AF574299

Graduate Texts in Physics

Graduate Texts in Physics publishes core learning/teaching material for graduate- and advanced-level undergraduate courses on topics of current and emerging fields within physics, both pure and applied. These textbooks serve students at the MS- or PhD-level and their instructors as comprehensive sources of principles, definitions, derivations, experiments and applications (as relevant) for their mastery and teaching, respectively. International in scope and relevance, the textbooks correspond to course syllabi sufficiently to serve as required reading. Their didactic style, comprehensiveness and coverage of fundamental material also make them suitable as introductions or references for scientists entering, or requiring timely knowledge of, a research field.

More information about this series at http://www.springer.com/series/8431

Keh Yung Cheng

III–V Compound Semiconductors and Devices

An Introduction to Fundamentals

Springer

Keh Yung Cheng
Department of Electrical
and Computer Engineering
University of Illinois at Urbana-Champaign
Urbana, Illinois, USA

Department of Electrical Engineering
National Tsing Hua University
Hsinchu, Taiwan

ISSN 1868-4513 ISSN 1868-4521 (electronic)
Graduate Texts in Physics
ISBN 978-3-030-51901-8 ISBN 978-3-030-51903-2 (eBook)
https://doi.org/10.1007/978-3-030-51903-2

This Springer imprint is published by the registered company Springer Nature Switzerland AG
The registered company address is: Gewerbestrasse 11, 6330 Cham, Switzerland

Preface

Since 1990, the creation of the World Wide Web and the strong surge of personal computer popularity, which promotes Internet usage, created a steep increase of Internet traffic. In the meantime, wireless communication systems also have evolved rapidly through several generations with ever-increasing bandwidth. To handle the vast amount of information traffic flow, high-speed and wide-bandwidth networks are needed, which require III-V compound semiconductor-based high-speed electronic and photonic devices. Furthermore, the realization of the *p*-type GaN not only led to widespread applications of III-nitride-based light-emitting diodes (LEDs) for solid-state lighting and full-color displays but also unlocked the opportunities for the development of next-generation power electronics using wide-bandgap semiconductors. On the other hand, miniaturization of the feature size of silicon-based integrated circuits is rapidly approaching the limit of Moore's law. Taking advantage of the very high electron mobility of some III-V alloys, the integration of compound semiconductor devices on silicon offers another opportunity for future developments. So the field has continued to expand rapidly. Although a number of books devoted to specific subjects on compound semiconductors and devices are available with the emphasis on the state of the art, a single monograph providing a complete discussion of the whole field is still lacking.

It is not the purpose of this book to give an exhaustive treatment of compound semiconductors and devices. Rather, this book conveys the relevant fundamental physics, and material and device concepts, of III-V compound semiconductors in a coherent manner within a single text, so as to provide the reader with basic concepts and ideas needed for understanding related literature and developments. The preliminary concepts gained also serve as foundations for more advanced studies of compound semiconductor-based optoelectronics and high-speed electronic devices. The content of this book was developed based on a one-semester course for beginning graduate students and advanced seniors in the Department of Electrical and Computer Engineering (ECE) at the University of Illinois at Urbana-Champaign (UIUC). Normally, the students in this course have the background in electromagnetic theory and have been exposed to semiconductors at the level of Streetman and Banerjee, *Solid-State Electronics Devices*, 6th ed. (Prentice Hall 2006). Certain students may have gained the knowledge of solid-state physics. Students from other engineering disciplines such as materials science and engineering also attended the class. Because of the diverse background of students interested in these subjects, introductory materials are included throughout the book to help bring all students to the same starting point.

The book is organized into four parts: Semiconductor fundamentals, Compound semiconductor materials, Properties of heterostructures, and Heterostructure devices. The book starts with a concise introduction to the historical development of the semiconductor industry and the relevance of III-V compound semiconductors. Chapters 2 and 3 introduce basic semiconductor concepts with supplementary materials on concepts beyond silicon. It also serves as a brief review for those who are familiar with the topics. Chapter 4 describes the physical properties of compound semiconductors including some distinctive features of multielement alloys. The material technologies of preparing compound semiconductor crystals in bulk or thin film forms and post-growth modification of material structures are presented in Chap. 5. The unique properties of heterostructures formed in compound semiconductors are then introduced in Chaps. 6–8. In Chap. 6, the alignments of energy bands at heterojunctions are derived first for the construction of heterostructures. Then strain effects on band-edge energies are discussed. The fundamental electrical and optical properties of heterostructures follow in Chaps. 7 and 8, respectively. Finally, important high-speed devices and photonic devices are presented in the last two chapters. Two of the widely used heterostructure high-speed devices—high electron mobility transistor (HEMT) and heterojunction bipolar transistor (HBT)—are discussed in Chap. 9. In addition, the key breakthroughs that led to the successful development of long sought after III-V MOSFETs are also discussed. Chapter 10 deals with photonic devices including various semiconductor injection lasers, LEDs, quantum-well infrared photodetectors, and an integrated optoelectronic device—the transistor laser. To assist students in probing further into the subject matter discussed, cited journal articles and a list of general references are added at the end of each chapter. Each chapter also contains an assortment of problems at its end to assess students' ability to apply the principles of the chapter.

Finally, it is my great pleasure to acknowledge a number of people who have encouraged me to take the charge of developing this book. I thank colleagues of the ECE Department at UIUC who have collaborated with me on various inspiring III-V compound semiconductor-related device research projects, particularly Profs. Milton Feng, Nick Holonyak Jr., K. C. Hsieh, and the late Shun-Lien Chuang. I am especially grateful to Nick Holonyak Jr. for many stimulating discussions during the coffee hours on technical issues as well as on his personal experience in early days of semiconductor developments. I am grateful to James Hutchinson for proofreading the whole manuscript. The continuous guidance and encouragement offered by Dr. Alfred Y. Cho of Bell Laboratories over the years are highly appreciated. The National Tsing-Hua University in Hsinchu, Taiwan, provided an ideal writing environment through the Ho Chin Tui chair professorship. Finally, I wish to thank my wife Kuo Ping for her encouragement and patience during the course of this endeavor.

Walnut Creek, California, USA Keh Yung (Norman) Cheng

Contents

Chapter 1
Introduction

Abstract Nowadays, people of the world take their vast benefits for granted—in mobile smartphones, laptop and tablet computers, smart watches, ATMs, electrical vehicles, and numerous other electronic devices. The core technology centered in all these device applications is the microchip, which consists of billions of transistors. The transistor was presciently called the 'nerve cell' of the Information Age by William Shockley. December 16, 1947, marks the beginning of the Information Age when John Bardeen and Walter Brattain invented the transistor at Bell Laboratories. It can be said that the transistor is the most important artifact of the twentieth century which deeply affects human civilization. In this chapter, the evolution of the transistor after Bardeen and Brattain's work from discrete device to very-large-scale integrated circuits, from elemental Ge, Si, to III-V compound semiconductors, and from transport device to photonic device applications is concisely discussed. The necessity of using III-V compound semiconductors for high-speed and high-power optoelectronic devices is also examined.

1.1 Historical Perspective

1.1.1 Rise of the Semiconductor Industry

On December 23, 1947, John Bardeen and Walter Brattain demonstrated the first transistor they invented at Bell Telephone Laboratories (BTL), Murray Hill, New Jersey (Fig. 1.1). The *p–n–p* point contact transistor was made on an *n*-type germanium semiconductor. Inspired by this invention, William Shockley conceived the bipolar junction transistor design in early 1948 by replacing two-point contacts of the transistor with *p–n* junctions. Due to the increased device structure complexity, until late 1950, a working junction transistor was finally made from a germanium crystal grown by the double-doped Czochralski method. To improve the device performance, other fabrication techniques such as alloying and diffusion were pursued. In late 1954 the first germanium transistor fabricated using diffusion was made with performance superior to that of the best commercially available transistors of the time. But the

K. Y. Cheng, *III–V Compound Semiconductors and Devices*, Graduate Texts in Physics,
https://doi.org/10.1007/978-3-030-51903-2_1

Fig. 1.1 This publicity photo accurately depicts the working relationship that led to the invention of the transistor at BTL: Brattain handles the apparatus, Bardeen enters data, and Shockley looks on. Courtesy AIP Emilio Segrè Visual Archives, Brattain Collection

germanium transistor has serious drawbacks such as leaky reverse currents and poor high-temperature characteristics. So, the crucial question at the time was whether to continue improving germanium-based transistors or concentrate on a different semiconductor material—silicon.

Silicon has much larger bandgap energy than germanium (1.124 vs. 0.664 eV), which means orders-of-magnitude smaller reverse leakage currents than germanium at normal operating temperatures. Therefore, John Moll, who led a group to develop *p–n–p–n* switches at BTL, insisted on using silicon instead of germanium. Since there are three *p–n* junctions in the switch structure, the innermost junction could not be easily formed using grown-junction or alloying techniques. Diffusion was the only real alternative. However, the high melting temperature of silicon made processing silicon at extremely high temperature a daunting problem. To solve the high-temperature diffusion problem, Moll's group collaborated with Carl Frosch from the BTL chemistry department. In early 1955, during a diffusion experiment, Frosch accidentally ignited hydrogen gas and introduced water vapor into the diffusion chamber, causing the growth of a silicon dioxide film on the silicon surface. The oxide formed a smooth, hard, protective layer that kept the silicon from degrading at high processing temperatures. Soon, transistor development within the Bell System, and later in the rest of the industry, would focus on silicon as the material and diffusion as the processing technology.

Although most of the early semiconductor technology advances took place within the Bell System, the integrated circuits concept was developed simultaneously in Texas and California. In 1958, Jack Kilby of Texas Instruments and Robert Noyce of Fairchild Semiconductor independently came up with the idea of integrating active and passive device components on a single slice of semiconductor material. Kilby concentrated on how to make a circuit that connected discrete silicon or germanium components using wires and solder. With the advent of diffusion and photolithography technologies developed at Bell Telephone Laboratories and Shockley Semiconductor Laboratory, Noyce focused on making electrical connections to all necessary *p–n* junctions and resistors on the same silicon wafer underneath a protective layer of silicon dioxide using the planar processing technique. This basic integrated circuit (IC) structure resembles today's most advanced very-large-scale integrated circuits. In the same year, a group led by M. M. Atalla at BTL found that by carefully cleaning the surface and applying a very pure oxide layer, it could drastically reduce the surface states at the silicon–oxide interface. With the surface states under control, Atalla and Dawon Kahng demonstrated the first metal-oxide-silicon field-effect transistor (MOSFET) in 1960—the kind of device that has come to dominate integrated circuits and microchips.

Through technology diffusion, new product development, and market expansion, the silicon semiconductor industry revenue rapidly reached $1.0 billion in 1961. But it took the industry another two decades to achieve revenue of $10 billion. The situation changed by the early 1980s. The advancements in integration technologies allowed Intel, founded by Noyce and Gordon Moore in 1968, to produce microprocessors in early-to-mid 1970s using the MOS process. The microprocessor is a CPU on a chip and could be used in a wide range of applications. For example, the Atari 8800 introduced in 1975 for hobbyists, using the Intel 8080 microprocessor, was essentially a rudimentary computer. Soon, many companies, including Apple and Microsoft, transformed the hobbyists' toy into a personal computer (PC) with more capabilities that was easier to use. Eventually, the PC became the worldwide driver for the semiconductor industry and the rapidly expanding information technology (IT) industry. The semiconductor industry revenue achieved $100 billion in 1993.

In 1969, the US Defense Department's Advanced Research Projects Agency (ARPA) initiated an experimental network project—ARPANET—to link computers at scientific laboratories across the country so that researchers might share computer resources. Additionally, it might be possible to connect computers in a network redundantly, so that if one line went down, a message could take another path. The network started out with four sites near the west coast. The ARPANET sites eventually extended to nineteen universities and research institutes from coast to coast. In 1989, the ARPANET was terminated, replaced by a network established by the National Science Foundation (NSF)—NSFNET. By that time anyone on a college campus with a connection to the campus network could become an Internet user. In 1990, the creation of the World Wide Web, a multimedia branch of the Internet, by researchers at CERN, the European Laboratory for Particle Physics near Geneva, Switzerland, made the Internet easier to navigate. At the same time, PC popularity surged, further promoting Internet use. The steep increase in Internet use helped further expansion of the silicon-based semiconductor industry. Meanwhile, to handle the vast information traffic, high-speed and wide-bandwidth networks are needed, which produces high demand for compound semiconductor-based high-speed electronic and photonic devices.

1.1.2 Development of III–V Compound Semiconductor Industry

Since the invention of integrated circuits, the semiconductor industry has been dominated by silicon due to several favorable key attributes including the formation of excellent dielectric layer of silicon dioxide, the high thermal conductivity, and the use of low-power complementary MOS (CMOS) circuitry that enables the integration of very-high-density circuits over large wafers ($\geq 12''$). However, due to the nature of an indirect energy band structure, silicon is not an efficient light-emitting material and its electron mobility is relatively low. The active devices supporting the high-speed networks, therefore, have to rely on III–V compound semiconductors, such as GaAs and InP. These compound semiconductors have several major advantages over their elemental semiconductor counterpart: First, efficient light emission is readily achievable from these direct energy band structure semiconductors. The direct energy gap also leads to the second advantage, high electron mobility and high saturation velocity. The carrier drift velocity in a direct bandgap semiconductor increases linearly with a moderate applied electric field. In indirect bandgap semiconductors, the multiplicity of the conduction band minima increases the electron scattering probability under high fields. Therefore, the drift velocity saturates under high fields. In contrast, the scattering probability of direct bandgap semiconductors is minimized, and a strong velocity overshoot is observed before the velocity saturation occurs. Third, some of the most popular crystals of III–V compounds can be prepared in semi-insulating forms. The use of these semi-insulating substrates can eliminate speed-degrading parasitic capacitance. It also simplifies the electrical isolation of devices. Next, by mixing two or three III–V binary compounds, one can prepare a series of ternary or quaternary compounds with greater freedom in the selection of crystal lattice constant and/or bandgap energy (Fig. 1.2). Finally, the most important attribute of III–V compounds is the creation of new function devices utilizing bandgap engineering in heterostructures, where semiconductor thin films with different energy bandgaps are stacked together forming devices with designed electrical and optical properties (Fig. 1.3). Quantum-well lasers and high-electron-mobility transistors are two examples.

In 1952, the semiconductor nature of III–V compounds was first reported independently by Heinrich Welker of Siemens, West Germany, and a group at the Ioffe Institute, Russia. Inspired by the demonstration of the first laser, a solid-state ruby laser was developed in May 1960 by Theodore Maiman of the Hughes Research

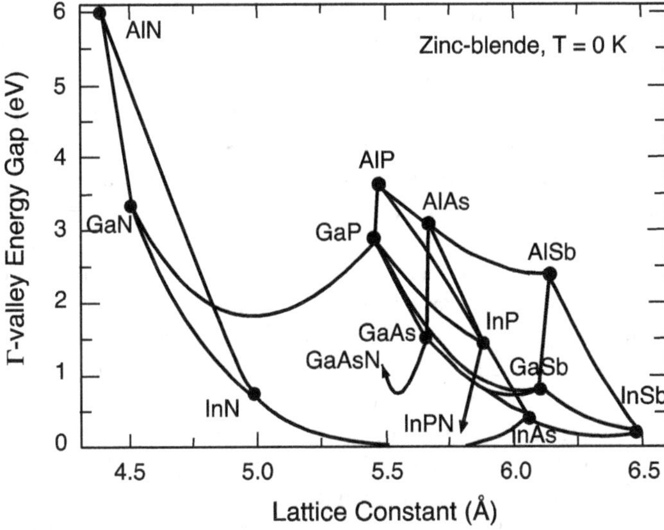

Fig. 1.2 Lattice constant as a function of Γ-valley bandgap energy of zinc-blende structure III–V compound semiconductors at zero temperature. The bandgap energy of ternaries follows the line connecting the constituent binaries. Reprinted with permission from [1], copyright AIP Publishing

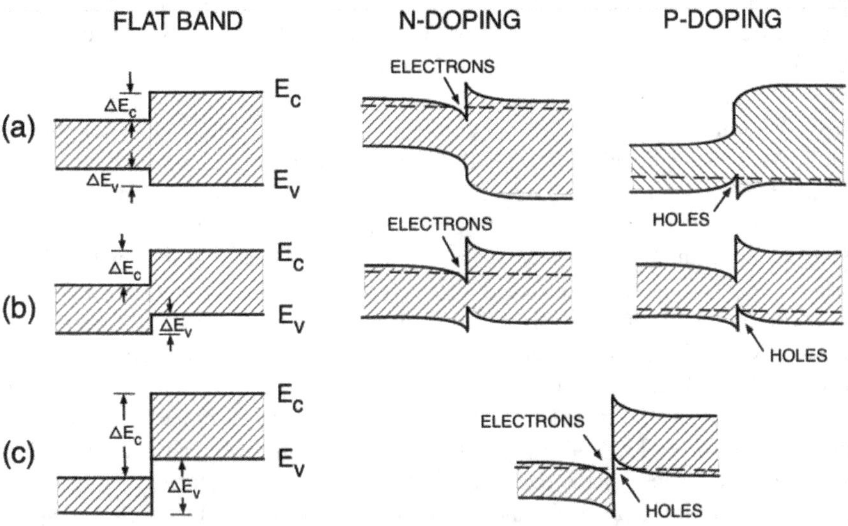

Fig. 1.3 **a** Type I, **b** staggered type II, and **c** broken-gap type II band-edge alignments in semiconductor heterostructures. The energy band diagrams of flat band, *n*-type heterostructures, and *p*-type heterostructures are shown in the left, middle, and right panels, respectively

Laboratory, and three groups (GE, IBM, and MIT Lincoln Laboratory) independently demonstrated GaAs homojunction injection lasers in 1962 using diffused *p–n* junction structures. The first ternary alloy (GaAsP) laser emitting in the visible spectrum (710 nm) was also reported at about the same time by Nick Holonyak Jr. at GE. The next year, J. G. Gunn discovered microwave oscillation due to the transferred electron effect in direct bandgap III–V compounds. These important early inventions precisely highlight the strength of III–V semiconductor alloys over silicon—their light-emitting and high-speed properties.

Compared to silicon, compound semiconductors are more complex in terms of material preparation. Heterostructure devices are entrenched in the grown layer structures, not formed by post-growth diffusion processing commonly used in silicon IC fabrication. Thus, further advance of compound semiconductor technologies relies on the successful development of hetero-epitaxial growth technologies. For example, independently suggested by Herb Kroemer and Zhores Alferov in 1963, the incorporation of the double-heterojunction (DH) structure can improve the device efficiency of injection lasers. At the time, the threshold current density of GaAs injection lasers was nearly 10^5 A/cm^2, which is too high even for room-temperature pulse operation. However, room-temperature cw DH laser operation was finally demonstrated in 1970 after the successful demonstration of the liquid phase epitaxy (LPE) technique to grow AlGaAs on GaAs in 1967. To improve the uniformity and growth rate control of LPE-grown layers, two major epitaxy techniques have been developed. Alfred Cho of Bell Laboratories pioneered the molecular beam epitaxy (MBE) technique and demonstrated MBE growth of AlGaAs and GaP in 1969. At Rockwell International, H. M. Manasevit first reported metalorganic chemical vapor deposition (MOCVD) of compound semiconductors on foreign substrates such as sapphire in 1968. With further developments, both MBE and MOCVD techniques have proved to have the characteristics of producing high-quality uniform multiple layered structures with excellent morphology, sharp interface, and precise doping and thickness control. These attributes ultimately formed the basis for successful growth of innovative bandgap-engineered devices with never-before-realized electrical and optical properties. In 1975, the quantum-well laser operation was demonstrated using MBE-grown AlGaAs/GaAs multilayer structures, where the quantum size effect occurred in very thin GaAs layers. The phrase *quantum-well laser* was first used in 1978 by Holonyak et al. to describe their MOCVD-grown laser structures. One additional advantage of these ultra-thin-layer structures is that high-quality strained layers can be achieved to further modify the energy band structure and enhance the laser performance. The ability to precisely make quantum wells has had a far-reaching impact, ranging from classroom physics to revolutions in electronic and optical devices for the consumer electronics, computers, and communications industries. Today, the global communication network relies on high-speed laser diodes and sensitive photodiodes to transmit and receive extremely high capacity information optically over hundreds of kilometers in low-loss optical fibers. High-electron-mobility transistors (HEMTs) are utilized as high-speed circuit components and in high-frequency, low-noise, direct broadcast satellite and wireless communications. The heterojunction bipolar transistor (HBT) is the key component to enabling the efficient performance and small size of cellular phones. Further development of these high-speed electronic devices will likely hasten the implementation of ultra-wide-bandwidth communication systems.

Another important development of compound semiconductor technology is the invention of the visible (red) light-emitting diode (LED) in 1962 by Holonyak using GaAsP ternary alloys. The performance (lumens/watt) of (As,P)-based LEDs increased 10 × per decade from 1970 to 2000 using GaP:Zn,O and GaP:N materials, and AlGaAs/GaAs and AlGaInP/GaAs heterostructures, progressively. However, the emission color of these LEDs was restricted by the bandgap energy of these alloys to between red and orange. Nevertheless, these LEDs have been used for many applications including instrument panels, displays, vehicle break lights, and traffic signals. In order to expand LED emissions to cover the full visible spectrum, specifically the blue–green band, GaN has been investigated since the early 1970s. However, the lack of suitable *p*-type dopants prevented it from forming useful *p*–*n* junction devices. In 1989, Isamu Akasaki and Hiroshi Amano of Nagoya University in Japan finally made the breakthrough; they obtained *p*-type conduction in Mg-doped GaN using post-growth annealing to remove hydrogen passivation and activate acceptors. This key discovery led Shuji Nakamura of Nichia in Japan to demonstrate the first InGaN/GaN blue LED grown by MOCVD. Now the LEDs with three primary colors are realized for making full-color displays. More important, white light LEDs for lighting applications are readily attainable using yellow-phosphorus-coated blue LEDs. As a general illumination lighting source, white light LED products surpass many conventional lighting technologies (including incandescent and fluorescent light sources) in energy efficiency, lifetime, and versatility and rival them in color and light quality. At the other end of the spectrum, innovative bandgap-engineered devices utilizing inter-subband transitions in quantum wells, such as quantum cascade (QC) lasers, extend the laser emission wavelength well into the mid-infrared (mid-IR), which is not available from bulk materials. Large area and uniform mid-IR quantum-well infrared photodetectors (QWIPs) based on the same inter-subband transitions were also developed. These mid-IR devices are critical for future planetary exploration, biomedical, and security applications.

1.2 Future Outlooks

Due to limitations in the material properties of silicon, III–V compound semiconductors have dominated photonic and high-speed electronic devices. This trend is likely to continue in the fast-growing wireless communications and Internet applications. The ability to form heterostructures further enhances the ability of compound semiconductors to form new quantum effect devices and efficient photonic devices.

It is expected that III–V compound semiconductors will continue to thrive in high-speed, wide-bandwidth, and low-power consumption devices and circuits. In addition, the realization of the *p*-type GaN not only led to widespread applications of III-N-based LEDs for solid-state lighting and full-color displays but also enabled the development of next-generation power electronics using wide-bandgap semiconductors. The inherent properties of the bulk GaN, including a wide bandgap, higher electron mobility, and excellent thermal conductivity, allow III-N material-based power electronic devices to operate efficiently at higher voltages and power densities, higher frequency and temperature than silicon semiconductors, yielding significant energy savings. However, the realization of the full potential of wide-bandgap semiconductor-based power devices depends on the development of cost-effective substrates.

On the other hand, miniaturization of the feature size of ICs to increase the density of components has been the foundation of the 50-year-long steady advance of silicon technology. In 1965 Moore wrote an *Electronics* article based on an analysis of the cost of IC production where he claimed that at any given time there was a component density that led to the lowest cost per component [2]. Moore further proclaimed that the minimum cost point had been moving out in density by a factor of two per year. This statement, commonly referred to as *Moore's law*, became the metronome of the industry advancement. The exponential increase of transistor count per unit area as a function of time has been faithfully delivered by the industry since 1970. The minimum transistor gate length also decreases with time exponentially (Fig. 1.4). However, the semiconductor industry faces increasingly difficult challenges in technology as well as cost as it moves into production at feature sizes below 100 nm. To keep up the CMOS performance with scaling, strained SiGe channel material

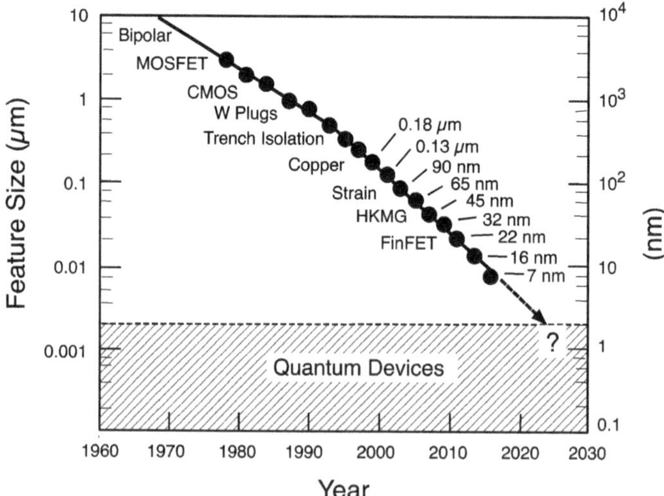

Fig. 1.4 Feature size versus time in silicon ICs. Major device technology milestones are also listed

(90/65 nm node) and high-κ gate dielectrics (45/32 nm node), even non-planar multi-gate structures (22/16 nm node) of MOSFETs have been incorporated. Although 5-nm-node CMOS circuits are on the horizon, further reduction of the feature size to below 2 nm will push the device structure into the quantum regime. At that moment, the silicon technology will reach the point at which significant materials and device innovations will be required to sustain further development. One possible situation for the future is the hybrid material system where silicon, germanium, and III–V compound semiconductors are integrated together on silicon wafers. For example, taking advantage of the very high electron mobility in GaInAs and high hole mobility in germanium, advanced CMOS circuits consisting of *n*-GaInAs channel and *p*-Ge channel MOSFETs might be fabricated together on the silicon wafer. One may also imagine the incorporation of optical interconnects within a silicon wafer using integrated compound semiconductor lasers and Si/SiO_2 waveguides to improve the speed of the circuits. Thus, the distinction between silicon and compound semiconductors in terms of their roles in device applications may become less well defined.

References

1. I. Vurgaftman, J.R. Meyer, L.R. Ram-Mohan, J. Appl. Phys. **89**, 5815 (2001)
2. G. E. Moore, Electronics **19**, 114–117 (1965, April); and Proc. IEEE **86**, 82 (1998)

Further Reading

1. M. Riordan, L. Hoddeson, *Crystal fire* (W. W. Norton & Co., New York, 1997)
2. R.K. Bassett, *To the Digital Age* (John Hopkins University Press, Baltimore, 2002)
3. J. Orton, *The Story of Semiconductors* (Oxford University Press, 2004)

Chapter 2
Atomic Bonding and Crystal Structures

Abstract On December 23, 1947, John Bardeen and Walter Brattain demonstrated the world's first transistor at Bell Telephone Laboratories. The point-contact transistor was made in the area of a large single-crystalline grain on a piece of reused polycrystalline germanium. Above the Ge slab was a polystyrene wedge structure where, at the bottom tip of the wedge, the gold-foil emitter and collector electrodes make point contacts, separated by ~50 μm, to the *n*-type Ge base. In this device, the injection of both electrons and holes into a semiconductor was first demonstrated. Shortly after that, the zone refining process was developed by William Pfann for the growth of silicon and germanium crystals of predetermined purity. Today, transistors, light-emitting diodes, injection lasers, integrated circuits, and many other semiconductor devices are made from single crystals of semiconductor material with extremely high purity. There are other semiconductor devices which are not made from single-crystal material, though they could be and their operation is usually obtained as if they were. Hence, a study of the basic properties of crystal structures is necessary for a meaningful explanation of the operation of semiconductor devices. In this chapter, the tetrahedron bond structure—the basic building block of major semiconductor crystals—is first derived from atomic sp^3 orbitals. The reciprocal lattice is then introduced as an alternative but powerful approach to investigate the lattice structure of the crystal. For example, crystal diffraction results are best discussed in the reciprocal lattice space. Finally, the Brillouin zone concept is introduced.

K. Y. Cheng, *III–V Compound Semiconductors and Devices*, Graduate Texts in Physics,
https://doi.org/10.1007/978-3-030-51903-2_2

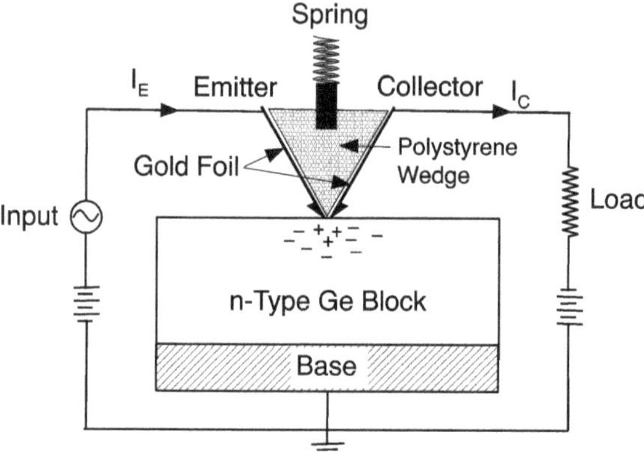

Fig. 2.1 Schematic of the point-contact bipolar transistor invented by Bardeen and Brattain of Bell Telephone Laboratories in 1947. The polystyrene wedge covered with gold foil on two sides was held against the Ge with a spring to form emitter and collector contacts. The *n*-type Ge slab forms the base of the transistor. Minority carrier (hole) injection was demonstrated for the first time

2.1 Crystal Structures and Symmetry

2.1.1 Crystal Structures

The solids of primary interest for us are semiconductors that have all their atoms arranged in a regular pattern in three dimensions (3D). Such solids are called single crystals, and the arrangement of atoms is termed the crystal structure. The basic 3D building blocks that pack together to fill all the space of the crystal are called the *unit cells*. A unit cell can be seen as a parallelepiped defined by three lengths a, b, and c and three angles α, β, and γ as shown in Fig. 2.2. By changing the lengths and angles of the parallelepiped, while making sure to fill all the space, a number of crystal structures can be built. Depending on the specifications about the lengths

Fig. 2.2 Definition of axes, dimensions and angles for a general unit cell

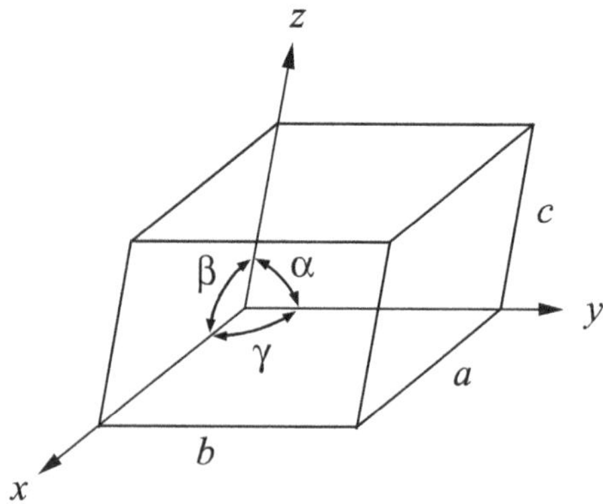

Table 2.1 Seven crystal classes and related Bravais lattices

System	Unit cell	Bravais lattice
Cubic	$a = b = c$, $\alpha = \beta = \gamma = 90°$	*P* (primitive), *I* (body-centered), *F* (face-centered)
Hexagonal	$a = b \neq c$, $\alpha = \beta = 90°$, $\gamma = 120°$	*P* (rhombohedral)
Tetragonal	$a = b \neq c$, $\alpha = \beta = \gamma = 90°$	*P*, *I*
Orthorhombic	$a \neq b \neq c$, $\alpha = \beta = \gamma = 90°$	*P*, *I*, *F*, *C* (base-centered)
Monoclinic	$a \neq b \neq c$, $\alpha = \beta = 90° \neq \gamma$	*P*, *C*
Trigonal	$a = b = c$, $120° > \alpha = \beta = \gamma \neq 90°$	*P* (rhombohedral)
Triclinic	$a \neq b \neq c$, $\alpha \neq \beta \neq \gamma$	*P*

and the angles, these crystal structures are classified into seven systems as listed in Table 2.1. Some structures can be derived from other crystal structures. For example, under a uniaxial stress, such as in a strained quantum well, a cubic crystal can be transformed into an orthorhombic structure. Additionally, these basic structures are further modified by four different 3D unit cells: (a) The *primitive* unit cell (symbol *P*) has a lattice point at each corner and forms a minimum volume. (b) The *body-centered* unit cell (symbol *I*) has a lattice point at each corner and one at the center of the cell. (c) The *face-centered* unit cell (symbol *F*) has a lattice point at each corner and one at the center of each face. (d) The *base-centered* unit cell (symbol *A*, *B*, or *C*) has a lattice point at each corner and one pair of opposite faces; e.g., a *C*-centered cell has lattice points in the centers of the *ab* faces. When combing these four types of lattice with seven possible unit cell forms, 14 *Bravais lattices* are produced. For example, the simple cubic is the primitive unit cell of the cubic system while the body-centered cubic and face-centered cubic correspond to the modified (*I* and *F*, respectively) cubic unit cells (Fig. 2.3). It is important to note that the *lattice point* represents equivalent positions in a crystal structure and not atoms. A lattice point could be occupied either by an atom, a complex ion, a single molecule, or a group of molecules.

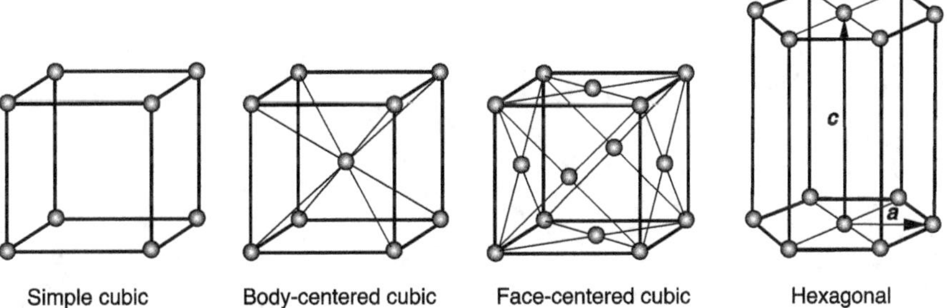

Fig. 2.3 Four basic 3D Bravais lattices most relevant to semiconductor crystals. The face-centered cubic and hexagonal crystal structures are particularly important in semiconductor physics

2.1.2 Crystallographic Notation—Miller Indices

In semiconductor technologies, we often make devices and circuits on a particular crystallographic plane or a particular direction of semiconductor crystals. This particular crystal plane is usually described by its *Miller indices*, which are established as follows. Consider a 3D lattice, a portion of which is shown in Fig. 2.4, where *A*, *B*, *C*, and *O* are lattice points; ***a***, ***b***, and ***c*** are translational unit vectors; and n_1, n_2, and n_3 are the proper integers. Thus the plane *ABC* can be defined by vectors ***OA*** (=$n_1\boldsymbol{a}$), ***OB*** (=$n_2\boldsymbol{b}$), and ***OC*** (=$n_3\boldsymbol{c}$). For reasons that will become clear when the properties of reciprocal lattices are explored in Sect. 2.5, we shall find it useful to refer to the *ABC* plane as the (*hkl*) plane, where *hkl* is a set of integers which expresses the ratio

$$h : k : l = \frac{1}{n_1} : \frac{1}{n_2} : \frac{1}{n_3} \tag{2.1}$$

Thus the plane *ABC* has the Miller indices of (*hkl*) which are a set of the smallest integers that satisfy (2.1). For example, for the plane whose intercepts are 2, 1, 3, the reciprocals are $^1/_2$, 1, and $^1/_3$; the smallest three integers having the same ratio are (362). If a plane cuts an axis (say, the *a*-axis) on the negative side of the origin, the corresponding index is negative, indicated by placing a bar (minus sign) above the index as $(\bar{h}kl)$. For an intercept at infinity, the corresponding index is zero. Figure 2.5 illustrates the Miller indices of some important plans in a cubic crystal. The indices (*h*00) may denote a plane parallel to both *b*- and *c*-axes, e.g., the (100) plane, and the indices (*hk*0) represent a plane parallel to the *c*-axis. The normal to the plane with indices (*hkl*) is the direction [*hkl*]. In a cubic crystal there are six cubic faces: (100), (010), (001), $(\bar{1}10)$, $(0\bar{1}0)$, and $(00\bar{1})$. This set of equivalent planes of a cubic crystal may be denoted as a group by {100}. The family of plane directions in the cubic crystal may be expressed as ⟨100⟩.

For hexagonal crystal structures, there are four coordinates, namely the a_1, a_2, a_3, and *c* axes as shown in Fig. 2.6. The angle between adjacent a_i's is 120°. Therefore, it is common to use a set of four Miller indices, expressed in terms of the reciprocal intercepts with respect to the vectors $\boldsymbol{a}_1$, $\boldsymbol{a}_2$, $\boldsymbol{a}_3$, and ***c*** as (*hkjl*). However, it can be

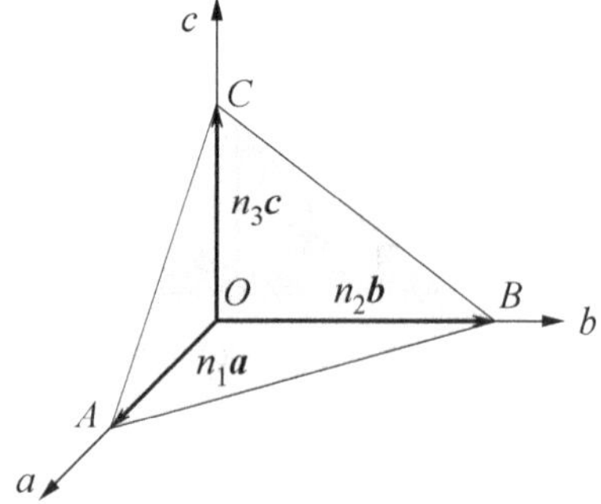

Fig. 2.4 Description of the construction of Miller indices for the plane passing through the lattice points *A*, *B*, and *C*. ***a***, ***b***, and ***c*** are unit vectors and n_1, n_2, and n_3 are integers

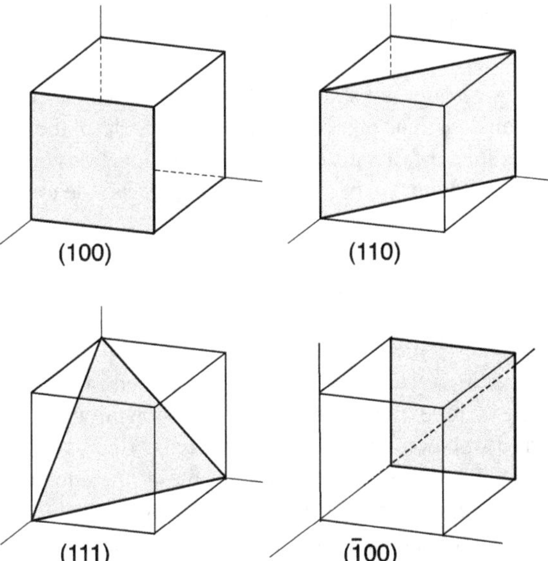

Fig. 2.5 Miller indices of some important planes in a cubic crystal

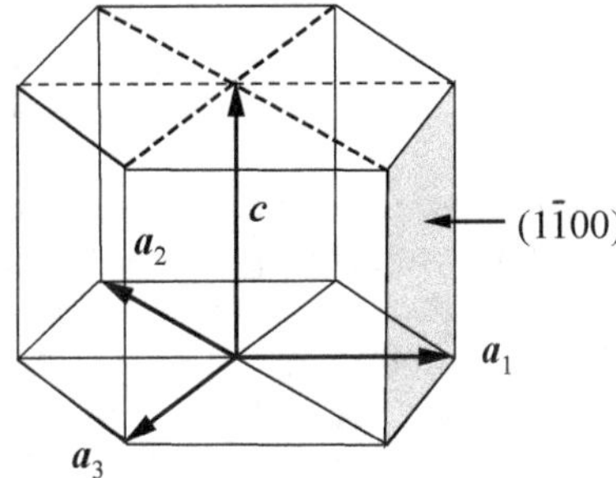

Fig. 2.6 Four unit vectors used to formulate Miller indices for a hexagonal crystal

proved that the sum of *h, k,* and *j* of a plane is always zero. Thus the added index *j* serves as a check and does not provide any new information. For example, the shaded surface shown in Fig. 2.6 cuts a_1 and a_2 at 1 and -1, respectively, and is parallel to a_3 and *c*. Thus this plane may be designated as $\left(1\bar{1}00\right)$ according to

$$h{:}\,k{:}\,j{:}\,l = \frac{1}{1} : \frac{1}{-1} : \frac{1}{\infty} : \frac{1}{\infty} = 1 : -1 : 0 : 0$$

where $h + k + j = 1 - 1 + 0 = 0$. The Miller indices of the base of the hexagonal crystal, or the basal plane, are (0001).

2.1.3 Lattice Symmetries

As discussed earlier, a single-crystal solid has all its atoms arranged in a regular pattern in three dimensions. The unit cells or building blocks of the crystal are shaped as 14 basic parallelepipeds of Bravais lattices. Any two locations in the crystal having identical atomic environment can be achieved through a simple translation action on the unit cells following

$$\boldsymbol{T} = n_1\boldsymbol{a} + n_2\boldsymbol{b} + n_3\boldsymbol{c} \tag{2.2}$$

where $\boldsymbol{T}$ is the translation vector; $\boldsymbol{a}$, $\boldsymbol{b}$, and $\boldsymbol{c}$ are the unit vectors; and n_1, n_2, and n_3 are integers. In addition, lattice points can be recurred into themselves by other operations including (a) reflection at a plane, (b) inversion through a point, (c) rotation about an axis, and (d) rotation-inversion about an axis.

Using simple cubic structure as an example, these operations are illustrated in Fig. 2.7. Through the operation of reflection at a mirror plane, the x–y plane in Fig. 2.7a, point P may transform to point Q. This symmetry is expressed mathematically by a coordinate transformation $x' = x$, $y' = y$, and $z' = -z$. The tensor representation of the mirror reflection operation in the x–y plane is expressed as

$$\begin{bmatrix} 1 & 0 & 0 \\ 0 & 1 & 0 \\ 0 & 0 & -1 \end{bmatrix}\begin{bmatrix} x \\ y \\ z \end{bmatrix} = \begin{bmatrix} x \\ y \\ -z \end{bmatrix} \tag{2.3}$$

The presence of a mirror plane in a crystal is represented by the symbol m. The inversion symmetry operation (Fig. 2.7b) may be described by the coordinate transformation $x' = -x$, $y' = -y$, and $z' = -z$. It may also be expressed by the transformation of the position vector $\boldsymbol{r}' = -\boldsymbol{r}$ at the inversion point. The symbol of the inversion symmetry is $\bar{I}$. The other symmetry operation is the rotation about an axis. The axis of rotation is named an n-fold axis if the angle ϕ of rotation to carry a unit cell into itself is equal to $\phi = 2\pi/n$. In this example, Fig. 2.7c, a 90° rotation about the axis will bring point P to point Q or a fourfold symmetry. All possible rotation operations

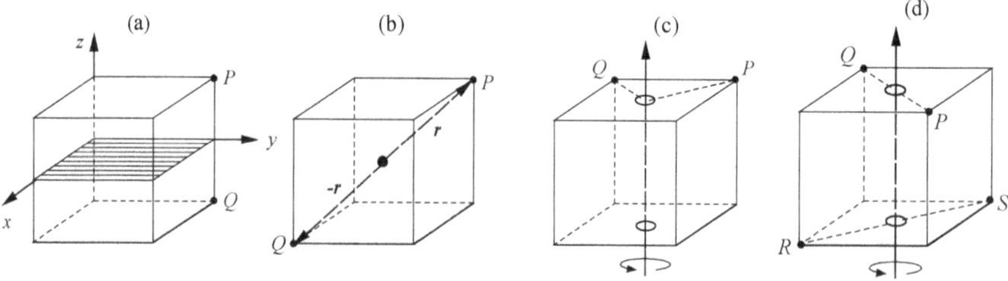

Fig. 2.7 Symmetry operations in a cubic crystal. **a** Reflection at a plane, **b** inversion through a point, **c** rotation about an axis, and **d** rotation-reflection about an axis

are one-, two-, three-, four-, and sixfold symmetry, which correspond to rotations by $2\pi, \pi, 2\pi/3, \pi/2$, and $\pi/3$ radians. These are the only five allowed rotation symmetry operations since they are consistent with a crystal's requirements for translational symmetry.

The symmetry operations may be combined with one another in various ways. For example, there is a rotation-reflection compound operation to carry the lattice into itself. As shown in Fig. 2.7d, a 90° (fourfold) rotation about an axis parallel to the z-axis through the center of the cube combining with a mirror reflection at the x–y plane will bring points P and Q to S and R, respectively. This combined operation is represented by the symbol $\bar{n}$ for the n-fold rotation symmetry. Conversely, every crystal may be described by a particular combination of symmetry elements that strongly influence its material properties. More specifically, this combination determines the nature of independent components of the tensors describing macroscopic material properties.

To illustrate this correlation, we examine the dielectric constant ϵ of the material using the symmetry concept and tensor representation. By definition, $\boldsymbol{D} = \epsilon \boldsymbol{E}$ or

$$\begin{bmatrix} D_x \\ D_y \\ D_z \end{bmatrix} = \begin{bmatrix} \epsilon_{11} & \epsilon_{12} & \epsilon_{13} \\ \epsilon_{21} & \epsilon_{22} & \epsilon_{23} \\ \epsilon_{31} & \epsilon_{32} & \epsilon_{33} \end{bmatrix} \begin{bmatrix} E_x \\ E_y \\ E_z \end{bmatrix} \tag{2.4}$$

For a cubic crystal, there are four threefold (along diagonals as shown in Fig. 2.8) and three fourfold (along three major axes) rotation axes. When rotating along [111] for 120°, it brings, for example, $x \to y$, $y \to z$ and $z \to x$. In tensor form, it is

$$\begin{bmatrix} D_y \\ D_z \\ D_x \end{bmatrix} = \begin{bmatrix} \epsilon_{11} & \epsilon_{12} & \epsilon_{13} \\ \epsilon_{21} & \epsilon_{22} & \epsilon_{23} \\ \epsilon_{31} & \epsilon_{32} & \epsilon_{33} \end{bmatrix} \begin{bmatrix} E_y \\ E_z \\ E_x \end{bmatrix} \tag{2.5}$$

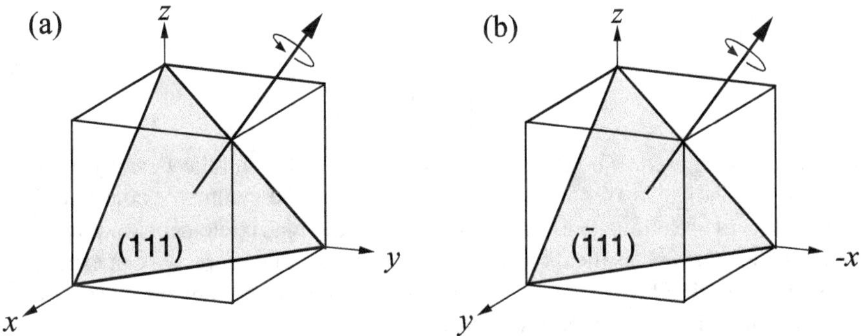

Fig. 2.8 Rotation symmetry operations along **a** [111] and **b** [$\bar{1}$11] axes of a cubic crystal

Equations (2.4) and (2.5) are essentially the same and their components can be compared. By examining D_i in both equations, we can deduce the relations of some tensor components

$$D_x = \epsilon_{11} E_x + \epsilon_{12} E_y + \epsilon_{13} E_z = \epsilon_{31} E_y + \epsilon_{32} E_z + \epsilon_{33} E_x \tag{2.6a}$$

and

$$D_y = \epsilon_{21} E_x + \epsilon_{22} E_y + \epsilon_{23} E_z = \epsilon_{11} E_y + \epsilon_{12} E_z + \epsilon_{13} E_x \tag{2.6b}$$

By comparison, we have $\epsilon_{11} = \epsilon_{22} = \epsilon_{33}$, $\epsilon_{13} = \epsilon_{21} = \epsilon_{32}$, and $\epsilon_{12} = \epsilon_{31} = \epsilon_{23}$. The threefold rotation symmetry exists along the other rotation axes. A rotation of 120° about $[\bar{1}11]$ will bring $x \to -y$, $y \to z$, $z \to -x$ (Fig. 2.8b), or

$$\begin{bmatrix} -D_y \\ D_z \\ -D_x \end{bmatrix} = \begin{bmatrix} \epsilon_{11} & \epsilon_{12} & \epsilon_{13} \\ \epsilon_{21} & \epsilon_{22} & \epsilon_{23} \\ \epsilon_{31} & \epsilon_{32} & \epsilon_{33} \end{bmatrix} \begin{bmatrix} -E_y \\ E_z \\ -E_x \end{bmatrix} \tag{2.7}$$

Comparing (2.4), (2.5), and (2.7), we find $\epsilon_{12} = -\epsilon_{23} = \epsilon_{31}$ and $\epsilon_{13} = \epsilon_{21} = -\epsilon_{32}$. In order to fulfill both these results obtained from the two different rotation axes, we conclude that $\epsilon_{ij} = 0$ for $i \neq j$ and $\epsilon_{11} = \epsilon_{22} = \epsilon_{33}$. Thus,

$$\epsilon = \begin{bmatrix} \epsilon_{11} & 0 & 0 \\ 0 & \epsilon_{11} & 0 \\ 0 & 0 & \epsilon_{11} \end{bmatrix} \tag{2.8}$$

Therefore, the simple cubic crystal has an isotropic dielectric constant ϵ.

Using a similar approach, we can calculate the dielectric constant tensors in other crystals. For hexagonal crystals, the dielectric constant tensor has the form

$$\epsilon = \begin{vmatrix} \epsilon_{11} & 0 & 0 \\ 0 & \epsilon_{11} & 0 \\ 0 & 0 & \epsilon_{33} \end{vmatrix} \tag{2.9}$$

if the z-axis is chosen to be the axis of sixfold symmetry and the x- and y-axes are the two perpendicular axes in the basal plane. Thus, the symmetry considerations help us find the relations that may exist between the matrix elements in a material parameter tensor such as the dielectric constant tensor. Therefore, we can expect the less symmetric zinc-blende crystal structure of GaAs to have an anisotropic dielectric constant and nonlinear optical properties.

2.2 Ionic Bond and Inter-atomic Forces

Ionic crystals are generally transparent insulators of inorganic compounds. The crystal structures of some ionic crystals are simple, such as NaCl and CsCl, while others are extremely complex. For the purpose of studying the properties of ionic bonding, we consider the simple compound NaCl. Sodium (Na), an alkali metal, has a single 3*s* valence electron outside the closed shell, whereas chlorine (Cl), a halogen, is one electron short of having a complete outer 3*p* shell. As shown in Fig. 2.9, an electron transfer from the alkali metal to the halogen results in closed-shell configurations in both Na^+ and Cl^- ions. During this electron transfer process, there are four major energy exchanges involved in forming the ionic bond. First, the energy given up when a neutral (Cl) atom gains an electron and becomes a negative ion (*anion*) is called the *electron affinity*. It increases the energy of the system and has a positive value. Second, the energy necessary to remove an electron from the (Na) atom creating a positive ion (*cation*) is simply the *ionization energy*—a negative energy value. Third, the Coulomb force between Na^+ and Cl^- ions keeps these ions together in a Na^+Cl^- ionic crystal, and a lowering of potential energy occurs. Therefore, the Coulomb attraction force has a positive energy value. Finally, when the electron shells of the neighboring ion cores start to overlap, a repulsive force must be considered. The repulsive energy is a result of the Pauli exclusion principle for a closed ion shell. Energy is required to prevent the collapse of the lattice and is considered a negative term. The balance of competing requirements for attractive and repulsive components of the electrostatic interaction warrants that highly symmetric crystal structures ensue with maximized volumes. Figure 2.10 shows two of the simplest ionic crystal structures of all—the rock salt structure of NaCl and the cesium chloride (CsCl) structure. The NaCl structure can be viewed as an FCC crystal with a basis containing one Na^+ ion at (0, 0, 0) and one Cl^- ion at (½, ½, ½). The other way of constructing the NaCl structure is to position alternatively with Na^+ or Cl^- ions at the lattice sites of a simple cubic crystal. Each ion is surrounded by six nearest-neighbor ions of the opposite type in $\langle 100\rangle$ directions. The CsCl structure is based on the simple cubic crystal structure with a basis containing one Cs^+ ion at (0,0,0)

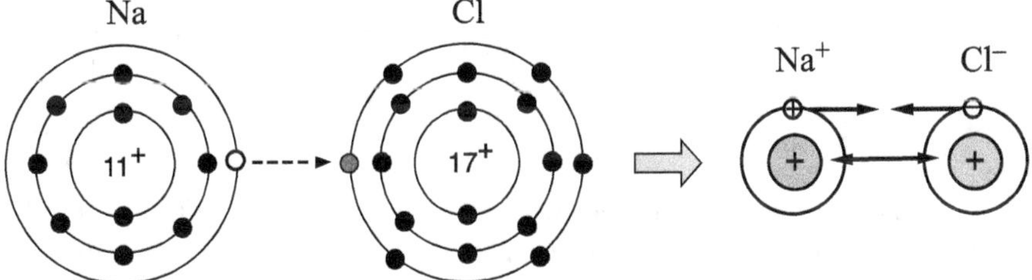

Fig. 2.9 Electron configuration in NaCl. The outer ring represents the shell of $n = 3$. The dashed arrow indicates the electron transfer from Na to Cl resulting in closed shell configurations in Na and Cl ions. The Coulomb attraction between two ions and repulsive force between cores are also shown

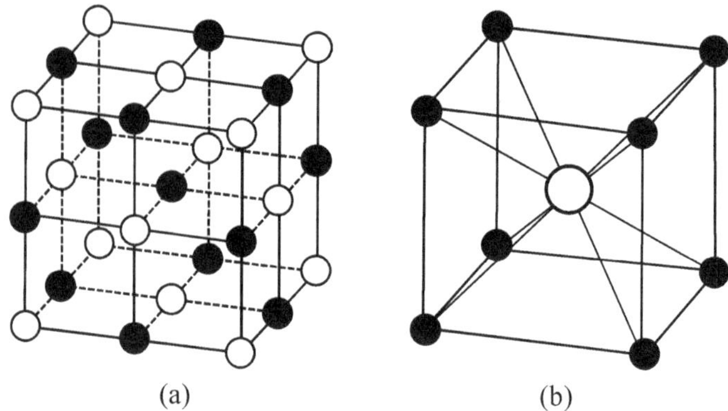

Fig. 2.10 Two typical ionic crystal structures: **a** NaCl and **b** CsCl. The ionic radii of Na^+, Cs^+, and Cl^- are 0.875 Å, 1.455 Å, and 1.475 Å, respectively

and one Cl^- ion at (1/2, 1/2, 1/2). The structure looks like BCC but, strictly speaking, is not since different ions occupy the two lattice sites at the body center and corners. There are eight nearest-neighbor ions of the opposite type in $\langle 111 \rangle$ directions.

The total energy required to separate the atoms in the solid into isolated neutral atoms and vice versa is called the *cohesive* energy or *binding* energy, E_0. Since the ionization energy of the positive ion and the electron affinity of the negative ion very nearly cancel each other, the major contribution to the binding energy is the electrostatic energies. We can formulate these relations in the following equation:

$$E_{\mathrm{T}}(r) \cong E_{\mathrm{Coul}}(r) + E_{\mathrm{Rep}}(r) \tag{2.10}$$

where E_{T} is the sum of the Coulomb attraction energy E_{Coul} and the energy of the repulsive force E_{Rep} at a positive ion–negative ion separation distance r. In a NaCl crystal structure, surrounding any Na^+ are six neighboring Cl^- ions, 12 next-nearest-neighboring Na^+ ions, and so forth. The Coulomb energy can be calculated as

$$E_{\mathrm{Coul}} = -\frac{q^2}{4\pi\epsilon_0 r}\left(\frac{6}{1} - \frac{12}{\sqrt{2}} + \frac{8}{\sqrt{3}} - \frac{6}{\sqrt{4}} + \frac{24}{\sqrt{5}} - \cdots\right) = -1.748\frac{q^2}{4\pi\epsilon_0 r} = -\alpha\frac{q^2}{4\pi\epsilon_0 r} \tag{2.11}$$

where α is the Madelung constant which has the values of 1.641, 1.638, and 1.763 for wurtzite, zinc-blende, and CsCl structures, respectively. The negative sign of E_{Coul} represents the net attractive force between ions. Since the repulsive force decreases rapidly as the distance between ions increases, it has a power-law form of

$$E_{\mathrm{Rep}} = \frac{A}{r^n} \propto r^{-n} \tag{2.12}$$

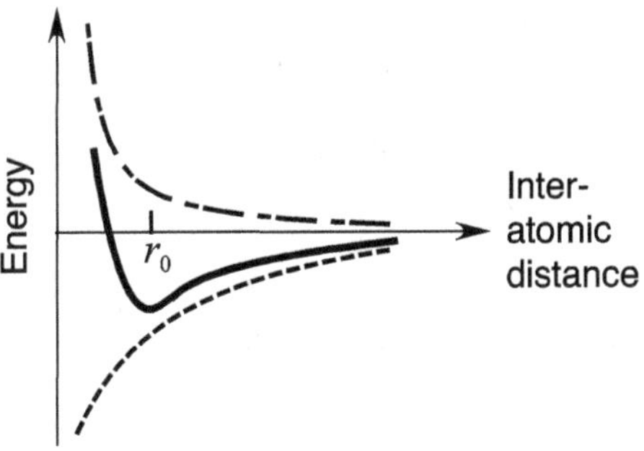

Fig. 2.11 Total potential energy as a function of the separation of two ions. The dot-dashed and dashed curves correspond to contributions of repulsive and Coulomb attraction forces, respectively. r_0 is the equilibrium separation of ions

where A and n are constants and E_{Rep} has a positive value. Thus the sum of (2.11) and (2.12) gives the total energy.

$$E_{\text{T}}(r) = -\frac{\alpha q^2}{4\pi\epsilon_0 r} + \frac{A}{r^n} \tag{2.13}$$

A qualitative expression of $E_{\text{T}}(r)$ is plotted in Fig. 2.11. At equilibrium, $r = r_0$,

$$\frac{\mathrm{d}E_{\text{T}}}{\mathrm{d}r} = \frac{\alpha q^2}{4\pi\epsilon_0 r_0^2} - \frac{nA}{r_0^{n+1}} = 0$$

where r_0 is the equilibrium inter-atomic distance of the ionic crystal. Thus,

$$r_0^{n-1} = \frac{4\pi\epsilon_0 nA}{\alpha q^2} \tag{2.14}$$

The cohesive or binding energy E_0 is obtained from the total energy $E_{\text{T}}(r)$ at $r = r_0$.

$$E_0 = E_{\text{T}}(r_0) = \left(E_{\text{Coul}} + E_{\text{Rep}}\right)_{r_0} = -\frac{\alpha q^2}{4\pi\epsilon_0 r_0}\left(1 - \frac{1}{n}\right) \tag{2.15}$$

For stability of the crystal, n must be greater than unity. Of course it should be possible to determine it from an exact quantum mechanical treatment of the problem. However, from a fit to experimental quantities, n has been determined to be in the range 6–10. Therefore, the Coulomb attraction force becomes the major contributor to the cohesive energy.

In general, the ionic bond is a non-directional bond. It forms a close atomic packing to maximize the number of bonds per unit volume and to minimize the bonding energy per unit volume. Due to the size difference between ions, it can form either a simple cubic or FCC structure to avoid anion–anion or cation–cation contact. The condition that determines the crystal structure can be understood by using a 'hard sphere' model. Imagine the atoms as small hard spheres. When identical atoms are assembled in a layer, the most compact arrangement is the close-packed structure in which every sphere has six touching neighbors. This plane array, as shown in Fig. 2.12, has a hexagonal symmetry and is assigned as layer *A*. To build up a close-packed solid in 3D we must now add a second layer, *B*. The spheres of the second layer will sit in half of the holes of the first layer as shown in Fig. 2.13. When we come to add the third layer, *C*, there are two possible positions where it can fit. First, it could be positioned on top of layer *B* so that its atoms sit in the holes of layer *B* which were also holes in layer *A*. A fourth layer identical to layer *A* can then be added and the *ABCABC* sequence continued. This is known as cubic close packing and produces a FCC lattice. The stacking layer surfaces simply correspond to the {111} surface of the FCC structure. An alternative form of close packing is to position the third layer *C* identically to layer *A*, to fit on top of layer *B*. When we repeat this sequence, the pattern of *ABABAB* produces a hexagonal close-packed (HCP) structure. For cations and anions of similar size, the simple cubic structure (CsCl) is preferred to achieve the highest packing density. Otherwise, it will form an FCC structure (NaCl). Due to their large cohesive energies, these alkali halides, which are among the most stable ionic crystals, have large bandgaps. For example, the bandgaps for NaCl and CsCl are 8.5 and 8.4 eV, respectively.

Fig. 2.12 Hard spheres arranged in a close-packed layer structure

Fig. 2.13 Two layers of close-packed spheres

2.3 Covalent Bond and sp^3 Hybrid Orbit

2.3.1 General Properties of Covalent Bond Formation

Atoms in molecule or solid can 'share' their electrons to form a covalent bond. The simplest example is the N_2 molecule. The nitrogen atom has an electron configuration of $1s^2 2s^2 2p^3$. The $2p$ shell needs six valence electrons to fill its outermost orbit. This can be accomplished by sharing six electrons (three from each atom) between two nitrogen atoms (Fig. 2.14). These shared electrons orbit the two nitrogen atoms, spending equal time with each. Therefore they do not specifically belong to a particular nitrogen atom, but are shared by both. These shared electrons 'bond' two nitrogen atoms into a nitrogen molecule. We shall discuss the physics of the bond formation in detail next.

To illustrate the relation of electron distribution behavior and bonding formation, we use the H_2^+ ion as an example, in which the two protons are held together by one electron. For an isolated hydrogen atom, the wave function has a spherical symmetric form of

$$\psi(1s) = \frac{1}{\sqrt{\pi a_0^3}} \exp(-r/a_0) \tag{2.16}$$

where a_0 is the Bohr radius. The wave functions of two nearby but independent hydrogen atoms, as described by (2.16), are shown in Fig. 2.15a. As the protons are brought closer together to form the H_2^+ ion, the interaction between the electron and two protons leads to two possible wave functions as derived by quantum mechanical calculation. The two lowest energy states have either *symmetric* (even) wave function or *anti-symmetric* (odd) wave function. The change in sign for the anti-symmetric wave merely means that the two wave functions associated with ions are 180° out of phase. From these wave functions, the corresponding probability densities of the electron distribution can be easily estimated as shown in Fig. 2.16. For the anti-symmetric wave function, there is no electron near the center of the system. In the case of symmetric wave function, there is a high probability of an electron existing in the middle of the two nuclei. Different electron distributions directly relate to the energy profile of the system. The potential energy of the system is given as

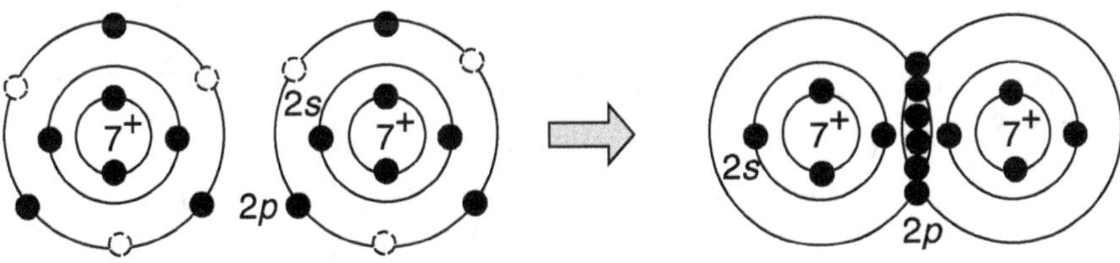

Fig. 2.14 Schematic of the formation of covalent bonding in a nitrogen molecule. The circles in each nitrogen atom represent the unfilled $2p$ valence electron states

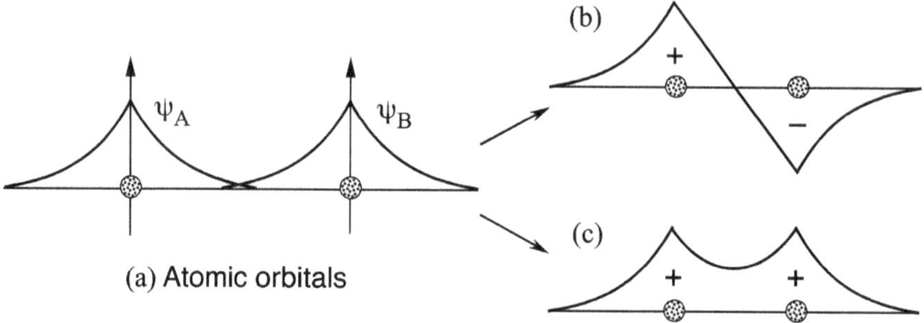

Fig. 2.15 Formation of bonding and anti-bonding combinations of atomic orbitals in diatomic hydrogen molecules with a single bonding electron. **a** The wave functions of two nearby hydrogen atoms. **b** The anti-symmetric wave function and **c** symmetric wave function formed in a H_2^+ ion

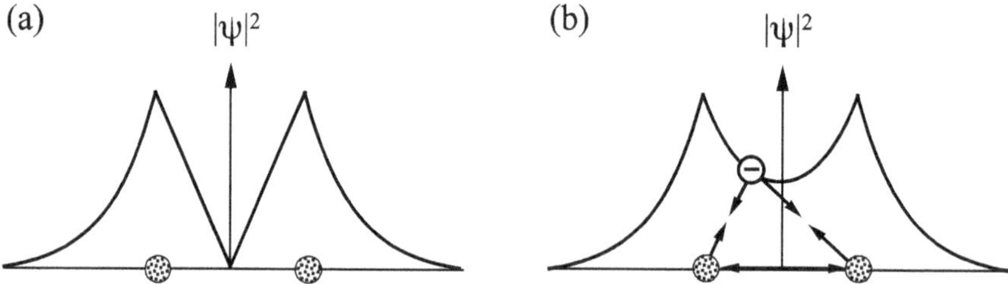

Fig. 2.16 Probability density of the electron distribution in a H_2^+ ion with **a** anti-symmetric and **b** symmetric wave functions

$$E_{\text{PE}} = -E_{e-n} + E_{n-n} \tag{2.17}$$

where $-E_{e\text{-}n}$ represents the electron–nucleus attraction energy and $E_{n\text{-}n}$ represents the nucleus–nucleus repulsion energy (Fig. 2.16b). The high probability of finding electrons in between two nuclei increases the attractive force contribution (more negative) to E_{PE}. As shown in Fig. 2.17, this decreases the potential energy distribution near the center and forms the primary mechanism for the covalent bonding

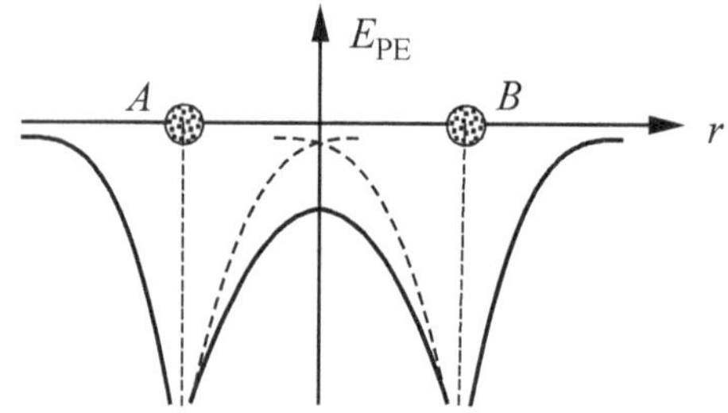

Fig. 2.17 Potential energy of the H_2^+ ion (solid curves). The dashed curve is the potential energy of a single hydrogen nucleus

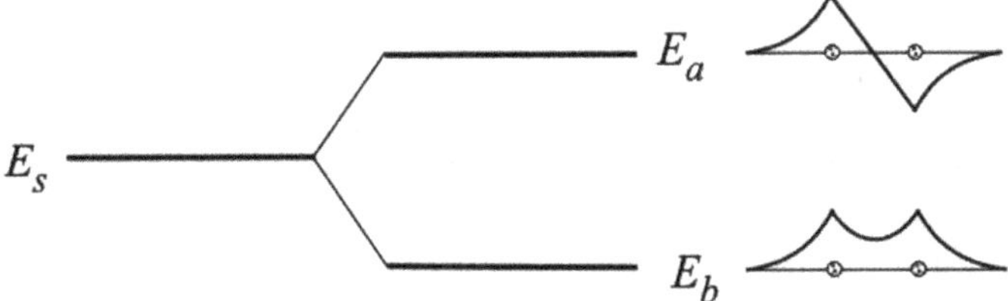

Fig. 2.18 Energy level diagram shows the relative position of bonding (E_b) and anti-bonding (E_a) energies with respect to isolated $\psi(1s)$ state energy (E_s). The related wave functions are also shown

(bonding state). In the anti-symmetry case, the lack of electrons in between two cores means it is unfavorable to form a covalent bond (anti-bonding state).

In other words, the spatial overlap of the wave functions ψ_A and ψ_B leads to the splitting of the original energy level E_s, associated with the $\psi(1s)$ state, into a higher and a lower molecular energy level (Fig. 2.18). The molecular orbital corresponding to the higher energy level is the anti-bonding state. The bonding state is where the electron occupies the lower-lying bonding orbital, thereby giving rise to a reduction in the total energy. This energy reduction corresponds to the binding energy of the covalent bond.

Now, let us turn our attention to hydrogen molecules where there are two valence electrons. When forming covalent bonding in H_2 molecules, the two valence electrons are shared equally by the two nuclei. The symmetric electron wave functions from the two electrons have the same quantum numbers *n, l,* and *m.* They, then, must have opposite spin orientations ($s = \pm$ 1/2) or *anti-parallel spins* to comply with the Pauli exclusion principle. Thus, in a covalent bond the two shared electrons must have anti-parallel spins, and this is also known as the *electron-pair bond*. This concept applies to other materials, such as N_2, which has been discussed earlier. The six 2*p* electrons in a N_2 molecule are bound in pairs with spins aligned anti-parallel in each pair as shown Fig. 2.19a.

We further examine this idea in water (H_2O) and phosphine (PH_3) (Fig. 2.19). The oxygen has an electron configuration $1s^22s^22p^4$, and phosphorous has an electron configuration $1s^22s^22p^63s^23p^3$. Since the *s* electrons in the outer shell already have

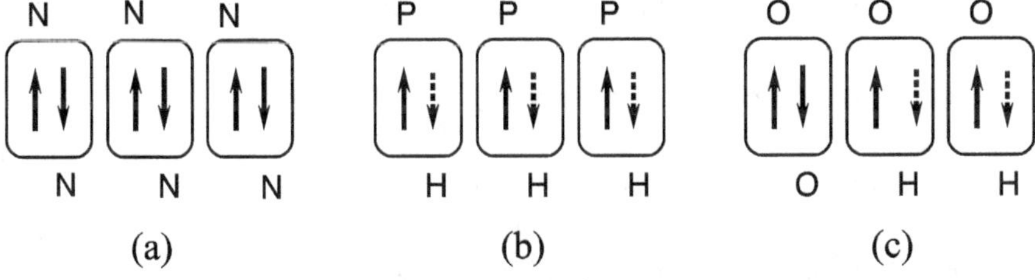

Fig. 2.19 Electron-pair bond formed in between a pair of electrons in *p* orbital with anti-parallel spin in **a** N_2, **b** PH_3, and **c** H_2O. The dotted arrows indicate hydrogen electrons which make pair-bonds with *p* electrons of phosphorous and oxygen

their spins paired, only the p electrons need to be considered. For a phosphorous atom, there are three electrons that have unpaired spins, whereas for an oxygen atom, two of the four electrons already have paired spins leaving two electrons having unpaired spins. Therefore, a phosphorous atom is capable of making three electron-pair bonds with three hydrogen atoms in PH_3, and an oxygen atom is capable of forming two electron-pair bonds in H_2O.

2.3.2 Directional Property of Covalent Bonds

In forming covalent bonds, the accumulated electrons in between the nuclei result in the overlapping of wave functions. A strong bond is formed if there is a maximum wave function overlap. The orbital wave functions of an atom can be derived using a simple model of a hydrogen atom consisting of a proton and a single electron that interact through electrostatic attraction. The wave function in 3D space is expressed as $\psi(r,\theta,\phi) = R(r)\Theta(\theta)\Phi(\phi)$ where R, Θ, and Φ are the radial, azimuthal, and angular wave functions, respectively. The algebraic expressions for some of the wave functions are given below.

$$1s^2\colon\ \psi_{100} = N_{100}\exp\left(-\frac{r}{a_0}\right) \tag{2.18a}$$

$$2s^2\colon\ \psi_{200} = N_{200}\left(2-\frac{r}{a_0}\right)e\left(-\frac{r}{2a_0}\right) \tag{2.18b}$$

$$2p^6\colon \begin{cases} \psi_{210} = N_{210}\left(\frac{r}{a_0}\right)\exp\left(-\frac{r}{2a_0}\right)\cos\theta \\ \psi_{211} = N_{211}\left(\frac{r}{a_0}\right)\exp\left(-\frac{r}{2a_0}\right)\sin\theta\exp(i\phi) \\ \psi_{21\bar{1}} = N_{21\bar{1}}\left(\frac{r}{a_0}\right)\exp\left(-\frac{r}{2a_0}\right)\sin\theta\exp(-i\phi) \end{cases}. \tag{2.18c}$$

where a_0 is the Bohr radius. The superscripts to s and p above indicate the allowed degeneracy of the state. The radial wave functions for the 2s state (ψ_{200}) and 2p state (ψ_{210}) are shown in Fig. 2.20a. The probability density distributions of the s states are always spherically symmetric, while the p, d, f states, etc., have angular dependence. For example,

$$\begin{cases} |\psi_{210}|^2 \propto r^2\exp\left(-\frac{r}{a_0}\right)\cos^2\theta \\ |\psi_{21\pm1}|^2 \propto r^2\exp\left(-\frac{r}{a_0}\right)\sin^2\theta \end{cases} \tag{2.19}$$

Figure 2.21 shows the shape of the probability density distributions for 2s and 2p states in space for a hydrogen atom. However, it is not $|\psi|^2$ itself but the integration of $|\psi|^2$ over the volume that represents the correct electron distribution in a 3D system. In a spherical system like the hydrogen atom model, the volume increases rapidly

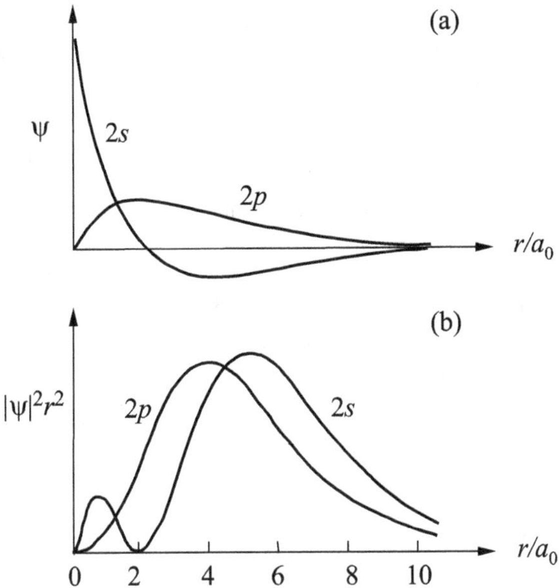

Fig. 2.20 **a** Radial wave functions $R_{nl}(r)$ and **b** radial probability density $|\psi|^2 r^2$ of the 2*s* (ψ_{200}) and 2*p* (ψ_{210}) states of the hydrogen atom as a function of radial distance *r* in the unit of Bohr radius a_0

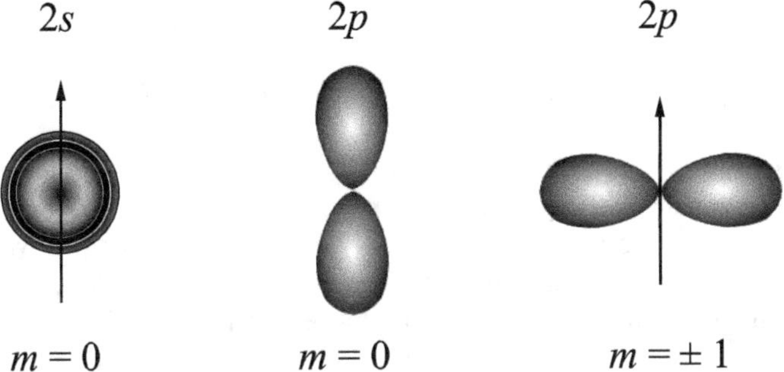

Fig. 2.21 Electron density distributions of 2*s* and 2*p* states for a hydrogen atom. The polar (*z*) axis is oriented vertically

with the radius *r* and the larger values of *r* should be weighted heavily. Therefore, it is more interesting to find the probability that the electron is between *r* and ($r + \mathrm{d}r$),

$$|\psi|^2 \mathrm{d}v = |\psi|^2 r^2 \sin\theta \mathrm{d}r \mathrm{d}\phi \mathrm{d}\theta = |\psi|^2 4\pi r^2 \mathrm{d}r \propto |\psi|^2 r^2. \tag{2.20}$$

Figure 2.20 shows the electron wave functions and the corresponding radial probability densities ($|\psi|^2 r^2$) of 2*s* and 2*p* states in a hydrogen atom. Note that the radial probability density distributions of s and p states overlap heavily. More $2s$ electrons than $2p$ electrons are occupying states at larger r. As we will see in the next section, these results have important consequences in determining crystal structures of semiconductors. Consider the case of mixing s and p orbital wave functions. The s orbits ($l = 0$) of an atom always have a spherical symmetry, i.e., $\psi_s = g(r)$. For p orbits ($l = 1$), the wave functions are angularly dependent as described in (2.18c). Consequently, the spatial distribution of the wave functions leads to a strong orientation dependency in covalent bonding.

Let us take a closer look at the spatial overlap between s and p wave functions. Since the linear combination of ψ_{210} and $\psi_{21\pm1}$ is also a solution to the Schrödinger wave equation, we can transform the polar components of wave functions to the Cartesian coordinates with

$$\begin{cases} x = r \sin\theta \cos\phi \\ y = r \sin\theta \sin\phi \\ z = r \cos\theta \end{cases} \tag{2.21}$$

Thus,

$$\begin{cases} \psi_{211} + \psi_{21\bar{1}} \propto x \\ \psi_{211} - \psi_{21\bar{1}} \propto y \\ \psi_{210} \propto z \end{cases} \tag{2.22}$$

Letting $R(r) = rf(r)$, we can express the three p wave functions as

$$\psi_{px} = xf(r),\ \psi_{py} = yf(r),\ \text{and}\ \psi_{pz} = zf(r) \tag{2.23}$$

The shapes of ψ_s and ψ_p's are shown in Fig. 2.22. Each p orbital extends along a specific coordinate axis, e.g., the x-axis for the ψ_{px} orbital. Therefore, when ψ_s and ψ_p overlap to form a covalent bond, only certain ψ_p orientations will provide a maximum overlap of electron distributions with ψ_s. The covalent bond will be formed along one of these orientations to achieve a bond of maximum strength.

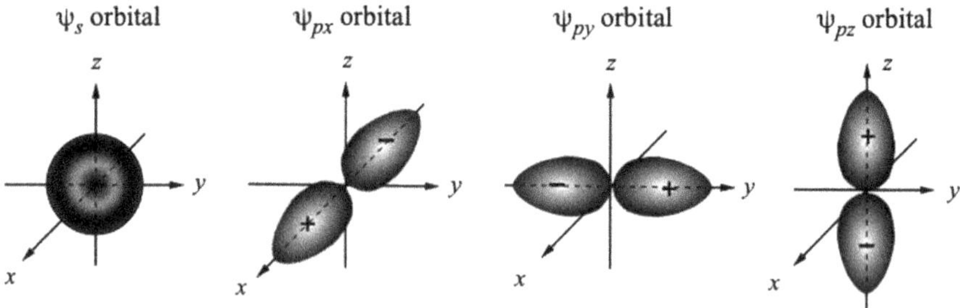

Fig. 2.22 Spherical s orbital and three p orbitals each directed along a different Cartesian axis

2.3.3 sp³ Hybrid Orbit

In semiconductor crystals, the atomic bonds are not structured in the form of pure s or p wave function but a mixture of them. We will use the electron configuration of the methane (CH_4) molecule to illustrate the hybrid orbitals. For a carbon atom, the electron configuration is shown in Fig. 2.23a ($1s^2 2s^2 2p^2$). The inner s orbitals are all full. For the $2p$ orbital, there are only two unpaired electrons available for four single valence electrons from hydrogen atoms to form electron-pair bonds. A closer examination reveals that the energy levels of $2s$ and $2p$ states are very close to each other, and, as shown in Fig. 2.20b, the probability density $|\psi(2p)|^2 r^2$ overlaps heavily with $|\psi(2s)|^2 r^2$. Therefore, one of the two electrons in the $2s$ state can be elevated to the empty $2p$ state. Promoting one of the two electrons to the empty $2p$ state increases the energy of the system. But this allows four electrons from hydrogen atoms to form more electron-pair bonds, which decreases the total energy of the system. Thus, the electron configuration of a carbon atom is changed into $1s^2 2s^1 2p^3$ in a methane molecule. This new configuration has a maximum of four unpaired electrons and is ready to form CH_4 molecules by bonding with electrons from four hydrogen atoms. The electron configuration of the methane molecule now looks like that shown in Fig. 2.23b.

The $2s^1 p^3$ configuration in methane is called sp^3 *hybridization.* If an s-orbital is mixed with three p-orbitals, four sp^3-hybrids are generated. The wave functions of the four sp^3 hybrid orbitals in a CH_4 molecule can be derived by linear combinations of s and p state (2.23) wave functions.

$$\begin{cases} \psi_{111} = \frac{1}{2}[\psi_s + \psi_x + \psi_y + \psi_z] \\ \psi_{1\bar{1}\bar{1}} = \frac{1}{2}[\psi_s + \psi_x - \psi_y - \psi_z] \\ \psi_{\bar{1}1\bar{1}} = \frac{1}{2}[\psi_s - \psi_x + \psi_y - \psi_z] \\ \psi_{\bar{1}\bar{1}1} = \frac{1}{2}[\psi_s - \psi_x - \psi_y + \psi_z] \end{cases} \tag{2.24}$$

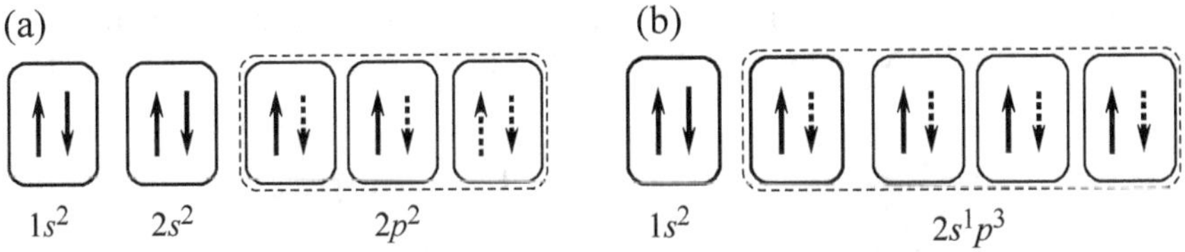

Fig. 2.23 Electron-pair bond formed in **a** carbon atom, and **b** CH_4. The dotted boundary indicates the bonding orbitals. The dotted arrows are sites available for electron-pair bonds

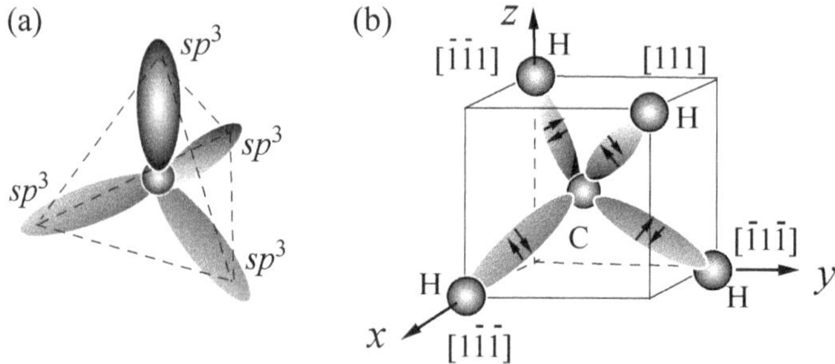

Fig. 2.24 **a** A set of four sp^3 orbitals forms a tetrahedron with each sp^3 orbital containing a large lobe and a small lobe. The large lobes point to the corners of the tetrahedron and can be used to share electrons in covalent bonds. **b** The sp^3 hybrid orbitals of carbon in a CH_4 molecule, each having an electron distribution directed toward one of the four alternating corners of a cube

Four sp^3 orbitals are identical in shape, each one having a large lobe and a minor lobe as shown in Fig. 2.24a. The four orbitals are oriented in the space such that the large lobes form a tetrahedral arrangement. These new hybrid orbitals on carbon are used to share electron pairs with the 1*s* orbitals from the four hydrogen atoms to form a methane molecule, as shown in Fig. 2.24b. Once the methane molecule is formed, there is no extra unpaired free bond to link with other methane molecules. Therefore, a CH_4 crystal can only be formed by van der Waals bonding force. The bonding structure of CH_4 is a *tetrahedron* structure that is determined by optimizing repulsive forces between hydrogen atoms. The tetrahedron bond structure represents the basic form of all semiconductor crystals. For example, the silicon atom has an electron configuration of $1s^22s^22p^63s^23p^2$. The four unpaired electrons in the 3*s* and 3*p* states form sp^3 hybrid orbits for crystal bonding. Each Si atom is bonded with four neighboring Si atoms in a tetrahedron structure shown in Fig. 2.25a. This basic tetrahedron structure can propagate indefinitely in all directions as illustrated in Fig. 2.26. As will be discussed in more detail in Sect. 2.4, the resultant crystal has a diamond structure. For GaAs, the Ga ($1s^22s^22p^63s^23p^63d^{10}4s^24p^1$) and As ($1s^22s^22p^63s^23p^63d^{10}4s^24p^3$) can also bind in the diamond-like (zinc-blende) structure with tetrahedron coordination (Fig. 2.25b). Each Ga atom is surrounded by four

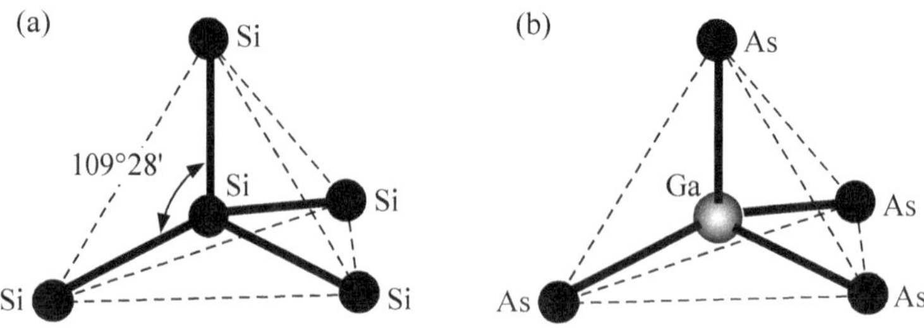

Fig. 2.25 Tetrahedron bond structure of **a** Si and **b** GaAs

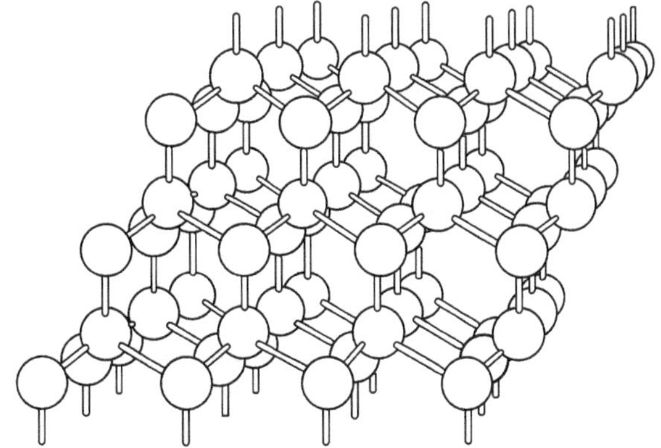

Fig. 2.26 An infinite tetrahedron arrangement of a covalent bonded diamond cubic structure. Reprinted with permission from [1], copyright Oxford University Press

As atoms and vice versa. The shared bonding electrons consist of five electrons from the As atom and three electrons from the Ga. The total number of electrons per atom is the same as in the case of the diamond structure of carbon or silicon.

2.3.4 *Mixed Ionic-Covalent Bonds*

In pure covalent crystals like Si and diamond, the binding orbital between the nearest neighbors is a linear combination of hybridized orbitals from each atom and the wave function of the binding orbital can be expressed as

$$\psi_b \sim \psi_1 + \psi_2 \tag{2.25}$$

where ψ_1 and ψ_2 are the wave functions of the two nearest-neighbor atoms. Due to symmetry, the electrons shared in a bond must spend, on the average, 'equal time' on each atom. This leads to a symmetrical distribution of the electron density and to a purely covalent bond.

In compound semiconductors, the bonding orbitals are less ideal than the pure covalent bond and the sharing of bonds is not so obvious. Consider GaAs as an example, where there are three and five valence electrons associated with Ga ($4s^2 4p^1$) and As ($4s^2 4p^3$) atoms, respectively. On average, there are four bonds per Ga atom sharing with four As atoms. In the simplest picture, each group V atom donates an electron to a group III atom so that each atom has four valence electrons with sp^3 hybrid orbitals. This situation can be expressed as

$$\mathrm{Ga}\left(4s^2 4p^1\right) + \mathrm{As}\left(4s^2 4p^3\right) \rightarrow \mathrm{Ga}\left(4s^1 4p^3\right)^- + \mathrm{As}\left(4s^1 4p^3\right)^+ \tag{2.26}$$

However, this also leads to the formation of III^- and V^+ ions. The situation can be seen partially as electron sharing and partially as electron transfer. Shared electrons

Table 2.2 Crystal structure and ionicity for III–V compound semiconductors. (W = wurtzite structure; ZB = zinc-blende structure) [2]

III–V	Structure	Ionicity	III–V	Structure	Ionicity	III–V	Structure	Ionicity
AlN	W	0.449	GaN	W	0.500	InN	W	0.578
AlP	ZB	0.307	GaP	ZB	0.374	InP	ZB	0.421
AlAs	ZB	0.274	GaAs	ZB	0.310	InAs	ZB	0.357
AlSb	ZB	0.426	GaSb	ZB	0.261	InSb	ZB	0.321

in the bonding states spend a greater fraction of time on the cation, similar to a partially ionic bond. Thus, the time-averaged wave function for a bonding electron is

$$\psi_b \sim \psi_{\text{cov}} + \lambda\psi_{\text{ion}} \tag{2.27}$$

where ψ_{cov} and ψ_{ion} are wave functions for completely covalent and ionic forms, respectively. The weighting factor λ, called the *ionicity parameter*, determines the degree of ionicity (f_{i}):

$$f_i = \frac{\lambda^2 - 1}{\lambda^2 + 1} \tag{2.28}$$

Another measure of ionicity in a compound is the difference in the electronegativity of its constituent elements as suggested by Linus Pauling. The *electronegativity* of an element represents the power of an atom to attract an electron to itself in a compound. Therefore, the electronegativity difference is well correlated with the ionicity of a compound. For purely covalent crystals such as Si and Ge, $f_{\text{i}} = 0$. For III–V compound semiconductors, the theoretical ionicity values along with their crystal structures are listed in Table 2.2. It is interesting to note that compounds with large ionicity value ($f_{\text{i}} \geq 0.45$), e.g., III-nitride compounds, tend to crystallize in wurtzite structure rather than zinc-blende crystal structure.

2.4 Major Semiconductor Crystal Structures

2.4.1 Diamond Structure

To generate a large silicon crystal, we must always fit silicon atoms together in the four-nearest-neighbor pattern suggested in Fig. 2.25a. A section of a crystal so formed, also called a *unit cell* of the crystal, is shown in Fig. 2.27. The atoms in the lower left-hand corner of this figure are accentuated to indicate the fundamental nature of Fig. 2.25a in the crystal construction. This type of lattice structure is characteristic of many important semiconductor crystals including diamond, Si, and Ge

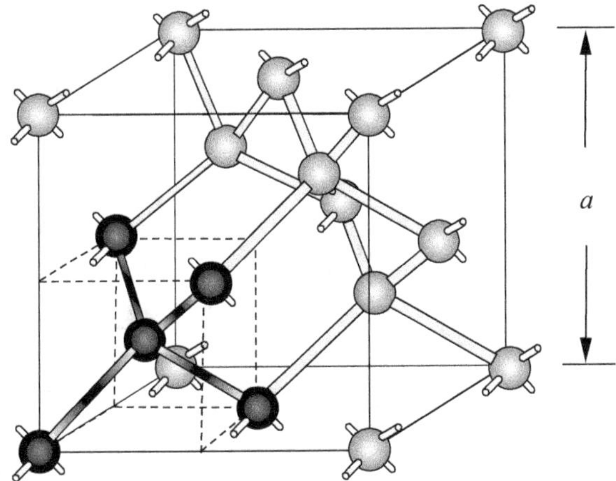

Fig. 2.27 Diamond structure. Each atom is symmetrically surrounded in an imaginary cube (enclosed by dotted lines) by four nearest neighbors

and is called the *diamond* structure. The diamond crystal structure can be commonly described as two interpenetrating FCC structures that are displaced relative to one another along the body diagonal. In a tetrahedral configuration, each atom in the structure is surrounded by four nearest-neighbor atoms. In Fig. 2.27, if we define one of the corners of the first FCC structure as the origin, the position of the origin of the second FCC structure is at [¼,¼,¼]a in terms of the basis unit vector where a is the lattice constant. The nearest-neighbor distance is $\left(\sqrt{3}/4\right)a$. The unit cell cube has four atoms in its interior, an atom in the center of each of its faces (each of which is to be thought of as 'shared' with an adjoining cube in a large crystal), and eight atoms at the cube vertices (each of which is shared with seven other elementary cubes). Thus, the number of atoms in a crystal is eight times the number of elementary cubes; or, equivalently, there are eight atoms in a cube of size a^3.

The other way of constructing the diamond lattice is to use four tetrahedral structures (Fig. 2.25a) as building blocks within a unit cell. As shown in Fig. 2.28, the bottom half of the unit cell is formed by joining two tetrahedral structures diagonally. The top half of the unit cell has the same arrangement but is rotated 90° with respect to the bottom half diagonal.

2.4.2 *Zinc-Blende Structure*

The zinc-blende crystal structure is closely related to the diamond structure. The only difference is that in zinc-blende structures, there are two different types of atoms in the lattice. As shown in Fig. 2.29, the two different atoms occupy alternating lattice sites. In each tetrahedral cubic structure, the center atom is different from the atoms located at corners.

The diamond structure of elemental semiconductors has the highest symmetry among all semiconductors. However, it lacks many of the symmetrical features

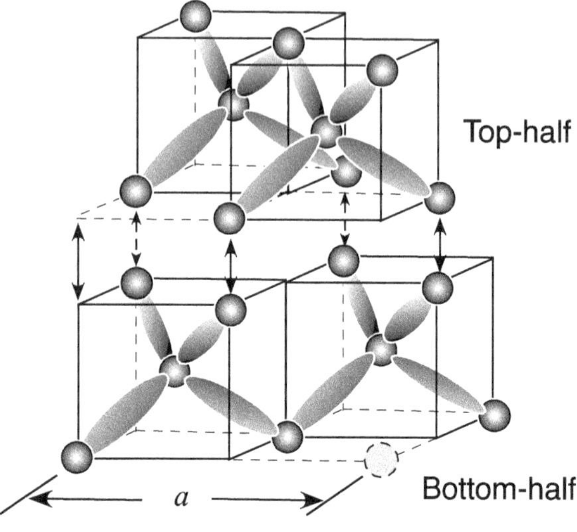

Fig. 2.28 Construction of a diamond lattice structure using tetrahedral structures as building blocks

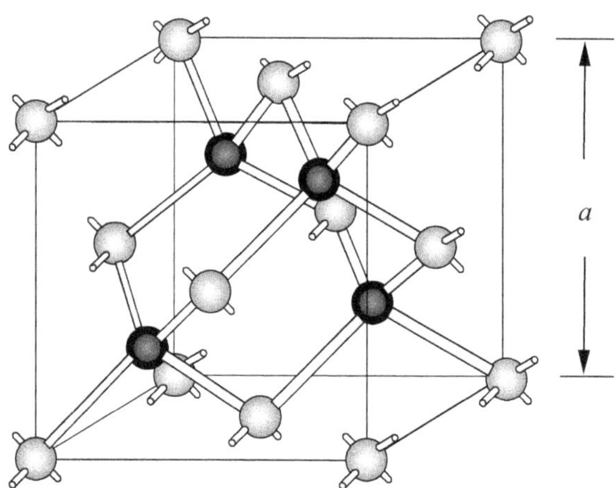

Fig. 2.29 Zinc-blende crystal structure. Two types of atoms are arranged alternately with each atom surrounded by four nearest neighbors of different types of atoms

enjoyed by primitive cubic. Compared to the diamond crystal structure, the symmetry of zinc-blende structure is further reduced. The physical structures of (111) and $(\bar{1}\bar{1}\bar{1})$ surfaces are the same for diamond structure but different for zinc-blende crystal. For example, in GaAs the (111) consists entirely of Ga atoms, while $(\bar{1}\bar{1}\bar{1})$ is a sheet of As atoms. Because of the different surface constituents, these two types of surface show very different chemical properties. Further, due to the nature of partially ionic bonding in zinc-blende structure, electric dipoles exist between the neighboring Ga and As atoms along $\langle 111 \rangle$. Although, at equilibrium, the net internal electric field is zero as the individual dipole moments get canceled out, compressions along $\langle 111 \rangle$ will cause atoms to move closer and induce a net internal electric field. This piezoelectric effect in compound semiconductors enhances carrier scattering which reduces the carrier mobility. More detailed discussions on the physical properties of III–V compounds with zinc-blende structure will be given in Chap. 4.

2.4.3 Wurtzite Structure

The wurtzite structure has the same tetrahedral coordination as the zinc-blende structure. It is the disposition of second-nearest neighbors which produces a hexagonal crystal. Figure 2.30 shows the unit cell of the hexagonal close-packed (HCP) structure. The basic HCP is formed by interlacing two basic hexagonal structures along the c-axis. The second hexagonal lattice is displaced from the first by $c/2$ along the c-axis and by $2a/3$ and $a/3$ in the $\boldsymbol{a}_1$- and $\boldsymbol{a}_2$-directions, respectively. Only three atoms from the base of the second hexagonal are enclosed inside the first hexagonal unit cell. Each atom has twelve nearest neighbors: six in its own plane and three each in the planes above and below. The primitive unit cell has a rhombohedral structure as outlined. In the HCP structure, the $\boldsymbol{c}$- and $\boldsymbol{a}$-axes are related by $c = 2a\sqrt{2/3}$.

The wurtzite structure is formed by interlacing two identical HCP structures containing different atoms, e.g., group-III (Ga) and group-V (N) atoms, respectively, along c-axis. The displacement of the two lattices is the bond length of $3c/8$. This arrangement of atoms can be seen from the primitive unit cell structure illustrated in Fig. 2.31. The hexagonal unit cell of a wurtzite crystal structure is shown in Fig. 2.32. It is clear that along the c-axis, layers containing atom A and atom B are staggered similarly to the (111) surface in the zinc-blende structure.

Next, we shall investigate why the nearest-neighbor atomic arrangement is the same for both zinc-blende and wurtzite structures but different for the second-nearest neighbors. In both structures, the bonds are covalent in nature with a certain fraction of ionicity. For zinc-blende crystals, the electron–electron repulsive force dominates the bond-forming mechanism, leading to the tetrahedral structure. The second nearest neighbors are also arranged by optimizing the repulsive forces between atoms. As shown in Fig. 2.33a, using zinc-blende $\langle 111 \rangle$ GaN as an example, the second-nearest

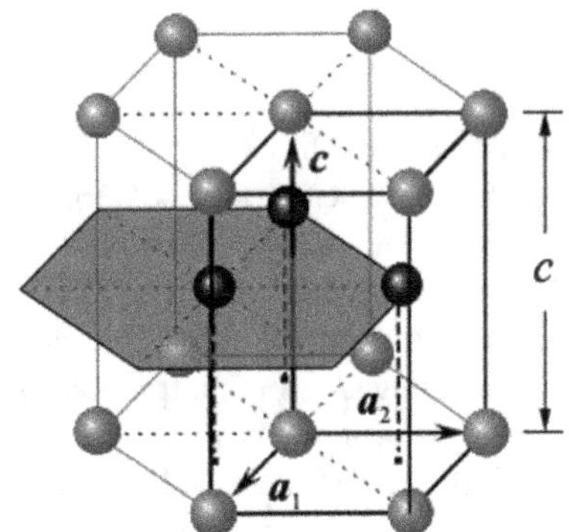

Fig. 2.30 Unit cell of the hexagonal close-packed (HCP) structure. The three atoms from the second hexagonal inside the unit cell are shown as dark spheres. The primitive cell is outlined and defined by unit vectors $\boldsymbol{a}_1$, $\boldsymbol{a}_2$, and $\boldsymbol{c}$

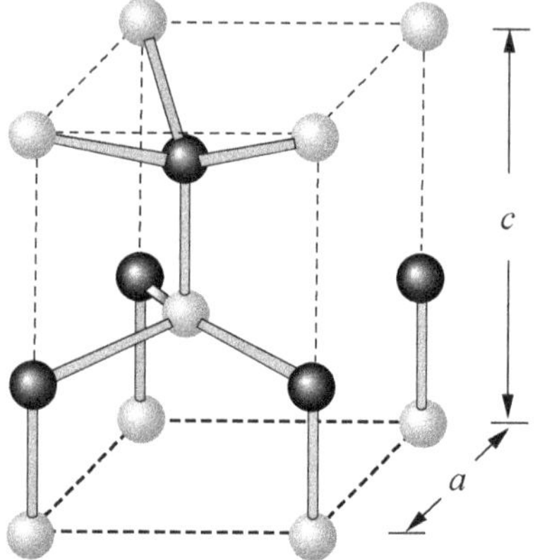

Fig. 2.31 Primitive unit cell of the wurtzite structure

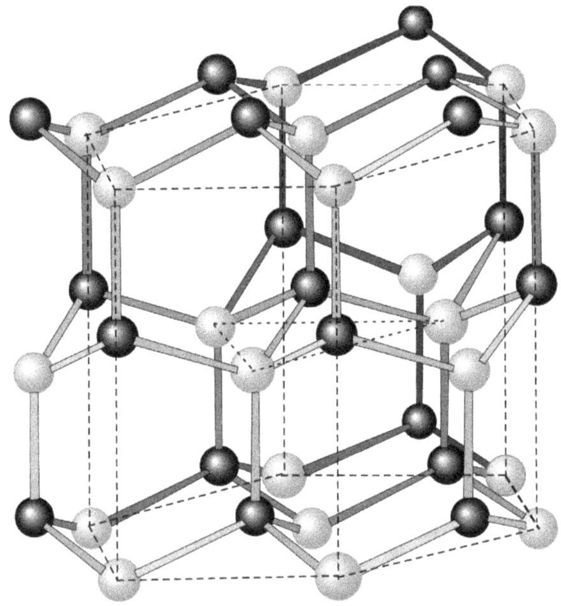

Fig. 2.32 Unit cell of the wurtzite structure. The hexagonal boundary is outlined with dashed lines

Ga atoms in the top (111) plane are positioned in between the N atoms in the lower (111) plane. The two types of atoms in the neighboring (111) planes tend to minimize the repulsive force by maximizing the separation. For the wurtzite GaN structure, the high fraction of ionicity (0.5) implies that atoms are more ionized into Ga^- and N^+. As shown in Fig. 2.33b, the cations and the anions tend to align along the c-axis by the strong Coulomb attraction force. Therefore, the preferred crystal structure of GaN is the wurtzite structure.

The layered structures shown in Fig. 2.33 containing atom A and atom B are similar in that they are staggered along the c-axis and the ⟨111⟩ axis in wurtzite structure and

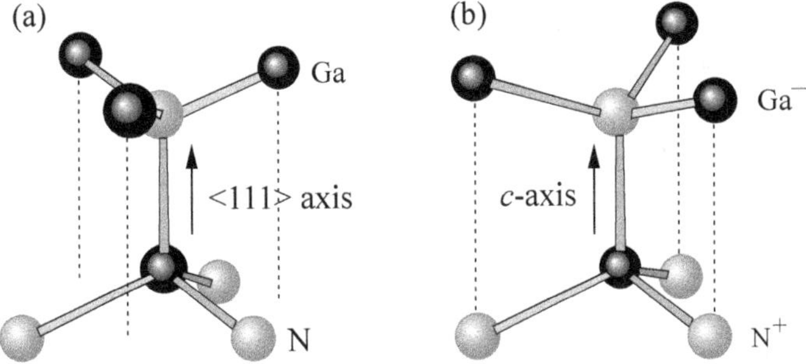

Fig. 2.33 Second-nearest-neighbor bond arrangements along **a** ⟨111⟩ axis in zinc-blende and **b** *c*-axis in wurtzite GaN crystals

zinc-blende structure, respectively. But the similarity stops here. The alignment of cations and anions along the *c*-axis in wurtzite crystals such as GaN can easily induce a piezoelectric field by a mechanical stress applied along the same direction. The relative movements between cations and anions, caused by lattice vibrations above finite temperatures, break the equilibrium of the total electric field of the system. A net spontaneous polarization electric field is induced in the crystal along the *c*-axis. Thus, the (0001) *c*-plane is a polar surface. However, for some crystal planes containing mixed anions and cations that have equal numbers of bonds pointing out of the plane, e.g., $(1\bar{1}00)$ *m*-plane and $(11\bar{2}0)$ *a*-plane, the net spontaneous polarization field is zero. These are non-polar planes with *c*-axis parallel to the layer surface as shown in Fig. 2.34. Any crystal plane inclined between a non-polar plane and *c*-plane is a semi-polar plane including $(10\bar{1}1)$, $(11\bar{2}2)$ and $(20\bar{2}1)$.

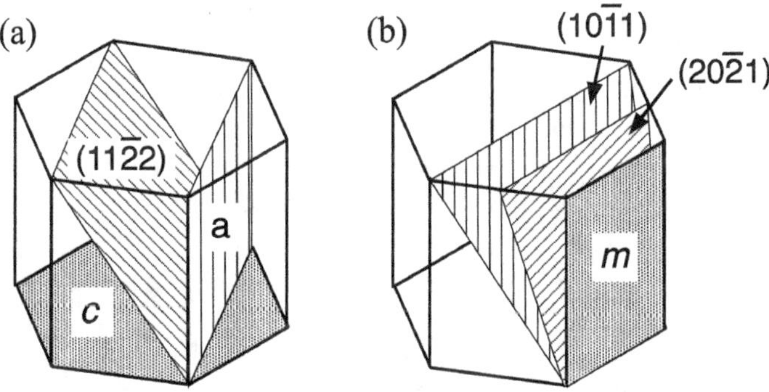

Fig. 2.34 Non-polar and semi-polar crystal planes in wurtzite III-N crystals. **a** $(11\bar{2}0)$ *a* plane and $(11\bar{2}2)$ semi-polar plane and **b** $(1\bar{1}00)$ *m*-plane and $(10\bar{1}1)$ and $(20\bar{2}1)$ semi-polar planes

2.5 Reciprocal Lattice, Diffraction Condition, and Brillouin Zone

2.5.1 Crystal Diffraction

The lattice structure of a crystal can be deduced by diffraction experiments using radiation of a wavelength comparable with lattice dimensions. X-rays with proper energy are widely used as the diffraction sources to study semiconductor crystals. A suitable combination of X-ray wavelength λ and incident angle θ will enhance the diffraction according to Bragg's law

$$n\lambda = 2d_{hkl} \sin \theta \tag{2.29}$$

where d_{hkl} is the lattice spacing and n is an integer (Fig. 2.35). Thus X-rays of a single wavelength directed on a crystal at an arbitrary angle will not be reflected. Currently, the two most widely used X-ray diffraction techniques are the following: The Laue method utilizes X-rays of many wavelengths at a single angle of incidence, and the rotating crystal method allows monochromatic rays to encounter the crystal at a variety of angles.

The Laue method uses a narrow beam of broadband X-rays to shine on a single crystal as shown in Fig. 2.36a. A diffraction beam will emerge for any X-ray wavelength satisfying the Bragg condition. All backscattered diffraction beams are recorded simultaneously in the photographic plate positioned between the crystal and the X-ray source. Each recorded diffraction 'spot' represents a specific wavelength that fits the Bragg condition. Therefore the crystal structure information recorded in the photographic plate is in the *frequency* domain. On the other hand, the rotating crystal method, shown in Fig. 2.36b, uses a monochromatic X-ray beam, shaped from a broadband source through multiple Bragg reflections from separate crystals, as the diffraction source on a crystal rotating at a constant speed. At any rotation angle θ that fulfills the Bragg condition, a diffraction beam emerges. Since the orientation (θ) of the crystal is varying as a function of time, the diffraction beams are recorded as 'peaks' in the *time* domain. For the same crystal subjected to diffraction

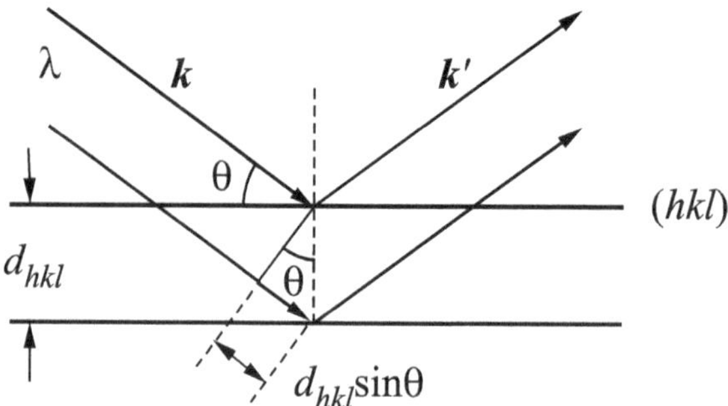

Fig. 2.35 Reflection of a parallel beam of X-rays from two adjacent *(hkl)* planes of atoms in a crystal. The path difference between two X-rays has to be an integer multiplication of the path difference of $2d_{hkl} \sin \theta$ for constructive reinforcement

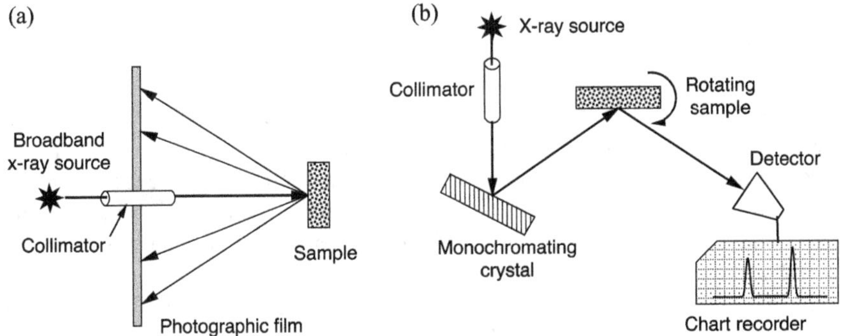

Fig. 2.36 **a** Laue diffraction method arrangement using broadband X-ray source. **b** A rotating crystal X-ray diffraction arrangement using monochromatic X-rays selected by Bragg reflection from a separate single crystal

measurements using both Laue and rotating crystal methods, the different diffraction results presented in the frequency and time domains are equivalent. In fact, because the frequency ($f = 1/t$) is the inverse of time (t), the Fourier transform allows us to convert diffraction results from time to frequency domains, and vice versa. Just as a time-varying quantity can be described as a sum of Fourier components in the frequency domain, so the spatial properties of a crystal can be described as the sum of Fourier components in Fourier space, known as the *reciprocal* space which has a dimension of length^{-1}. This leads to the concept of the *reciprocal lattice.* In the discussion that follows, we shall see that the reciprocal lattices are associated with the Bragg diffraction condition in a crystal. Later in Chap. 3 we will apply the Fourier transformation to associate wave functions in real space (in length) and momentum space (in length^{-1}).

2.5.2 Reciprocal Lattice

So far we have described the crystal structure in terms of the real-space lattice points of atomic positions. However, to understand the wave diffractions by crystals, as described in the last section, it is more convenient to describe the crystal structure from the point of view of crystal planes. A set of (hkl) planes of the crystal structure is completely specified by the unit vector of the plane $\boldsymbol{n}$ and the inter-plane spacing d_{hkl}. The overall structure of the crystal is then completely specified by the set of values of $\boldsymbol{n}$ and d_{hkl}. This specific way of relating crystal structures to values of $\boldsymbol{n}$ and d_{hkl} is cumbersome. Conversely, a much more useful representation is to define a vector which uniquely relates to each set of $\boldsymbol{n}$ and d_{hkl} of the real-space lattice points

$$\boldsymbol{G}_{hkl} = 2\pi \boldsymbol{n}/d_{hkl} \tag{2.30}$$

It is obvious that this vector has the dimension of inverse length and can be used to define the so-called *reciprocal lattice.* Thus, every crystal structure has both a real-space lattice and a reciprocal lattice associated with it.

To further explore the concept of reciprocal lattice vectors, the interactions of a wave with a periodic physical quantity of a crystal are illustrated below. In crystal diffraction measurements, the condition for the constructive interference of a wave scattered from the lattice points depends on the detailed unit cell structure of the crystal. Thus, the scattering intensity of the incident wave is determined by the spatial distribution of atoms within each cell. In periodic lattice structures, Fourier transformation can be applied to relate the direct lattice and wave vector domains. In a 1D system of a periodic lattice with a period a along x, the scattering wave density ρ satisfies the condition

$$\rho(x + a) = \rho(x) \tag{2.31}$$

Using Fourier transformation, the 1D wave density becomes

$$\rho(x) = \frac{a_0}{2} + \sum_{n=1}^{\infty}\left(a_n \cos\frac{2n\pi}{a}x + b_n \sin\frac{2n\pi}{a}x\right) = \sum_{-\infty}^{\infty} c_n \exp\left(i\frac{2n\pi}{a}x\right) \tag{2.32}$$

where

$$c_n = \frac{1}{a}\int_{-a/2}^{a/2} \rho(x) \exp\left(-i\frac{2n\pi}{a}x\right)\mathrm{d}x \tag{2.33}$$

is the Fourier coefficient. A displacement of $x = ma$, where m is an integer, leads to an identical $\rho(x)$, hence satisfying the periodic requirement. Assume $2\pi/a \equiv |\boldsymbol{k}|$ is the wave vector associated with the direct lattice spacing a. The 1D scattering wave density is expressed in terms of the wave vector $\boldsymbol{k}$ and the lattice vector $\boldsymbol{x}$ by

$$\rho(x) = \sum_{-\infty}^{\infty} c_n \exp(in\boldsymbol{k}\cdot\boldsymbol{x}) \tag{2.34}$$

In 3D periodic systems,

$$\rho(\boldsymbol{r} + \boldsymbol{T}) = \rho(\boldsymbol{r}) \tag{2.35}$$

where $\boldsymbol{T} = h'\boldsymbol{a} + k'\boldsymbol{b} + l'\boldsymbol{c}$ is the translation vector. We can use the same equation with a minor modification as shown below:

$$\rho(r) = \sum_{n} c_G \exp(i\boldsymbol{G}\cdot\boldsymbol{r}_n) \tag{2.36}$$

The wave vector $\boldsymbol{k}$ and lattice vector $\boldsymbol{x}$ are replaced by $\boldsymbol{G}$ and $\boldsymbol{r}_n$, respectively. In order to preserve the translation invariance of $\rho(r)$ with respect to the lattice vector $\boldsymbol{r}_n$, the vector $\boldsymbol{G}$ must fulfill certain conditions. Just like the 1D case, the product of $\boldsymbol{G}$ and $\boldsymbol{r}_n$ has to preserve the periodic property of the lattice. Thus,

$$\boldsymbol{G}\cdot\boldsymbol{r}_n = 2m\pi \text{ and } \boldsymbol{r}_n = n_1\boldsymbol{a} + n_2\boldsymbol{b} + n_3\boldsymbol{c} \tag{2.37}$$

where m is an integer for all n_1, n_2, and n_3, and $\boldsymbol{a}$, $\boldsymbol{b}$, $\boldsymbol{c}$ are the unit vectors of the lattice. Assume vector $\boldsymbol{G}$ can be expressed as

$$\boldsymbol{G} = h\boldsymbol{a}^* + k b^* + l\boldsymbol{c}^* \tag{2.38}$$

where h, k, l are integers and $\boldsymbol{a}^*$, $\boldsymbol{b}^*$, $\boldsymbol{c}^*$ are unit vectors of a different coordination system. Then

$$\boldsymbol{G}\cdot\boldsymbol{r}_n = \left(h\boldsymbol{a}^* + k\boldsymbol{b}^* + l\boldsymbol{c}^*\right)\cdot(n_1\boldsymbol{a} + n_2\boldsymbol{b} + n_3\boldsymbol{c}) = 2m\pi$$

The relations between $\boldsymbol{a}$, $\boldsymbol{b}$, $\boldsymbol{c}$ and $\boldsymbol{a}^*$, $\boldsymbol{b}^*$, $\boldsymbol{c}^*$ are derived as follows. For example, in the case of $n_2 = n_3 = 0$, we have

$$\boldsymbol{G}\cdot\boldsymbol{r}_n = n_1\left(h\boldsymbol{a}\cdot\boldsymbol{a}^* + k\boldsymbol{a}\cdot\boldsymbol{b}^* + l\boldsymbol{a}\cdot\boldsymbol{c}^*\right) = 2m\pi$$

For an arbitrary choice of n_1, this can only be satisfied by

$$\boldsymbol{a}\cdot\boldsymbol{a}^* = 2\pi \text{ and } \boldsymbol{a}\cdot\boldsymbol{b}^* = \boldsymbol{a}\cdot\boldsymbol{c}^* = 0$$

This implies a general expression of

$$\boldsymbol{G}_i\cdot\boldsymbol{r}_j = 2\pi\delta_{ij} \tag{2.39}$$

where $\boldsymbol{G}_i$ and $\boldsymbol{r}_j$ are the ith and jth components of vectors $\boldsymbol{G}$ and $\boldsymbol{r}_n$, respectively, and

$$\delta_{ij} = \begin{cases} 1 \text{ for } i = j \\ 0 \text{ for } i \neq j \end{cases} \tag{2.40}$$

The conditions stated above of $\boldsymbol{a}\cdot\boldsymbol{a}^* = 2\pi$ and $\boldsymbol{a}\cdot\boldsymbol{b}^* = \boldsymbol{a}\cdot\boldsymbol{c}^* = 0$ imply $\boldsymbol{a}^* = k_1(\boldsymbol{b}\times\boldsymbol{c})$, or $\boldsymbol{a}^*$ is perpendicular to the plane containing $\boldsymbol{b}$ and $\boldsymbol{c}$. Therefore

$$\boldsymbol{a}\cdot\boldsymbol{a}^* = 2\pi = k_1(\boldsymbol{a}\cdot\boldsymbol{b}\times\boldsymbol{c})$$

and

$$k_1 = \frac{2\pi}{(\boldsymbol{a}\cdot\boldsymbol{b}\times\boldsymbol{c})} \tag{2.41}$$

This leads to the definition of the *reciprocal lattice* unit vectors $\boldsymbol{a}^*$, $\boldsymbol{b}^*$, and $\boldsymbol{c}^*$,

$$\begin{cases} \boldsymbol{a}^* = 2\pi \dfrac{\boldsymbol{b} \times \boldsymbol{c}}{(\boldsymbol{a} \cdot \boldsymbol{b} \times \boldsymbol{c})} \\ \boldsymbol{b}^* = 2\pi \dfrac{\boldsymbol{c} \times \boldsymbol{a}}{(\boldsymbol{a} \cdot \boldsymbol{b} \times \boldsymbol{c})} \\ \boldsymbol{c}^* = 2\pi \dfrac{\boldsymbol{a} \times \boldsymbol{b}}{(\boldsymbol{a} \cdot \boldsymbol{b} \times \boldsymbol{c})} \end{cases} \tag{2.42}$$

and $\boldsymbol{G} = h\boldsymbol{a}^* + k\boldsymbol{b}^* + l\boldsymbol{c}^*$ is the reciprocal lattice vector.

Whether this $\boldsymbol{G}$ satisfies the periodicity requirement can be examined by using the following example. Assuming $\boldsymbol{G}$ is the solution, it should possess the periodic property of

$$\rho(\boldsymbol{r} + \boldsymbol{T}) = \sum_G c_G \exp(i\boldsymbol{G} \cdot \boldsymbol{r}) exp(i\boldsymbol{G} \cdot \boldsymbol{T}) = \rho(\boldsymbol{r})$$

$$\exp(i\boldsymbol{G} \cdot \boldsymbol{T}) = \exp\left[i\left(hh' + kk' + ll'\right)\right] = \exp(iu) = \sin(2u\pi) + \cos(2u\pi) = 1$$

where $u = (hh' + kk' + ll') =$ integer. Thus,

$$\rho(\boldsymbol{r} + \boldsymbol{T}) = \sum_G c_G \exp(i\boldsymbol{G} \cdot \boldsymbol{r}) = \rho(\boldsymbol{r}) \tag{2.43}$$

The periodicity is maintained in the 3D reciprocal lattice.

2.5.3 Properties of the Reciprocal Lattice Vector

As shown in Fig. 2.37, in direct lattice, the (hkl) plane intersects with ***a*-, *b*-,** and ***c***-axes at $1/h$, $1/k$, and $1/l$, respectively. The three position vectors extending from the

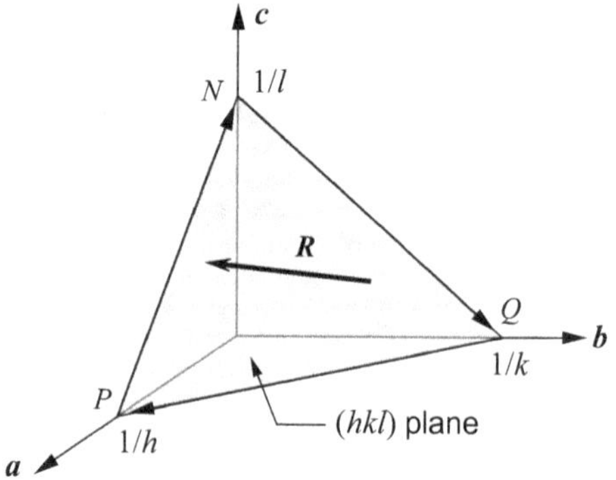

Fig. 2.37 Geometry of the direct lattice (hkl) plane and related vectors

origin to these three intersections are $\boldsymbol{a}/h$, $\boldsymbol{b}/k$, and $\boldsymbol{c}/l$. Using these three vectors, we can define three vectors lying along the edges of the (hkl) plane as indicated by

$$\boldsymbol{QP} = \left(\frac{\boldsymbol{a}}{h} - \frac{\boldsymbol{b}}{k}\right),\ \boldsymbol{PN} = \left(\frac{\boldsymbol{a}}{h} - \frac{\boldsymbol{c}}{l}\right),\ \text{and}\ \boldsymbol{NQ} = \left(\frac{\boldsymbol{b}}{k} - \frac{\boldsymbol{c}}{l}\right)$$

A linear combination of the above three vectors ($\boldsymbol{QP} \times D + \boldsymbol{PN} \times E + \boldsymbol{NQ} \times F$) forms a general vector $\boldsymbol{R}$ lying in the (hkl) plane:

$$\boldsymbol{R} = \frac{\boldsymbol{a}}{h}(D - E) + \frac{\boldsymbol{b}}{k}(F - D) + \frac{\boldsymbol{c}}{l}(E - F) \tag{2.44}$$

where D, E, and F are integers. It can be further simplified by assuming $A = D - E$, $B = F - D$, $C = E - F$, and $A + B + C = 0$. Thus,

$$\boldsymbol{R} = A\left(\frac{\boldsymbol{a}}{h}\right) + B\left(\frac{\boldsymbol{b}}{k}\right) + C\left(\frac{\boldsymbol{c}}{l}\right) = \left(\frac{A}{h}\right)\boldsymbol{a} + \left(\frac{B}{k}\right)\boldsymbol{b} + \left(\frac{C}{l}\right)\boldsymbol{c} \tag{2.45}$$

$$\begin{aligned} \boldsymbol{R} \cdot \boldsymbol{G}_{hkl} &= \left[\left(\frac{A}{h}\right)\boldsymbol{a} + \left(\frac{B}{k}\right)\boldsymbol{b} + \left(\frac{C}{l}\right)\boldsymbol{c}\right] \cdot (h\boldsymbol{a}^* + k\boldsymbol{b}^* + l\boldsymbol{c}^*) \\ &= \left(\frac{A}{h}\right)h + \left(\frac{B}{k}\right)k + \left(\frac{C}{l}\right)l = A + B + C = 0 \end{aligned} \tag{2.46}$$

Therefore, $\boldsymbol{R} \perp \boldsymbol{G}_{hkl}$. The reciprocal lattice vector defined by $\boldsymbol{G}_{hkl} = h\boldsymbol{a}^* + k\boldsymbol{b}^* + l\boldsymbol{c}^*$ lies perpendicular to the direct lattice (hkl) plane.

We now show that with the aid of Fig. 2.38 the separation of the planes in the direct lattice d_{hkl} is equal to $2\pi/\boldsymbol{G}_{hkl}$.

$$d_{hkl} = \boldsymbol{r} \cdot \boldsymbol{n} = \boldsymbol{r} \cdot \frac{\boldsymbol{G}_{hkl}}{|\boldsymbol{G}_{hkl}|} = \frac{\boldsymbol{a}}{h} \cdot \frac{h\boldsymbol{a}^* + k\boldsymbol{b}^* + l\boldsymbol{c}^*}{|\boldsymbol{G}_{hkl}|} = \frac{2\pi}{|\boldsymbol{G}_{hkl}|} \tag{2.47}$$

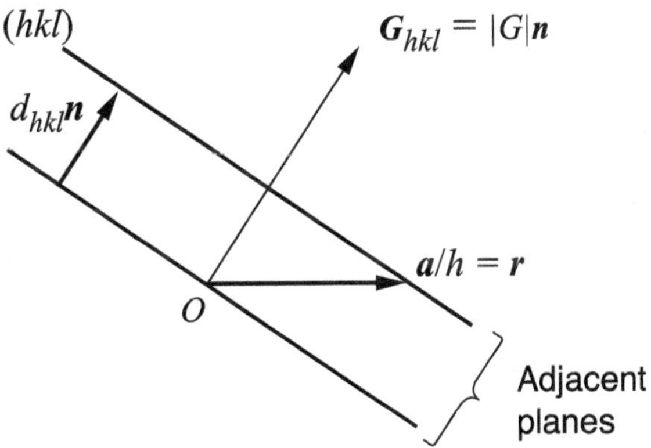

Fig. 2.38 Construction for the determination of the spacing between successive (hkl) planes. $\boldsymbol{n}$ is the surface normal unit vector

The perpendicular distance of the adjacent (hkl) planes is

$$d_{hkl} = \frac{2\pi}{|G_{hkl}|} \text{ or } d_{hkl} = \frac{2\pi}{\sqrt{h^2 + k^2 + l^2}} \tag{2.48}$$

2.5.4 Diffraction Condition

From Fig. 2.35, the Bragg law of diffraction (2.29) requires $n\lambda = 2d_{hkl} \sin\theta$. It is also possible to express the diffraction condition with the reciprocal lattice vector. Instead of considering the wavelength λ of the radiation which interacts with the crystal, we shall now consider the initial and final wave vectors $\boldsymbol{k}$, $\boldsymbol{k}'$ of a reflected diffraction. Assume a wave vector $\boldsymbol{k}$ associated with a wavelength λ is incident on the (hkl) plane and encounters an elastic diffraction or elastic scattering where there is no change in λ. Therefore, the diffracted wave vector

$$|\boldsymbol{k}'| = |\boldsymbol{k}| = 2\pi/\lambda \tag{2.49}$$

The change of $\boldsymbol{k}$, shown in Fig. 2.39, is

$$\Delta\boldsymbol{k} = \boldsymbol{k}' - \boldsymbol{k} = 2|\boldsymbol{k}|\boldsymbol{n}\sin\theta = \left(\frac{4\pi}{\lambda}\sin\theta\right)\boldsymbol{n} = \left(\frac{4\pi}{\lambda}\sin\theta\right)\frac{\boldsymbol{G}_{hkl}}{|\boldsymbol{G}_{hkl}|} \tag{2.50}$$

The change $\Delta\boldsymbol{k}$ is in the direction of the surface normal and perpendicular to the (hkl) planes.

Since $d_{hkl} = 2\pi/|\boldsymbol{G}_{hkl}|$ and $2d\sin\theta = n\lambda$, then

$$\Delta\boldsymbol{k} = \left(\frac{2d_{hkl}}{\lambda}\sin\theta\right)\boldsymbol{G}_{hkl} \tag{2.51}$$

$$\Delta\boldsymbol{k} = \boldsymbol{G}_{hkl} \tag{2.52}$$

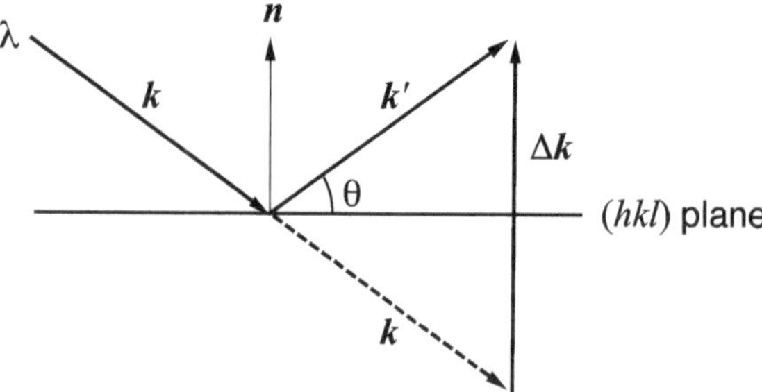

Fig. 2.39 The change in wave vector when undergoing an elastic scattering at the (hkl) plane

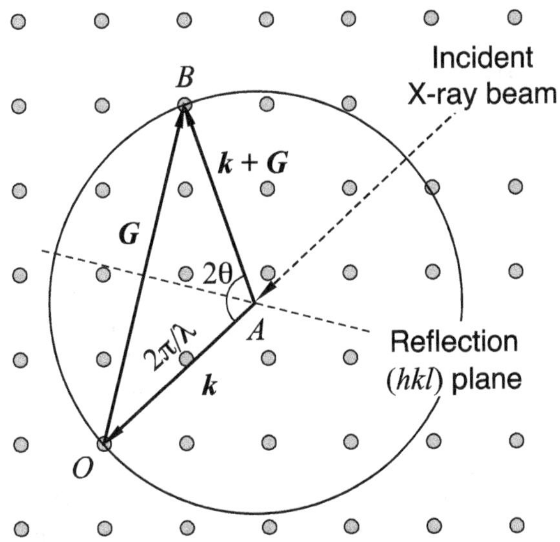

Fig. 2.40 Ewald construction for the Laue diffraction condition from a crystal in the reciprocal lattice. The incident X-ray beam has a wavelength λ

or

$$\boldsymbol{k}' = \boldsymbol{k} + \boldsymbol{G}_{hkl} \tag{2.53}$$

Thus the scattering vector $\Delta\boldsymbol{k}$ is equal to a reciprocal lattice vector $\boldsymbol{G}_{hkl}$. This is the 3D Bragg diffraction condition.

In the Laue method of Fig. 2.36a, the array of diffraction spots from a crystal, recorded on the photographic film, thus corresponds to the set of points generated by $\boldsymbol{G}_{hkl}$ in the reciprocal lattice space. This can be represented graphically by the *Ewald sphere construction* shown in Fig. 2.40. We start by drawing a vector corresponding to the incident X-ray beam, whose length is $2\pi/\lambda$ and which terminates at a point O of the reciprocal lattice. Note that the tail of the vector $\boldsymbol{AO}$ does not necessarily have to rest on a reciprocal lattice point. A sphere with a radius $|\boldsymbol{k}|$ or $2\pi/\lambda$ is constructed about point A as the center. The diffraction condition is satisfied whenever the spherical surface coincides with points of the reciprocal lattice where $\boldsymbol{k}' = \boldsymbol{k} + \boldsymbol{G}_{hkl}$. The vector $\boldsymbol{OB}$ or $\boldsymbol{G}_{hkl}$ that connects the two reciprocal lattice points coinciding with the sphere must be normal to the (hkl) plane of the direct lattice. Using simple trigonometric relations, we can verify in the following that $\boldsymbol{G}_{hkl}$ satisfies the Bragg condition. From (2.48), $\boldsymbol{G}_{hkl}$ has a length $2n\pi/d_{hkl}$, where n is an integer. From Fig. 2.40, the length OB can also be expressed as $4\pi \sin\theta/\lambda$. Equating these two expressions for the length of the vector $\boldsymbol{OB}$ leads to the Bragg condition.

The diffraction condition of a real crystal surface can also be correlated to the reciprocal lattice using the Ewald sphere construction scheme. For example, the reflection high-energy electron diffraction (RHEED) method is commonly used to study crystal surfaces as shown in Fig. 2.41. Electrons having energy of 5–30 keV, corresponding to an electron wavelength of less than the lattice constant, are incident

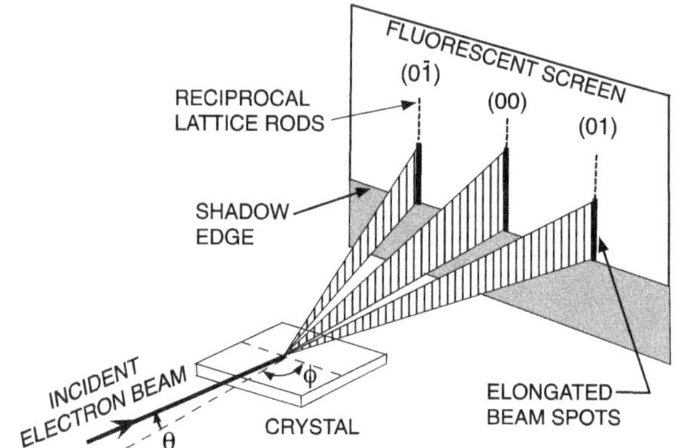

Fig. 2.41 Schematic diagram of the RHEED arrangement showing the incident electron beam at an angle $\theta \leq 5°$ to the sample surface. The elongated spots indicate the intersection of the Ewald sphere with diffraction rods. (Reprint with permission from [3], copyright AIP Publishing.)

on the crystal surface at a low glancing angle of ≤5°. The diffraction of the incoming primary electron beam by the two-dimensional (2D) atomic network of surface atoms leads to the appearance of streaks normal to the shadow edge on the fluorescence screen. The conditions for constructive interference of the elastically scattered electrons may be inferred using the Ewald sphere construction in the reciprocal lattice. Due to the loss of periodicity in the dimension above the crystal surface, the surface layer is represented by rods in a direction normal to the real surface in the reciprocal space (Fig. 2.42). The reciprocal lattice rods are labeled with only two Miller indices, h and k. Diffraction occurs where the Ewald sphere cuts a reciprocal lattice rod and the diffraction beam is labeled with the Miller index (hk) of the rod causing it. It should be noted that due to the high electron acceleration energy, the radius of the Ewald sphere is much larger than the separation of the rods. The wavelength of the incident electron beam λ is related to the energy by

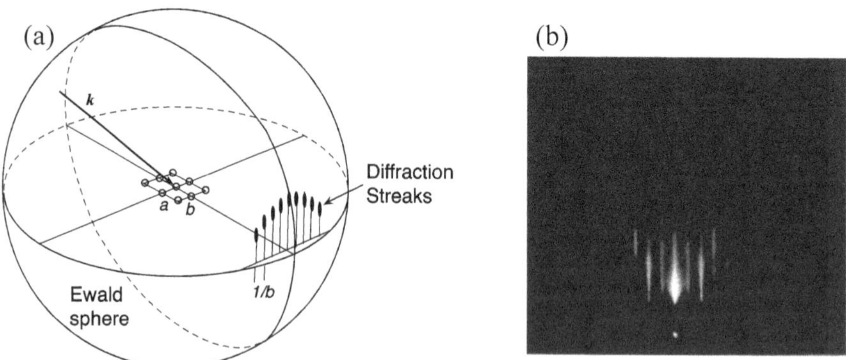

Fig. 2.42 **a** Ewald sphere construction for a reconstructed semiconductor surface in reciprocal lattice space. The real-space lattice atoms are shown as circles. (Reprint with permission from [4], copyright Elsevier) **b** Diffraction pattern from the GaAs (001) − (2 × 4) surface along the $[\bar{1}\bar{1}0]$ azimuth

$$E = \frac{\hbar^2}{2m}\left(\frac{2\pi}{\lambda}\right)^2 \tag{2.54}$$

Thus, at 5 keV, $\lambda = 0.0174$ nm. The radius of the Ewald sphere is $|\boldsymbol{k}| = 2\pi/\lambda = 362\ \text{nm}^{-1}$. If the surface lattice network has a lattice constant a of 5 Å, then the distance between adjacent rods in reciprocal space will be $2\pi/a = 12.57\ \text{nm}^{-1}$. As a result, the intersection of the nearly flat Ewald sphere and rods occurs along their length, resulting in a streaked diffraction pattern as illustrated in Fig. 2.42a. However, due to the finite surface roughness, the short diffraction streaks transform into elongated lines. Figure 2.42b is a typical RHEED pattern for a smooth (001) surface of GaAs.

2.5.5 *The Brillouin Zone*

An alternative expression of the Bragg diffraction condition can be derived by rewriting (2.53) as

$$(\boldsymbol{k}')^2 = (\boldsymbol{k} + \boldsymbol{G})^2 \tag{2.55}$$

or

$$(k')^2 = k^2 + G^2 + 2\boldsymbol{k}\cdot\boldsymbol{G} \tag{2.56}$$

Since $|\boldsymbol{k}| = |\boldsymbol{k}'|$, (2.56) can be expressed as

$$2\boldsymbol{k}\cdot\boldsymbol{G} + G^2 = 0 \tag{2.57}$$

or

$$\boldsymbol{k}\cdot(\boldsymbol{G}/G) = \boldsymbol{k}\cdot\boldsymbol{n} = G/2 \tag{2.58}$$

Equation (2.58) corresponds to a geometric structure that describes the locus of all points satisfying the Bragg condition, called the *Bragg plane* (Fig. 2.43). This is a plane that is the perpendicular bisector of a reciprocal lattice vector $\boldsymbol{G}$. The Bragg law is only satisfied for those incident wave vectors whose tips lie on the Bragg plane in reciprocal space when the vector origin is tied to a reciprocal lattice point. Since the reciprocal lattice of a periodic direct lattice is also periodic and infinite in extent, it is most useful only to require the use of a limited volume of reciprocal lattice, or $\boldsymbol{k}$-space. This volume is called the first Brillouin zone. The Bragg plane simply forms one of the faces of the first Brillouin zone. As shown in Fig. 2.44, the smallest 2D polyhedron centered at the origin and enclosed by perpendicular bisectors of reciprocal lattice vectors, i.e., Bragg planes, forms the (first) Brillouin zone.

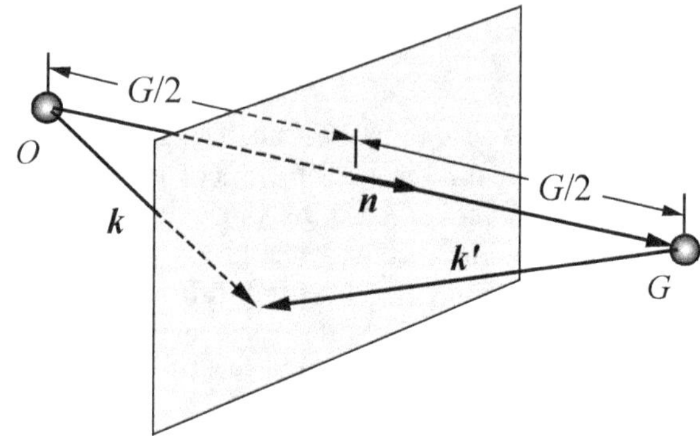

Fig. 2.43 Geometrical representation of the Bragg diffraction condition. Diffraction only occurs for an incident wave vector whose tip lies on the Bragg plane if its origin is tied to a reciprocal lattice point

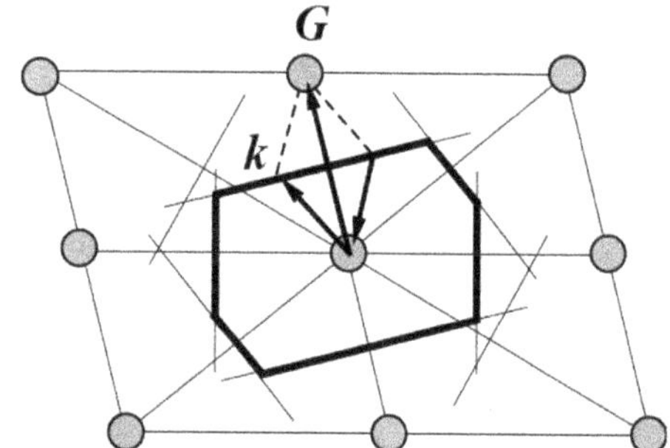

Fig. 2.44 First Brillouin zone formed in the reciprocal space for a 2D oblique lattice. It is shown bisecting the midpoints of six shortest reciprocal lattice vectors from the center of this lattice structure

The points on the Brillouin zone boundary are special because every wave with $\boldsymbol{k}$ extending from the origin to the zone boundary gives rise to a Bragg reflected wave. It is obvious from Fig. 2.43 that $\boldsymbol{k}' = \boldsymbol{k} + \boldsymbol{G}$ is satisfied on these boundaries. The interference of the incident primary waves and the Bragg reflected waves produces a 'standing wave.' This is important for the understanding of X-ray diffraction and the formation of a bandgap in the electronic structure of semiconductor materials.

(a) **Example for a 2D square lattice**

The reciprocal lattice of a square direct lattice, whose lattice spacing is a, is also a square lattice of spacing $1/a$ (Fig. 2.45). The reciprocal lattice vector pointing from the origin to a lattice point (u, v) can be expressed as $\boldsymbol{G} = (2\pi/a)(u\boldsymbol{i}_x + v\boldsymbol{i}_y)$ where u and v are integers. According to the Bragg condition, for $\boldsymbol{k} = k_x\boldsymbol{i}_x + k_y\boldsymbol{i}_y$

$$2\boldsymbol{k} \cdot \boldsymbol{G} + G^2 = \frac{4\pi}{a}\left(uk_x + vk_y\right) + \frac{4\pi^2}{a^2}\left(u^2 + v^2\right) = 0 \qquad (2.59\text{a})$$

or

$$f\left(k_x, k_y\right) = uk_x + vk_y + \frac{\pi}{a}\left(u^2 + v^2\right) = 0 \qquad (2.59\text{b})$$

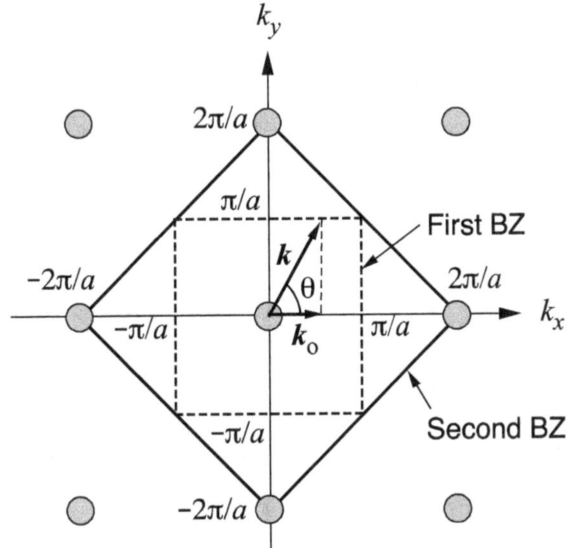

Fig. 2.45 Construction of the first (dashed line boundary) and second (solid line boundary) Brillouin zones (BZ) in a 2D reciprocal lattice space. The square direct lattice has a lattice spacing a

This is a straight line of k_x and k_y. The intersections on the x- and y-axes are

$$\begin{cases} k_x = -\dfrac{\pi}{a}\left(\dfrac{u^2+v^2}{u}\right) \text{ on } x\text{-axis} \\ k_y = -\dfrac{\pi}{a}\left(\dfrac{u^2+v^2}{v}\right) \text{ on } y\text{-axis} \end{cases} \tag{2.60}$$

To construct the smallest rectangle, we calculate k_x and k_y with the smallest possible u and v. At $u = 0$ and $v = \pm 1$, we have $k_y = \pm \pi/a$. Similarly, at $v = 0$ and $u = \pm 1$, we have $k_x = \pm \pi/\text{a}$. The four constant k's form a square. This is the first Brillouin zone as shown in Fig. 2.45. Alternatively, bisecting the midpoints of four shortest reciprocal lattice vectors from the origin can also form the first Brillouin zone. For the second Brillouin zone, $k_x \pm k_y = \pm 2\pi/a$ for $u = \pm 1$ and $v = \pm 1$. The higher-order Brillouin zones can be constructed through the same procedures. For example, using $u = 0$, $v = \pm 2$ and $v = 0$, $u = \pm 2$, the third zone boundaries are defined.

On the Brillouin zone boundary, the wave vectors must follow the Bragg condition. For example, the Bragg condition of the first Brillouin zone, $n = 1$, becomes

$$2a \sin\theta = \lambda \tag{2.61}$$

For an incident wave of a wavelength λ, as seen in Fig. 2.45, the wave vector is

$$|\boldsymbol{k}| = \sqrt{k_0^2 + \left(\frac{\pi}{a}\right)^2} = \frac{2\pi}{\lambda} \tag{2.62}$$

and

$$\sin\theta = \frac{k_y}{\sqrt{k_x^2 + k_y^2}} = \frac{\pi/a}{\sqrt{k_0^2 + (\pi/a)^2}} = \frac{2\pi}{\sqrt{k_0^2 + (\pi/a)^2}} \cdot \frac{1/a}{2} = \frac{\lambda}{2a} \tag{2.63}$$

Thus, $2a \sin\theta = 2a \cdot (\lambda/2a) = \lambda$, fulfilling the Bragg law.

Therefore the Brillouin zone construction exhibits all the wave vectors that can be Bragg diffracted by the crystal. Other waves get weakened and eventually cease to exist.

(b) 3D Brillouin zone of a FCC crystal

As we will discuss in Chap. 3, the Brillouin zone concept is essential for the development of the energy band structure of semiconductors. The construction of Brillouin zones in 3D semiconductor crystals follows a procedure similar to that for the 2D case. Since most semiconductor crystals have either zinc-blende or wurtzite lattice structure, we shall investigate the Brillouin zones of the basic FCC and simple hexagonal structures.

First, we need to transform the primitive unit cell of the FCC structure into reciprocal lattice spacing. As shown in Fig. 2.46, the translation vectors which define the primitive unit cell of the FCC lattice are

$$\boldsymbol{a}' = (a/2)(\boldsymbol{b} + \boldsymbol{c}), \boldsymbol{b}' = (a/2)(\boldsymbol{c} + \boldsymbol{a}), \boldsymbol{c}' = (a/2)(\boldsymbol{a} + \boldsymbol{b}) \tag{2.64}$$

where $a/2$ defines the size of the unit cell. The translation vectors $\boldsymbol{a}'$, $\boldsymbol{b}'$, and $\boldsymbol{c}'$ are vectors pointing from the origin to three neighboring face-centered surface lattice points, respectively. The formed cubic volume of the primitive cell is $\boldsymbol{a}' \cdot \boldsymbol{b}' \times \boldsymbol{c}' = a^3/2\sqrt{2}$.

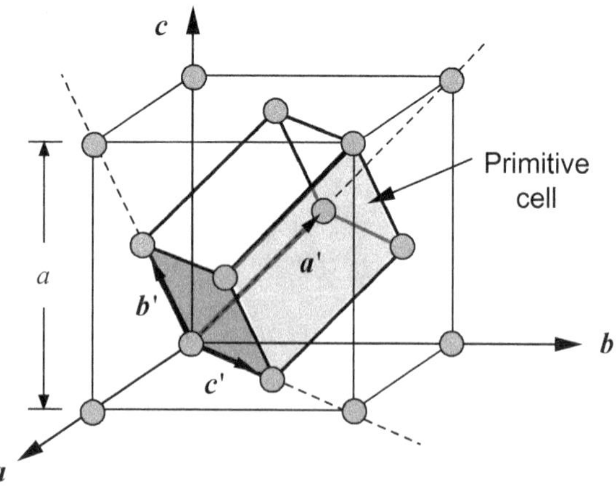

Fig. 2.46 Primitive unit cell of an FCC crystal structure is formed by three translation vectors $\boldsymbol{a}'$, $\boldsymbol{b}'$, and $\boldsymbol{c}'$. These translation vectors point from the origin toward the surface atoms diagonally

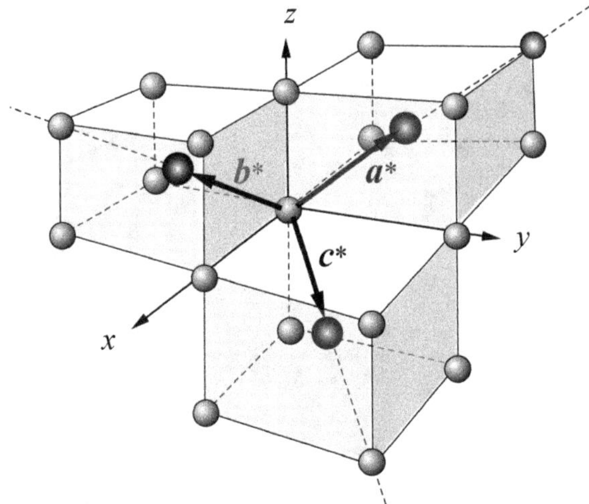

Fig. 2.47 Reciprocal translation vectors $\boldsymbol{a}^*$, $\boldsymbol{b}^*$, and $\boldsymbol{c}^*$ of the FCC crystal structure. The lattice points at the body centers are drawn as larger dark spheres

Then the primitive translation vectors in the reciprocal lattice are calculated following (2.42) as

$$\begin{cases} \boldsymbol{a}^* = 2\pi\left(\dfrac{\boldsymbol{b}'\times\boldsymbol{c}'}{\boldsymbol{a}'\cdot\boldsymbol{b}'\times\boldsymbol{c}'}\right) = \dfrac{2\pi a^2(-\boldsymbol{a}+\boldsymbol{b}+\boldsymbol{c})/4}{a^3/4} = \dfrac{2\pi}{a}(-\boldsymbol{a}+\boldsymbol{b}+\boldsymbol{c}) \\ \boldsymbol{b}^* = 2\pi\left(\dfrac{\boldsymbol{c}'\times\boldsymbol{a}'}{\boldsymbol{a}'\cdot\boldsymbol{b}'\times\boldsymbol{c}'}\right) = \dfrac{2\pi a^2(\boldsymbol{a}-\boldsymbol{b}+\boldsymbol{c})/4}{a^3/4} = \dfrac{2\pi}{a}(\boldsymbol{a}-\boldsymbol{b}+\boldsymbol{c}) \\ \boldsymbol{c}^* = 2\pi\left(\dfrac{\boldsymbol{a}'\times\boldsymbol{b}'}{\boldsymbol{a}'\cdot\boldsymbol{b}'\times\boldsymbol{c}'}\right) = \dfrac{2\pi a^2(\boldsymbol{a}+\boldsymbol{b}-\boldsymbol{c})/4}{a^3/4} = \dfrac{2\pi}{a}(\boldsymbol{a}+\boldsymbol{b}-\boldsymbol{c}) \end{cases} \tag{2.65}$$

These are vectors oriented along the $[\bar{1}11]$, $[1\bar{1}1]$, and $[11\bar{1}]$] directions of a cubic lattice, respectively. We can plot (2.65) as shown in Fig. 2.47 and realize that these are the primitive translation vectors of a BCC lattice! Thus, the reciprocal of an FCC lattice is a BCC lattice. The volume of the primitive cell in the reciprocal lattice is $\boldsymbol{a}^*\cdot\boldsymbol{b}^*\times\boldsymbol{c}^* = 4(2\pi/a)^3$.

To construct the first Brillouin zone from these reciprocal translation vectors, we shall derive a general form of $\boldsymbol{G}_{hkl}$:

$$\boldsymbol{G}_{hkl} = h\boldsymbol{a}^* + k\boldsymbol{b}^* + l\boldsymbol{c}^* = \left(\frac{2\pi}{a}\right)[(-h+k+l)\boldsymbol{a} + (h-k+l)\boldsymbol{b} + (h+k-l)\boldsymbol{c}] \tag{2.66}$$

In a BCC lattice, there are eight nearest neighbors from the body-centered lattice point along $\langle 111\rangle$ (Fig. 2.48). For $h = k = l = \pm 1$, there are eight reciprocal lattice vectors pointing to each corner of the BCC structure. The boundaries of the first Brillouin zone are located at $\boldsymbol{G}/2$, and

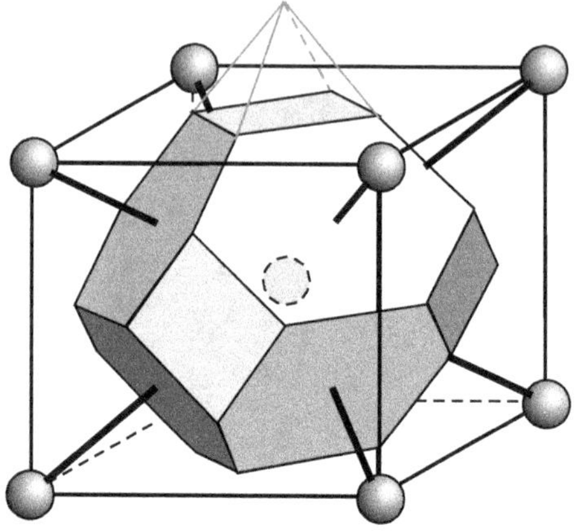

Fig. 2.48 First Brillouin zone of the face-centered cubic lattice in reciprocal space. Six pyramid-shaped volumes (one shown in gray lines) are excluded since they are intersected by second nearest neighbors

$$\frac{1}{2}\boldsymbol{G}_{\langle 111\rangle} = \frac{\pi}{\boldsymbol{d}_{\langle 111\rangle}} = \frac{\pi}{a}(\pm\boldsymbol{a} \pm \boldsymbol{b} \pm \boldsymbol{c}) \tag{2.67}$$

These eight vectors generate eight Brillouin zone surfaces halfway between the origin and the nearest points of the lattice, which form a regular octahedron. However, this is not the first Brillouin zone since the planes arising from the second nearest neighbors between body-center lattice points intersect it. Therefore, the six pyramid-shaped corners in the $\langle 100\rangle$ directions extending outside the first zone are excluded. The final form of the first Brillouin zone for an FCC lattice becomes a truncated octahedron (Fig. 2.48). The volume enclosed by the first Brillouin zone is also known as the *Wigner–Seitz* primitive cell of the reciprocal lattice. The eight hexagonal faces arise from the planes halfway to the atoms at the corners, while the six smaller square faces are halfway to the atoms in the middle of the next cells. As expected, these unit cells can fit together to fill the whole space as demonstrated in Fig. 2.49.

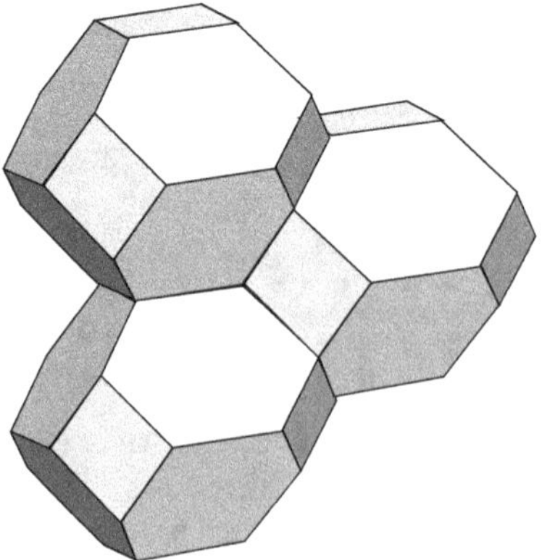

Fig. 2.49 Way in which the first Brillouin zone cells of an FCC lattice can fill the whole space

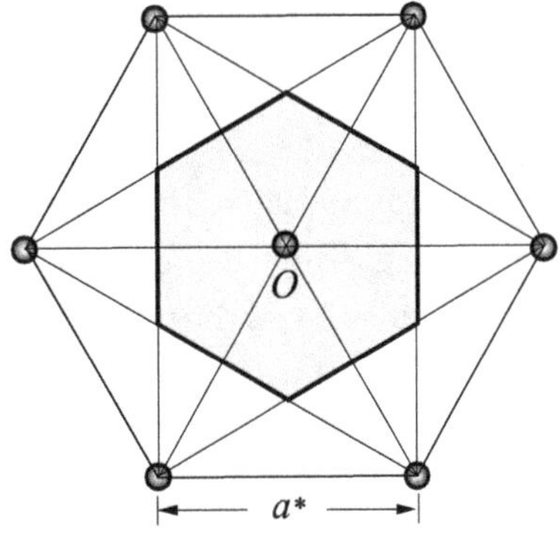

Fig. 2.50 Construction of the first Brillouin zone for the plane simple hexagonal lattice. The reciprocal lattice points are shown at six corners and the origin

(c) 3D Brillouin zone of a simple hexagonal crystal

The reciprocal lattice of a simple hexagonal lattice is another simple hexagonal lattice rotated 30° about the c-axis. If the basic lattice translations for the direct lattice in the basal plane and along the c-axis are, respectively, a and c, then for the reciprocal lattice they are a^* and c^*, where $a^* = 4\pi/\sqrt{3}a$ and $c^* = 2\pi/c$. Figure 2.50 shows seven points of the reciprocal lattice of a simple hexagonal lattice. If we take the central point O as origin, the other six lie at the corners of a regular hexagon of side $a^* = 4\pi/\sqrt{3}a$ and are the six nearest neighbors. The first Brillouin zone is formed by connecting bisectors of the lines joining O to the other six nearest neighbors. It is a hexagon whose sides have length $a^*/\sqrt{3} = 4\pi/3a$ and hence covers a unit cell area of $(a^*)^2\sqrt{3}/2 = 8\pi^2/\sqrt{3}a^2$. It should be noted that the hexagon forming the first Brillouin zone has the same orientation as the hexagons of the direct lattice.

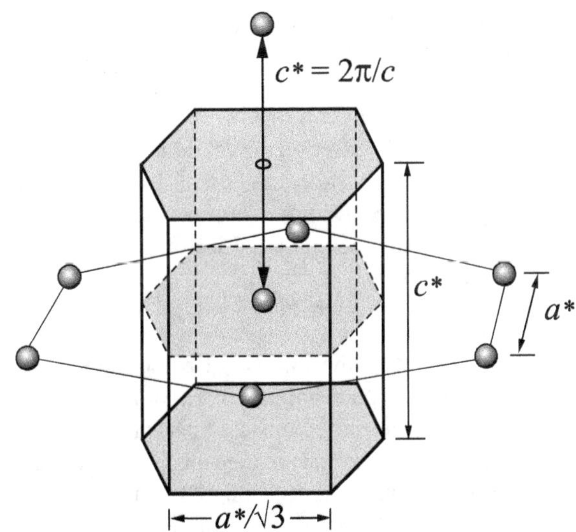

Fig. 2.51 First Brillouin zone for the simple hexagonal lattice. Some lattice points of the reciprocal lattice are also shown

The 3D first Brillouin zone of the simple hexagonal lattice is obtained simply by introducing planes parallel to the c-axis through the sides of the hexagon and two planes perpendicular to the c-axis, each at distance $c^*/2$ from O. The first Brillouin zone is therefore a hexagonal prism of height $c^* = 2\pi/c$ whose axis is parallel to the c-axis of the direct lattice. The hexagonal of side $4\pi/3a$ is oriented in the same direction as the hexagons of the direct lattice. It occupies a volume of $\left(16\pi^3/\sqrt{3}\right)/a^2c$. The first Brillouin zone is shown in Fig. 2.51, together with some points of the reciprocal lattice.

Problems

1. Both AsH_3 and $SiCl_4$ are common gases for semiconductor growth and processing. Discuss their molecular bonding properties.
2. The ionic radii of Na, Cs, and Cl are 0.875 Å, 1.455 Å, and 1.475 Å, respectively.
   (a) For NaCl, the FCC structure is the preferred crystal structure. Verify that this structure prevents anion–anion or cation–cation contact.
   (b) Verify that the simple cubic structure has a higher atomic packing density than the FCC structure in CsCl crystals.
3. Calculate the atomic packing density for the following crystal structures:
   (a) Simple FCC
   (b) Simple hexagonal close-packed (HCP)
   (c) Diamond
   (d) Wurtzite.
4. Assume the cube containing a tetrahedron bond structure has a length a on each side. Calculate the bond length of the tetrahedron bond in terms of length a. Also, determine the angle between two tetrahedron bonds.
5. The primitive unit cell of a hexagonal close-packed (HCP) structure can be defined by unit vectors a and c. Prove that, using the incompressible spheres model, the ratio between c and a is $\sqrt{8/3}$.
6. The primitive unit cell of a wurtzite structure can be defined by unit vectors a and c (Fig. 2.31). Derive the relations between a, c, and the bond length.
7. Prove that the reciprocal lattice of a simple hexagonal lattice is another simple hexagonal lattice. Determine the new lattice constants along the a-axis and c-axis.
8. (a) Derive the reciprocal lattice of the BCC structure. What is the new crystal structure?
   (b) What is the angle between $\boldsymbol{G}_{110}$ and $\boldsymbol{G}_{101}$?
9. Make a plot of the first two Brillouin zones of a primitive rectangular two-dimensional lattice with axes a, $b = 2a$.
10. Derive the dielectric constant tensor for a simple tetragonal crystal.
11. The hexagonal crystal is called an optically uniaxial crystal and has two principal dielectric constants: ϵ_{33} along the c-axis and ϵ_{11} in the basal plane. Show that the dielectric constant tensor of hexagonal crystals can be expressed as

$$\epsilon = \begin{bmatrix} \epsilon_{11} & 0 & 0 \\ 0 & \epsilon_{11} & 0 \\ 0 & 0 & \epsilon_{33} \end{bmatrix}.$$

Note: This can be done by considering a rotation of $\theta = \pi/3$ about the c-axis. The new coordinates (x' and y') are related to the original coordinates through

$$\begin{cases} x' = x\cos\theta + y\sin\theta \\ y' = y\cos\theta - x\sin\theta \end{cases}.$$

References

1. M. Prutton, *Introduction to Surface Physics* (Oxford University Press, 1994)
2. J.C. Phillips, Rev. Mod. Phys. **42**, 317 (1970)
3. B. Bolger, P.K. Larsen, Rev. Sci. Instrum. **57**, 1363 (1986)
4. P.K. Larsen, P.J. Dobson, J.H. Neave, B.A. Joyce, B. Bolger, J. Zhang, Surf. Sci. **169**, 176 (1986)

Further Reading

1. C. Kittel, *Introduction to Solid State Physics*, 6th edn. (Wiley, 1986)
2. H. Ibach and H. Lüth, *Solid State Physics*, 2nd edn. (Springer, Berlin, 1996)
3. S. Wang, *Fundamentals of Semiconductor Theory and Device Physics* (Prentice-Hall, 1989)

Chapter 3
Electronic Band Structures of Solids

Abstract The property that distinguishes semiconductors from other materials concerns the behavior of their charged carriers, in particular the existence of gaps in their electronic and photonic transition spectra. The microscopic behavior of charged carriers in a solid is most conveniently specified in terms of the electronic band structure. The metals are good conductors with overlapped conduction and valence bands, while insulators such as ionic crystals have large bandgaps of over 8 eV. The semiconductor crystals have a finite energy bandgap between metals and insulators. The purpose of this chapter is to develop the basic understanding of energy bandgap formation in solids. We start our discussion by first developing the free electron theory based on quantum theory in which the Fermi–Dirac statistics are used instead of the classical Boltzmann statistics. Then the spatial periodic potential of a solid is introduced according to Bloch's theorem. Finally, the interference at Brillouin zone boundaries leads to constructive and/or destructive interferences of electron waves forming the energy bandgap.

3.1 Free Electron Theory and Density of States

The simplest model for describing electrons in solids is to assume that the valence electrons of an atom are free to move anywhere throughout the volume of the material. This is an appropriate model for a metallic solid. Of course, for an isolated atom, the electrons are confined by the large barrier of the Coulomb potential well and cannot move away from the nucleus. But when atoms bond together to form crystals, the potential barriers between atoms are lowered by the bonding energy. The simple model further assumes that the periodically arranged nuclei or the cores of the atoms are stationary and the core electrons shield the nuclear charge completely. Thus the potential appears qualitatively as shown in Fig. 3.1 where a constant electrostatic potential exists everywhere in the solid. Any potential 'bump' associated with individual atoms is smoothed out. This allows valence electrons to move freely within the crystal as free electrons. Therefore, we can use a 'particle in a box' approach to treat the problem. Furthermore, we neglect any electron–electron interactions by allowing only one free electron in the system.

K. Y. Cheng, *III–V Compound Semiconductors and Devices*, Graduate Texts in Physics,
https://doi.org/10.1007/978-3-030-51903-2_3

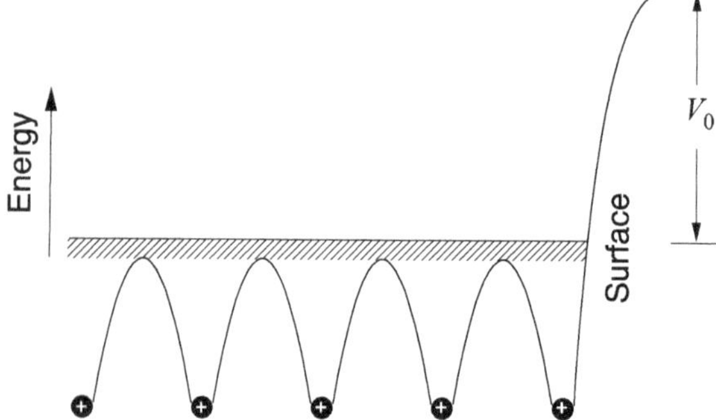

Fig. 3.1 Schematic of the potential within a perfectly periodic crystal lattice of positive cores. The vacuum level V_0 is the energy that the electron must acquire in order to leave the crystal

3.1.1 One-Dimensional System

The simplified 'one free electron theory' in 1D can be pictured as a finite potential well with a well depth V_0 and width L shown in Fig. 3.2. The only electron in the well has energy of $E < V_0$. Since the barrier height of the well is finite, there are bound to be leakage waves outside the well. The 1D Schrödinger wave equation of this system *outside* the well is

$$-\frac{\hbar^2}{2m}\frac{d^2\psi}{dz^2} + V_0\psi = E\psi \tag{3.1}$$

Since $E < V_0$, the solution of the wave function is

$$\psi = A\exp(-\alpha z) + B\exp(\alpha z) \tag{3.2a}$$

and

$$\alpha = \sqrt{2m(V_0 - E)/\hbar^2} \tag{3.2b}$$

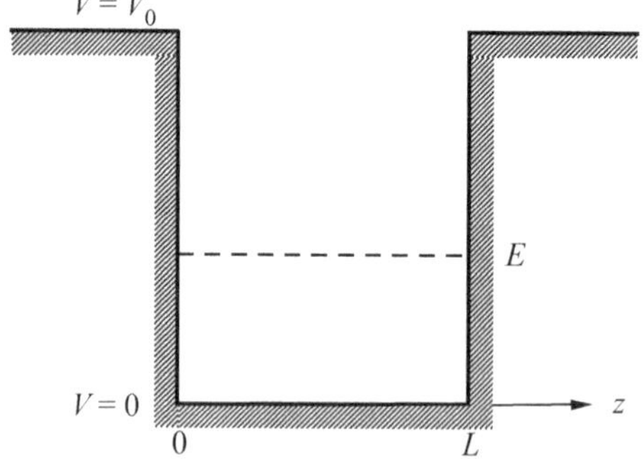

Fig. 3.2 Illustration of a finite square potential well experienced by an electron in the 1D free electron model

where $V_0 - E$ is the work function, and E is the kinetic energy of the electron. At $z = \infty$, ψ approaches 0, and $B = 0$. Thus, for $z > L$, the decay wave function outside the crystal is

$$\psi(z) = A\exp(-\alpha z) \tag{3.3}$$

In GaAs, $V_0 - E \sim 4$ eV and using the electron effective mass of $0.063m_0$, we have $\alpha \sim 2.57 \times 10^7\ \text{cm}^{-1}$. This means the amplitude of the wave function $\psi(z)$ drops to $1/e$ in 3.89 Å or about a monolayer of atomic thickness. Therefore, under practical situations, we can neglect the penetration depth outside the crystal and use the infinite well approximation instead. The solution within the well becomes

$$\psi(z) = \begin{cases} A\sin(k_z z), & 0 \le z \le L \\ 0, & \text{elsewhere} \end{cases} \tag{3.4}$$

where the wave vector $k_z = n\pi/L$ and $n = 1, 2, 3\ldots$. The first three wave functions are plotted in Fig. 3.3. Replacing the solutions (3.4) in the wave equation (3.1), we have

$$-\frac{\hbar^2}{2m}\frac{\mathrm{d}^2}{\mathrm{d}z^2}\left[A\sin(k_z z)\right] = \frac{\hbar^2}{2m}k_z^2\psi = E\psi \tag{3.5}$$

The energy eigenvalue is related to the wave vector by

$$E = \frac{\hbar^2 k_z^2}{2m} \tag{3.6}$$

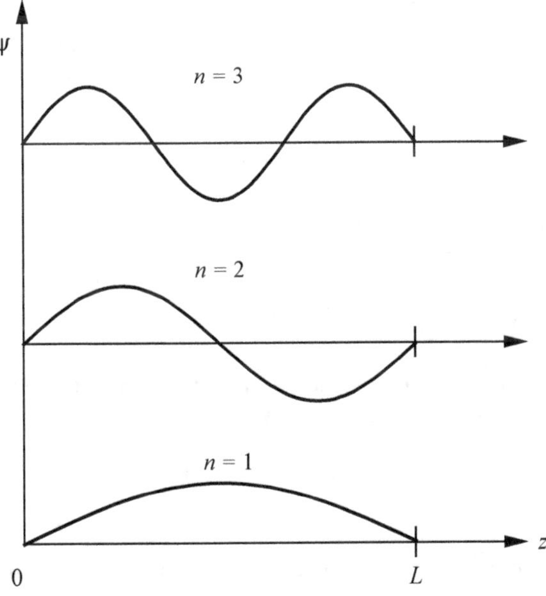

Fig. 3.3 First three wave functions of a free electron in a square potential well of length L

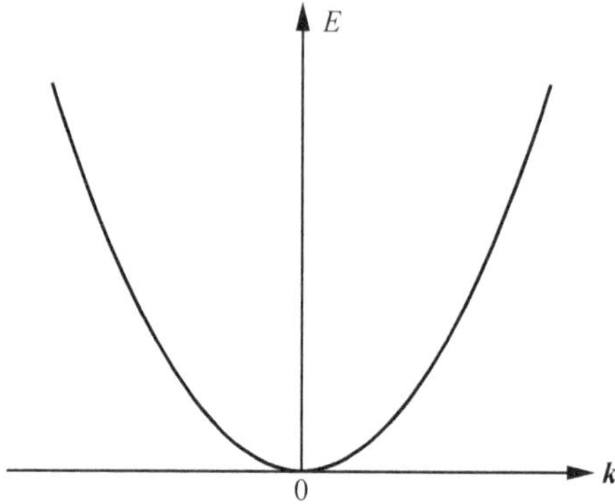

Fig. 3.4 Energy E plotted as a function of k according to (3.6)

In the 1D free electron model, as shown in Fig. 3.4, the energy eigenvalue is a parabolic function of the wave vector k_z.

3.1.2 *Three-Dimensional System*

(a) **Wave function $\psi(x, y, z)$**

The general expression of the 3D wave function is

$$-\frac{\hbar^2}{2m}\nabla^2\psi(x, y, z) + V_0\psi(x, y, z) = E\psi(x, y, z) \tag{3.7}$$

Using the infinite well approximation, we only have to consider $\psi(x, y, z)$ inside the crystal (a cube of side L), where $V = 0$. Solving the wave equation by separation of variables, assuming $\psi(x, y, z) = X(x)\,Y(y)\,Z(z)$, we get

$$-\frac{\hbar^2}{2m}\left(YZ\frac{\partial^2 X}{\partial x^2} + XZ\frac{\partial^2 Y}{\partial y^2} + XY\frac{\partial^2 Z}{\partial z^2}\right) = EXYZ \tag{3.8a}$$

or

$$-\frac{\hbar^2}{2m}\left(\frac{1}{X}\frac{\partial^2 X}{\partial x^2} + \frac{1}{Y}\frac{\partial^2 Y}{\partial y^2} + \frac{1}{Z}\frac{\partial^2 Z}{\partial z^2}\right) = E \tag{3.8b}$$

This equation is equivalent to three independent equations in X, Y, and Z. For example, the X-related equation is

$$-\frac{\hbar^2}{2m}\frac{\partial^2 X}{\partial x^2} = XE_x \tag{3.9}$$

where E_x is a constant. The solution of the wave equation in x is

$$X = A_x \sin k_x x + B_x \cos k_x x \tag{3.10a}$$

and

$$k_x = \sqrt{\frac{2mE_x}{\hbar^2}} \tag{3.10b}$$

Again, similar to the 1D system, we can use the condition of $V_0 = \infty$ at $x = 0$ and $x = L$ such that $\psi(0) = \psi(L) = 0$. Thus, $B_x = 0$ and

$$X = A_x \sin k_x x \tag{3.11a}$$

and

$$k_x = \frac{n_x \pi}{L} \tag{3.11b}$$

The wave functions of Y and Z can be solved in the same manner. We come to the final solution of the wave function in 3D as

$$\psi(x, y, z) = A\sin(k_x x)\sin(k_y y)\sin(k_z z) \tag{3.12a}$$

and

$$k_x = \frac{n_x \pi}{L},\ k_y = \frac{n_y \pi}{L},\ k_z = \frac{n_z \pi}{L} \tag{3.12b}$$

where n_x, n_y, and n_z are positive integers or quantum numbers which specify the relationship of ψ in the x-, y-, and z-directions. The amplitude A is calculated through the normalization process using $\int_V \psi^* \psi \mathrm{d}x\mathrm{d}y\mathrm{d}z = 1$.

$$A^2 \int_0^L \sin^2(k_x x)\mathrm{d}x \int_0^L \sin^2(k_y y)\mathrm{d}y \int_0^L \sin^2(k_z z)\mathrm{d}z = 1 \tag{3.13}$$

Since $\int \sin^2(ax)\mathrm{d}x = \frac{x}{2} - \frac{\sin(2ax)}{4a}$, for $a = k_i = n_i \pi/L$,

$$\left.\frac{\sin(2ax)}{4a}\right|_0^L = \frac{\sin(2n_i\pi) - \sin 0°}{4a} = 0$$

Therefore,

$$\int_0^L \sin^2(k_x x)\mathrm{d}x = \left.\frac{x}{2}\right|_0^L = \frac{L}{2} \tag{3.14a}$$

and

$$A = (2/L)^{3/2}. \tag{3.14b}$$

The final solution of the wave function becomes

$$\psi(x, y, z) = \left(\frac{2}{L}\right)^{3/2} \sin(k_x x) \sin(k_y y) \sin(k_z z) \tag{3.15a}$$

and

$$k_j = \frac{n_i \pi}{L} \text{ for } j = x, y, z \text{ and } n_i = 1, 2, 3, \ldots \tag{3.15b}$$

(b) **Energy eigenvalues (*E*)**

The allowed energy levels can be derived by inserting the wave function solution (3.15a) into the 3D wave function (3.7)

$$-\frac{\hbar^2}{2m}\nabla^2\psi = E\psi \text{ and } \nabla^2\psi = \frac{\partial^2\psi}{\partial x^2} + \frac{\partial^2\psi}{\partial y^2} + \frac{\partial^2\psi}{\partial z^2} \tag{3.16}$$

The second derivative terms in (3.16) are calculated using the wave function solution (3.15a) as follows:

$$\frac{\partial^2\psi}{\partial x^2} = -\psi k_x^2, \frac{\partial^2\psi}{\partial y^2} = -\psi k_y^2, \frac{\partial^2\psi}{\partial z^2} = -\psi k_z^2 \tag{3.17}$$

Then the wave equation becomes

$$-\frac{\hbar^2}{2m}\left(\frac{\partial^2\psi}{\partial x^2} + \frac{\partial^2\psi}{\partial y^2} + \frac{\partial^2\psi}{\partial z^2}\right) = \frac{\hbar^2}{2m}\left(k_x^2 + k_y^2 + k_z^2\right)\psi = E\psi \tag{3.18}$$

which leads to the energy eigenvalues of

$$E = \frac{\hbar^2 k^2}{2m} = \frac{\hbar^2}{2m}\left(k_x^2 + k_y^2 + k_z^2\right) \tag{3.19}$$

Thus,

$$E \propto k^2 = k_x^2 + k_y^2 + k_z^2 \tag{3.20}$$

In classical theory, the kinetic energy of a particle with momentum p can be expressed as

$$K.E. = \frac{p^2}{2m} = \frac{1}{2m}\left(p_x^2 + p_y^2 + p_z^2\right) \tag{3.21}$$

Since our discussions of the total energy of the free electron involve kinetic energy only, the expressions of energy in (3.19) and (3.21) are identical. The allowed energy eigenvalues indicate that

$$\boldsymbol{p} = \hbar \boldsymbol{k} \tag{3.22}$$

This is the de Broglie relationship, and $\boldsymbol{p}$ is the *crystal momentum*. However, it should be stressed that the wave vector $\boldsymbol{k}$ for electron wave functions in a periodic potential is not a measure of true physical momentum. As will be discussed later, the physically distinct values of the wave vector (crystal momentum) of an electron in a crystalline lattice—as opposed to those of a free particle, which can take any value—are restricted by the first Brillouin zone.

Further, the energy functions also contain the wave nature of the electron motion in the potential well. For $n = 1$ in (3.15b), $k = \pi/L$, and the allowed wavelength is $\lambda = 2\pi/k = 2L$ or $k = 2\pi/\lambda$. The wavelength is related to the particle nature of the electron through

$$\lambda = \frac{2\pi}{k} = \frac{2\pi\hbar}{\sqrt{2mE}} = \frac{h}{p} \tag{3.23}$$

Therefore, the electron with a momentum p will be diffracted like a wave with a wavelength λ in a crystal!

Since $k = n\pi/L$, the energy eigenvalue can also be expressed as a function of n as

$$E = \frac{\hbar^2}{2m}\left(\frac{\pi}{L}\right)^2 n^2 = \frac{\hbar^2}{2m}\left(\frac{\pi}{L}\right)^2 \left(n_x^2 + n_y^2 + n_z^2\right) \tag{3.24}$$

or

$$E \propto \left(n_x^2 + n_y^2 + n_z^2\right) = n^2 \tag{3.25}$$

The results of (3.19) and (3.24) indicate that the 3D constant energy surface in n or k-space is a sphere. For each k_i determined by n_i there is a corresponding λ_i or mode. Each allowed energy state corresponding to a point n_i can be displayed in the n-space with positive integer coordinates n_x, n_y, and n_z. It is obvious that the energy eigenvalues are degenerate for $n_x \neq n_y \neq n_z$. For example, for (n_x, n_y, n_z) equals (1, 2, 3), (1, 3, 2), (2, 3, 1), (2, 1, 3), (3, 1, 2), or (3, 2, 1), there are six degenerate energy states with the same energy. The number of degenerate states increases dramatically with increasing energy.

3.1.3 Density of States (DOS)

In many situations, such as when computing the electron distribution, it is necessary to know how the electrons are distributed in the energy spectrum. This could be done through the use of (3.24) to derive a density of states.

Consider a 2D crystal lattice in n-space where $n_z = 0$, as shown in Fig. 3.5. Each lattice point is defined by a pair of (n_x, n_y). Since $k_i = n_i\pi/L = 2\pi/\lambda_i$, each point corresponds to a specific mode of wave oscillation. The number of normal modes of oscillation inside a finite area defined by a rectangular box is simply the number of lattice points within $\Delta n_x \Delta n_y$. In the 3D case, the number of normal modes of oscillation is $\Delta N' \equiv \Delta n_x \Delta n_y \Delta n_z$, which forms an orthorombic volume. In a crystal the spacing between these points is extremly close. For example, the wavelength of, say, a 3 eV electron is $h/(2\,\mathrm{m}E)^{1/2} \cong 7$ Å and $k = 2\pi/\lambda \cong 9\times10^7\ \mathrm{cm}^{-1}$. For a 1 cm^3 cube, the number of oscillation modes in one coordinate direction is $kL/\pi \sim 3\times10^7$! Therefore, we shall replace Δn_i by dn and $\Delta N' \approx \mathrm{d}N' = \mathrm{d}n_x\mathrm{d}n_y\mathrm{d}n_z$. Due to the large number of modes, they form a continuous band instead of discrete states.

Since n_i are all positive integers, as shown in Fig. 3.6, one need only count the n_i in the first octant of the constant energy sphere. The number of normal modes in the first octant is

$$N' = \frac{1}{8}\left(\frac{4\pi n^3}{3}\right) \tag{3.26a}$$

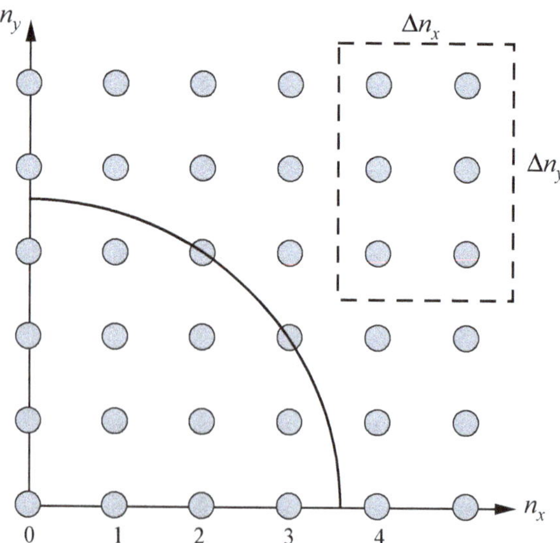

Fig. 3.5 Schematic of the states of an electron in a 2D infinte well. The circle corresponds to a constant energy

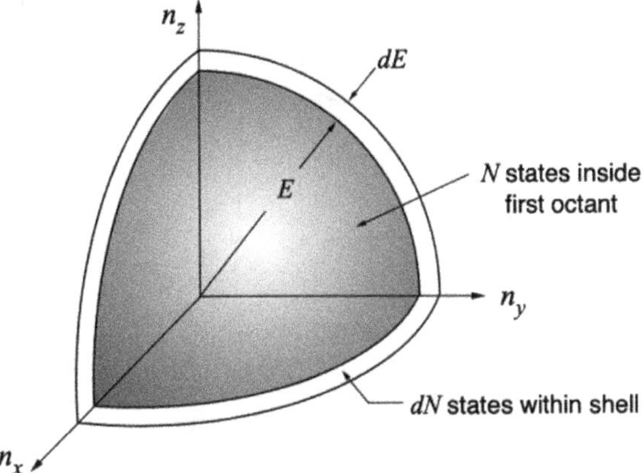

Fig. 3.6 Spherical surface corresponding to constant energies E and $E + \mathrm{d}E$ plotted in the momentum (n) space of a particle

where

$$n = \frac{kL}{\pi} = \frac{L}{\pi}\sqrt{\frac{2mE}{\hbar^2}} \tag{3.26b}$$

Then, including the two possible spins into the total number, we have

$$N = 2N' = \frac{2}{8}\left(\frac{4\pi}{3}\right)\left(\frac{2mE}{\hbar^2}\right)^{3/2}\left(\frac{L}{\pi}\right)^3 = \left(\frac{1}{3\pi^2}\right)L^3\left(\frac{2mE}{\hbar^2}\right)^{3/2} \tag{3.27}$$

The number of electron states within an energy interval dE, between E and $E +$ dE, is

$$\mathrm{d}N = \left(\frac{\mathrm{d}N}{\mathrm{d}E}\right)\mathrm{d}E = \left(\frac{L^3}{2\pi^2}\right)\left(\frac{2m}{\hbar^2}\right)^{3/2}\sqrt{E}\mathrm{d}E \tag{3.28}$$

The density of states (DOS) is defined as the number of states per unit energy interval near E per unit volume. It is expressed as

$$D_{3\mathrm{D}}(E) = \left(\frac{\mathrm{d}N}{\mathrm{d}E}\right)\frac{1}{L^3} = \frac{1}{2\pi^2}\left(\frac{2m}{\hbar^2}\right)^{3/2}\sqrt{E}\,(\mathrm{cm}^{-3}\mathrm{eV}^{-1}) \tag{3.29}$$

In the 2D case, such as in a quantum well,

$$N' = \frac{\pi n^2}{4} = \frac{\pi}{4}\cdot\frac{L^2}{\pi^2}\cdot\frac{2mE}{\hbar^2} = \frac{N}{2} \tag{3.30}$$

Thus,

$$\frac{\mathrm{d}N}{\mathrm{d}E} = \frac{mL^2}{\pi\hbar^2} \tag{3.31}$$

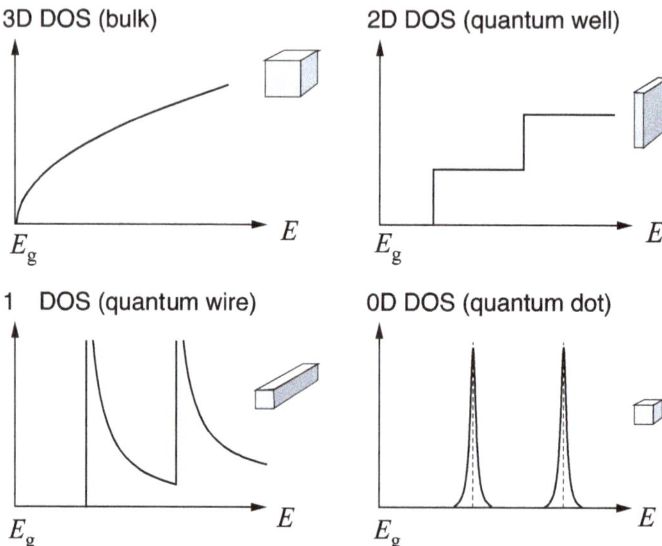

Fig. 3.7 Density of states distributions in bulk (3D), quantum well (2D), quantum wire (1D), and quantum dot (0D) structures

and

$$D_{2D}(E) = \left(\frac{\mathrm{d}N}{\mathrm{d}E}\right)\frac{1}{L^2} = \frac{m}{\pi\hbar^2} = \text{constant}\left(\text{cm}^{-2}\text{eV}^{-1}\right) \quad (3.32)$$

In the 1D case, such as in a quantum wire, $N' = n = kL/\pi$.

$$N = 2N' = \frac{2L}{\pi}\sqrt{\frac{2mE}{\hbar^2}} \quad (3.33)$$

Then

$$\frac{\mathrm{d}N}{\mathrm{d}E} = \frac{L}{\pi}\sqrt{\frac{2m}{\hbar^2 E}} \quad (3.34)$$

and

$$D_{1D}(E) = \left(\frac{\mathrm{d}N}{\mathrm{d}E}\right)\frac{1}{L} = \frac{1}{\pi\hbar}\sqrt{\frac{2m}{E}} \propto \frac{1}{\sqrt{E}}\left(\text{cm}^{-1}\text{eV}^{-1}\right) \quad (3.35)$$

Finally, when the volume of the 3D confinement of a potential well reduces to contain only one energy state, the structure becomes a quantum dot. Since there are two electrons allowed per state E_0, one spin-up and one spin-down, the DOS of zero-dimension (0D) is given as

$$D_{0D}(E) = 2\delta(E - E_0) \tag{3.36}$$

The delta function indicates that only when $E = E_0$ is there a state. The DOS as a function of energy for 3D, 2D, 1D, and 0D distributions is plotted in Fig. 3.7.

3.2 Periodic Crystal Structure and Bloch's Theorem

The simple model of the free electron theory using one-electron approximation of a square potential well has been successful in accounting for some physical properties of solids, in particular those of metals. However, in crystals, the 'free' electron situation is not very accurate. Therefore, this overly simplified model needs refinements in order to improve its ability to explain the properties of semiconductors. One obvious addition to the model is to add the spatial dependence of the potential experienced by a valence electron in a crystal. The effect of the added potential associated with the ion cores is particularly pronounced in periodic ion-core potentials. However, this also imposes a concomitant constraint on the wave functions that describing the motion of an electron in the periodic potential. Fortunately, this added constraint can be treated easily using the theory that Felix Bloch proposed in 1928.

3.2.1 *Bloch's Theorem*

The free electron theory developed so far has used the square well approximation that leads to the sinusoidal wave function solutions of (3.15a). In our present assumption, to simplify the discussions, all deviations from a perfect periodicity will be neglected. This means that even the surface effects are removed by assuming an infinitely extended potential. For an infinite 1D system with a uniform potential energy of V (=constant), the wave function is now represented by a traveling plane wave of the form

$$\psi_k(z) = A\exp(ikz) \tag{3.37}$$

This leads to a uniform distribution function of $\psi_k^*\psi_k = \psi_k^2 = A^2 = \text{constant}$. Therefore, the system is translationally invariant.

Now a periodic potential energy is introduced to the system. This potential is the sum of the electrostatic potential due to ion cores of lattice atoms and the potential due to all other outer electrons. The charge density from outer electrons would have the same average value in every unit cell of the crystal and would also be periodic. Thus the total potential has the periodicity of the lattice. Based on this reasoning, Bloch argued that the total potential energy $V(z)$ with a periodicity of the lattice could be substituted into the 1D Schrödinger equation.

$$-\frac{\hbar^2}{2m}\frac{d^2\psi}{dz^2} + V(z)\psi = E\psi \quad (3.38)$$

He concluded that the wave functions, which satisfy (3.38), subject to such a periodic potential, must be of the form

$$\psi_k(z) = u_k(z)\exp(ikz) \quad (3.39)$$

where $u_k(z)$ are functions with the same periodicity of the lattice. The theory leading to (3.39) is known as *Bloch's theorem*, and functions $u_k(z)$ are called *Bloch functions*.

To discuss Bloch's theorem, let us first examine a 1D crystal, shown in Fig. 3.8, which has a linear array of atoms with interatomic (lattice) spacing a. Since the potential has the periodicity of the lattice, then

$$V(z) = V(z + a) \quad (3.40)$$

for any value of z. For such a periodic potential system, the wave function solutions shall repeat after N unit cell length, where N is an arbitrary number. Then

$$\psi(z) = \psi(z + Na) \quad (3.41)$$

for any value of z. In view of the translational symmetry of the system, the wave functions of the neighboring unit cells can be related by an expression of the form

$$\psi(z + a) = \lambda\psi(z) \quad (3.42)$$

where λ is a function to be determined. This means that for the location separated by N unit cells,

$$\psi(z + Na) = \lambda^N\psi(z) = \psi(z) \quad (3.43)$$

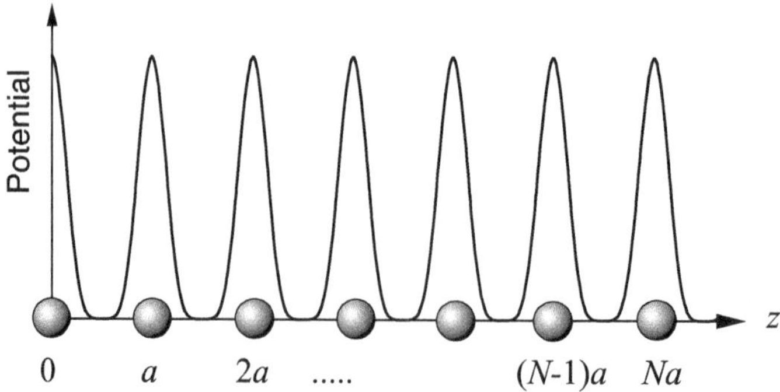

Fig. 3.8 Periodic potential for a linear atomic lattice. We may assume that the wave functions of all eigenstates must repeat after some arbitrary number N of unit cells

Comparing (3.41) and (3.43) shows that $\lambda^N = 1$ and λ must be one of the N roots of unity:

$$\lambda = \exp(i2\pi P/N) \tag{3.44}$$

where P is an integer representing the number of complete cycles in the distance Na. Therefore, we can relate the wave vector k to P by $k = 2\pi P/Na$. Then, for $\psi(z+a)$,

$$\lambda = \exp(ika) \tag{3.45}$$

Thus, one form of the wave function is

$$\psi(z+a) = \exp(ika)\psi(z) \tag{3.46}$$

An equivalent and general form of the above equation is that the wave function is the modulated plane wave

$$\psi(z) = \exp(ikz)u_k(z) \tag{3.47}$$

where $u_k(z)$ is a periodic function which also has the translational periodicity of the lattice. This is the one-dimensional Bloch function.

It is clear from (3.42), (3.45), and (3.47) that

$$\begin{aligned} u_k(z+a) &= \exp[-ik(z+a)]\psi(z+a) \\ &= \exp[-ik(z+a)]\lambda\psi(z) \\ &= \exp[-ik(z+a)]\exp(ika)\psi(z) \\ &= \exp(-ikz)\psi(z) = u_k(z) \end{aligned} \tag{3.48}$$

The function $u_k(z)$ is thus periodic with a period a.

We can expand this 1D result to 3D crystal systems with a periodic potential function proportional to the periodicity of the lattice, shown as

$$V(\boldsymbol{r}) = V(\boldsymbol{r}+\boldsymbol{T}) \text{ and } \boldsymbol{T} = n_1\boldsymbol{a} + n_2\boldsymbol{b} + n_3\boldsymbol{c} \tag{3.49}$$

The wave functions which satisfy the corresponding Schrödinger wave equation have the form of a plane wave with a propagation constant $\boldsymbol{k}$ modulated by a function $u_k(\boldsymbol{r})$ whose periodicity is that of the crystal lattice.

$$\begin{cases} \psi_k(\boldsymbol{r}) = u_k(\boldsymbol{r})\exp(i\boldsymbol{k}\cdot\boldsymbol{r}) \\ u_k(\boldsymbol{r}) = u_k(\boldsymbol{r}+\boldsymbol{T}) \end{cases} \tag{3.50}$$

3.2.2 Reduced Zone Representation

The 1D Bloch function of (3.39) can be written as

$$\psi(z) = u_k(z)\exp(i2\pi Pz/Na) \tag{3.51}$$

In this equation, as will become clear next, it is not necessary to use the value for P which is larger than $\pm(N/2)$, corresponding with $k = \pm(\pi/a)$. Any excess periodicity can be transferred into the $u_k(z)$ factor of the Bloch function. We notice that the range of wave vector from $k = -(\pi/a)$ to $k = +(\pi/a)$ corresponds to the first Brillouin zone in reciprocal lattice of the 1D lattice with a lattice spacing of a. Thus it is useful to examine the Bloch wave function near the origin of the k-space.

For the 3D case, the Bloch wave function whose wave vectors differ from $\psi_k(\boldsymbol{r})$ by a reciprocal lattice vector $\boldsymbol{G}$ according to

$$\boldsymbol{k} + \boldsymbol{G} = \boldsymbol{k}' \tag{3.52}$$

is also equivalent in terms of the electron energy. It shows

$$\begin{aligned}\psi_{k+\mathrm{G}}(\boldsymbol{r}) &= u_{k+\mathrm{G}}(\boldsymbol{r})\exp[i(\boldsymbol{k}+\boldsymbol{G})\cdot\boldsymbol{r}]\\ &= \left[u_{k+\mathrm{G}}(\boldsymbol{r})\exp(i\boldsymbol{G}\cdot\boldsymbol{r})\right]\exp(i\boldsymbol{k}\cdot\boldsymbol{r})\\ &= u_k(\boldsymbol{r})\exp(i\boldsymbol{k}\cdot\boldsymbol{r})\end{aligned}$$

Thus,

$$\psi_{k+\mathrm{G}}(\boldsymbol{r}) = \psi_k(\boldsymbol{r}) \tag{3.53}$$

Here, we used $u_{k+\mathrm{G}}(\boldsymbol{r}) = u_k(\boldsymbol{r})$ and $\exp(i\boldsymbol{G}\cdot\boldsymbol{r}) = \exp(i2m\pi) = 1$.

The Schrödinger wave equations for ψ_k and $\psi_{k+\mathrm{G}}$ can be expressed, respectively, as

$$H\psi_k = E_k\psi_k \tag{3.54}$$

and

$$H\psi_{k+\mathrm{G}} = E_{k+\mathrm{G}}\psi_{k+\mathrm{G}} \tag{3.55}$$

Using (3.53), we can combine these wave equations into

$$E_k\psi_k = E_{k+\mathrm{G}}\psi_{k+\mathrm{G}} \tag{3.56}$$

or

$$E_k = E_{k+\mathrm{G}} \tag{3.57}$$

Thus, the translational real-space periodicity of the lattice potential imposes periodicity on the energy eigenvalues and wave functions in reciprocal k-space. To examine the energy–momentum relationships of the electron, one needs only know these functions for k- values in the first Brillouin zone, k'. For example, as shown in Fig. 3.9, assume $\pi/a \le k \le 3\pi/a$ in a 1D system with a lattice constant a, or

$$k = \frac{2\pi}{a} + k' \text{ and } -\frac{\pi}{a} < k' < \frac{\pi}{a} \tag{3.58}$$

Then,

$$\begin{aligned}\psi_k(z) &= u_k(z)\exp(ikz)\\ &= \{u_k(z)\exp[i(2\pi/a)z]\}\exp(ik'z)\end{aligned} \tag{3.59}$$

where $\exp[i(2\pi/a)z]$ is a periodic function with the same periodicity as $u_k(z)$. We can rewrite the term inside the curly bracket as a new periodic function $u_{k'}(z)$ with the same periodicity. Thus,

$$\psi_k(z) = u_{k'}(z)\exp(ik'z). \tag{3.60}$$

This reduces the Bloch wave number to the range $-\pi/a \le k \le \pi/a$, which is the first Brillouin zone. Other higher-order wave vectors can also be reduced to the first Brillouin zone using the reduced zone representation.

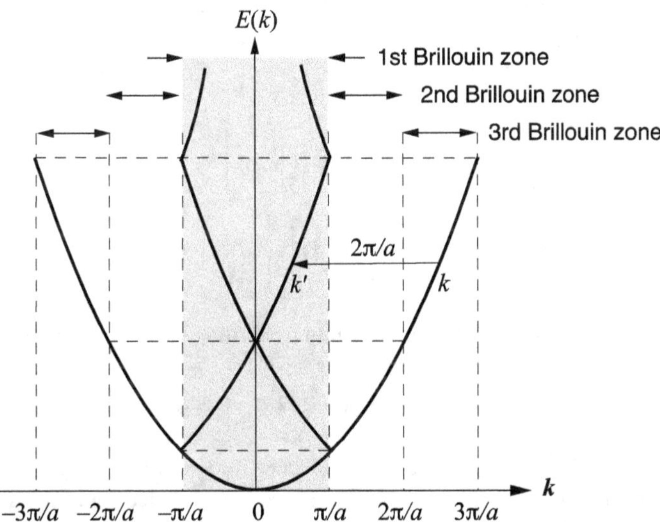

Fig. 3.9 Energy as a function of wave vector for electrons in a one-dimensional crystal of lattice constant a, where the amplitude of the periodic potential is set to zero. The continuous energy function is shown as a multiple-valued function of k in the shaded first Brillouin zone using the reduced zone representation

3.2.3 Empty Lattice Model—Energy Band Calculation for an FCC Crystal

In the free electron model, it is simple to discuss electron energy and wave vector without regard for the crystallography of the solid. As soon as we take the periodic potential into consideration, the Brillouin zone boundaries (surfaces) in k-space become particularly important. Therefore, the following discussion of the relationship between energy and momentum for electrons in crystals will be in k-space.

For the free electron model, the allowed energy values are distributed continuously in a parabolic form of

$$E = \frac{\hbar^2 k^2}{2m} \tag{3.19}$$

When a finite periodic potential distribution is added, the periodic lattice information needs to be incorporated. Since we consider one electron in the crystal lattice only, without including electron–electron wave interactions, this approach is also called the *empty lattice model*. The 3D parabolic E-k relationship is modified with the incorporation of the reciprocal lattice vector $\boldsymbol{G}$ as

$$E = \frac{\hbar^2}{2m}|\boldsymbol{k} + \boldsymbol{G}|^2 \tag{3.61}$$

where $\boldsymbol{k} = k_x\boldsymbol{a} + k_y\boldsymbol{b} + k_z\boldsymbol{c}$ and $\boldsymbol{G} = h\boldsymbol{a}^* + k\boldsymbol{b}^* + l\boldsymbol{c}^*$. As we have discussed in Sect. 2.5, $\boldsymbol{G}$ contains the lattice information for a specific crystal structure. For FCC crystals, using the translation vectors of the $\boldsymbol{G}$-vector from (2.65) leads to

$$\begin{aligned} E_{\mathrm{FCC}} = \frac{\hbar^2}{2m}\Bigg\{&\left[k_x + \frac{2\pi}{a}(-h + k + l)\right]^2 + \left[k_y + \frac{2\pi}{a}(h - k + l)\right]^2 \\ &+ \left[k_z + \frac{2\pi}{a}(h + k - l)\right]^2\Bigg\} \end{aligned} \tag{3.62}$$

By properly adjusting the system of units used, the expression can be reduced to

$$E = (k_x + G_x)^2 + \left(k_y + G_y\right)^2 + (k_z + G_z)^2 \tag{3.63}$$

where G_x, G_y, and G_z are integers to be selected for different energy states. Based on this equation, we can calculate the E-k curves, or the electronic band structure, along the $\langle 100\rangle$ and $\langle 111\rangle$ axes in the reduced zone representation.

(a) ⟨100⟩

As shown in Fig. 3.10, along [100], $k_y = k_z = 0$, and $E = (k_x + G_x)^2 + G_y^2 + G_z^2$. The lowest energy state is located at $G_x = G_y = G_z = 0$ and $E = k_x^2$. The E-k curve follows a parabola and $E = 1$ at $k_x = 1$. Figure 3.11 shows the lowest three E-k curves along [100].

To continue, using the reduced zone scheme, the next E needs to start at $E = 1$ ($k_x = 1$) and increase as k_x approaches zero. By trial and error, there are two possible G's that match this condition. However, $G(-1, -1, 0)$ is not selected due to the lack of continuation in the third Brillouin zone and other diffraction directions. The appropriate G's are $(G_x, G_y, G_z) = (-2, 0, 0)$, and $E = (k_x - 2)^2$. Started from $k_x = 1$, $E = 1$, the E-k curve increases toward $k_x = 0$, $E = 4$.

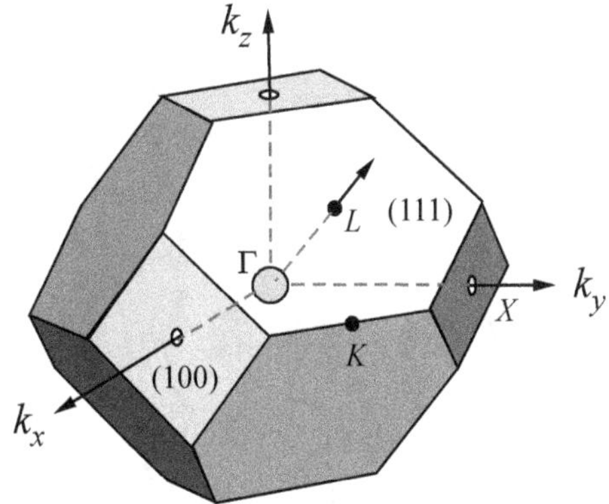

Fig. 3.10 Brillouin zone of the face-centered cubic lattice. The high symmetry points Γ, L, X, and K correspond to the zone center, ⟨111⟩, ⟨100⟩, and ⟨110⟩ directions, respectively

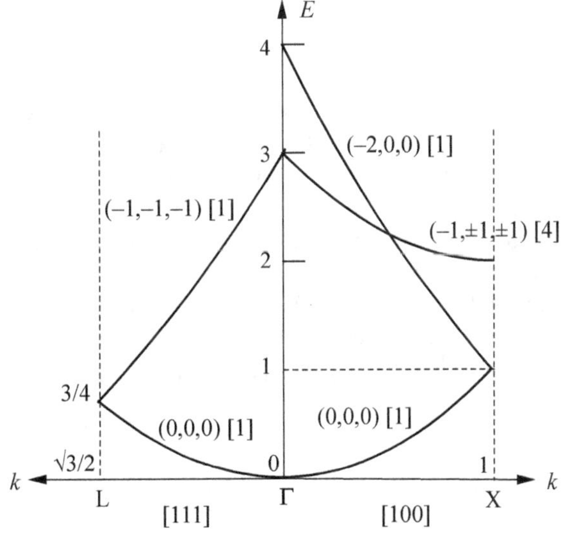

Fig. 3.11 Band structure for a free electron in an FCC lattice along [100] and [111] in the first Brillouin zone. For each curve, the selected ***G***'s and degeneracy (in brackets) are indicated

The next higher energy level is $E = 3$ at $k_x = 0$. This energy level matches the second lowest energy state along [111] shown next. This can be calculated by setting $(G_x, G_y, G_z) = (-1, \pm 1, \pm 1)$, which has a four-fold degeneracy. The E-k dispersion relationship is $E = (k_x - 1)^2 + 2$. At $k_x = 1$, $E = 2$, and at $k_x = 0$, $E = 3$.

(b) $\langle 111 \rangle$

In the reciprocal lattice of FCC structures, assuming the first Brillouin zone boundary along $\langle 100 \rangle$, measured from the zone center Γ, has a length of $k_{\max} = 1$. From the geometry of the Wigner–Seitz cell shown in Fig. 3.10, the Brillouin zone boundary along $\langle 111 \rangle$ has a length of $(\sqrt{3}/2)\, k_{\max}$ and $k_x = k_y = k_z$. The lowest energy is at $(G_x, G_y, G_z) = (0,0,0)$. $E = k_x^2 + k_y^2 + k_z^2 = k^2$. At $k = 0$, $E = 0$, and at $k = \left(\sqrt{3}/2\right) k\text{max}$, $E = 3/4$.

The next energy level should have an energy of 3/4 at $\left(\sqrt{3}/2\right) k_{max}$. This corresponds to a set of $(G_x, G_y, G_z) = (-1, -1, -1)$. Then

$$\begin{aligned} E &= (k_x - 1)^2 + \left(k_y - 1\right)^2 + (k_z - 1)^2 \\ &= \left(k_x^2 + k_y^2 + k_z^2\right) - 2\left(k_x + k_y + k_z\right) + 3 \\ &= k^2 - 2\sqrt{3}k + 3 \end{aligned} \tag{3.64}$$

where we recognized that

$$\begin{aligned} \left(k_x + k_y + k_z\right)^2 = {} & \left(k_x^2 + k_y^2 + k_z^2\right) \\ & + 2\left(k_x k_y + k_y k_z + k_z k_x\right) = 3k^2 \end{aligned}$$

and

$$k_x + k_y + k_z = \sqrt{3}k \tag{3.65}$$

According to (3.64), at $k = 0$, $E = 3$, and at $k = 0.868\ k_{\max}$, $E = 3/4$. This E-k curve is also shown in Fig. 3.11. One can continue the process along [110] and other major directions of the FCC crystal to complete the E-k diagram shown in Fig. 3.12.

In the free electron theory, where the potential induced by the core ions is neglected, the energy band is a continuous parabola. However, the above results indicate that the motion of electrons is not free, but is constrained by the spatial arrangement of the periodic potential associated with the lattices. Nevertheless, in this simple 'one' electron model, it allows the Bloch function wave of all energies to extend without attenuation through a crystal.

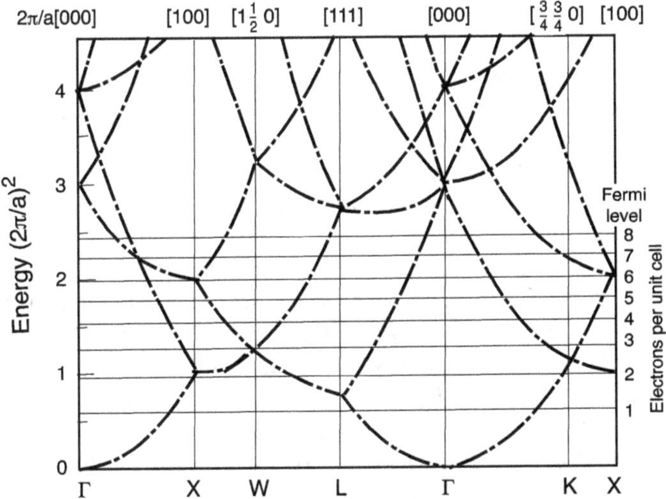

Fig. 3.12 Band structure for a free electron in an FCC lattice in the first Brillouin zone. Reprinted with permission from [1], copyright Wiley

3.3 Nearly Free Electron Approximation and the Energy Gap

3.3.1 Origin of Bandgaps

The E-k dispersion curve of 'one' free electron in a 1D system with a small periodic potential function, equivalent to the lattice constant of semiconductors, can be represented by a parabolic curve centered on a reciprocal lattice point. The possible electron states are not restricted to that single parabola in k-space, but are equally as likely to be found on a parabola shifted by any $\boldsymbol{G}$-vector as shown in Fig. 3.13. This is due to the periodic property of E, i.e.,

$$E(\boldsymbol{k}) = E(\boldsymbol{k} + \boldsymbol{G}) = \left(\hbar^2/2m\right)|\boldsymbol{k} + \boldsymbol{G}|^2 \tag{3.66}$$

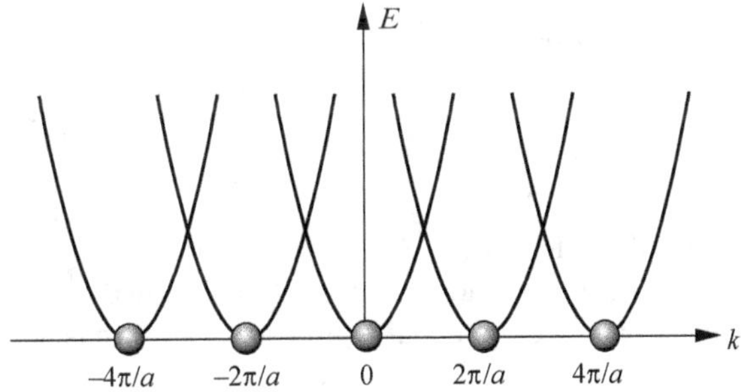

Fig. 3.13 Energy bands for an electron in a 1D periodic array of potential with a periodicity of a plotted in the repeated-zone scheme

However, when there is only one electron in the system, the electron–electron wave interactions are missing, and the resulting energy band structure shows only the crystal structural property, e.g., as illustrated in Fig. 3.12 for FCC lattices. Of course, in reality, there are extremely large quantities of electrons available in semiconductors. To improve the model, we shall allow more electrons in the crystal lattice such that the electron–electron wave interactions could be included.

When an electron wave propagates through the crystal lattice, it gets scattered in all directions from the lattice atoms. Because of the periodic nature of the lattice, in certain directions, waves scattered by many lattice points interfere constructively and a strong scattered beam results. For example, the Bragg condition for an incident wave normal to the crystal plane with a lattice periodicity of a is $n\lambda = 2a$ or $k = n\pi/a$. The wave vector that fulfills the Bragg condition simply equals one half of the reciprocal lattice vector $\boldsymbol{G}$, i.e., $k = \pm G/2$. In general, since the scattering from the lattice involves all wave numbers, a plane wave of wave number k and energy $E(k)$ will mix with other electron waves with wave number $k + G_n$ for all n's. The mixing becomes strong only when the electron waves propagating through the crystal lattice are of equal energy where $E(\boldsymbol{k}) = E(\boldsymbol{k} \pm \boldsymbol{G})$ or $|\boldsymbol{k}| = |\boldsymbol{k} \pm \boldsymbol{G}|$ or $k = \pm G_n/2$. Again, the Bragg condition for strong reflection occurs at $k = \pm G_n/2$. Therefore, the backscattering becomes very strong, and the electron is unable to propagate through. The forward and back reflected waves establish a standing wave in the crystal. Assuming the contributions from the nearest neighboring waves dominate the resulting waves, the contribution from other reciprocal lattice vectors can be neglected.

The wave functions at the edge of the first Brillouin zone at $\pm G/2 = \pm\pi/a$ are

$$\psi(+\pi/a) = A\exp(ik \cdot r) = A\exp(iGz/2) \tag{3.67a}$$

and

$$\psi(-\pi/a) = A\exp[i(G/2 - G)z] = A\exp(-iGz/2) \tag{3.67b}$$

The resultant forward (ψ_+) and reflected (ψ_-) plane waves are

$$\begin{cases} \psi_+ \sim \exp(iGz/2) + \exp(-iGz/2) \sim \cos(\pi z/a) \\ \psi_- \sim \exp(iGz/2) - \exp(-iGz/2) \sim \sin(\pi z/a) \end{cases} \tag{3.68}$$

The corresponding electron probability densities are

$$\begin{cases} \rho_+ = \psi_+^*\psi_+ \sim \cos^2(\pi z/a) \\ \rho_- = \psi_-^*\psi_- \sim \sin^2(\pi z/a) \end{cases} \tag{3.69}$$

As illustrated in Fig. 3.14, the probability density of ψ_+ (solid curve) peaks at the valleys of the potential energy function, and the associated electrons spend more time at the lower potential region. Thus, electrons associated with ψ_+ have a lower

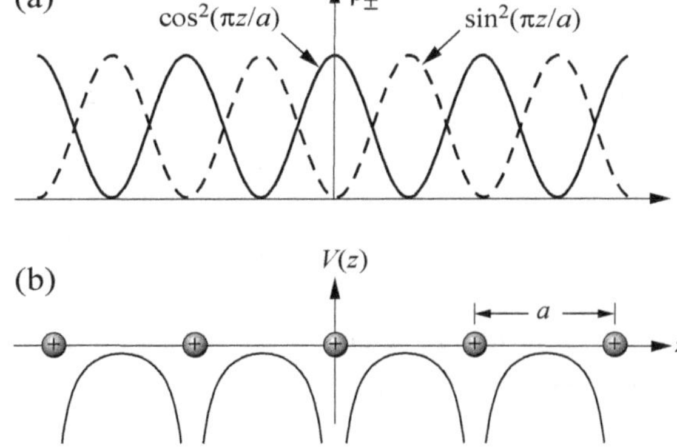

Fig. 3.14 Schematic illustration of the relationship between **a** electron charge density distribution and **b** the ion core positions in a 1D lattice

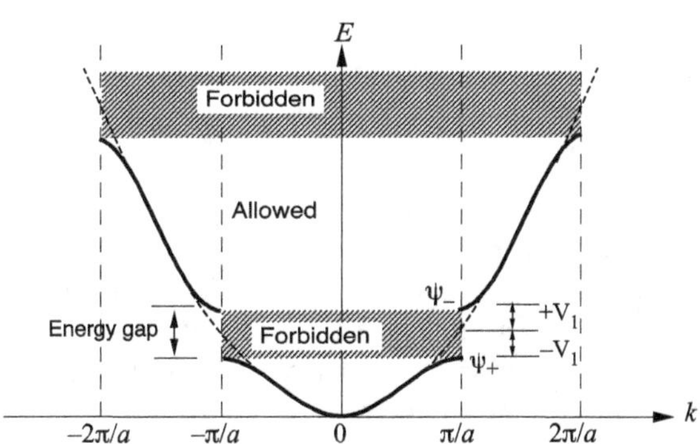

Fig. 3.15 Electronic band structure for a 1D crystal with periodicity a in the nearly free electron approximation. Many electrons are presented. The dashed curve shows the continuous E-k curve of the free electron model

total energy. The peaks of the probability density of ψ_- (dashed curve) correspond to the peaks of the potential energy function and have a higher total energy. The degeneracy of the two states, ψ_+ and ψ_-, with different energies at certain wave numbers, breaks the continuity of the parabolic E-k dispersion curve and forms the energy gap (Fig. 3.15). At $k = \pm\, n\pi/a$, the wave functions of the upper and lower boundaries of the energy gap are ψ_- and ψ_+, respectively. There is no allowed energy value in between ψ_+ and ψ_- at the zone boundary, and an energy bandgap exists in this forbidden region.

3.3.2 Energy Gap—A Quantitative Approach

The behavior of the energy bands in the vicinity of the bandgaps can be analyzed by considering the Schrödinger equation for a translational periodic potential energy expressed in terms of Fourier expansion. In a 1D lattice similar to that shown in Fig. 3.14b, the periodic potential with a periodicity a is expressed as $V(z) = V(z + a)$. The Fourier transformation of the symmetric periodic function is

$$V(z) = \sum_n V_n \exp(i2n\pi z/a) \tag{3.70}$$

Since $V(z)$ is an even function, by neglecting the sine term, $V(z)$ becomes

$$\begin{aligned} V(z) &\cong V_0 + \sum_n [2V_n(z)\cos(2n\pi z/a)] \\ &= V_0 + V_1'(z) + V_2'(z) + \cdots \end{aligned} \tag{3.71}$$

Further neglecting higher-order terms, the lattice potential function can be represented as

$$V(z) \approx V_0 + V_1'(z) \tag{3.72}$$

and

$$\begin{aligned} V_1'(z) &= 2V_1\cos(2\pi z/a) \\ &= V_1\left[\exp(i2\pi z/a) + \exp(-i2\pi z/a)\right] \end{aligned} \tag{3.73}$$

where V_1 is a constant.

The wave functions of this lattice structure can be expressed as $\psi(z) = u(z)\exp(ikz)$. Due to Bragg diffraction at zone boundaries, the incident and reflected waves form constructive interference. At the first Brillouin zone boundary, $k = \pm \pi/a$, the wave functions are

$$\begin{aligned} \psi(z) &= A_1\exp(ikz) + A_2\exp(-ikz) \\ &= A_1\exp(i\pi z/a) + A_2\exp(-i\pi z/a) \end{aligned} \tag{3.74}$$

where A_1 and A_2 are constants, and

$$\frac{\partial^2\psi(z)}{\partial z^2} = -k^2\psi(z) \tag{3.75}$$

Replacing (3.72) and (3.75) in the Schrödinger wave equation

$$\frac{\hbar^2}{2m}\frac{\partial^2\psi}{\partial z^2} + [E - V(z)]\psi = 0 \tag{3.76}$$

it becomes

$$-\frac{\hbar^2k^2}{2m}\psi(z) + \left[E - V_0 - V_1'(z)\right]\psi(z) = 0 \tag{3.77}$$

Letting $E_k \equiv \dfrac{\hbar^2 k^2}{2m}$ and using $\psi(z) = A_1 \exp(ikz) + A_2 \exp(-ikz)$, the wave equation becomes

$$\begin{aligned}&(E - E_k)\left[A_1 \exp(ikz) + A_2 \exp(-ikz)\right] \\ &= \left[V_0 + V_1^{'}(z)\right]\left[A_1 \exp(ikz) + A_2 \exp(-ikz)\right]\end{aligned} \tag{3.78}$$

Regrouping, we have

$$\begin{aligned}&(E - E_k - V_0)\left[A_1 \exp(ikz) + A_2 \exp(-ikz)\right] \\ &= \left\{V_1\left[\exp(i2kz) + \exp(-i2kz)\right]\right\}\left[A_1 \exp(ikz) + A_2 \exp(-ikz)\right] \\ &= V_1\left\{\left[A_1 \exp(-ikz) + A_2 \exp(ikz)\right] + \left[A_1 \exp(i3kz) + A_2 \exp(-i3kz)\right]\right\}\end{aligned} \tag{3.79}$$

By neglecting the higher-order terms and comparing exponential terms on each side, we obtain the following two equations as functions of A_1 and A_2:

$$(E - E_k - V_0)A_1 - V_1 A_2 = 0 \tag{3.80}$$

$$-V_1 A_1 + (E - E_k - V_0)A_2 = 0 \tag{3.81}$$

The non-trivial solutions of these equations are when the determinant equals zero, yielding

$$(E - E_k - V_0)^2 - (-V_1)^2 = 0 \tag{3.82}$$

This second order equation has two energy eigenvalues:

$$E_{\pm} = E_k + V_0 \pm V_1 \text{ at } k = \pm\frac{\pi}{a} \tag{3.83}$$

The E-k curve at the first Brillouin zone boundary breaks open by a value of $2V_1$ with equal amounts above and below the value for the free electron model. Recall that V_i is the Fourier coefficient of the small periodic lattice potential. Therefore, as shown in Fig. 3.15, the width of the forbidden energy bands is small compared to that of the allowed bands. Equation (3.83) can be used to calculate the values of A_1 and A_2. By replacing $E_{\pm}$ into (3.80), we find

$$A_1 = A_2 \text{ for } E_+ \text{ and } A_1 = -A_2 \text{ for } E_- \tag{3.84}$$

This leads to the solutions of the wave functions in the first Brillouin zone (3.74) as

$$\begin{cases} \psi_+(z) = A\big[\exp(ikz) + \exp(-ikz)\big] = 2A\cos(\pi z/a) \\ \psi_-(z) = A\big[\exp(ikz) - \exp(-ikz)\big] = 2iA\sin(\pi z/a) \end{cases} \tag{3.85}$$

These are the same qualitative results we obtained in (3.68). The $\psi_+(z)$ and $\psi_-(z)$ functions near the zone boundary represent two traveling waves of equal amplitude. The interference of these two waves generates two standing waves which represent two different electron distributions with respect to the lattice potential. The difference in potential energy associated with these two standing waves leads to a splitting of the energy bands at the zone boundary as shown in Fig. 3.15. Thus, in the nearly free electron approximation, by considering *many electrons* propagating through the periodic crystal lattice, bandgaps are formed in the otherwise continuous parabolic energy band.

3.4 The Kronig–Penney Model

In the free electron approximation discussed above, the periodic potential energy of the electron was assumed to be small compared to its total energy, and the wave functions for electrons on neighboring atoms overlap heavily. This model led to the energy band structure where the width of the forbidden energy bands was found to be small compared to that of allowed bands. The other important approach, the tight-binding approximation, proceeds from the opposite point of view, where the atoms of the crystal are so far apart that the wave functions for electrons associated with neighboring atoms overlap only to a small extent. Thus the wave functions and allowed energy levels of the crystal will be closely related to the wave functions and energy levels of isolated atoms. The resulting allowed energy bands are narrow in comparison with the forbidden bands. Depending on the particular solid to be studied, we will use the free electron approximation, the tight-binding approximation, or a mixture of the two. Of course, there are very many advanced methods for full calculations of semiconductor energy bands, but the topic is too complex for an adequate treatment in this book. Fortunately, we still can learn some interesting properties about the appearance of energy gaps when an 'adjustable' periodic potential is added by studying a simple 1D model suggested by R. de L. Kronig and W. G. Penney. In the Kronig–Penney model, the depth, width, and periodicity of the potential wells can be varied to simulate the environment that electrons will experience under the condition of the nearly free electron model, the tight-binding approximation, or a mixture of the two.

3.4.1 Theoretical Model

The Kronig–Penney model uses a simplified rectangular periodic potential energy for a 1D lattice potential as shown in Fig. 3.16. The total interatomic distance is $(a + b)$, of which the potential energy is V_0 over the range b. The one-dimensional lattice potential energy function is

$$V(z) = \begin{cases} 0, & \text{in region I} \quad (0 < z < a) \\ V_0, & \text{in region II} \quad (-b < z < 0) \end{cases} \tag{3.86}$$

Using this periodic potential $V(z)$ in the wave equation of the system

$$\frac{\hbar^2}{2m}\frac{\partial^2\psi}{\partial z^2} + [E - V(z)]\psi = 0 \tag{3.87}$$

the corresponding solutions are

$$\begin{cases} \psi_{\mathrm{I}}(z) = A\exp(i\alpha' z) + B\exp(-i\alpha' z) \\ \psi_{\mathrm{II}}(z) = C\exp(\beta' z) + D\exp(-\beta' z) \end{cases} \tag{3.88}$$

and

$$\alpha' = \sqrt{\frac{2mE}{\hbar^2}},\ \beta' = \sqrt{\frac{2m(V_0 - E)}{\hbar^2}} \tag{3.89}$$

However, this approach is not suitable for such a potential distribution where the periodic property of the lattice structure is not properly addressed. The Kronig–Penney model incorporates Bloch's theorem in the wave equation to overcome this problem. Assume the solution of the wave equation has the form of

$$\psi(z) = u(z)\exp(ikz). \tag{3.90}$$

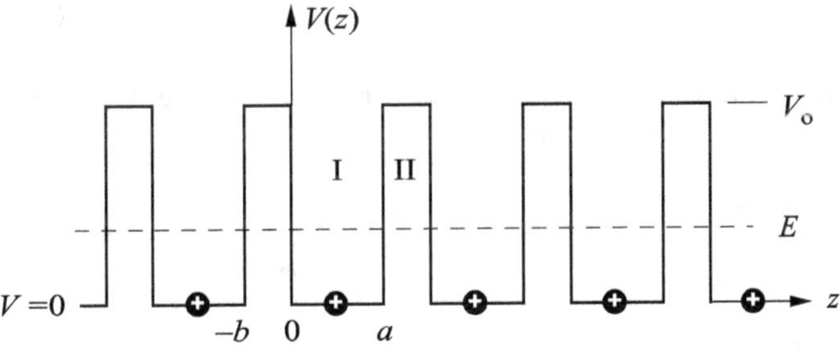

Fig. 3.16 One-dimensional periodic square well potential energy used in Kronig–Penney model

where k has a real value. Using this in the Schrödinger wave equation, the general form of the wave equation becomes

$$\frac{d^2u}{dz^2} + 2ik\frac{du}{dz} - \left(k^2 - \frac{2mE}{\hbar^2} + \frac{2mV}{\hbar^2}\right)u = 0 \tag{3.91}$$

Equation (3.91) reduces to the following two equations in these two regions:

(1) In region I, $V = 0$:

$$\frac{d^2u_I}{dz^2} + 2ik\frac{du_I}{dz} - (k^2 - \alpha^2)u_I = 0 \text{ and } \alpha = \sqrt{\frac{2mE}{\hbar^2}} \tag{3.92}$$

(2) In region II, $V = V_0$:

$$\frac{d^2u_{II}}{dz^2} + 2ik\frac{du_{II}}{dz} - (k^2 - \beta^2)u_{II} = 0 \text{ and}$$

$$\beta = \sqrt{\frac{2m(E - V_0)}{\hbar^2}} \tag{3.93}$$

Note that β becomes an imaginary number for $E < V_0$.
These two equations are a differential equation of the form

$$\frac{d^2f}{dz^2} + a\frac{df}{dz} + bf = 0 \text{ or } m^2 + am + b = 0 \tag{3.94}$$

The solutions of this equation are

$$f = A\exp(m_1z) + B\exp(m_2z) \tag{3.95}$$

This leads to the solutions of the wave equations:

$$\begin{cases} u_I(z) = A\exp[i(\alpha - k)z] + B\exp[-i(\alpha + k)z],\ 0 < z < a \\ u_{II}(z) = C\exp[i(\beta - k)z] + D\exp[-i(\beta + k)z],\ -b < z < 0 \end{cases} \tag{3.96}$$

The factors A, B, C, and D are derived by using the continuity rules at $-b$, 0, and a. It leads to the following conditions:

(i) At $z = 0$, $u_I = u_{II}$,

$$A + B = C + D \tag{3.97}$$

(ii) At $z = 0$, $u'_I = u'_{II}$,

$$(\alpha - k)A - (\alpha + k)B = (\beta - k)C - (\beta + k)D \tag{3.98}$$

(iii) At $x = a$ and $-b$, $u_{\mathrm{I}} = u_{\mathrm{II}}$,

$$A\exp[i(\alpha - k)a] + B\exp[-i(\alpha + k)a] = C\exp[-i(\beta - k)b] + D\exp[i(\beta + k)b] \tag{3.99}$$

(iv) At $x = a$ and $-b$, $u'_{\mathrm{I}} = u'_{\mathrm{II}}$,

$$(\alpha - k)A\exp[i(\alpha - k)a] - (\alpha + k)B\exp[-i(\alpha + k)a] = (\beta - k)C\exp[-i(\beta - k)b] - (\beta + k)D\exp[i(\beta + k)b] \tag{3.100}$$

The coefficients A, B, C, and D can be determined as the solution of a set of four simultaneous linear equations in those quantities. For a non-trivial solution, the determinant of the coefficients must be zero, that is,

$$\begin{bmatrix} 1 & 1 & -1 & -1 \\ \alpha - k & -(\alpha + k) & -(\beta + k) & \beta + k \\ \exp[i(\alpha - k)a] & \exp[-i(\alpha + k)a] & -\exp[-i(\beta - k)b] & -\exp[i(\beta + k)b] \\ (\alpha - k)e^{i(\alpha-k)a} & -(\alpha + k)e^{-i(\alpha+k)a} & -(\beta + k)e^{-i(\beta-k)b} & (\beta + k)e^{i(\beta+k)b} \end{bmatrix} = 0 \tag{3.101}$$

Thus, for $E > V_0$, after expanding the determinant and rearranging, (3.101) can be expressed as

$$-\left(\frac{\alpha^2 + \beta^2}{2\alpha\beta}\right)\sin(\alpha a)\sin(\beta b) + \cos(\alpha a)\cos(\beta b) = \cos[k(a + b)] \tag{3.102}$$

However, in the range $0 < E < V_0$, according to (3.93), β is imaginary. It is common to express (3.102) in a different form. Letting

$$\beta = i\gamma = i\sqrt{\frac{2m(V_0 - E)}{\hbar^2}} \tag{3.103}$$

in this region and noting that $\cosh(z) = \cos(iz)$ and $i\sinh(z) = \sin(iz)$, (3.102) can be rewritten as

$$\left(\frac{\gamma^2 - \alpha^2}{2\alpha\gamma}\right)\sin(\alpha a)\sinh(\gamma b) + \cos(\alpha a)\cosh(\gamma b) = \cos[k(a + b)] \tag{3.104}$$

We may use (3.102) or (3.104) to find the energy eigenvalues for various periodic lattice potential settings. Nevertheless, since $\cos(z)$ on the right-hand side has a real value in the range of $\pm$ 1, any result of $|\cos k(a + b)|$ larger than one is not a solution.

3.4.2 Analysis

Both (3.102) and (3.104) represent a rather involved problem unless we consider some set of simplified conditions. In the following, we shall discuss models with the potential function designed to replicate the nearly free electron approximation or the tight-binding approximation.

(a) **Nearly free electron model**

In the limit of the nearly free electron model, the weak periodic potential function is achieved by using the tall but thin barriers in the Kronig–Penney model. We assume that $V_0 \approx \infty$, $b \approx 0$, and $V_0 b$ remains finite and small. The electron in this model is nearly free except near the extremely thin periodic barriers, which can be easily tunneled through. Under these conditions, for $V_0 > E$, and

$$\lim_{\substack{V_0 \to \infty \\ b \to 0}} \sinh(\gamma b) \approx \gamma b \text{ and } \lim_{\substack{V_0 \to \infty \\ b \to 0}} \cosh(\gamma b) \approx 1 \tag{3.105}$$

Equation (3.104) is reduced to

$$\left(\frac{m V_0 b}{\hbar^2} \cdot a \right) \frac{\sin(\alpha a)}{\alpha a} + \cos(\alpha a) = \cos(ka) \tag{3.106}$$

It can also be written as a function of energy,

$$f(E) = P\left[\frac{\sin(\alpha a)}{\alpha a} \right] + \cos(\alpha a) = \cos(ka) \tag{3.107}$$

where

$$P = \frac{m(V_0 b)a}{\hbar^2} \propto (V_0 b)a \tag{3.108}$$

is a measure of the 'effective area' of each periodic barrier.

The left-hand side of (3.107) is clearly a function of α which is proportional to $\sqrt{E}$. Therefore, we can plot $f(E)$ in (3.107) as a function of $\sqrt{E}$ and impose a limit of $-1 \leq f(E) \leq 1$ to yield the allowed solutions shown in Fig. 3.17. We note that the ordinate is $(1 + P)$ for $\alpha = 0$, where $\sin(\alpha a) \approx \alpha a$, and it is ± 1 for α equal to any multiple of (π/a), regardless of the value of P. The left side of (3.107) contains a $[\sin(\alpha a)]/(\alpha a)$ term, which is a sinc function, with the same periodicity and phase as $\cos(\alpha a)$. Since the amplitude of the sinc function decreases quickly with increasing α, its contribution diminishes after a few periods. Therefore, for a small P, the resultant $f(E)$ has a shape similar to $\cos(\alpha a)$ beyond several times of (π/a). Note that, consistent with our earlier discussions of the free electron model, the forbidden bands are narrower than the allowed energy bands.

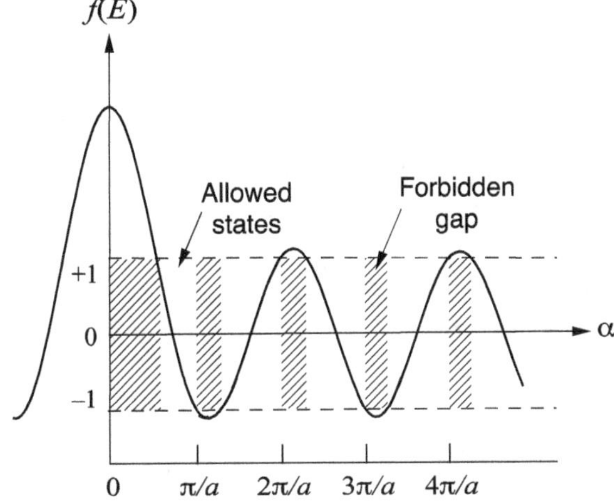

Fig. 3.17 Plot of the left side of (3.107) for $P = 2$ as a function of α. The allowed states are located within $1 \geq f(E) \geq -1$, and the shaded areas are the forbidden regions

In the following, we shall discuss two extreme cases of the nearly free electron model.

(i) *Case I*: $P \to \infty$

Since P is a product of $(V_0 b)$ and a, where $(V_0 b)$ is finite and small, an infinite P means that the lattice constant $(a+b)$ is near infinity. This is equivalent to the case of discrete atoms. In the range of $-1 \leq f(E) \leq 1$, the only possible solution of (3.107) is when $\sin(\alpha a) \approx 0$ or $\alpha a = \pm\, n\pi$ for $n = 1, 2, 3, \ldots$ We can deduce the allowed energy states as

$$\alpha = \pm \frac{n\pi}{a} = \sqrt{\frac{2mE}{\hbar^2}} \tag{3.109}$$

or

$$E = \frac{\hbar^2}{2m}\left(\frac{n\pi}{a}\right)^2 \quad n = 1, 2, 3, \ldots \tag{3.110}$$

The allowed E's are discrete levels as seen in an isolated atom.

(ii) *Case II*: $P \to 0$

Since $(V_0 b)$ is a small and finite, the condition of P approaching zero implies that the lattice constant $(a + b)$ is also near zero. The electron wave functions can easily tunnel through the thin barriers and act like a pool of free electrons. The solution (3.107) reduces to

$$\cos(\alpha a) = \cos(ka) \tag{3.111}$$

or

$$\alpha = k = \sqrt{\frac{2mE}{\hbar^2}} \tag{3.112}$$

Therefore, $E = \hbar^2 k^2/2m$, and the allowed energy states form a continuous band.

(b) **Tight-binding approximation**

This approach takes a point of view opposite to that of free electron model. It is based on the assumption that the atoms of the crystal are so far apart that the wave functions for neighboring electrons overlap only to a small extent. If we keep V_0 and a fixed and vary $d = (a + b)$, we see that as b becomes large,

$$\sinh(\gamma b) \approx \cosh(\gamma b) \approx \exp(\gamma b)/2. \tag{3.113}$$

Thus the general solution of the Kronig–Penney model becomes

$$\left[\left(\frac{\gamma^2 - \alpha^2}{2\alpha\gamma}\right)\sin(\alpha a) + \cos(\alpha a)\right]\frac{\exp(\gamma b)}{2} = \cos[k(a + b)] \tag{3.114}$$

Since b is a large number and both b and γ are positive, it is clear that $|\cos[k(a + b)]| \leq 1$ only if

$$\left(\frac{\gamma^2 - \alpha^2}{2\alpha\gamma}\right)\sin(\alpha a) + \cos(\alpha a) = 0 \tag{3.115}$$

or

$$\tan(\alpha a) \approx \frac{2\alpha\gamma}{\alpha^2 - \gamma^2} \tag{3.116}$$

This equation has an equivalent form of

$$(\alpha a)\tan(\alpha a) \cong \gamma' a \tag{3.117}$$

for $V_0 > E$, where γ' is a function of α and γ. This expression is similar to the solution of an isolated finite depth quantum well with a well depth of V_0 and a width of $2a$. This implies that the allowed energy bands are very narrow in the tight-binding approximation, which leads to a nearly linear relationship between $f(E)$ and E in each allowed narrow energy band. However, as shown in Fig. 3.18, the solution (3.115) still maintains the shape of a sinc function.

We know that in a two-dimensional system, the line equation of a linear function between point 1 at (x_1, y_1) and point 2 at (x_2, y_2) can be written as

$$\frac{y - y_1}{x - x_1} = \frac{y_2 - y_1}{x_2 - x_1} \tag{3.118}$$

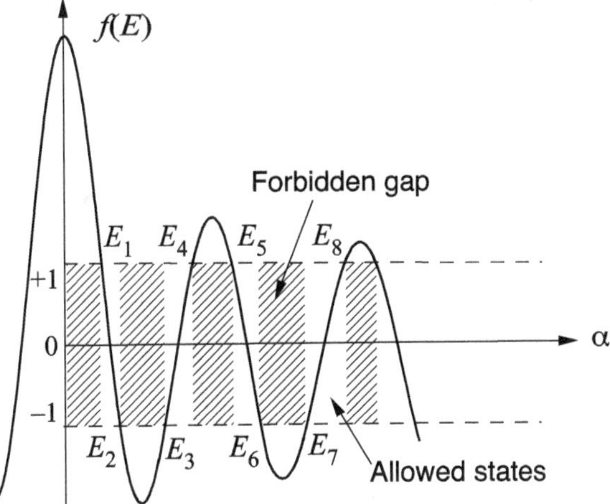

Fig. 3.18 Plot of the left side of (3.115) as a function of α. The allowed states are located within $1 \geq f(E) \geq -1$, and the shaded areas are the forbidden regions. The allowed energy bands are between E_1 and E_2, E_3 and E_4, E_5 and E_6, etc

We can apply (3.118) to derive the allowed energy states. In the lowest (first) allowed energy band, between E_1 and E_2, the line equation can be expressed as

$$\frac{f(E) - f(E_1)}{E - E_1} = \frac{f(E_2) - f(E_1)}{E_2 - E_1} \tag{3.119}$$

For the allowed energy states, $f(E_1) = f(E_4) = f(E_5) = f(E_8) = \cdots = +1, f(E_2) = f(E_3) = f(E_6) = f(E_7) = \cdots = -1$, and $f(E) = \cos[k(a + b)]$, which lead to the following results.

In the first allowed energy band, according to (3.119),

$$\cos[k(a + b)] - 1 = -2\left(\frac{E - E_1}{E_2 - E_1}\right) \tag{3.120}$$

If we assume the first allowed energy band width is

$$\Delta E_1 = E_2 - E_1 \tag{3.121}$$

then (3.120) can be expressed in terms of ΔE_1 as

$$E = E_1 + \frac{\Delta E_1}{2}\{1 - \cos[k(a + b)]\} \tag{3.122}$$

The result of (3.122) is plotted in Fig. 3.19a. Note that, in contrast to the free electron theory, the minimum allowed energy state is a nonzero value.

The second allowed energy band is between $f(E_3) = -1$ and $f(E_4) = +1$. The line equation has the form of

$$\cos[k(a + b)] + 1 = 2\left(\frac{E - E_3}{E_4 - E_3}\right) \tag{3.123}$$

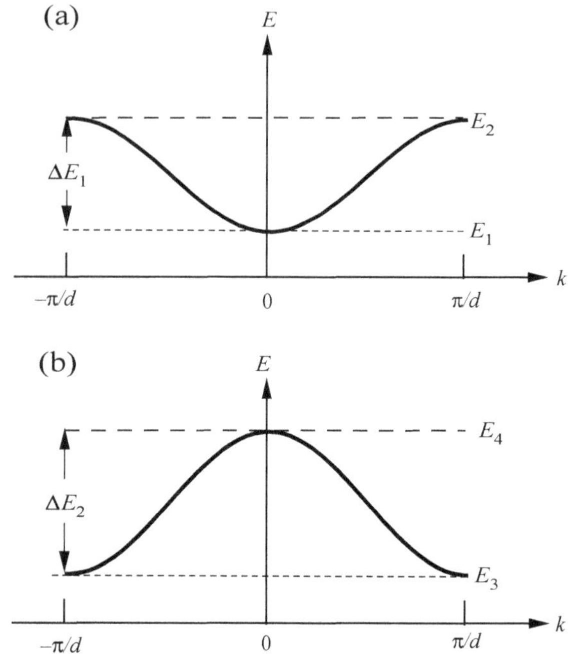

Fig. 3.19 Plot of the E-k curve of the **a** first and **b** second allowed energy bands using tight-binding approximation

This is rearranged below to show the allowed energy function.

$$E = E_3 + \frac{\Delta E_2}{2}\{1 + \cos[k(a + b)]\} \tag{3.124}$$

where $\Delta E_2 = E_4 - E_3$. This cosine function (3.124) is plotted in Fig. 3.19b. Using this process for higher energy bands, we can construct the complete energy band diagram. In Fig. 3.20, we note that there are discontinuities at $k = \pm n\pi/a,$ where energy gaps exist. The minimum allowed energy state is not zero and has a finite value. At higher energies, the allowed states become wider.

3.5 Effective Mass

Before we move on to examine the realistic energy band structure of common semiconductors, let us now consider the motion of an electron in the crystal under the influence of an applied electric field. First, if the Bloch function $u_k(z)$ in (3.39) is set to be a constant, then the wave function has the form of $\exp(\pm ikz)$, corresponding to a perfectly free electron of momentum $p = \hbar k$ whose energy would be

$$E = \frac{\hbar^2 k^2}{2m} \tag{3.125}$$

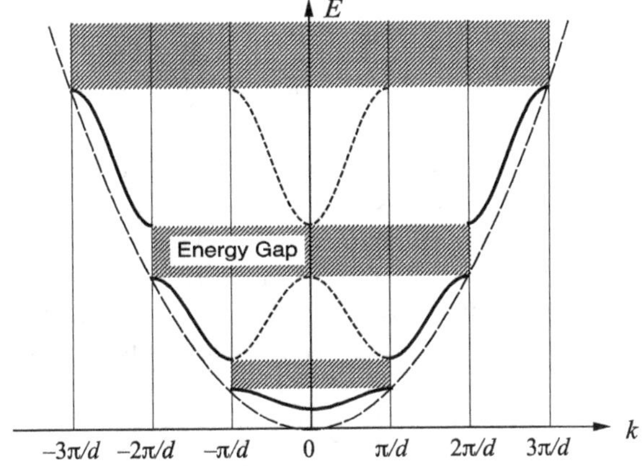

Fig. 3.20 Plot of the E-k curve according to the tight-binding approximation (solid curves) compared with the free electron mode (long-dashed curves). The reduced zone representation is plotted as short-dashed curves

where k is a propagation constant. This relation is shown as the long-dashed curve in Fig. 3.20. Under this condition, $p = \hbar k$ is the crystal momentum and has a value close to the real momentum of a moving particle. However, in a real crystal, the nonzero periodic potential function leads to discontinuities in the E-k curve at $k = \pm n\pi/d$. The true instantaneous momentum of an electron is no longer a constant of the motion, while the crystal momentum $\hbar k$ is a perfectly well-defined constant value for a given energy state. Since there exist allowed wave functions with different k's, we shall use the *group velocity* to relate the electron 'wave packet' following

$$\upsilon_g = \frac{d\omega}{dk} = \frac{1}{\hbar}\frac{dE}{dk} \tag{3.126}$$

where $E = \hbar\omega$. Again, this equation shows that the instantaneous momentum of the traveling electrons at the Brillouin zone boundaries is no longer a constant of the motion.

Suppose an electric field F is applied such that, over a distance dz in a period dt, the group velocity of the electron increases by dυ_g. The rate of change of electron energy due to the applied field is

$$dE = -qF dx = -qF\upsilon_g dt \tag{3.127}$$

Then, using (3.126) and (3.127) we see that

$$dE = \frac{dE}{dk}dk = -\frac{qF}{\hbar}\frac{dE}{dk}dt \tag{3.128}$$

Therefore,

$$dk = -\frac{qF}{\hbar}dt \tag{3.129}$$

or

$$\hbar \frac{\mathrm{d}k}{\mathrm{d}t} = \frac{\mathrm{d}p}{\mathrm{d}t} = -qF = \mathcal{F} \tag{3.130}$$

where p is the crystal momentum. Equation (3.130) indicates simply that the time rate of change of crystal momentum equals the force, $\mathcal{F}$. Thus it is the analogue of Newton's law, showing that the crystal momentum of the electron in a periodic lattice changes under the influence of an applied electric field in the same way as does the true momentum of a free electron in free space.

Differentiating (3.126) with respect to time, the result is

$$\frac{\mathrm{d}\upsilon_g}{\mathrm{d}t} = \frac{1}{\hbar}\frac{\mathrm{d}}{\mathrm{d}t}\left(\frac{\mathrm{d}E}{\mathrm{d}k}\right) = \frac{1}{\hbar}\frac{\mathrm{d}^2E}{\mathrm{d}k^2}\left(\frac{\mathrm{d}k}{\mathrm{d}t}\right) = \frac{1}{\hbar^2}\frac{\mathrm{d}^2E}{\mathrm{d}k^2}\left(\hbar\frac{\mathrm{d}k}{\mathrm{d}t}\right) \tag{3.131}$$

Using (3.130), the magnitude of acceleration $\boldsymbol{a}$ can be written as

$$a = \frac{\mathrm{d}\upsilon_g}{\mathrm{d}t} = \frac{1}{\hbar^2}\frac{\mathrm{d}^2E}{\mathrm{d}k^2}\left(\hbar\frac{\mathrm{d}k}{\mathrm{d}t}\right) = \frac{\mathcal{F}}{m^*} \tag{3.132}$$

where the *effective mass* m^* is defined as

$$m^* \equiv \frac{\hbar^2}{\left(\mathrm{d}^2E/\mathrm{d}k^2\right)} \tag{3.133}$$

or

$$m^* = \left(\frac{1}{\hbar^2}\frac{\partial^2 E}{\partial k_i \partial k_j}\right)^{-1} \tag{3.134}$$

for the non-spherical constant energy surface case. We can check this concept in a free electron system with mass m_0 where

$$E = \frac{\hbar^2 k^2}{2m_0}, \ \frac{\mathrm{d}E}{\mathrm{d}k} = \frac{\hbar^2 k}{m_0}, \ \text{and} \ \frac{\mathrm{d}^2E}{\mathrm{d}k^2} = \frac{\hbar^2}{m_0}$$

Then the effective mass simply equals the free electron mass as expected, i.e.,

$$m^* = \frac{\hbar^2}{\left(\mathrm{d}^2E/\mathrm{d}k^2\right)} = m_0 \tag{3.135}$$

To further illustrate the relationship between the E-k curve, group velocity, and m^*, a tight-binding limit model is used for two specific simple $E(k)$ variations in Fig. 3.21.

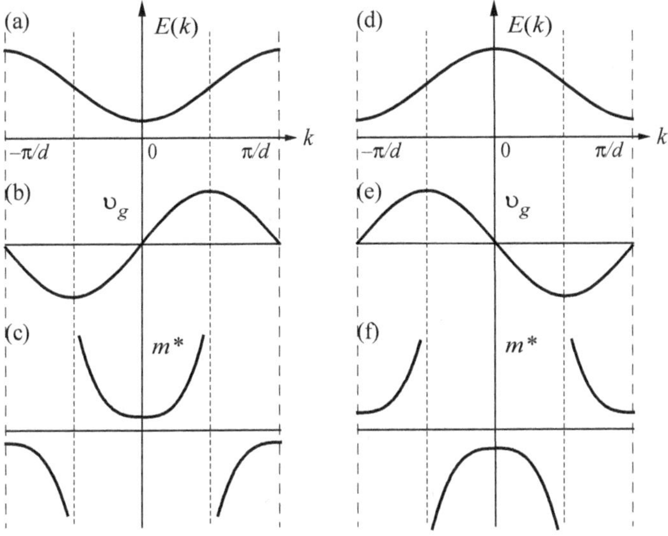

Fig. 3.21 Variation of **a**, **d** electron energy, **b**, **e** electron velocity, and **c**, **f** effective mass with reduced wave vector for states in a band of one-dimensional crystal. The *E*-*k* curve is a cosine function for the left panel, and in the form of $(1 + \cos ka)$ for the right panel

Assume, in one case, that

$$E = (\Delta E/2)(1 - \cos kd) \tag{3.136}$$

It represents the *E*-*k* dispersion curve of the lowest band, where ΔE is the allowed energy bandwidth. Under an applied electric field, the group velocity and effective mass are calculated as

$$\upsilon_g = \frac{1}{\hbar}\left(\frac{\mathrm{d}E}{\mathrm{d}k}\right) = \frac{d\Delta E}{2\hbar}\sin kd \tag{3.137}$$

and

$$m^* = \frac{\hbar^2}{\left(\mathrm{d}^2 E/\mathrm{d}k^2\right)} = \frac{2\hbar^2}{d^2\Delta E}\sec kd \tag{3.138}$$

As shown in Fig. 3.21b, the resultant group velocity is not a constant within the first Brillouin zone. Rather, it oscillates from zero to a maximum, then going through a minimum and back up to maximum again. Since the zone boundaries $\pm\pi/d$ are equivalent using the periodicity of $\boldsymbol{k}$-space, the zero velocity at the zone boundaries corresponds to the standing wave nature of the electron waves due to Bragg reflections. Furthermore,

$$x(t)=\int v_g(t)\mathrm{d}t=\frac{\Delta E}{2qF}\left[1-\cos\left(\frac{qFtd}{\hbar}\right)\right]$$
$$=\frac{\Delta E}{2qF}(1-\cos kd) \tag{3.139}$$

where $k=(qFt)/\hbar$. The results of (3.137) and (3.139) show that electrons oscillate in both k- and real space, rather than accelerating uniformly as they would do the absence of a periodic potential.

This velocity oscillation is equivalent to a mass change at critical points in k-space shown in Fig. 3.21c. The electron mass has a constant positive value near the zone center but becomes negative on the zone boundary. One unusual property of the effective mass is that at wave vectors corresponding to the maximum and minimum velocities, the effective mass goes to infinite. This marks the point where a deceleration must begin to slow down the electron. The other peculiar property is that the effective mass has regions in which it is negative. A negative m^* means that the acceleration resulting from the externally applied electric field is in a direction opposite to that of the applied field; this is the effect of electron reflection due to the periodic crystal lattice.

In the second tight-binding limit model, as shown in Fig. 3.21d, the $E(k)$ variation can be represented by

$$E=\left(\Delta E'/2\right)(1+\cos kd) \tag{3.140}$$

It resembles the E-k curve of the valence band in a semiconductor with an allowed energy bandwidth of $\Delta E'$. By the same procedures, we can derive the group velocity and effective mass as displayed in Fig. 3.21e, f, respectively.

The E-k dispersion curve shown in Fig. 3.21a is very much like the conduction band edge of a direct bandgap semiconductor such as GaAs. We shall use it to discuss some properties related to semiconductor materials. Since the electron effective mass is a function of k within the first Brillouin zone, what value of k should be used for determining m^*? Ideally, in a perfect crystal, under the influence of an applied electric field, the electron takes a periodic motion indefinitely within the first Brillouin zone as described by (3.137) and (3.139). This is known as the *Bloch oscillation* with a frequency $\omega=qFd/\hbar$, and there have been many attempts to observe and utilize it as a possible terahertz (THz) radiation source. However, all attempts to observe Bloch oscillation in conventional bulk solids have failed. For all reasonable applied fields, the Bloch oscillation is destroyed by scattering with impurities, defects, and phonons, before a single oscillation cycle is completed. The existence of such periodic oscillations in THz (1 THz $=10^{12}$ Hz) range was finally experimentally observed in 1993 in a semiconductor superlattice structure [2]. The multiple periods of GaAs/AlGaAs thin layers form superlattices that have a lattice periodicity much larger than the lattice constant of GaAs. This large lattice periodicity reduces the width of the Brillouin zone in k-space such that the scattering loss is suppressed. Therefore, in real semiconductors, the electron only travels a short range of the periodic E-k curve due to scattering events. The two dominant scattering processes are phonon scattering due to lattice vibration, and impurity and defect scattering. In high quality semiconductor materials, phonon scattering dominates the carrier relaxation process near room temperature and above. Through these processes, an electron can lose its energy quickly and move to the zone center of the E-k curve where the potential energy is the lowest. Near the zone center, the energy can be approximated by a parabolic function of k or $E = A(k - k_0)^2$. The effective mass has a constant value of $m^* = \hbar^2/2A$, and the dynamical behavior of the electron will be the same as that of a free particle with this effective mass. For indirect bandgap semiconductors like Si and Ge, although the minimum conduction band edge is not located at the Γ point, the energy is still a parabolic function of k. The effective mass will also be a constant similar to that of the direct bandgap material.

3.6 Band Structures of Common Semiconductors

The introduction of the lattice influence in terms of a small periodic potential function in a 1D nearly free electron system leads to the development of the energy bandgap concept. When we move from the 1D to the 3D nearly free electron system incorporating the effect of a small periodic potential, we encounter not only the increased dimension but also new possibilities that arise with the availability of more dimensions. For example, for an electron moving through a 3D crystal, an energy bandgap in one $\boldsymbol{k}$-direction may be bridged by an allowed band in another $\boldsymbol{k}$-direction. This possibility of the 3D system gives rise to the existence of overlapping bands, which is not possible in a 1D system. Typical energy bands for a diamond structure crystal are shown in Fig. 3.22. These E-k curves may be compared with the empty lattice model of a FCC crystal given in Fig. 3.12.

3.6.1 General Trend of Energy Band Structure in Semiconductors

The tetrahedral bonds of diamond and zinc-blende semiconductor structures are derived from the sp^3 hybrid orbit. By linear combination of spherically symmetric s-like waves with three directional p-like waves, as expressed in (2.24), four sp^3 hybrids are produced. These four sp^3 hybrid orbitals form a tetrahedron bond structure, which is the basic arrangement of all semiconductor crystals. Figure 3.23 shows schematically the evolution of the s- and p-like atomic states to the conduction and valence bands in hybridized elemental semiconductor systems. Atomic s- and p-states are hybridized to form sp^3 hybrid bonds which then can form bonding and anti-bonding combinations between orbitals on neighboring atoms pointing along the common bond. Finally, these molecular orbitals associated with bonding and

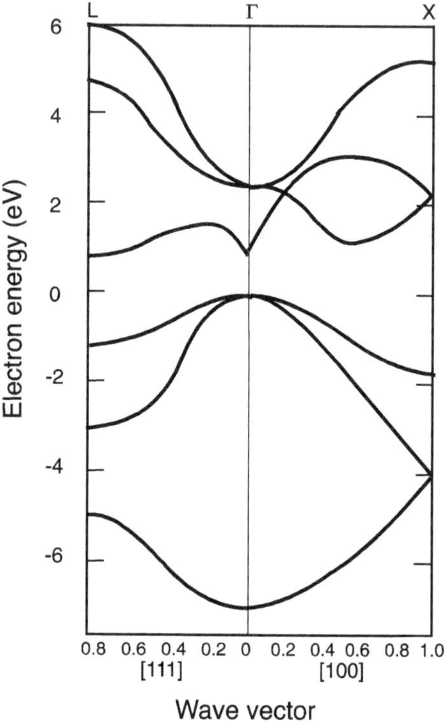

Fig. 3.22 Typical energy bands for a diamond-like crystal structure. Reprinted with permission from [3], copyright Wiley

anti-bonding levels become broadened into bands by interactions between hybrid orbitals on the same atom, the valence band being formed from bonding orbitals and the conduction band from anti-bonding orbitals. Therefore, the states are purely *s*-like or *p*-like. At the zone center (Γ-point), the bottom of the valence band is therefore *s*-like, and the top is *p*-like and triply degenerated when spin-orbit coupling is neglected. Similarly, at Γ-point, the bottom of the conduction band is purely *s*-like. The development of the energy band structure for a polar (compound) semiconductor proceeds in the same way. Hybrids are formed on both atom types and then form bonding and anti-bonding states with the other hybrid in the bond. Finally, the valence and conduction bands are formed from the broadened bonding and anti-bonding levels.

Using Ge as an example, the true energy bands and the nearly free electron bands are compared in Fig. 3.24. The true bands are obtained by combining pseudopotential calculations with experimental optical measurement data. The nearly free electron-like bands in Fig. 3.24b are remarkably similar to those at the bottom of the valence band and the upper part of the conduction band of the 'true' band structure (Fig. 3.24a), except that bandgaps at certain critical points in the Brillouin zone do not occur in the nearly free electron band structure. The other feature displayed in Fig. 3.24 is that Ge is an indirect bandgap semiconductor. Its maximum of the valence band is located at Γ, and the minima in the conduction band are located

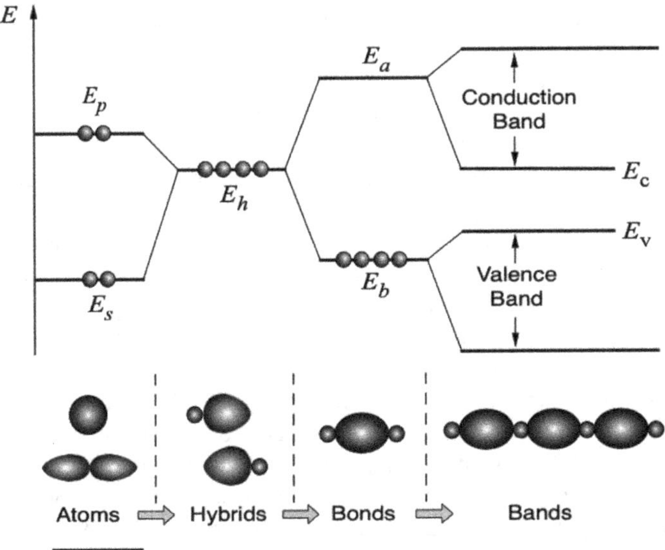

Fig. 3.23 Development of energy band structure of silicon. The single *s*-states (E_s) and *p*-states (E_p) on each atom are transformed to four hybrid states (E_h), which are combined with neighboring hybrids to form bonds. The bonding (E_b) and anti-bonding (E_a) levels finally broaden into energy bands in the crystal

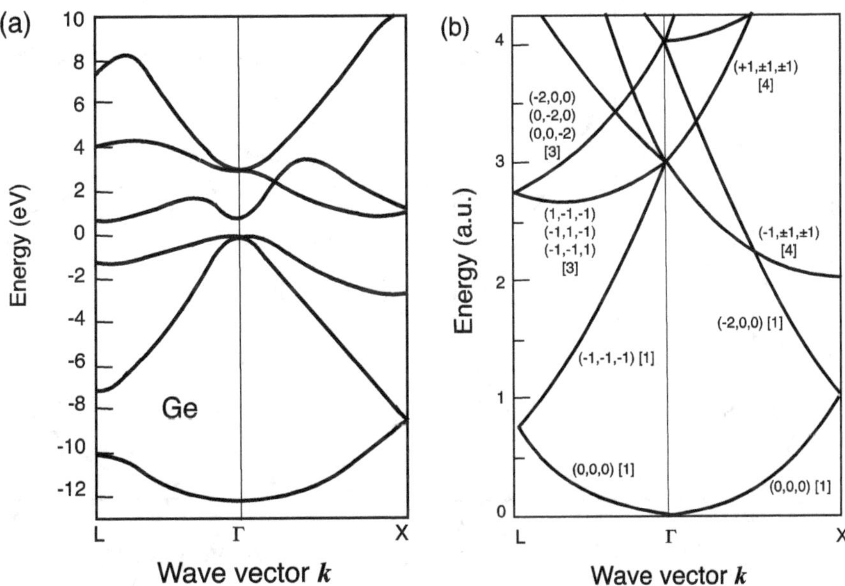

Fig. 3.24 Energy bands of germanium. **a** Energy band obtained by pseudopotential calculations. Reprinted with permission from [4], copyright American Physical Society, **b** energy band calculated using the nearly free electron model

in the ⟨111⟩ directions at the L-point. On the other hand, GaAs is a direct bandgap semiconductor with both the maximum of the valence band and the minimum of the conduction band located at the Γ-point. Therefore, it is instructive to examine the trends in the band structures of tetrahedrally bonded semiconductors, specifically the III–V compounds exhibiting mixed ionic-covalent bonding.

For group IV elements, there is an increase in metallic tendency (metallicity) on going from C to Sn when increasing the atomic number. With increasing metallicity, as expected, the conduction band minimum at the Γ-point drops faster than at X or L. At some point, the semiconductor turns into direct bandgap material. Indeed the indirect bandgaps of diamond, silicon and germanium turn into direct in α-Sn. This trend is also true for III–V compounds. The lowest conduction band drops more quickly at the Γ-point than at X or L with increasing metallicity, so the III–V compounds become direct bandgap semiconductors. The Al-compounds and GaP are the exceptions with indirect bandgaps. The energy band structures of Si and GaAs are shown in Fig. 3.25. On the other hand, for III–V compounds, there is an increase in ionicity which has a tendency to increase the bandgap energy. For example, the III–V nitride compounds with wurtzite crystal structure have bandgap energies much larger than 2 eV.

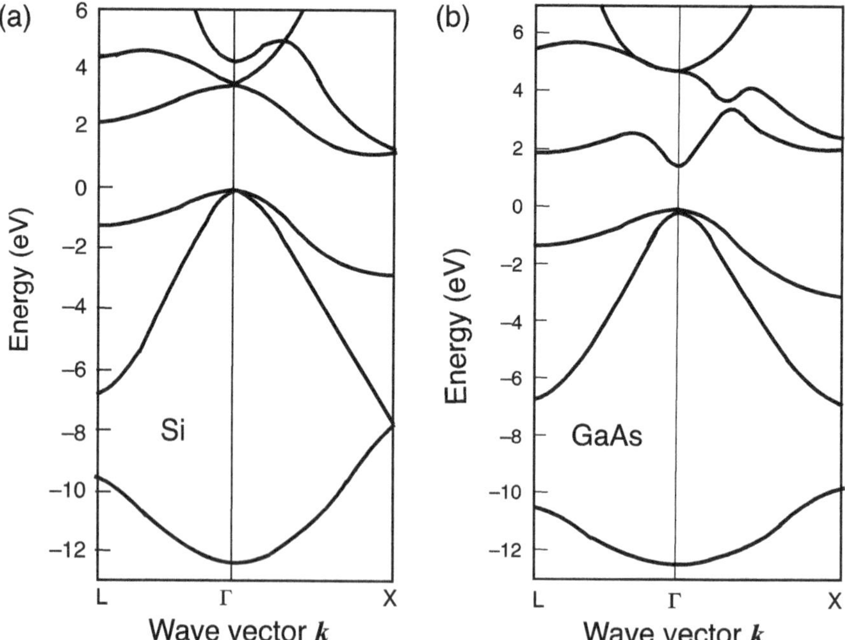

Fig. 3.25 Detailed energy band structures of **a** Si and **b** GaAs. Reprinted with permission from [4], copyright American Physical Society

3.6.2 Valence Band

Consider the p_z-orbit in a cubic crystal. As shown in Fig. 3.26a, the wave functions are polarized along the z-direction. Electrons move easily along the z-direction. This corresponds to a light effective mass shown as the E-k curve with a small radius of curvature (Fig. 3.26b). On the other hand, electrons moving in x- and y-directions are very restricted and have a heavy effective mass, shown as the slow varying E-k curve. The same property can be said about p_x- and p_y-like orbits. The combined 3D E-k diagram of the valence band turns out to be a complex composition. As shown in Fig. 3.27, there are heavy-hole, light-hole, and split-off bands. The maximum of the heavy- and light-hole bands degenerate at $k = 0$. Away from the Γ-point, the heavy- and light-hole bands with different radii of curvature in the E-k curve result. Near the zone center, they can be described as

$$E(k) = E_v - \frac{\hbar^2 k^2}{2m_0 m_\mathrm{h}} \tag{3.141}$$

where m_h equals m_hh or m_lh for heavy holes or light holes, respectively. Although widely used for its simplicity and convenience, it must be stressed that this approximation is valid only near the vicinity of the zone center. In realistic situations, the bands are both non-parabolic and anisotropic, depending on the direction of k as

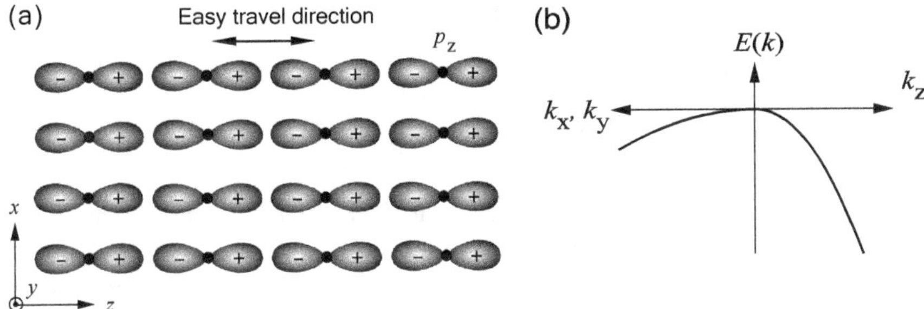

Fig. 3.26 Valence band constructed from p_z orbitals. **a** Regular arrangement of p_z orbitals. **b** E-k curves of p_z orbitals only; the band is 'light' along k_z and 'heavy' along k_x or k_y

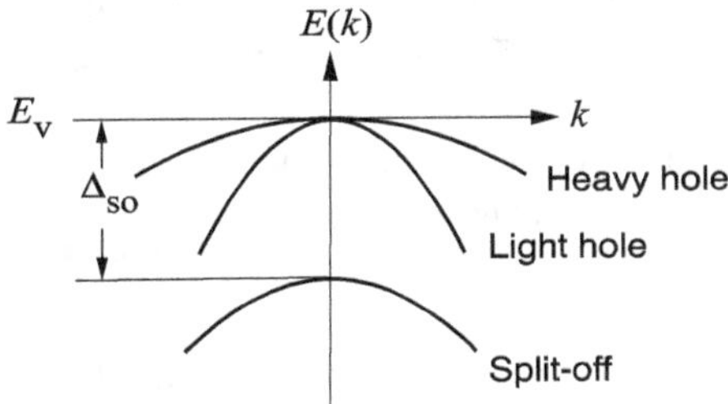

Fig. 3.27 Total band structure at the top of the valence band for tetrahedral semiconductors

well as its location from the zone center. The simple spherical surfaces predicted by (3.141) are modified into complex warped spheres. The non-spherical hole dispersion relation near the zone center can be described by Luttinger parameters. The situation becomes even worse if the cubic symmetry is broken by introducing strain in the material.

The decoupling of the spin-orbit split-off band from the top of the valence band is caused by the interaction of the electron spins ($s = \pm 1/2$) of the valence electrons with the magnetic moment ($l = 1$ for p-state) arising from their orbital motion, which yield a total angular momentum $j = l + s = 3/2$ and 1/2. The separation between the top of the heavy-/light-hole bands and the spin-orbit split-off band, Δ_{so}, increases with atomic number, essentially because the greater the nuclear charge, the greater the internal magnetic field experienced by the orbiting electron. Thus, Δ_{so} is very small for Si (0.04 eV), fairly large for GaAs (0.34 eV), and quite large for InSb (0.81 eV).

3.6.3 Conduction Band

(a) **Direct bandgap materials**

The conduction band originates from s-like orbitals which have a spherical symmetry. For III–V compounds such as GaAs, the conduction band minimum is located at the zone center (Γ-point), and these are called direct bandgap semiconductors. Near the conduction band minimum, the band can be described by

$$E(k) = E_{\mathrm{c}} + \frac{\hbar^2 k^2}{2m_0 m_{\mathrm{e}}} \tag{3.142}$$

Thus the constant energy surface of the conduction band has a spherical shape. For direct bandgap semiconductors, the radius of the E-k curve of the lowest conduction band is relatively small, leading to a light electron effective mass. In GaAs, $m_{\mathrm{e}}^* \cong 0.063 m_0$ is the electron effective mass.

(b) **Indirect bandgap materials**

For indirect bandgap semiconductors, the shape of the lowest conduction band minimum is more complex. In Si (and diamond), the minima are located along $\langle 100 \rangle$ about 85% of the way to the X-zone boundary. As illustrated in Fig. 3.28, there are six Δ-directions and, therefore, six equivalent minima in a cubic crystal. Due to their off-center location, these minima (valleys) have anisotropic E-k relationships along k_x, k_y, and k_z.

$$E(k) = E_{\mathrm{c}} + \frac{\hbar^2}{2m_0}\left[\frac{(k_x - k_0)^2}{m_{\mathrm{L}}} + \frac{k_y^2}{m_{\mathrm{T}}} + \frac{k_z^2}{m_{\mathrm{T}}}\right] \tag{3.143}$$

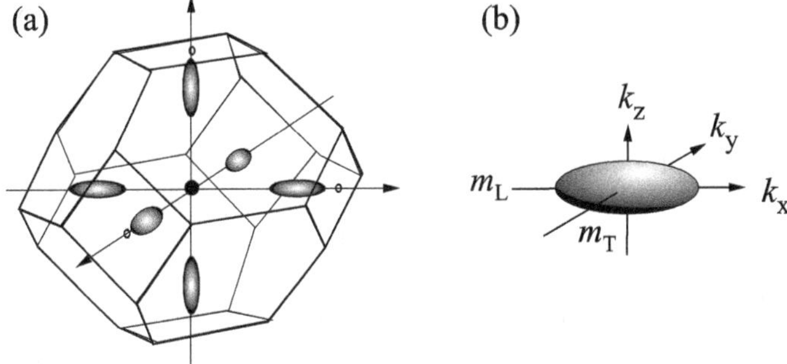

Fig. 3.28 **a** The Brillouin zone of Si with six equivalent constant energy surfaces along Δ-direction. **b** Enlarged view of one of the constant energy volumes, showing longitudinal and doubly degenerated transverse masses m_L and m_T, respectively

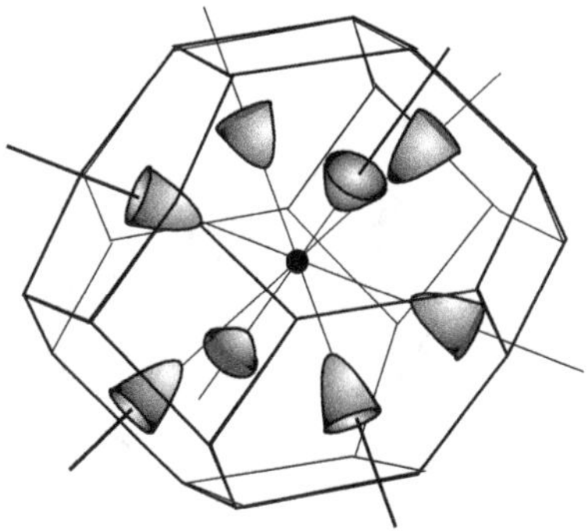

Fig. 3.29 Brillouin zone of Ge with constant energy surfaces for the eight equivalent *L*-valleys

where the subscripts *L* and *T* indicate longitudinal and transverse components, respectively. In Si, $m_L = 0.92$ and $m_T = 0.19$. As mentioned earlier, Ge also has indirect bandgap with the conduction band minima along $\langle 111 \rangle$ in the *L*-point. In Ge, as shown in Fig. 3.29, there are eight ellipsoidal equivalent valleys located at the zone boundaries where $m_L = 1.59$ and $m_T = 0.082$. Since only half of the constant energy surface is located inside the first Brillouin zone, the degenerate factor *g* is 4, not 8.

3.6.4 Band Structures of Wurtzite Crystals

Although the crystal structure of most III–V compounds is zinc-blende, the III-nitride alloys prefer the formation of wurtzite crystal structure. Since both structures are characterized by the tetrahedral lattice arrangement, as discussed in Sect. 2.4, the nearest-neighbor environments are identical with only slight differences in the next nearest-neighbor arrangements. It is necessary to go beyond next nearest neighbors before both positional and directional differences occur. One can regard the wurtzite structure along the c-axis as a cubic zinc-blende lattice slightly deformed along the body diagonal of [111]. Figure 3.30 compares the Brillouin zone of zinc-blende and wurtzite crystal structures. A correspondence is obtained between the [111] direction ΓL in zinc-blende structure and the [0001] direction $\Gamma\Gamma'$ in wurtzite. Therefore, a strong similarity between the band structure of zinc-blende and wurtzite crystals can be expected. However, in contrast to the more symmetric cubic zinc-blende crystal, the in-plane (basal plane) behavior of the energy bands is different from the behavior along the [0001] axis (the c-axis) in wurtzite crystal. The anisotropic hexagonal structure, when including spin-orbit interaction, leads to energy splitting in the valence band edges. The valence band structure for GaN is shown in Fig. 3.31. The heavy-hole (HH) and light-hole (LH) bands no longer degenerate at $k = 0$ and separated by the spin-orbit interaction (Δ_{so}). The crystal-hole (CH) valence band

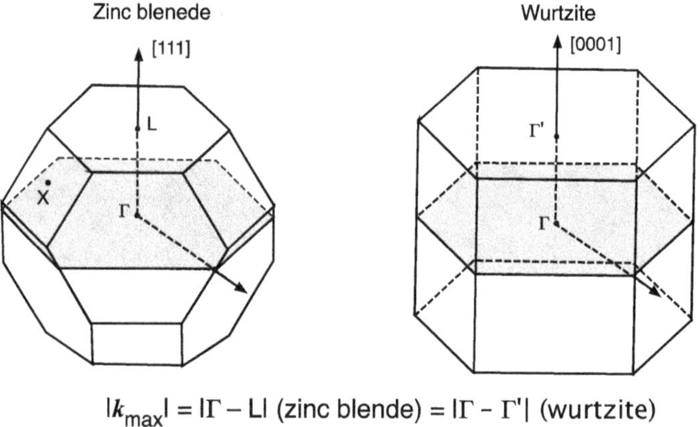

Fig. 3.30 Brillouin zone of zinc-blende and wurtzite-type materials. Reprinted with permission from [5], copyright American Physical Society

Fig. 3.31 Valence band structure of wurtzite GaN. Reprinted with permission from [6], copyright AIP Publishing

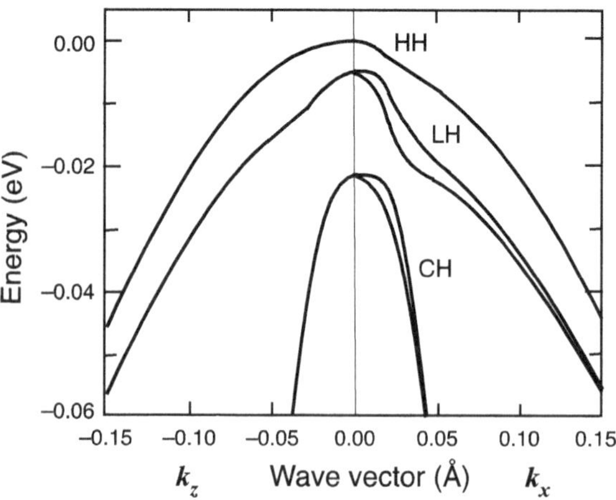

results from the crystal-field splitting energy (Δ_{cr}). The topmost three valence band energy levels at $k = 0$ (E_1, E_2, and E_3, in decreasing order) are

$$E_1 = \Delta_1 + \Delta_2 \tag{3.144a}$$

$$E_2 = \frac{\Delta_1 - \Delta_2 + \sqrt{(\Delta_1 - \Delta_2)^2 + 8\Delta_3^2}}{2} \tag{3.144b}$$

$$E_3 = \frac{\Delta_1 - \Delta_2 - \sqrt{(\Delta_1 - \Delta_2)^2 + 8\Delta_3^2}}{2} \tag{3.144c}$$

where E_1 is labeled as HH band, E_2 is labeled as LH band, and E_3 is labeled as CH band. We also assumed

$$\Delta_1 = \Delta_{cr}, \ \Delta_2 = \Delta_3 = \Delta_{so} \tag{3.145}$$

In general, Δ_{cr} and Δ_{so} have positive values for III-nitrides except that AlN has a negative Δ_{cr}. Thus, the top valence band of AlN is E_2 instead, and the valence bands from top to bottom are CH, HH, and LH, respectively. Note that the valence band energy splitting $E_1 - E_2$ and $E_1 - E_3$ are measurable quantities through inter-band optical transition measurements whereas Δ_{cr} and Δ_{so} are parameters of theory, obtainable indirectly by fitting experimental energy splitting.

All III-nitride semiconductor alloys have direct bandgaps. The bottom of the conduction band in GaN and InN is well approximated by a parabolic dispersion relationship, while a greater anisotropy in AlN than GaN is predicted due to the reduced lattice symmetry. The conduction band-edge energy is deduced by adding the bandgap energy to E_1.

$$E_c = E_g + \Delta_1 + \Delta_2 \tag{3.146}$$

Problems

1. Calculate the first two E-k curves along the [110] axis for a BCC crystal using the 3D free electron model.
2. For an artificial two-dimensional square lattice structure shown below, calculate the first two energy bands for free electrons along the Γ-X, X-K, and Γ-K directions.

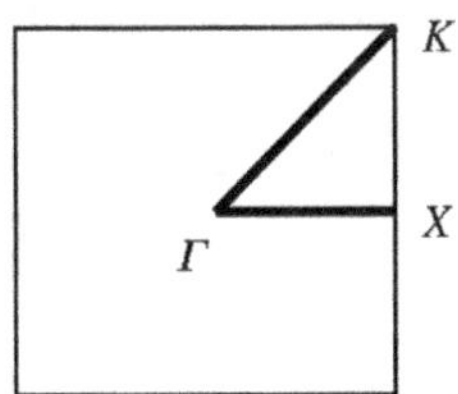

3. Calculate and plot the density of states as a function of energy in the conduction band of a GaAs/AlGaAs quantum well for the first three bound states located at 25, 100, and 225 meV above the conduction band edge. The electron effective mass of GaAs in the conduction band is $m^* = 0.067m_0$.
4. Derive the density of states for a quantum dot system. In a quantum dot system, the quantum confinement is three-dimensional.
5. Suppose the *E*-*k* relation is given by $E = E_0 + (\Delta E/2)(1 - \cos ka)$. For very small k. prove that the parabolic approximation holds and determines the effective mass.
6. For the delta function periodic potential ($V_0 \sim \infty$, $b \sim 0$ and $V_0 b$ is small but finite or $P \ll 1$), use the Kronig–Penney model to calculate

   (a) The energy of the lowest energy band at $k = 0$.
   (b) The bandgap at $k = \pi/a$.

7. Derive the following solution of the Kronig–Penney model for a crystal:

$$-\left(\frac{\alpha^2+\beta^2}{2\alpha\beta}\right)\sin(\alpha a)\sin(\beta b)+\cos(\alpha a)\cos(\beta b)=\cos[k(a+b)]$$

   by setting the determinant of wave functions (3.101) to zero.
8. Consider a simple cosine approximation to the shape of an energy band as

$$E(k)=\frac{\Delta E}{2}(1-\cos ka).$$

   (a) For small k, prove that the parabolic approximation holds and $\upsilon(k) \cong \hbar k/m^*$.
   (b) Many semiconductors have a saturation velocity for electrons of around 10^7 cm/s. How much of the first Brillouin zone do electrons need to explore to reach this velocity? Assume the initial position of electrons is at the zone center. For calculations, use the velocity equation of part (a) and the following material parameters of GaAs: $m^* = 0.067m_0$ and lattice constant $a_0 = 5.653$ Å.

9. Consider a single electron in a perfect crystal. Starting from $k = 0$ in an empty band, the electron is accelerated by a constant electric field $-F$ toward $+k$. Assume $E = (\Delta E/2)(1 - \cos ka)$ and $\hbar k = (qF)t = \mathcal{F}t$.

   (a) Find the velocity of the electron.
   (b) Prove that the position of the electron in real space as a function of time is

$$x(t)=\frac{\Delta E}{2qF}\left[1-\cos\left(\frac{qFa}{\hbar}\right)t\right]$$

The electron oscillates in real space, rather than accelerating uniformly as in the free electron model. This periodic motion of an electron in a periodic crystal structure is called the *Bloch oscillation*.

(c) The Bloch oscillation frequency is $\omega = qFa/\hbar$. For $F = 10^8$ V/m, calculate the Bloch oscillation frequency in GaAs.

(d) In semiconductors, due to scattering with defects, the Bloch oscillation of electrons in the Brillouin zone cannot be observed. Assuming the average electron can only travel one tenth of the first Brillouin zone, design a GaAs/AlAs superlattice structure such that the Bloch oscillation can be measured. The lattice constants of GaAs and AlAs are 5.653 Å and 5.660 Å, respectively.

References

1. F. Herman, in *An Atomistic Approach to the Nature and Properties of Materials*, ed. by J.A. Pask (Wiley, New York, 1967)
2. C. Waschke, H.G. Roskos, R. Schwedler, K. Leo, H. Kurz, K. Kohler, Phys. Rev. Lett. **70**, 3319 (1993)
3. D. Long, *Energy Bands in Semiconductors* (Wiley-Interscience, New York, 1968)
4. J.P. Chelikowsky, M.L. Cohen, Phys. Rev. **B14**, 556 (1976)
5. J.L. Birman, Phys. Rev. **115**, 1493 (1959)
6. I. Vurgaftman, J.R. Meyer, J. Appl. Phys. **94**, 3675 (2003)

Further Reading

1. C. Kittel, *Introduction to Solid State Physics*, 6th edn. (John Wiley & Sons, 1986)
2. H. Ibach, H. Lüth, *Solid State Physics*, 2nd edn. (Springer, 1996)
3. W.A. Harrison, *Elementary Electronic Structures* (World Scientific, 1999)

Chapter 4
Compound Semiconductor Crystals

Abstract As we learned from previous chapters, silicon has a diamond crystal structure with pure covalent bonding. Due to its nature as a single element, the Si crystal structure is highly symmetric. Since there are two different elements in binary III–V compounds, the basic diamond structure turns into zinc-blende crystal structure, which destroys some of the symmetry observed in Si. This lower symmetry in crystal structure leads to band structures with different features which, in turn, give the unique properties not available to Si. Using GaAs as an example, the main advantages over Si are (a) Larger bandgap: The larger bandgap leads to a low intrinsic carrier concentration (n_i) suitable for high-temperature operation. (b) Smaller effective mass: The effective mass is inversely proportional to the carrier mobility. The higher mobility contributes to high-speed operation. (c) Direct bandgap: The location of the lowest conduction band minimum also varies with a decreasing degree of crystal symmetry in III–V compounds. Some of them have both the conduction band minimum and the valence band maximum located at the zone center. These are called *direct bandgap* semiconductors. This allows for their use in photonic device and transfer electron device applications. (d) Heterostructures: Additionally, binary III–V compounds can form ternary and quaternary compounds, which leads to an expanded selection of lattice constants and bandgap energies. New heterostructure and quantum effect devices then become possible.

In this chapter, the basic structural, transport, doping, and surface properties of III–V binaries are introduced first. The variation of lattice constant and energy gap in ternary and quaternary alloys are discussed next.

The history of preparation of III–V compounds goes back to the late nineteenth century. Since no naturally occurring III–V crystals have been found, the early work was to prepare these compounds. During the early twentieth century, before WWII, a wide range of the compounds, such as AlSb, AlN, AlP, AlAs, InP, InAs, InSb, and GaN, was studied for their crystal chemistry and structure without realizing the semiconducting nature of these compounds. After the war, work continued in both Russia and the Western world. In 1950, N. A. Goryunova, a Ph.D. student at Leningrad State University, came to the conclusion that compounds with the zinc-blende structure are not only structural analogs of group IV elements, but they are

K. Y. Cheng, *III–V Compound Semiconductors and Devices*, Graduate Texts in Physics,
https://doi.org/10.1007/978-3-030-51903-2_4

also semiconductors. She substantiated the idea with exploratory experiments and reported the results in her Ph.D. thesis in 1951. These ideas are contained in a summary of intermetallic compounds reported by A. F. Ioffe, where he pointed out that InSb, HgSe, and HgTe are semiconductors like silicon. Ioffe's work was included in papers by A. I. Blum, N. P. Mokrovski, and A. R. Ragel [1].

At about the same time, in 1952, H. Welker in West Germany published his classical paper on new semiconducting compounds, including the important III–V compound semiconductor GaAs and applied for patents on III–V compound semiconductor devices and their method of manufacturing [2]. He was recognized in the Western world as the pioneer of III–V compound semiconductors. In the USA, the first discussion of the unique properties of III–V compound semiconductors was held at the American Physical Society March Meeting at Durham, North Carolina, in 1953. The materials discussed at this meeting were mainly Sb-related compounds, e.g., InSb and AlSb. Research on III–V compounds got more attention after the demonstration of semiconductor injection lasers in 1962 and the discovery of coherent microwave oscillation in transfer electron devices in 1963. These important early inventions precisely highlight the strength of III–V compound semiconductors over silicon—their light-emitting and high-speed properties.

4.1 Structural Properties

4.1.1 Lattice Constant

The lattice constant of a III–V compound zinc-blende crystal can be estimated by using the tetrahedron covalent radii defined by L. Pauling. Assuming the atoms pack in as hard spheres, the lattice constant (a) is simply related to the tetrahedron covalent radii (r_{III} and r_{V}) by

$$a = \frac{4}{\sqrt{3}}(r_{\mathrm{III}} + r_{\mathrm{V}}) \tag{4.1}$$

For all III–V compounds, except AlSb, as shown in Table 4.1, the error between the calculated and experimental results is less than 1%. Although this is a remarkably close fit, the accuracy is not good enough for predicting lattice constants of new compounds. In heterostructures, a lattice-mismatch of $\leq$0.1% is required to avoid the formation of misfit dislocations. In most III–V compounds, this tolerance is less than 0.0005 Å. Therefore, we have to rely on experimental lattice constant values.

For III-nitride compounds, their crystal structure is wurtzite. The lattice constants a and c are related to the tetrahedron covalent radii by

$$a = \frac{2\sqrt{2}}{\sqrt{3}}(r_{\mathrm{III}} + r_{\mathrm{V}}) \text{ and } c = \frac{8}{3}(r_{\mathrm{III}} + r_{\mathrm{V}}) \tag{4.2}$$

Table 4.1 Measured lattice constant (in Å) of III–V compounds as compared to calculated values from covalent radii (value in parenthesis) of each pair of III and V elements

	r_V							
	P (1.10Å)		As (1.18Å)		Sb (1.36Å)		N (0.70Å)	
r_{III}	Cal.	Exp.	Cal.	Exp.	Cal.	Exp.	Cal.	Exp.
Al (1.26Å)	5.450	5.464	5.635	5.660	6.051	6.136	*a*: 3.20 *c*: 5.227	*a*: 3.112 *c*: 4.982
Ga (1.26Å)	5.450	5.451	5.635	5.653	6.051	6.096	*a*: 3.20 *c*: 5.227	*a*: 3.189 *c*: 5.185
In (1.44Å)	5.866	5.869	6.051	6.058	6.466	6.479	*a*: 3.495 *c*: 5.706	*a*: 3.54 *c*: 5.70

For nitride compounds, *a* and *c* indicate the lattice constants in the (0001) plane and along the *c*-axis, respectively

4.1.2 Cleavage Properties

The side views and 3D structures of a zinc-blende crystal along the [111], [110], and [100] directions are shown in Fig. 4.1. For silicon, the cleavage plane is in {111} due to its slightly larger separation between the two planes. For compound semiconductors, {111} is not the favored cleavage plane. This is because the two opposite surfaces

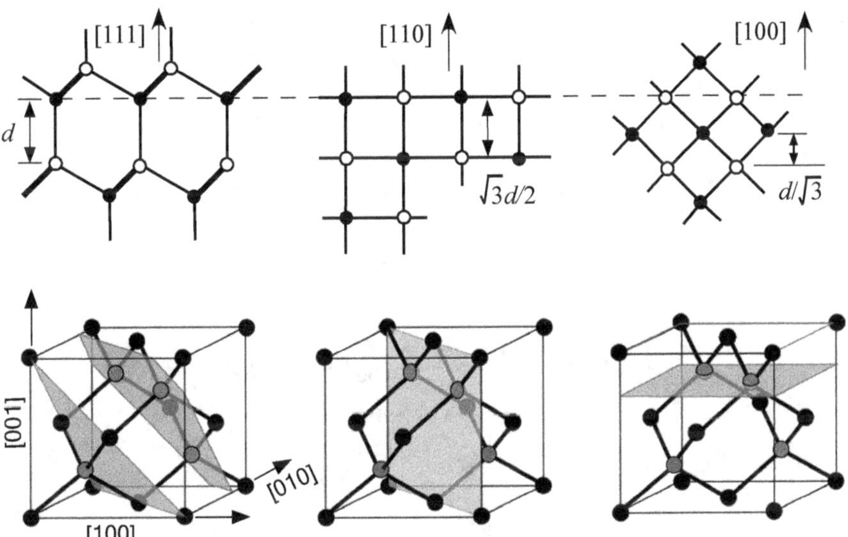

Fig. 4.1 (Top row) Side views of bond structures of (111), (110), and (100) surfaces of a zinc-blende crystal. The dashed line indicates a plane parallel to the surface and *d* is the bond length. The thicker lines in (111) side view represent double bonds. (Second row) The 3D structures of [111], [110], and [100] planes in zinc-blende structure. The shaded plane corresponds to the dashed line in the top panel

are filled with different atoms where an electrostatic attraction force is formed upon cleaving. The (111) surface occupied by group III atoms is called the (111)A surface. The one filled with group V atoms is the (111)B surface. Because of this chemical difference, these two surfaces show very different characteristics in etching and epitaxial growth processes. On each side of the (110) cleaved plane, there are equal numbers of group III and group V atoms. Because there is no electrostatic attraction between the two sides, this is the easy cleavage plane among III–V compounds. The atoms on the [100] plane have double bonds and have the strongest bonding among three directions discussed here.

4.1.3 Lattice Vibration—Phonons

The vibration of crystal lattice at a finite temperature is represented, in quantum mechanics, by *phonons*. As discussed in Sect. 2.5 and in later sections, phonon scattering due to lattice vibration plays a critical role in the carrier relaxation process, which determines many important basic properties of the semiconductor including effective mass, carrier recombination, and thermal conductivity, etc. The periodicity of the lattice means that phonons have band structure in much the same way as electrons, as discussed later in this section. The crystal lattice can be visualized as an array of point masses connected by springs. The masses are just the core ions. The length of the spring can be seen as the equilibrium bond length. It resists any attempt to change its length. The force between the neighboring atoms linked with the spring is governed by Hooke's law, $\mathcal{F} = \alpha x$, where α is the Hooke's law constant. To examine the lattice vibration property of III–V compound semiconductors, a 1D diatomic linear chain model shown in Fig. 4.2 will be used to represent the diatomic primitive unit cell structure. The lighter atoms with a mass m and the heavier atoms with a mass M are arranged alternately with an equal separation of a in between.

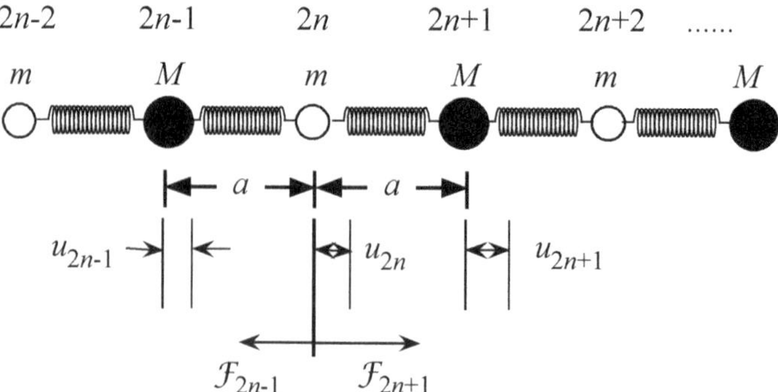

Fig. 4.2 One-dimensional row of springs and masses in a linear chain diatomic model of III–V compounds. These spring-linked masses are allowed to vibrate longitudinally through displacements u_i

Let us first discuss the 1D motion along the linear chain only, i.e., longitudinal vibration, without concerning the sideways motion. It is assumed that the only significant force interactions between atoms are direct nearest-neighbor interactions. Therefore, the net force acting upon the 2*n*th atom can be seen as interactions through two springs connecting to atoms $(2n + 1)$ and $(2n - 1)$. The force equation is given as

$$\mathcal{F}_{2n} = \alpha(u_{2n+1} - u_{2n}) - \alpha(u_{2n} - u_{2n-1}) = \alpha(u_{2n+1} + u_{2n-1} - 2u_{2n}) \tag{4.3}$$

where u_i's are the displacements of the *i*th atoms. In obtaining (4.3), we note that the force acting on the 2*n*th atom from the $(2n-1)$th atom is opposite to that from the $(2n + 1)$th atom. The force between nearest neighbors can be expressed by Hooke's law and Newton's second law as

$$\begin{cases} \mathcal{F}_{2n} = m\dfrac{d^2u_{2n}}{dt^2} = \alpha(u_{2n+1} + u_{2n-1} - 2u_{2n}) \\ \mathcal{F}_{2n+1} = m\dfrac{d^2u_{2n+1}}{dt^2} = \alpha(u_{2n+2} + u_{2n} - 2u_{2n+1}) \end{cases} \tag{4.4}$$

The periodic solutions to these two equations are in the forms of

$$\begin{cases} u_{2n} = A\exp[i(\omega t - 2nka)] \\ u_{2n+1} = B\exp\{i[\omega t - (2n+1)ka]\} \end{cases} \tag{4.5}$$

where ω is the oscillation frequency of atoms. The displacement of the $(2n + 2)$th and $(2n - 2)$th atoms can be expressed as

$$\begin{cases} u_{2n+2} = A\exp\{i[\omega t - (2n+2)ka]\} = u_{2n}\exp(-2ika) \\ u_{2n-1} = B\exp\{i[\omega t - (2n-1)ka]\} = u_{2n+1}\exp(2ika) \end{cases} \tag{4.6}$$

Replacing all the relevant *u*'s in the force equation for the $(2n + 1)$th atom, we find

$$-M\omega^2 u_{2n+1} = \alpha\{[1 + \exp(-2ika)]u_{2n} - 2u_{2n+1}\} \tag{4.7}$$

Solving for u_{2n+1}, we have

$$u_{2n+1} = \frac{\alpha[1 + \exp(-2ika)]}{2\alpha - M\omega^2}u_{2n} \tag{4.8}$$

The force equation for the $(2n)$th atom has the form

$$-m\omega^2 u_{2n} = \alpha\{[1 + \exp(-2ika)]u_{2n+1} - 2u_{2n}\} \tag{4.9}$$

Combining (4.8) and (4.9), it is found that

$$(2\alpha - m\omega^2)(2\alpha - M\omega^2) - 4\alpha^2\cos^2 ka = 0 \tag{4.10}$$

Rearranging the equation in terms of ω, we obtain

$$\omega^4 - \frac{2\alpha(m+M)}{mM}\omega^2 + \frac{4\alpha^2 \sin^2 ka}{mM} = 0 \tag{4.11}$$

The solution of the ω-k dispersion relation has the form

$$\omega_{\pm}^2 = \frac{\alpha(m+M)}{mM}\left[1 \pm \sqrt{1 - \frac{4mM\sin^2 ka}{(m+M)^2}}\right] \tag{4.12}$$

The positive sign group of the ω-k dispersion curve is called the *optical branch*, and the negative sign group is called the *acoustical branch* because the energies associated with these two modes of vibrations are in the optical and acoustical range of the frequency spectrum, respectively. The dispersion curves of phonons vibrating longitudinally are shown in Fig. 4.3.

Near the zone center, at $k = 0$, ω_+ has a maximum value of

$$\omega_+(0) = \sqrt{\frac{2\alpha(m+M)}{mM}} \text{and } \omega_-(0) = 0 \tag{4.13}$$

The oscillation frequency of the acoustical branch for small values of k, $\sin ka \approx ka$, becomes

$$\omega_-(0) \approx ka\sqrt{\frac{2\alpha}{(m+M)}} \tag{4.14}$$

Here we assumed $(1+x)^{1/2} = 1 + x/2 - x^2/8 + \ldots \approx 1 + x/2$. The frequency of the acoustical branch near the zone center is a linear function of k.

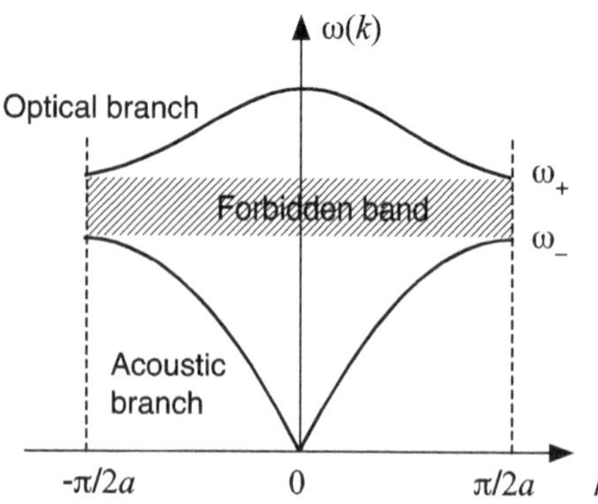

Fig. 4.3 Phonon dispersion curves for a 1D linear diatomic crystal with longitudinal vibrations

The smallest allowed wavelength in this diatomic system should equal twice the lattice constant ($2a$), which is $\lambda = 4a$. Therefore, the first zone boundary is located at $k = \pm 2\pi/\lambda = \pm\pi/2a$, and where the two roots become

$$\begin{cases} \omega_+ = \sqrt{2\alpha/m} \\ \omega_- = \sqrt{2\alpha/M} \end{cases} \quad (4.15)$$

For $m \neq M$, a forbidden frequency band is formed.

The relative movement of the two atoms in the unit cell can be calculated by taking the ratio of u_{2n+1} and u_{2n}.

$$\frac{u_{2n+1}}{u_{2n}} = \frac{\alpha\left[1 + \exp(-2ika)\right]}{2\alpha - M\omega^2} = \frac{B}{A}\exp(-ika) \quad (4.16)$$

or

$$\frac{B}{A} = \frac{\alpha\left[\exp(-ika) + \exp(ika)\right]}{2\alpha - M\omega^2} = \frac{2\alpha}{2\alpha - M\omega^2}\cos ka \quad (4.17)$$

For the acoustical branch, $\omega = \omega_-(0) = 0$ at $k = 0$, thus (4.17) leads to $A = B$. As shown in Fig. 4.4, the displacements in the longitudinal acoustic (LA) branch are like sound waves, where the two atoms in each unit cell move in the same direction by almost the same distance. It appears as if the whole crystal has been compressed or stretched over a small region. This creates alternating zones of compression and dilation. In a crystal, these waves can propagate in three directions. If we choose the $\langle 100 \rangle$ direction in a cubic crystal, one acoustic mode is longitudinal and the other two are transverse. In the transverse acoustic (TA) modes, the atoms vibrate in the plane normal to the direction of wave propagation, similar to an electromagnetic wave in free space. The atoms and their center of mass move together as in long-wavelength acoustical vibrations.

For the optical branch, using the values $k = 0$ and $\omega = \omega_+(0)$ in (4.17), we find that $B/A = -m/M$. As shown in Fig. 4.5, the heavy and light atoms move in opposite

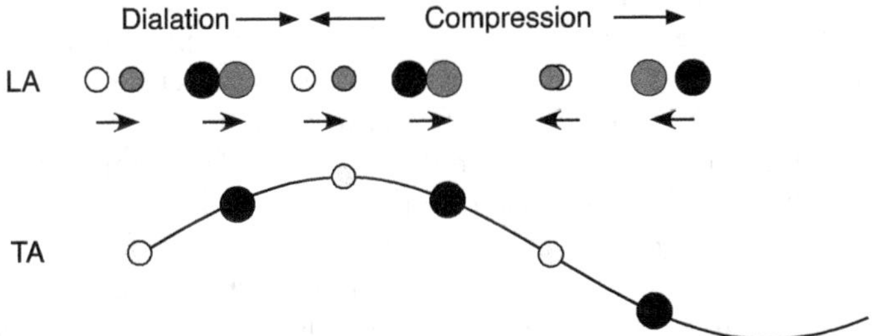

Fig. 4.4 Motion of the atoms in longitudinal acoustic (LA) mode for a 1D chain of alternating light and heavy atoms. The two types of atoms move together in transverse acoustical (TA) mode of the same crystal system. Displaced positions of moving atoms are shown as gray circles

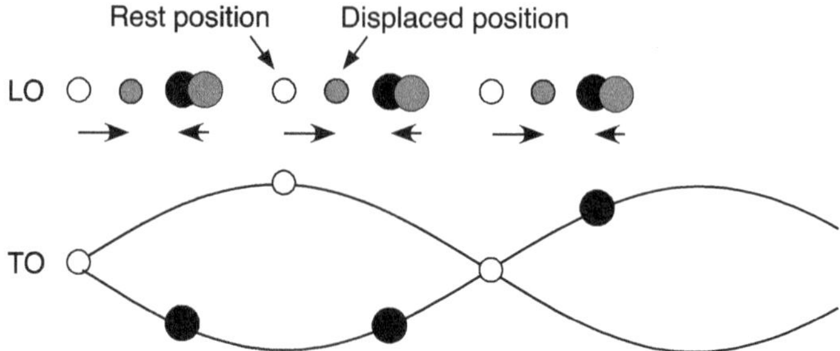

Fig. 4.5 Motion of the atoms in longitudinal optical (LO) mode for a 1D chain of alternating light and heavy atoms. The two types of atoms move in opposite directions in the transverse optical (TO) mode

directions and the displacement is inversely proportional to the mass of the atom. In the longitudinal optical (LO) mode, both the light and heavy atoms vibrate back and forth in the same direction in which the wave travels. Similar to the TA modes, in the transverse optical (TO) modes, the atoms vibrate in the plane normal to the direction of wave propagation, but the atoms move in opposite directions.

Both Si and GaAs have two atoms in each primitive cell, allowing two different modes of vibrations close to that of Fig. 4.3. The experimental phonon dispersion curves of GaAs, shown in Fig. 4.6, are very similar to those predicted by the 1D model. The doubly degenerated transverse branches are also displayed. The LO phonon energy in the zone center of GaAs is 36 meV. We also notice that the oscillation frequency for TO phonons is smaller than that of LO phonons at the zone center. In compound semiconductors, the relative displacement of the two atoms, as in the optical mode, sets up an electric dipole. This leads to the formation of a polarization field and an electric field, which can interact with electromagnetic waves. Electrons can be scattered rapidly by the electric field associated with the polar LO phonons, raising their frequency. Purely transverse vibrations do not generate dipoles, so this effect is absent. This polar nature is not seen in elemental semiconductors because all atoms are identical and there is no transfer of charge.

Since phonons are the quantum mechanical equivalents of lattice vibrations, there are abundant phonons available in semiconductor crystals at room temperature. For a wave of frequency ω, the crystal momentums of photons and phonons are both given by $\hbar k$ and $k = \omega/\upsilon$, where υ is the velocity of the wave. For an electromagnetic wave, $\upsilon = c = 3 \times 10^8$ m/s. The velocity of a lattice wave is similar to that of a sound wave, and its amplitude is smaller by about five orders of magnitude. Therefore, for the same frequency, the momentum of a phonon is much higher than that of a photon. This finite momentum associated with the phonon becomes very important in connecting electrons and photons in a semiconductor. In an indirect bandgap semiconductor, as discussed in Chap. 8, phonons can provide high enough momentum to satisfy

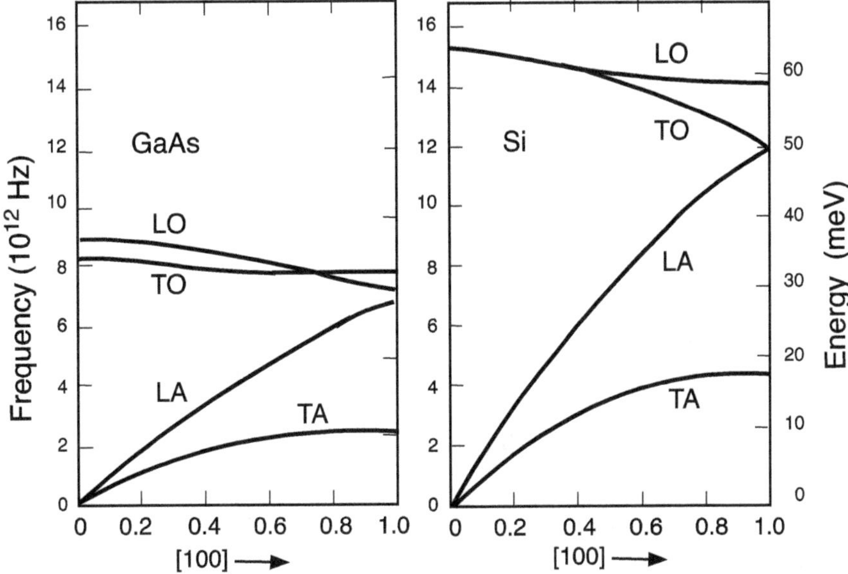

Fig. 4.6 Left: Phonon dispersion curves along [100] from Γ to X for a GaAs crystal at room temperature. The transverse mode branches are doubly degenerated. (Reprinted with permission from [3], copyright American Physical Society) Right: Measured dispersion curves for phonons in Si along [100]. (Reprinted with permission from [4], copyright American Physical Society)

momentum conservation required to excite an electron from the valence band to the conduction band by absorbing or emitting a photon with the proper energy. Phonons also play an equally important role in the light emission process in direct bandgap semiconductors. Electrons in higher states inside the conduction band need to relax to the band minimum before they can recombine with holes residing near the valence band maximum for photon generation. The relaxation process relies on phonons, with energy $\hbar\omega$, to dissipate extra energy into heat and to conserve momentum.

4.2 Electrical Properties

4.2.1 Effective Mass

The electron concentration in a non-degenerate semiconductor is calculated by integrating the electron distribution above the conduction band edge and formulated as

$$n = \int_{E_c}^{\infty} D(E) f(E) \mathrm{d}E \tag{4.18}$$

where $D(E)$ is the density of states of a bulk semiconductor and has the form

$$D(E) = \frac{1}{2\pi^2}\left(\frac{2m_{\mathrm{DOS}}}{\hbar^2}\right)^{3/2}\sqrt{E} \tag{4.19}$$

where m_{DOS} stands for the DOS effective mass. In a non-degenerated semiconductor, where $E_c - E_F > 3kT$, the Fermi–Dirac distribution function $f(E)$ can be approximated by the Boltzmann distribution function to simplify the calculation.

$$f(E) = \left[1 + \exp\left(\frac{E_c - E_F}{kT}\right)\right]^{-1} \approx \exp\left(-\frac{E_c - E_F}{kT}\right) \tag{4.20}$$

This leads to a non-degenerate electron concentration of

$$n = N_c \exp\left(-\frac{E_c - E_F}{kT}\right) \tag{4.21}$$

where N_c is the effective density of states in the conduction band and has a value of

$$N_c = 2\left(\frac{2\pi m_{\mathrm{DOS}} kT}{h^2}\right)^{3/2} = 2.509 \times 10^{19}\left(\frac{m_{\mathrm{DOS}}}{m_0}\frac{T}{300}\right)^{3/2}\left(\mathrm{cm}^{-3}\right) \tag{4.22}$$

For a parabolic conduction band with spherical energy surface, like GaAs, m_{DOS} simply equals m_e^*. For non-spherical energy surface, the DOS effective mass used for carrier concentration calculation is modified according to the procedures outlined below. For example, in Si and AlAs, the conduction band is non-spherical (ellipsoidal) and has multiple valleys. For an ellipsoidal energy surface along k_x, it can be expressed as

$$E - E_c = \frac{\hbar^2 (k_x - k_{x0})^2}{2m_L} + \frac{\hbar^2 k_y^2}{2m_T} + \frac{\hbar^2 k_z^2}{2m_T} = \frac{(p_x - p_{x0})^2}{2m_L} + \frac{p_y^2 + p_z^2}{2m_T} \tag{4.23}$$

This ellipsoidal surface can transform into a spherical surface through a coordinate transformation process and is expressed as

$$E - E_c = \frac{\left(p_x'\right)^2 + \left(p_y'\right)^2 + \left(p_z'\right)^2}{2m'} = \frac{\left(p'\right)^2}{2m'} \tag{4.24}$$

where $p_x' = (p_x - p_{x0})\sqrt{m'/m_L}$, $p_y' = p_y\sqrt{m'/m_T}$, and $p_z' = p_z\sqrt{m'/m_T}$. The volume of p' space in the spherical shell bounded by radii p' and $\left(p' + \mathrm{d}p'\right)$ for a single ellipsoidal is $\mathrm{d}V_{p'} = 4\pi p'^2 \mathrm{d}p'$ and $p'\mathrm{d}p' = m'\mathrm{d}E$. Thus,

$$\mathrm{d}V_{p'} = 4\sqrt{2}\pi\left(m'\right)^{3/2}\sqrt{E - E_\mathrm{c}}\mathrm{d}E \tag{4.25}$$

Since $\mathrm{d}V_\mathrm{p} = \mathrm{d}p_x\mathrm{d}p_y\mathrm{d}p_z$ and $\mathrm{d}V_{p'} = \mathrm{d}p_x'\mathrm{d}p_y'\mathrm{d}p_z'$, $\mathrm{d}V_\mathrm{p}$ and $\mathrm{d}V_{p'}$ are related through

$$\mathrm{d}V_p = \frac{\sqrt{m_\mathrm{L}m_\mathrm{T}^2}}{(m')^{3/2}}\mathrm{d}V_{p'} = 4\sqrt{2}\pi\sqrt{m_\mathrm{L}m_\mathrm{T}^2}\sqrt{E - E_\mathrm{c}}\mathrm{d}E \tag{4.26}$$

for one ellipsoidal energy surface. For g equivalent valleys, the total DOS effective mass becomes

$$(m_\mathrm{DOS})^{3/2} = \mathrm{g}\left(m_\mathrm{L}m_\mathrm{T}^2\right)^{1/2} \tag{4.27}$$

In the valence band, there is a single peak containing multiple bands, i.e., light-hole and heavy-hole bands. Similarly, the DOS function is also modified to include the multiband effect.

$$\begin{aligned} p = p_\mathrm{lh} + p_\mathrm{hh} &= (N_\mathrm{vl} + N_\mathrm{vh})\exp\left(-\frac{E_F - E_v}{kT}\right) \\ &= 2\left(\frac{2\pi kT}{h^2}\right)\left(m_\mathrm{lh}^{3/2} + m_\mathrm{hh}^{3/2}\right)\exp\left(-\frac{E_F - E_v}{kT}\right) \end{aligned} \tag{4.28}$$

where N_vl and N_vh are the effective density of states in the light-hole band and heavy-hole band, respectively. Therefore, the DOS effective mass of the valence band becomes

$$m_\mathrm{DOS}^{3/2} = m_\mathrm{lh}^{3/2} + m_\mathrm{hh}^{3/2} \tag{4.29}$$

The calculated effective DOS and intrinsic carrier concentration (n_i) in Si and GaAs are listed in the Table 4.2.

In addition, there is another type of effective mass associated with the conductivity. The conductivity is related to the effective mass in the form of

$$\sigma = \frac{nq^2\tau}{m_\mathrm{c}^*} \tag{4.30}$$

Table 4.2 Effective density of states and intrinsic carrier concentration of Ge, Si, and GaAs

(cm^{-3})	Ge	Si	GaAs
N_c	1.04×10^{19}	3.2×10^{19}	4.7×10^{17}
N_v	5×10^{18}	1.8×10^{19}	9.4×10^{18}
n_i	2.33×10^{13}	1.02×10^{10}	2.1×10^{6}

where n is the carrier concentration, q is the electron charge, τ is the carrier relaxation time, and m_c^* is the conductivity effective mass. For semiconductors with a spherical energy surface, such as GaAs, the conductivity effective mass has a simple form of

$$m_c^* = m_e^* \tag{4.31}$$

For non-spherical energy surfaces of the conduction band, as for Si and AlAs, the conductivity has a tensor form of

$$\sigma_i = \frac{n_i q^2 \tau}{\hbar^2}\begin{bmatrix} \partial^2 E/\partial k_x^2 & \partial^2 E/\partial k_x k_y & \partial^2 E/\partial k_x k_z \\ \partial^2 E/\partial k_y k_x & \partial^2 E/\partial k_y^2 & \partial^2 E/\partial k_y k_z \\ \partial^2 E/\partial k_z k_x & \partial^2 E/\partial k_z k_y & \partial^2 E/\partial k_z^2 \end{bmatrix} \tag{4.32}$$

for each ellipsoidal surface. Since $\left(\partial^2 E/\partial k_i \partial k_j\right) = 0$ for $i \neq j$, so that all the off-diagonal elements of the tensor are zero. Using Si as an example, the major axes of the ellipsoidal energy surface lie along the coordinate axes of k-space. The conductivity tensor for, say, [001] and [00$\bar{1}$] ellipsoids along k_z becomes

$$\sigma_{(001)} = \sigma_{([00\bar{1}])} = n_{(001)} q^2 \tau \begin{bmatrix} 1/m_T & 0 & 0 \\ 0 & 1/m_T & 0 \\ 0 & 0 & 1/m_L \end{bmatrix} \tag{4.33}$$

where $m_x = m_y = m_T$ and $m_z = m_L$. Since all six ellipsoids are equivalent energy minima (Fig. 3.28), $n_{(100)} = n_{(\bar{1}00)} = n_{(010)} = n_{(0\bar{1}0)} = n_{(001)} = n_{(00\bar{1})} = n/6$. By summing up conductivities of all six ellipsoids, one can write the total conductivity as

$$\sigma = nq^2\tau \begin{bmatrix} \frac{1}{3}\left(\frac{2}{m_T} + \frac{1}{m_L}\right) & 0 & 0 \\ 0 & \frac{1}{3}\left(\frac{2}{m_T} + \frac{1}{m_L}\right) & 0 \\ 0 & 0 & \frac{1}{3}\left(\frac{2}{m_T} + \frac{1}{m_L}\right) \end{bmatrix} \tag{4.34}$$

This leads to the conductivity and the conductivity effective mass of

$$\sigma = \frac{nq^2\tau}{3}\left(\frac{1}{m_L} + \frac{2}{m_T}\right) = \frac{nq^2\tau}{m_c^*} \tag{4.35}$$

$$\frac{1}{m_c^*} = \frac{1}{3}\left(\frac{1}{m_L} + \frac{2}{m_T}\right) \tag{4.36}$$

In general, the carrier effective mass is a constant for non-degenerate semiconductors. However, the conduction band effective mass of a degenerated semiconductor becomes dependent on doping concentration. At high doping levels, in n-type semiconductor the Fermi level moves close to, or even into, the conduction band. Since the non-parabolicity of the band becomes significant, the effective mass increases with doping level for carrier density larger than 10^{18} cm^{-3}.

4.2.2 Mobility

The mobility (μ) is defined as the drift velocity (v_d) per unit electric field intensity (F).

$$\mu \equiv \frac{v_d}{F} = \frac{q\tau}{m^*}\left(\mathrm{cm}^2/\mathrm{V\text{-}s}\right) \tag{4.37}$$

where τ is the relaxation time or the inverse of scattering probability ($1/\tau$). The equilibrium drift velocity is determined from the balance between the electric field-induced acceleration and scattering generated deceleration. If the scattering mechanisms are assumed to act independently, the total probability of a scattering event occurring in the differential time dt is the sum of the individual events. Thus,

$$\frac{1}{\tau} \equiv \sum_j \frac{1}{\tau_j} \tag{4.38}$$

where τ_j is the relaxation time for each scattering mechanism. Since the mobility is proportional to τ, according to Matthiessen's rule, the total carrier mobility is the inverse sum of each individual contribution.

$$\frac{1}{\mu_{\mathrm{total}}} \equiv \sum_j \frac{1}{\mu_j} = \frac{1}{\mu_{\mathrm{i}}} + \frac{1}{\mu_{\mathrm{l}}} + \frac{1}{\mu_{\mathrm{po}}} + \frac{1}{\mu_{\mathrm{pe}}} + \frac{1}{\mu_{\mathrm{al}}} \tag{4.39}$$

The major scattering mechanisms included in compound semiconductors are impurity scattering (μ_{i}), lattice scattering (μ_{l}), polar scattering (μ_{po}), piezoelectric scattering (μ_{pe}), and alloy scattering (μ_{al}). The total mobility is thus limited by the dominant scattering mechanisms occurring in the semiconductor, as illustrated in Fig. 4.7.

For holes, the mobility is low compared to electrons, mainly due to the heavy effective mass associated with the heavy-hole band. Due to the degeneracy of the heavy and light holes, the interaction between these two bands also contributes to the extra scattering.

In the following, we briefly discuss the mobility due to major scattering mechanisms.

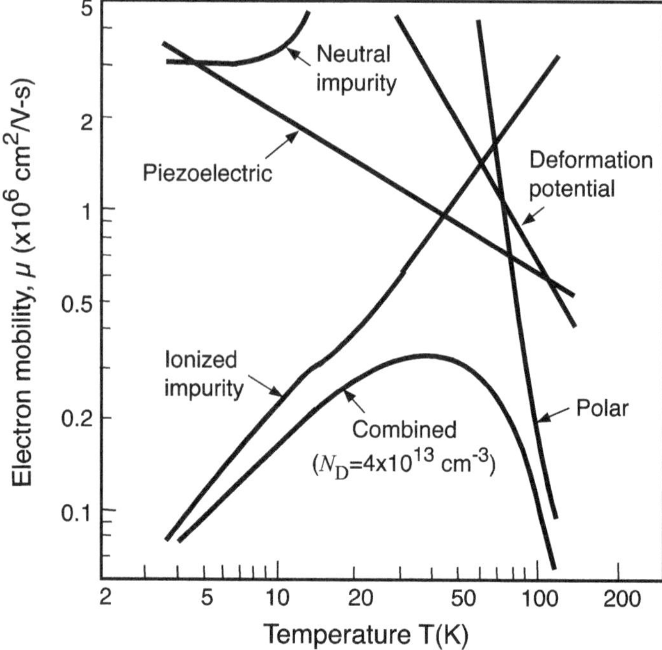

Fig. 4.7 Temperature dependence of the mobility for high-purity n-type GaAs showing the separate and combined scattering processes. Reprinted with permission from [5], copyright AIP Publishing

(a) **Impurity scattering (μ_i)**

The calculation of the mobility for ionized impurity scattering is based upon the scattering of charged particles by Coulomb potential of nuclei, which was originally developed by E. Rutherford to explain the scattering of α-particles. The resultant mobility has a temperature dependence of $T^{3/2}$.

$$\mu_i \propto N_i^{-1} \left(m^*\right)^{-1/2} T^{3/2} \tag{4.40}$$

where N_i is the ionized impurity concentration. The impurity scattering is the major mobility limiting mechanism at low temperatures for both elemental and compound semiconductors.

(b) **Lattice (deformation potential) scattering (μ_l)**

The vibrations of lattice atoms at finite temperatures cause a local variation in the energies of the conduction and valence band edges. The abrupt changes in the band edges further cause increased electron deflections. The energy step height is related to the degree of lattice compression/dilation defined by the deformation potential. At low temperatures, the thermal energy available to excite optical-mode lattice vibrations is quite limited. So we shall consider only the scattering by LA phonons. The temperature dependence of the lattice scattering has a form of $T^{-3/2}$.

$$\mu_1 \propto \Xi^{-2}(m^*)^{-5/2}T^{-3/2} \tag{4.41}$$

where Ξ is the deformation potential. This is the major scattering mechanism in the high-temperature region for elemental semiconductors, but not for III–V compounds.

However, experimental data show that, in both Si and Ge, the temperature dependences deviate quite considerably from the $T^{-3/2}$ law in the region dominated by lattice scattering. As discussed in the previous section, lattice vibrations generate both acoustic and optical branches of phonons. In (4.41), only the low-frequency acoustical mode is considered. In silicon or AlAs, there are six equivalent conduction band minima in the first Brillouin zone. Electrons scatter between these valleys, forming an inter-valley scattering process. The large momentum change involved requires the participation of high-energy phonons, both acoustical and optical. Consider the optical-mode phonon contribution, the inter-valley scattering mobility shows a different temperature dependence of the following form:

$$\mu_{op} \propto (m^*)^{-5/2}T^{-1}\exp(\hbar\omega_{op}/kT) \tag{4.42}$$

where $\hbar\omega_{op}$ is the optical phonon energy.

(c) **Polar scattering (μ_{po})**

In compound semiconductors, when responding to lattice vibrations, the two neighboring atoms move in opposite directions in the optical phonon modes. Because of the nonzero ionicity, the relative movement of neighboring atoms constitutes an electric polarization, which in turn produces an electric field. Deflection of the motion of free carriers by the field limits the mobility of carriers. Thus, the electron mobility in III–V compounds is limited by polar scattering instead of acoustical mode scattering in the high-temperature region. Polar scattering becomes the most dominant scattering mechanism in compound semiconductors at high temperatures.

$$\mu_{po} \propto \begin{cases} (m^*)^{-3/2}(T/\theta_l)^{-1/2}, T > \theta_l \\ (m^*)^{-3/2}\left[\exp(\theta_l/T) - 1\right], T < \theta_l \end{cases} \tag{4.43}$$

where $\theta_l = \hbar\omega_l/k$ is the equivalent temperature associated with longitudinal optical phonons. Since the polar scattering mobility is inversely proportional to $(m^*)^{3/2}$, the large difference in electron mobility and hole mobility in III–V compounds is explainable by (4.43).

(d) **Piezoelectric scattering (μ_{pe})**

A polarization electric field induced in an ionic crystal, such as II–VI compounds and III-N alloys, by the applied mechanical stress, or vice versa, is called the piezoelectric effect. The electric field accompanying the lattice vibrations can interact with electron motion through the lattice. Since zinc-blende III–V compounds are not pure covalent crystals, but mixtures of covalent and ionic bonding, there exist finite piezoelectric fields. The piezoelectric scattering mobility is

$$\mu_{pe} \propto \left(m^*\right)^{3/2} T^{-1/2} \tag{4.44}$$

This is the main scattering mechanism which limits the mobility in ionic II–VI compounds, but it is less important for zinc-blende III–V compounds.

(e) **Alloy scattering (μ_{al})**

When mixing binary compounds into ternary or quaternary alloys, the lattice constant, according to Vegard's law, is proportional to the composition ratio of the binaries. A close examination indicates that the bond lengths associated with each individual binary do not change significantly, as shown in Fig. 4.8. Using an extended x-ray absorption fine structure (EXAFS) technique, in InGaAs alloys, the measured Ga-As and In-As near-neighbor distances change by only 0.04 Å over the whole alloy composition range. However, the virtual crystal model that follows Vegard's law requires a change in average near-neighbor spacing of 0.17 Å. Therefore, the compound can be seen as a mixture of clusters of binary alloys instead of a completely random material. The carriers moving through regions containing different proportions of binaries may suffer scattering as

$$\mu_{al} \propto \left(m^*\right)^{-5/2} S^{-1} \Delta u^{-2} T^{-1/2} \tag{4.45}$$

where S represents the degree of randomness ($S = 1$ for complete random composition), and Δu is the alloy scattering potential. It turns out that this is the limiting

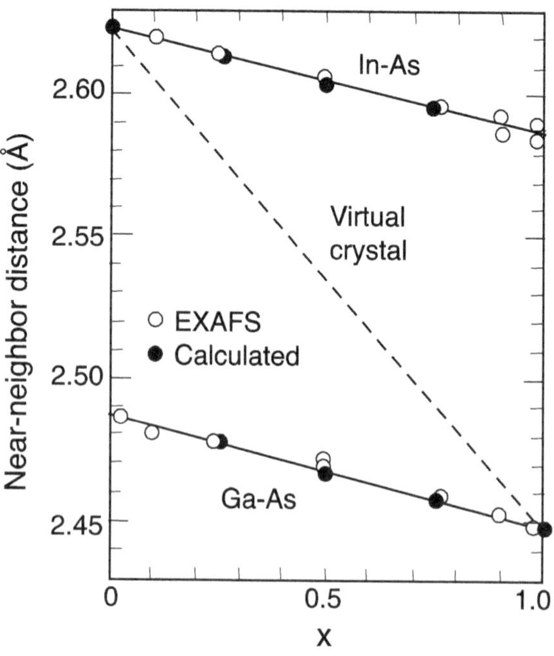

Fig. 4.8 Near-neighbor distance versus InAs mole fraction x in the $In_xGa_{1-x}As$ ternary alloy measured by EXAFS. The dashed line represents the average cation–anion spacing according to Vegard's law. Reprinted with permission from [6], copyright American Physical Society

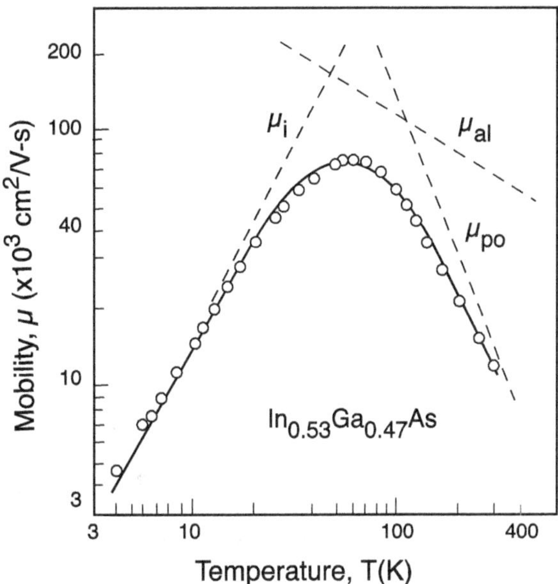

Fig. 4.9 Temperature dependence of the electron Hall mobility in $In_{0.53}Ga_{0.47}As$. The contribution due to alloy scattering dominates the total mobility near 100 K. Reprinted with permission from [7], copyright Elsevier

scattering mechanism in the middle temperature range for ternary and quaternary compounds. Figure 4.9 shows the dominance of alloy scattering in electron Hall mobility near 100 K in an $In_{0.53}Ga_{0.47}As$ ternary alloy lattice-matched to InP.

(f) **High-field phenomena**

As shown in (4.37), the carrier mobility is determined by the total scattering events occurring inside the semiconductor. If scattering rates were fixed, mobility would remain constant and carrier velocity would be limited by the magnitude of the applied electric field. In fact, under the increasing applied electric field, new scattering mechanisms come into play as carrier energies increase. Under a moderately high-field (~10^4 V/cm) condition, carriers may acquire sufficient energy to excite acoustic phonons within the semiconductor lattice. Increased phonon scattering is particularly pronounced between multiple conduction band minima in indirect bandgap materials. This leads to a sublinear rate of the mobility change with increasing applied field. For further increases of the field to above 10^5 V/cm, the energy gained by carriers is transferred to lattice heating through optical phonon emission (~60 meV in Si and ~36 meV in GaAs), resulting in a saturated drift velocity (or mobility) with a value of about 10^7 cm/s. The velocity–field characteristics for electrons in Si, GaAs, and InP are shown in Fig. 4.10.

Between the initial roll-off and final velocity saturation, the velocity–field behavior of electrons in the direct bandgap semiconductor is quite different from that in Si. As shown in Fig. 4.10, GaAs and InP display an anomalous peak in drift velocity corresponding to a negative differential resistance. This is the basis of microwave oscillation in compound semiconductors, known as the *Gunn effect*. To

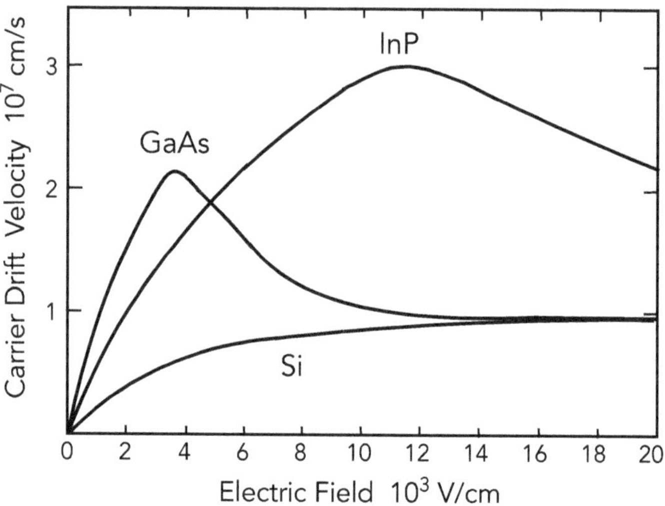

Fig. 4.10 Velocity–field characteristics for GaAs, InP, and Si [8]. Reprint with permission from [9], copyright American Physical Society

understand this, one must look into semiconductor band structures, e.g., Fig. 3.25. For GaAs, the conduction band has a number of satellite valleys at L and X minima with 0.31 meV and 0.51 meV above the Γ valley minimum, respectively. At low field, electrons are in the Γ valley, characterized by a light effective mass of $0.063m_0$ so that the low-field mobility is high. At intermediate field, electrons may gain enough energy to be transferred into the lowest satellite valleys. For GaAs, the lowest satellite energy band minima are located at L valleys with a heavy effective mass of $0.85m_0$ and a high density of states ($\times 50$ than Γ valley) for scattering electrons, leading to lower mobility. Thus the electron transfer between Γ and L valleys results in a downturn in electron velocity as shown in Fig. 4.10. The transfer electron effect has been utilized in devices to generate negative resistance and microwave oscillations.

4.2.3 *Intentional Impurity*

In III–V compounds, intentional impurities are selected from columns adjacent to column III or V of the periodic table. Elements from group II predominantly occupy group III sites to form p-type impurities. Substitution of group V elements by group VI impurities would result n-type materials. Group IV impurities can occupy either cation or anion sites and are amphoteric dopants. The energy required to ionize a dopant is called the *activation energy*, which is mainly determined by the Coulombic interaction between the impurity atom and the attached charge carrier. The hydrogen atom model can be applied to understand the shallow donors in III–V compounds. We derived the ionization energy for a hydrogen atom as

$$E_n = \left[\frac{m_0 q^4}{2(4\pi\epsilon_0\hbar)^2}\right]\frac{1}{n^2} = \frac{R_y}{n^2} = \frac{13.6}{n^2} \quad (4.46)$$

where R_y is the Rydberg energy. To apply this model to shallow donors, we have to make two modifications. First, the effective mass of electrons in semiconductors is very different from the free electron mass. Second, the dielectric constant of semiconductors is much larger than the free space value. Then the donor ionization energy required for the $n = 1$ level is

$$E_1 = \frac{m_e^* q^4}{2(4\pi\epsilon_r\hbar)^2} = \left[\frac{m_e^*/m_0}{(\epsilon_r\epsilon_0/\epsilon_0)^2}\right]R_y = 13.6\left(\frac{m_e^*/m_0}{\epsilon_r^2}\right) \text{ (eV)} \quad (4.47)$$

The donor Bohr radius is given by

$$a_e = \frac{4\pi\epsilon\hbar^2}{m_e^* q^2} = 0.53\left(\frac{\epsilon_r}{m_e^*/m_0}\right) \text{ (Å)} \quad (4.48)$$

The calculated ionization energy and Bohr radius for a hydrogen-like donor in GaAs with $\epsilon_r = 12.85$ and $m_e^* = 0.067\ m_0$ are 5.5 meV and 102 Å, respectively. The bound state wave function of the impurity extends much farther than the unit cell of the lattice (~5 Å). This result provides the justification for using effective mass and the dielectric constant of the semiconductor in these calculations. The calculated activation energy is smaller than or comparable to the thermal energy kT at room temperature. Therefore, all shallow donor impurities can be considered as activated at room temperature.

The application of the hydrogen model to acceptors in III–V compounds is complicated by their degenerate valence band structure. In the limit of strong spin-orbit interaction—that is, for spin-orbit splitting (Δ_{so}) much larger than the acceptor energy—only the heavy-hole and light-hole bands are included for acceptor binding energy calculation. The calculated results of the effective Bohr radius and effective Rydberg energy are found to be functions of Luttinger parameters, γ_1, γ_2, and γ_3, which describe the valence band shapes near the center of the Brillouin zone. Table 4.3 lists the calculated acceptor ionization energy for the major III–V compounds, which agrees reasonably well with experimental results. The measured values for each compound are given below.

Following is a summary of the most commonly used dopants in III–V compound semiconductors.

(a) **Group II** impurities incorporate on the anion sites of III–V compounds to form shallow acceptors.

Table 4.3 Calculated acceptor ionization energy of selected III–V compounds [8]

Material	AlSb	GaP	GaAs	GaSb	InP	InAs	InSb
E_a	42.4	47.5	25.6	12.5	35.3	16.6	8.6

Beryllium (Be) is a major shallow acceptor in III–V compounds. It has an ionization energy of 28 and 41 meV in GaAs and InP, respectively. Except at the highest concentration $\geq 10^{19}$ cm^{-3}, Be diffuses slowly in GaAs and the diffusion coefficient as a function of temperature has the form

$$D = D_0 \exp(-E_a/kT) \tag{4.49}$$

with the activation energy $E_a = 1.95$ eV and $D_0 = 2\times10^{-5}$ cm^2/s. However, at very high doping concentrations of ~10^{20} cm^{-3}, Be redistributes severely which can degrade the device performance. Be is widely used in molecular beam epitaxy (MBE)-growth, ion implantation, and as Au–Be alloy for *p*-type ohmic contact metallization.

Zinc (Zn) is another major shallow acceptor in III–V compounds besides Be. The ionization energy is 31 meV in GaAs and 47 meV in InP. Due to its high vapor pressure, Zn is not suitable for MBE. In chemical vapor deposition (CVD), Zn is a popular dopant. Zn diffuses rapidly at high concentration and its diffusion rate is concentration dependent. The Au-Zn alloy is commonly used as a *p*-type ohmic contact metallization.

Magnesium (Mg) is a shallow acceptor impurity in III–V compounds with an ionization energy of 28 meV in GaAs and 41 meV in InP. Mg has a strong affinity for oxygen. Elemental Mg can form MgO easily during the doping process, which leads to a low doping incorporation coefficient at high growth temperature (>500 °C). Recently, it was found that Mg is an efficient *p*-type dopant in III-nitride compounds grown by both MBE and metalorganic chemical vapor deposition (MOCVD). Nevertheless, the doping efficiency is low with deep ionization energy of $\geq$150 meV.

Cadmium (Cd) is a shallow acceptor in GaAs and InP with ionization energies of 35 and 56 meV, respectively. It has a very high vapor pressure, approximately 1–2 orders of magnitude higher than Zn. It is a useful dopant in liquid phase epitaxy (LPE) growth of InP for temperature $\leq$550 °C with a maximum doping concentration of ~10^{18} cm^{-3}.

(b) **Group VI** impurities incorporate on the cation sites of III–V compounds to form shallow donors.

Sulfur (S) is a shallow donor in III–V compounds with an ionization energy of 6 meV in GaAs. It has a high vapor pressure, and the incorporation depends strongly on the epitaxial growth temperature. This makes the accurate control of doping concentration from a solid sulfur source difficult. However, hydrogen sulfide (H_2S) is the most common doping precursor used in CVD for *n*-type doping of III–V compounds.

Selenium (Se) is also a high vapor pressure shallow donor in III–V compounds similar to sulfur. It has an ionization energy of 6 meV in GaAs. Hydrogen selenide (H_2Se) is a commonly used gaseous doping source of Se. However, the doping memory effect of using H_2Se is the major drawback. The memory effect manifests itself by a lack of abruptness of doping profiles.

Tellurium (Te) is an efficient shallow donor in III–V compounds. It has an ionization energy of 30 meV in GaAs. This is much larger than the calculated 5 meV due to a correction associated with the big Te atom. Elemental Te has been used as an *n*-type doping source in LPE growth of GaAs. Diethyltellurium [$(C_2H_5)_2Te$, DETe] has been used as a Te doping precursor for MOCVD and metalorganic MBE (MOMBE) growth of GaAs and InP.

(c) Group IV dopants

Group IV impurities incorporated into III–V compounds share some interesting characteristics. First, they all are *amphoteric*, i.e., they can occupy either the group III or the group V sites to become donors or acceptors, respectively. Second, they autocompensate. A group IV impurity that predominantly occupies the cation sites as a donor can also occupy the anion sites. This causes compensation and lowers the doping efficiency. Third, the electrically activated impurities saturate at high doping concentration. The autocompensation increases at high doping level and leads to the saturation of the free carrier concentration. Fourth, the free carrier concentration depends on growth temperature and V/III flux ratio during growth. For example, as shown in Fig. 4.11, when growing silicon-doped GaAs by liquid phase epitaxy, the type of conductivity depends on the growth temperature and the Si concentration in the liquid. At a fixed Si concentration above 4×10^{-4} atomic fraction in the growth solution, the GaAs shows *n*- and *p*-type conductivities when grown at high and low temperatures, respectively.

Carbon (C) is a shallow acceptor in GaAs (26 meV) but a donor in InP (43 meV) and InAs. Carbon is a stable impurity and diffuses very slowly. Its temperature-dependent diffusion coefficient in GaAs follows (4.49) with an activation energy

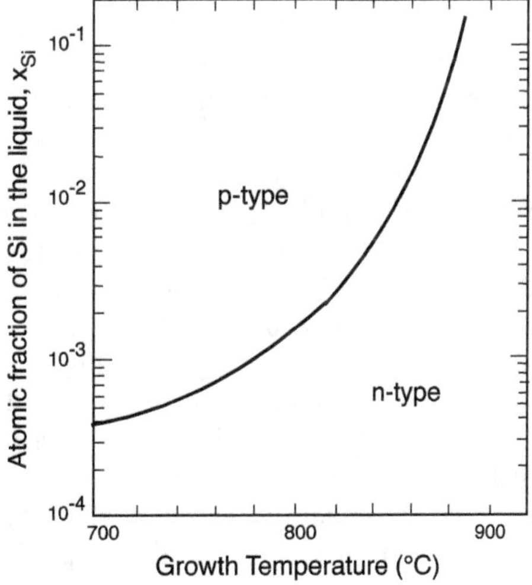

Fig. 4.11 Effect of growth temperature and Si concentration in the liquid on the doping behavior of Si in LPE grown GaAs

E_a= 1.75 eV and $D_0 = 5 \times 10^{-8}$ cm^2/s. Compared to Be, at 800 °C, the diffusion coefficient of C is almost two orders of magnitude smaller. It is an unintentional dopant in MOCVD when using metalorganic Ga and Al precursors. CCl_4 and CBr_4 are commonly used gaseous carbon doping sources. A very high acceptor level (~10^{20} cm^{-3}) is easily achieved in many III–V compounds. These behaviors make C the most ideal *p*-type dopant for the base region of III–V heterojunction bipolar transistors (HBT).

Silicon (Si) is a stable shallow donor (4–6 meV) in most III–V compounds at low concentration. The temperature-dependent diffusion coefficient in GaAs has an activation energy $E_a = 2.45$ eV and $D_0 = 4 \times 10^{-4}$ cm^2/s. It strongly compensates in the high concentration regime (~5×10^{18} cm^{-3}). It has been widely used as an *n*-type dopant for epitaxial growth and ion implantation. Gaseous silicon dopant precursors including silane (SiH_4) and disilane (Si_2H_6) have been widely used in MOCVD and MOMBE growth.

Germanium (Ge) is a strongly amphoteric impurity in III–V compounds. It is a *p*-type dopant for LPE growth of GaAs, but not popular for MBE and MOCVD growth. Au-Ge alloy is used for *n*-type ohmic contact metallization.

Tin (Sn) is a weakly compensating shallow donor in III–V compounds with 4–6 meV ionization energy in GaAs and InP. The major problem of Sn impurity during growth is the tendency to strongly redistribute due to the Sn diffusion and surface segregation. Nevertheless, high electron mobilities in GaAs:Sn are expected due to its low compensation.

(d) Hydrogen passivation

In III–V semiconductors, hydrogen is not an electrically active impurity. However, during the doping process, hydrogen generated from the carrier gas and/or dopant flux can passivate or neutralize shallow donor or acceptor in semiconductors. Using Si in (Al)GaAs as an example, under normal conditions, the substitutional silicon atom replaces a Ga and bonds with four nearest neighboring As atoms. When hydrogen atoms are incorporated into the lattice, a hydrogen can bond to the Si donor by breaking one of the Si-As bonds and sit in an interstitial site, as shown in Fig. 4.12a. The broken Si-As bonds of the As atom opposite the hydrogen atom are replaced with a lone pair of electrons—one provided by the donor atom and the other from the dangling bond of the As atom. The Si-H complex yields a neutral electron count, and the hydrogen has passivated the donor atom. The active donor concentration becomes very low compared to the impurity concentration. Hydrogen passivation of donor atoms has been found for a large number of donors including Si, Sn, Se, and Te. The dissociation of donor hydrogen bond by thermal annealing at temperatures above 400 °C allows hydrogen to diffuse out of the hydrogenated semiconductor and recover the doping concentration. Hydrogen also passivates shallow acceptors in III–V compounds as shown in Fig. 4.12b. The acceptor As bonds are broken, and the hydrogen atom is bonded to the arsenic atom and occupies a site between the acceptor impurity and the neighboring arsenic. The entire hydrogen acceptor complex is neutral, and the hydrogen passivates the acceptor.

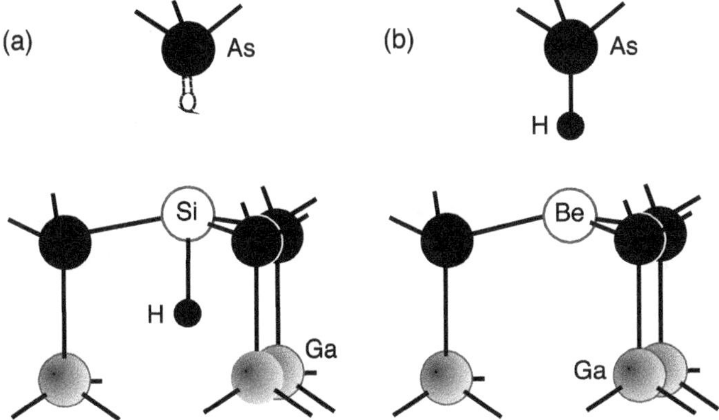

Fig. 4.12 **a** The group IV donor (Si)-hydrogen complex with hydrogen in an interstitial site and bound to the donor atom. **b** The group II acceptor (Be)-hydrogen complex with hydrogen bound to an As atom. Reprinted with permission from [11], copyright American Physical Society

4.2.4 Deep Levels

During the growth or doping of semiconductors, chemical impurities, native defects, or the combination of both, can form deep energy level(s) in the forbidden gap. These deep levels act like traps to capture and annihilate electrons and holes, making the material electrically inactive. Therefore, the deep traps may have many negative effects on the properties of semiconductors, including compensation of intentional impurities, minority carrier lifetime reduction, and reduced mobility and luminescence efficiency. In most cases, we should avoid deep levels in the fabrication of semiconductor materials and devices. Under certain circumstances, we utilize deep levels to control the conductivity of semiconductors. In this section, two particularly important deep levels in III–V compound semiconductors are discussed.

(a) **DX Centers**

During the 1970s, the rapid development of compound semiconductors was mostly focused on the $Al_xGa_{1-x}As$ system due to the availability of quality GaAs substrates, among other reasons. It was found that the *n*-type $Al_xGa_{1-x}As$ with $0.2 \leq x \leq 0.4$ has some unusual properties as compared to GaAs. Most noticeable are the large donor activation energy and a strong sensitivity of the conductivity to illumination. These properties have profoundly negative effects on the device performance of $Al_xGa_{1-x}As$/GaAs high-electron-mobility transistors (HEMT), as will be discussed in Chap. 9. The origin of the deep center observed in *n*-type $Al_xGa_{1-x}As$ involves a donor atom and another constituent to form a complex called the *donor-complex* (DX) *center*. The DX center formation mechanism and its properties are discussed below.

In $Al_xGa_{1-x}As$, donors such as Si, Te, or Sn can exist in two different configurations—substitutional and interstitial—as shown in Fig. 4.13. When the incorporated

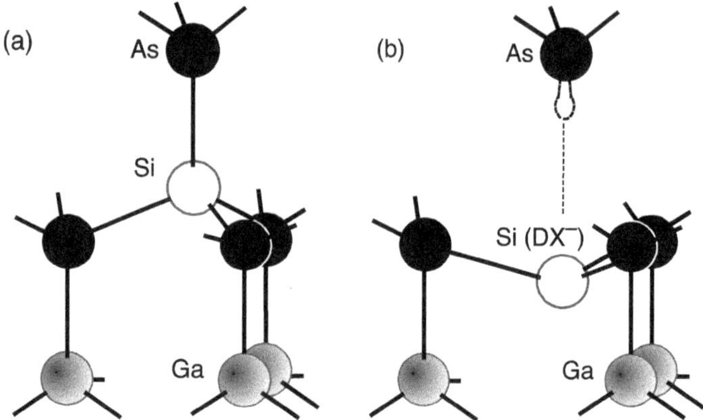

Fig. 4.13 **a** Substitutional and **b** DX configuration of a Si donor in (Al)GaAs. In DX configuration, the Si occupies an interstitial site and has three cations in its immediate vicinity. Reprinted with permission from [11], copyright American Physical Society

Si atom takes the substitutional site, as shown in Fig. 4.13a, it replaces one of the anions of $Al_xGa_{1-x}As$, e.g., Al or Ga. The Si atom is ~0.8 Å above the tetrahedron base formed by three adjacent As atoms. Due to the covalent bond length difference between Si (1.16 Å) and Al or Ga (1.26 Å), the substitutional configuration is the unrelaxed state for the Si atom. On the other hand, if the donor atom occupies an interstitial site, one of the four arsenic donor bonds is broken (Fig. 4.13b) and the interstitial configuration is the relaxed state. In this configuration, the Si atom is 0.2 Å below the tetrahedral plane defined by three As atoms. When the Si atom relaxes from the substitutional to interstitial configuration, the total displacement of the Si atom between the two configurations is approximately 1 Å. This displacement is large compared to the equilibrium bond length of the host crystal (~2.44 Å). This structural configuration change leads to a change in the electronic configuration of the defect.

To illustrate the interplay between structural and electronic configurations, consider the hydrogen molecule as an example. The total energy and nuclei distance of bonding and anti-bonding states of two hydrogen atoms are shown in Fig. 4.14. The equilibrium distance between the two protons corresponds to a minimum total energy which depends on the electronic state of the molecule. The bonding state has a shorter core distance and lower total energy minima than the anti-bonding state. For small lattice relaxation, e.g., between bonding and anti-bonding states of a hydrogen molecule, the direct transitions between the ground (bonding) state and the excited (anti-bonding) state through absorption and emission are readily allowed. The electron of the bonding state can be excited to the anti-bonding state vertically and then relaxed to a new core position (configuration) before returning to

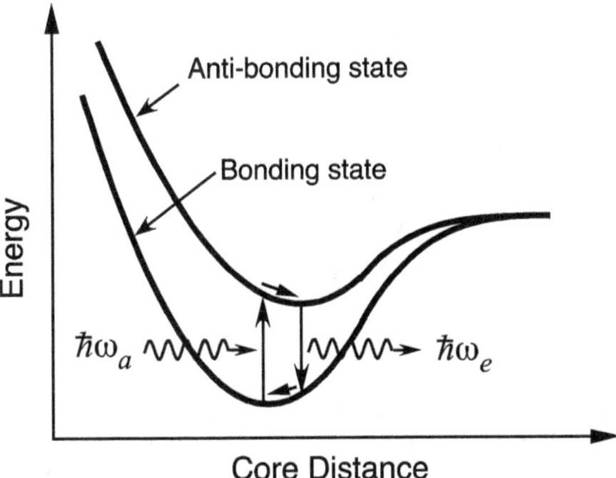

Fig. 4.14 Bonding and anti-bonding state energy of a hydrogen molecule as a function of distance between two core atoms. Note, $\hbar\omega_a > \hbar\omega_e$

the bonding state. This process leads to different absorption energy $\hbar\omega_a$ and emission energy $\hbar\omega_e$ for electron transitions between bonding and anti-bonding states, known as the *Franck-Condon shift*. However, due to the heavier mass of the nuclear as compared to the electronic mass, the two nuclei do not move during the short time of the electronic transition but the molecule slowly moves to the new configuration, i.e., new core distance. This situation is changed dramatically for large lattice relaxation. Figure 4.15 shows two configurations of such defects. Photon energy of $\hbar\omega_a$ is required to excite the defect from the ground (relaxed) state to the excited (unrelaxed) state. Once the defect has minimized its energy in the excited state, the defect becomes metastable and cannot return to the ground state by means of an optical (vertical) transition. In order for the defect to return from the metastable state to the

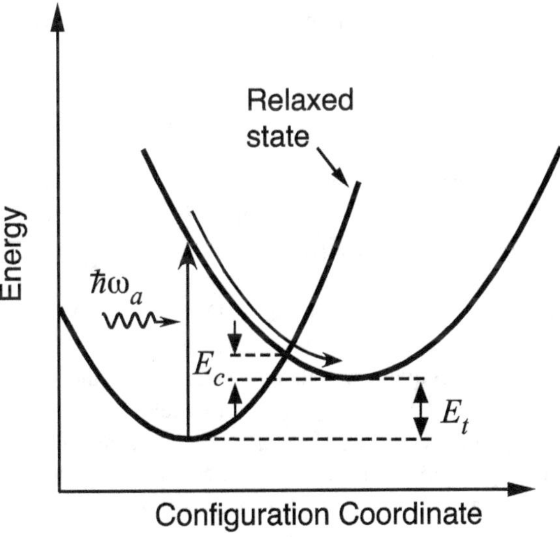

Fig. 4.15 Schematic configuration coordinate diagram for large lattice relaxation

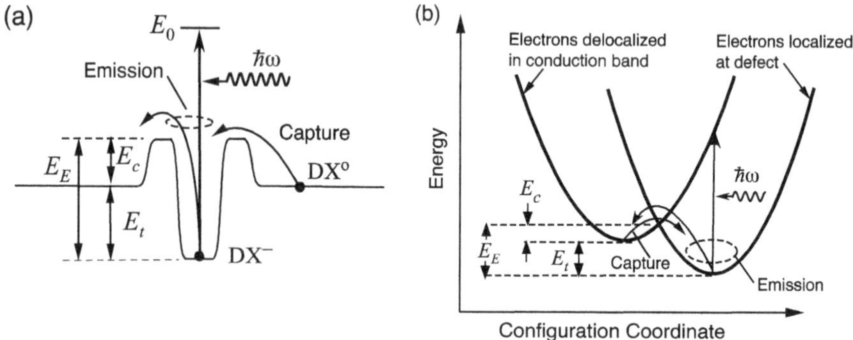

Fig. 4.16 **a** Energy diagram model of a DX center in n-type AlGaAs; E_c—capture energy barrier, E_E—transit emission energy, E_t—activation energy, and E_0—absorption energy of the DX center. **b** Configuration coordinate diagram of the DX center. Reprinted with permission from [12], copyright American Physical Society

ground state, it has to overcome the capture energy barrier E_c through other allowed paths such as thermally excited transitions. Thermal emissions are also allowed to excite the defect from the ground state to the excited state. The thermal emission energy $E_E = E_c + E_t$ is measured from the minimum of the ground state to the top of the thermal barrier.

We can use the same analysis to understand the physical mechanism of the DX center. The schematic representations of the relevant energies of the DX center and their representation in the configuration coordinate diagram are shown in Fig. 4.16. Experimentally, these relevant energies have been determined for Si-doped $Al_xGa_{1-x}As$: $E_0 \geq 1$ eV, $E_t \approx 150$ meV, $E_c \approx 150$ meV, and $E_E \approx 300$ meV. Apparently, this is a very deep level, as expected. Since the DX interstitial configuration in Si-doped $Al_xGa_{1-x}As$ is a stable (relaxed) configuration, the substitutional configuration is the excited (unrelaxed) configuration, and a neutral substitutional donor can capture an electron and transform into a DX center, i.e.,

$$D^0 + e \leftrightarrow DX^- \tag{4.50}$$

where D^0 and DX^- represent a neutral substitutional donor and a negatively charged DX center, respectively. An ionized shallow donor can also capture two electrons by assuming the DX configuration, i.e.,

$$D^+ + 2e \leftrightarrow DX^- \tag{4.51}$$

As a consequence, both deep DX^- donors and shallow D^0 donors coexist in n-type $Al_xGa_{1-x}As$ ($0.2 \leq x \leq 0.4$) at low temperatures. As an example, Fig. 4.17 shows the dependence of the Hall carrier concentration of Si-doped n-type $Al_{0.32}Ga_{-.68}As$ on temperature. When measured in dark, the electron concentration remains constant for temperature below ~150 K, with a value of 3.5×10^{17} cm^{-3}, only a fraction of the

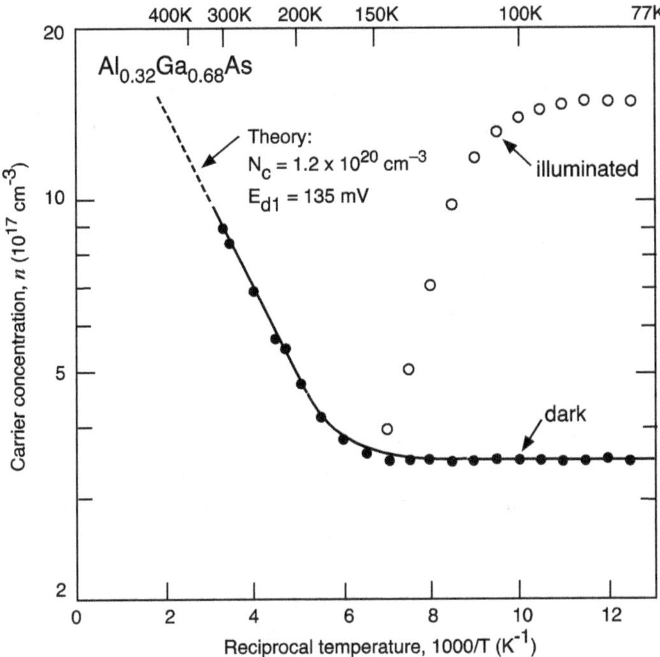

Fig. 4.17 Temperature dependence of the Hall electron concentration in *n*-type $Al_{0.32}Ga_{0.68}As$ with a Si doping concentration of 1.5×10^{18} cm^{-3}. Solid and open circles indicate experimental data measured in the dark and after illumination at low temperatures, respectively. Reprinted with permission from [13], copyright American Physical Society

Si impurity concentration of 1.5×10^{18} cm^{-3}. The electron concentration increases at temperatures higher than 150 K indicating the thermal ionization of the deep donor center. However, the saturation of the free carrier concentration does not occur even at 300 K confirming that the DX center has large thermal activation energy. Upon illumination with an infrared radiation having a wavelength of 820 nm, which is below the bandgap of $Al_xGa_{1-x}As$ at low temperature, the carrier concentration increases. The increase in carrier concentration persists even after the illumination has been turned off hours or days. This phenomenon is known as the *persistent photoconductivity* (PPC) effect, a key characteristic of DX center in *n*-type $Al_xGa_{1-x}As$. The model of two atomic configurations can explain the phenomenon of PPC. The negatively charged DX center resumes a neutral state, DX^0, under photoionization. This state is unstable, and the Si donor assumes the lower-energy substitutional D^0 configuration. Accompanying the neutral Si donors are the extra free carriers released from the DX^- state. At sufficiently low temperatures, the Si cannot assume the DX^- configuration due to the large capture barrier (E_c). Thus, the extra carriers remain free persistently. At temperatures higher than ~150 K, the neutral donor can capture an electron and return to the state of a DX center. Therefore, the high PPC concentration can be quenched by heating the sample to a temperature >150 K.

(b) **EL2 Level**

When growing bulk III–V semiconductors, it is unavoidable that some unintentional impurities (Si, C, B, Fe, S, and Mn) are incorporated into crystals. The dominant residual impurity in bulk GaAs crystal grown by liquid encapsulated Czochralski (LEC) technique is the shallow acceptor C with a concentration in the low 10^{15} cm^{-3} range. This impurity comes from the graphite heating elements and/or the wall of the stainless steel growth chamber. The residual conductivity would be *p*-type with a resistivity of 0.1–10 Ω cm. For high-speed device applications, it is necessary to minimize the parasitic capacitance and inductance by using semi-insulating (SI) substrates with a resistivity $>10^7$ Ω cm. Many deep impurities have been intentionally introduced in order to achieve high resistivity in compound semiconductors. To achieve SI GaAs, the most noticeable dopant used was Cr. The major drawback is the Cr diffusion and redistribution during high-temperature processing and epitaxial growth. Later, the EL2 defect was identified as the key to achieve undoped SI GaAs grown by the LEC technique. The incorporation of a sufficiently high concentration of EL2 defects allows one to obtain semi-insulating GaAs materials with a resistivity higher than 10^7 Ω cm.

In GaAs there are many deep levels whose chemical origins are unknown. The EL2 defect is one of these deep levels (forming an electron trap) in undoped GaAs. The EL2 level has been identified to be the native As anti-site defect, As_{Ga}. The arsenic atom has two excess electrons if occupying a Ga site, i.e., a double donor. The main EL2 donor level, located approximately 0.75 eV below the conduction band edge, compensates for the shallow C acceptor impurities. For high EL2 concentrations, due to their large ionization energy, only a fraction of the EL2 centers will be ionized, while the remaining EL2 centers will be neutral. As a result, the Fermi level is pinned at the EL2 level which is approximately in the mid-bandgap of the GaAs. Thus, GaAs with sufficiently high concentration of EL2 centers has near-intrinsic characteristics. The EL2 concentration can be controlled during bulk crystal growth via the As/Ga composition in the growth melt (Fig. 4.18). For high As/Ga ratios (As atom fraction $\geq$0.475) in the growth melt, the concentration of EL2 defects increases, which compensates the shallow C acceptors. At low As/Ga ratios in the growth melt, the shallow C acceptor impurities are not fully compensated, leading to residual *p*-type conductivity of the bulk GaAs.

4.3 Free Carrier Concentration and the Fermi Integral

4.3.1 Free Carrier Concentrations in 3D Semiconductors

The electron and hole concentrations in 3D semiconductor structures are calculated by integrating carrier distributions over appropriate energy ranges.

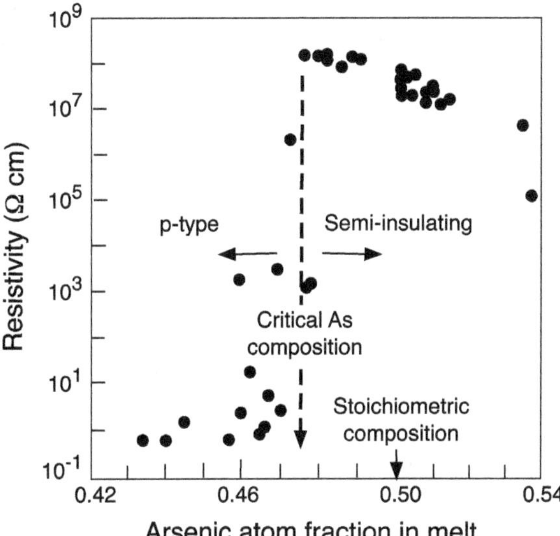

Fig. 4.18 Resistivity of unintentional doped GaAs grown by LEC technique as a function of As atom fraction in the growth melt. An As-rich melt produces semi-insulating GaAs with resistivity >10^7 Ω cm and a Ga-rich melt gives *p*-type conductivity. Reprinted with permission from [14], copyright AIP Publishing

$$\begin{cases} n = \int_{E_c}^{\infty} D_c(E) f(E) \mathrm{d}E \\ p = \int_{-\infty}^{E_v} D_v(E)[1 - f(E)] \mathrm{d}E \end{cases} \tag{4.52}$$

where $D(E)$ and $f(E)$ are the DOS and the Fermi–Dirac distribution function, respectively. The DOS of the conduction band and the valence band are

$$\begin{cases} D_c(E) = \frac{1}{2\pi^2}\left(\frac{2m_e^*}{\hbar^2}\right)^{3/2}\sqrt{E - E_c} \\ D_v(E) = \frac{1}{2\pi^2}\left(\frac{2m_h^*}{\hbar^2}\right)^{3/2}\sqrt{E_v - E} \end{cases} \tag{4.53}$$

In a semiconductor, the carriers are indistinguishable and have to obey the Pauli exclusion principle. The occupation behavior in the energy space is governed by the Fermi–Dirac distribution function

$$f(E) = \frac{1}{1 + \exp[(E - E_F)/kT]} \tag{4.54}$$

where E_F is the Fermi level. For non-degenerated semiconductors where $E - E_F > 3kT$, the Fermi–Dirac distribution function is simplified to the Boltzmann distribution function and (4.52) is solved analytically. For degenerated semiconductors where $|E - E_F|$ is less than a few kT, the Fermi–Dirac distribution function cannot be simplified. Using the DOS and $f(E)$ expressed above, one obtains

$$\begin{cases} n = \dfrac{1}{2\pi^2}\left(\dfrac{2m_e^*}{\hbar^2}\right)^{3/2} \displaystyle\int_{E_c}^{\infty} \dfrac{\sqrt{E - E_c}}{1 + \exp[(E - E_F)/kT]} dE \\ p = \dfrac{1}{2\pi^2}\left(\dfrac{2m_h^*}{\hbar^2}\right)^{3/2} \displaystyle\int_{-\infty}^{E_v} \dfrac{\sqrt{E_v - E}}{1 + \exp[(E_F - E)/kT]} dE \end{cases} \tag{4.55}$$

By replacing $\eta = (E_F - E_c)/kT$ and $\xi = (E - E_c)/kT$, the electron concentrations has the form of

$$n = \frac{1}{2\pi^2}\left(\frac{2m_e^* kT}{\hbar^2}\right)^{3/2} \int_0^{\infty} \frac{\sqrt{\xi}}{1 + \exp[\xi - \eta]} d\xi \tag{4.56}$$

We can reduce (4.56) to a more convenient form by using the Fermi–Dirac integrals defined as

$$F_j(\eta) = \frac{1}{\Gamma(j+1)} \int_0^{\infty} \frac{\xi^j}{1 + \exp[\xi - \eta]} d\xi \tag{4.57}$$

where $\Gamma(j + 1)$ is the Gamma function. With $j = 1/2$ and $\Gamma(3/2) = \sqrt{\pi}/2$ in (4.56) and (4.57), then

$$n = N_c F_{1/2}(\eta) \tag{4.58}$$

where N_c is the effective DOS and $F_{1/2}(\eta)$ is the Fermi–Dirac integral of the order $j = 1/2$. Their expressions are

$$N_c = 2\left(\frac{2\pi m_e^* kT}{h^2}\right)^{3/2}$$ and

$$F_{1/2}(\eta) = \frac{2}{\sqrt{\pi}} \int_0^{\infty} \frac{\sqrt{\xi}}{1 + \exp[\xi - \eta]} d\xi$$

Table 4.4 lists the Fermi–Dirac integral of orders 1/2 as a function of the reduced Fermi energy η.

Table 4.4 Fermi–Dirac integral of order +1/2, $F_{1/2}(\eta)$, as a function of reduced Fermi energy η

η	$F_{1/2}$	η	$F_{1/2}$	η	$F_{1/2}$	η	$F_{1/2}$	η	$F_{1/2}$
− 4.0	0.01820	− 2.0	0.12930	1.0	1.5756	4.0	6.5115	7.0	14.290
− 3.9	0.02010	− 1.8	0.15642	1.2	1.7900	4.2	6.9548	7.2	14.886
− 3.8	0.02220	− 1.6	0.18889	1.4	2.0221	4.4	7.4100	7.4	15.491
− 3.7	0.02451	− 1.4	0.22759	1.6	2.2720	4.6	7.8769	7.6	16.104
− 3.6	0.02706	− 1.2	0.27353	1.8	2.5393	4.8	8.3550	7.8	16.725
− 3.5	0.02988	− 1.0	0.32780	2.0	2.8237	5.0	8.8442	8.0	17.355
− 3.4	0.03299	− 0.8	0.39154	2.2	3.1249	5.2	9.3441	8.2	17.993
− 3.3	0.03641	− 0.6	0.46595	2.4	3.4423	5.4	9.8546	8.4	18.639
− 3.2	0.04019	− 0.4	0.55224	2.6	3.7755	5.6	10.375	8.6	19.293
− 3.1	0.04435	− 0.2	0.65161	2.8	4.1241	5.8	10.906	8.8	19.954
− 3.0	0.04893	0.0	0.76515	3.0	4.4876	6.0	11.447	9.0	20.624
− 2.8	0.05955	0.2	0.89388	3.2	4.8653	6.2	11.997	9.2	21.301
− 2.6	0.07240	0.4	1.0387	3.4	5.2571	6.4	12.556	9.4	21.986
− 2.4	0.08794	0.6	1.2003	3.6	5.6623	6.6	13.125	9.6	22.678
− 2.2	0.10671	0.8	1.3791	3.8	6.0806	6.8	13.703	9.8	23.378
− 2.0	0.12930	1.0	1.5756	4.0	6.5115	7.0	14.290	10	24.085

For the hole concentration in the valence band, it is convenient to set a new parameter $\xi_p = (E_v - E)/kT = -\xi - (E_g/kT) = -(\xi + \chi)$. The integral becomes

$$p = \frac{1}{2\pi^2}\left(\frac{2m_h^* kT}{\hbar^2}\right)^{3/2}\int_0^\infty \frac{\sqrt{\xi_p}}{1 + exp[\xi_p + \chi + \eta]} d\xi_p = N_v F_{1/2}(-\chi - \eta) \quad (4.59)$$

or

$$p = N_v F_{1/2}\left(\frac{E_v - E_F}{kT}\right) \text{ and } N_v = 2\left(\frac{2\pi m_h^* kT}{h^2}\right)^{3/2} \quad (4.60)$$

We assume $E_v = 0$. For non-degenerate n-type semiconductors, where $E_c - E_F \gg kT$ or $F_{1/2}(\eta) \ll 1$ or $n \ll N_c$, the Fermi–Dirac distribution can be approximated by Boltzmann distribution function.

$$f(E) \approx \exp[-(E - E_F)/kT] \qquad \text{and}$$

$$n = N_c \exp[-(E_c - E_F)/kT] \quad (4.21)$$

For p-type semiconductors, carrier concentration equation analogous to this is expressed as

$$p = N_v \exp[-(E_F - E_v)/kT] \quad (4.61)$$

For example, the calculated carrier concentration of a non-degenerate n-type semiconductor with $E_F - E_c = -2kT$, using Boltzmann approximation, is

$$n = N_c \exp(-2) = 0.13534 N_c$$

The Fermi–Dirac integral approach using $\eta = -2$ leads to $n = 0.12390 N_c$. The calculated electron concentrations using both approaches are quite comparable with a small error of ~4.7%. Now, if the Fermi level is moving closer to E_c, say $E_F - E_c = -kT$, the calculated carrier concentration error using Boltzmann approximation increases to 12%.

4.3.2 Free Carrier Concentrations in 2D Semiconductor Structures

For semiconductors with only 2D spatial freedom, e.g., in a quantum well, the carrier density is

$$n_{2D} = \int_{E_c}^{\infty} D_{2D}(E) f(E) \mathrm{d}E = \frac{m_e^*}{\pi \hbar^2} \int_{E_c}^{\infty} \frac{1}{1 + \exp[(E - E_F)/kT]} \mathrm{d}E \tag{4.62}$$

By replacing

$$\eta = \frac{E_F - E_c}{kT} \text{and } \xi = \frac{E - E_c}{kT} \tag{4.63}$$

the electron concentrations has the form of

$$n_{2D} = \frac{m_e^* kT}{\pi \hbar^2} \int_0^{\infty} \frac{1}{1 + \exp[\xi - \eta]} \mathrm{d}\xi \tag{4.64}$$

The integral is the Fermi–Dirac integral of zero order ($j = 0$)

$$F_0(\eta) = \frac{1}{\Gamma(1)} \int_0^{\infty} \frac{1}{1 + \exp[\xi - \eta]} \mathrm{d}\xi \tag{4.65}$$

where $\Gamma(1) = 1$ is the Gamma function. The Fermi–Dirac integral of zero order can be solved analytically and has a solution

$$F_0(\eta) = \ln\left[1 + \exp(\eta)\right] \tag{4.66}$$

Thus the 2D carrier density has a form of

$$n_{2D} = N_c^{2D} \ln\{1 + \exp[(E_F - E_c)/kT]\} \tag{4.67}$$

where $N_c^{2D} = m^* \, kT/\pi\hbar^2$. For a non-degenerate 2D system, the Fermi–Dirac distribution can be approximated by Boltzmann distribution function. Thus,

$$n_{2D} = N_c^{2D} \exp[(E_F - E_c)/kT] \tag{4.68}$$

In the highly degenerate 2D case, $E_F > E_c$, $\exp(n_{2D}/N_c^{2D}) \gg 1$ and

$$n_{2D} \approx N_c^{2D} \left(\frac{E_F - E_c}{kT} \right) \tag{4.69}$$

4.3.3 Carrier Concentration in the Multiple Valley Limit

When the energy differences between conduction band minimums along different crystallography directions are small, electron distribution is not limited to the lowest band. Depending on the energy separation between E_F and the conduction band minima in Γ, *L*, and *X*-directions, the electron distribution in each valley can be calculated. Let

$$\eta_i = \frac{E_F - E_c^i}{kT} \text{ for } i = \Gamma, L, \text{ and } X \text{ valley} \tag{4.70}$$

In a non-degenerate semiconductor the free carrier concentration including contributions from all valleys is

$$n = N_c^X \exp(\eta_X) + N_c^L \exp(\eta_L) + N_c^\Gamma \exp(\eta_\Gamma) \tag{4.71}$$

where $N_c^i = 2\left(\frac{2\pi m_{ei}^* kT}{h^2}\right)^{3/2} = 2.5 \times 10^{19}\left(\frac{m_{ei}^*}{m_0}\frac{T}{300}\right)^{3/2} (\text{cm}^{-3})$.

In the case of degenerate semiconductors, one obtains $n = \sum_i N_c^i F_{1/2}(\eta_i)$.

4.4 Surface States in Compound Semiconductors

In bulk compound semiconductor, the atomic bonds take the sp^3 configurations to form the zinc-blende crystal structure. A side view of the (110) surface along the [001] direction is shown in Fig. 4.19a where both group III and group V atoms are presented. When cleaving, the surface atoms take a relaxed configuration. In GaAs, due to their bond length difference and to minimize energy, on the (110) surface, the As bonds (~1.18 Å) move outward and the Ga bonds (~1.26 Å) move inward after the bonds are broken from the surface (Fig. 4.19b). To minimize their bond energies, the surface bonds of As and Ga atoms are taking s^2p^3 and sp^2 configurations, respectively, rather than the sp^3 configuration of the bulk bonds, and form 'dangling' surface bonds. These bonds are available for strong interactions with atoms and molecules of the ambient. The surface states are formed near the center of the forbidden gap as a result of interactions between the 'dangling' surface bonds and contamination

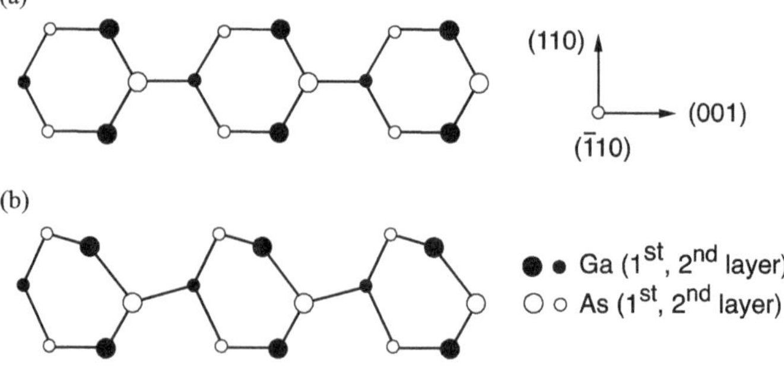

Fig. 4.19 **a** Bulk structure of the (110) surface, and **b** the relaxed configuration of (110) surface of a zinc-blende semiconductor

and surface imperfections. The surface imperfections are generated by deposited metal, which induces anti-site defects or vacancies near the interface. In silicon, the exposed surface atoms are the same and impose less severe surface problems. On polar (100) and (111) surfaces of III–V compounds, the surface atoms do not retain their ideal bulk structure, but undergo surface reconstruction. Depending on the surface condition, whether it is group III- or group V-rich, the surface can have different atomic arrangements that are more complex than on a (110) surface. For example, under As-rich conditions, the (100) GaAs surface atoms have periodicities twice and four times larger than the bulk atomic arrangement in the $[\bar{1}10]$ and [110] directions, respectively.

As a result of the surface states, the energy bands are modified near the surface. For an ideal metal–semiconductor interface, the Schottky barrier height (ϕ_b) is determined by the difference between the work function of the metal (ϕ_m) and the electron affinity of the semiconductor (χ_s) as shown in Fig. 4.20a.

$$\phi_b = \phi_m - \chi_s \tag{4.72}$$

In 1947, John Bardeen first recognized the importance of the semiconductor surface while studying metal–semiconductor contacts. He found that a surface space charge layer (surface states) is formed by the presence of surface defects and contaminations (Fig. 4.20b) and characterized by a mid-gap energy level ϕ_0. If the density of surface states is sufficiently high (Bardeen limit: $>10^{12}$ cm^{-2}), it can pin the surface Fermi level near the middle of the bandgap independent of the doping density. He further showed that if the density of the surface states is sufficiently high ($\geq 10^{13}$ cm^{-2}), the Schottky barrier height becomes a fixed value for each semiconductor independent of the metal work function.

$$\phi_b = E_g - \phi_0 \tag{4.73}$$

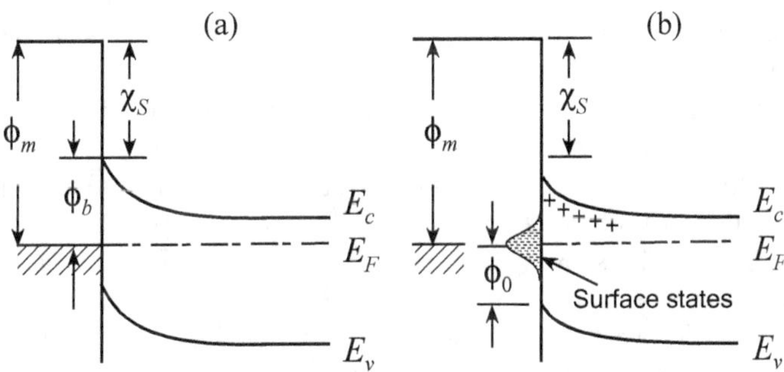

Fig. 4.20 Energy band diagram of **a** ideal Schottky barrier, and **b** metal–semiconductor barrier with high density of surface states. The Fermi level at the semiconductor surface is pinned near the mid-gap and represented by ϕ_0

Thus, the surface states play an important role in the physical properties of carrier transport near the semiconductor surface. Due to the nature of high surface state density, this observation has been verified in III–V compounds where the pinning of Schottky barrier heights is commonly observed. Because of this high density of surface states and the lack of robust native oxides to unpin the surface Fermi level, a high performance inversion-mode metal–oxide–semiconductor field-effect transistor (MOSFET) based on III–V compounds was not demonstrated until late 1990s. Using in situ electron beam evaporated $Ga_2O_3(Gd_2O_3)$ dielectric film on a clean as-grown GaAs in an ultra-high vacuum connected multiple chamber MBE growth system, MOS structures have been successfully fabricated on GaAs with a low interface trap density for the first time. The unpinning of the GaAs Fermi level results from the Gd_2O_3 restoring the surface As and Ga atoms to near-bulk charge. Later, high performance MOSFETs were demonstrated using ex situ atomic layer deposited (ALD) high-κ Al_2O_3 and HfO_2 films as the gate dielectrics on GaAs and other III–V materials. The metal alkyls used in ALD dielectric process, in particular, trimethyl aluminum for Al_2O_3 deposition, enable unpinning the Fermi levels on III–V semiconductors. A more detailed discussion of the development of III–V MOSFETs can be found in Chap. 9.

In semiconductors, the dangling-bond energy is typically located in the middle of the bandgap, and the interface trap density (D_{it}) increases exponentially in the energy ranges close to the band edges. Over the years, a variety of models of semiconductor surface pinning energy were developed. Among others, the charge neutrality level (CNL)-based model is more related to the band structures of III–V channel materials and offers a realistic explanation of all experimental results on III–V MOSFETs. The CNL energy level represents a weighted average value over the density of states (DOS). CNL is pushed away by the large DOS of the conduction and valence bands, as shown in Fig. 4.21. Therefore, the semiconductor surface pinning energy CNL is located inside the forbidden gap. If the Fermi level E_F is above CNL, the states are of acceptor type and negatively charged if the states are occupied. If the Fermi level E_F is below CNL, the states are of donor type and positively charged if the states are occupied. The CNL values above the valence band edges are determined by averaging values derived from various theoretical models as shown in Table 4.5. In general, the calculated CNL is in close agreement with Schottky barrier heights (ϕ_{bv}) on *p*-type III-V compounds.

In general, the surface pinning energies (ϕ_0) calculated from different models are in close agreement with Schottky barrier heights (ϕ_{bv}) on *p*-type III–V compounds (Table 4.5). Figure 4.22 shows the measured Fermi levels on cleaved (110) GaAs, GaSb, and InP surfaces deposited with different metals and overlayers in vacuum. Indeed, the Fermi level is pinned on these compounds and the surface pinning energy matches well with ϕ_{bv}. Note that in GaSb and InP, both *n*- and *p*-type materials, the pinning energies are closer to the valence band and conduction band, respectively.

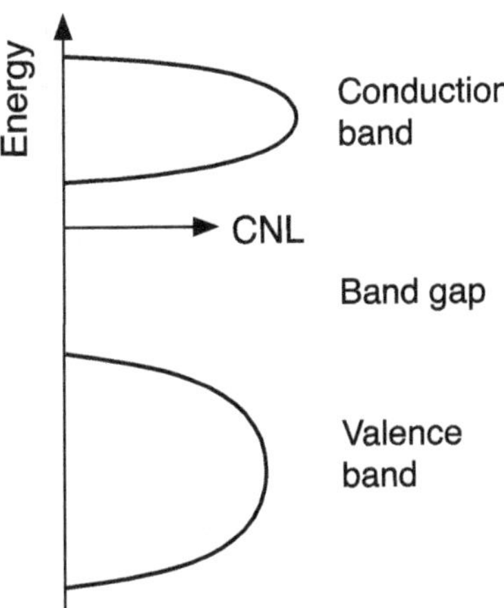

Fig. 4.21 CNL is a weighted average of DOS. A high DOS in the valence band tends to push the CNL toward the conduction band and vice versa. Reprinted with permission from [15], copyright AIP Publishing

Table 4.5 Schottky barrier height (p-type semiconductor) and charge neutrality level (CNL) data of selected III–V compounds [15]

Material	AlP	AlAs	AlSb	GaP	GaAs	GaSb	InP	InAs	InSb
ϕ_{bv} (eV)	1.27	1.002	0.47	0.797	0.562	0.07	0.857	0.58	0.04
CNL (eV)	1.3	0.92	0.4	0.8	0.55	0.06	0.6	0.50	0.15

This indicates that the nature and density of surface states are different for each material.

4.5 III–V Compound Semiconductors

4.5.1 *Lattice Constant*

For binary compounds, all physical parameters of the material are fixed. Thus, they have zero degrees of freedom. By mixing two binaries with a common element, one can form a ternary compound. Two group III or group V elements can share the same sublattice to form III–III′–V or III–V–V′ ternary alloy, respectively. Ternary compounds have one degree of freedom in selecting lattice constant or energy bandgap. The lattice constant, a_0, varies linearly with the alloy composition, following Vegard's law. Thus, the lattice constant of a ternary compound $A_xB_{1-x}C$ can be expressed as a linear combination of lattice constants of binaries AC and BC.

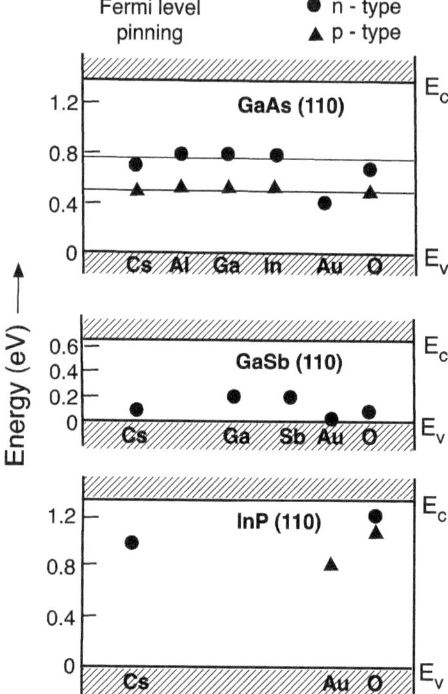

Fig. 4.22 Fermi pinning level positions for a range of overlayers on GaAs, GaSb, and InP (110) surfaces. Circles represent n-type and triangles represent p-type materials. Reprinted with permission from [16], copyright AIP Publishing

$$a_0(x) = xa_{\mathrm{AC}} + (1-x)a_{\mathrm{BC}} = a_{\mathrm{BC}} + (a_{\mathrm{AC}} - a_{\mathrm{BC}})x \tag{4.74}$$

where x is the composition fraction of binary AC and a_{MN} is the lattice constant of binary MN. The lattice constant and bandgap energy relationship of a number of III–V binary compounds are shown in Fig. 4.23.

Furthermore, the quaternary compounds have two degrees of freedom in selecting the lattice constant and energy bandgap independently. The quaternary region in Fig. 4.23 is bounded by either three or four ternaries. In the former case, a type-II quaternary $A_xB_yC_{1-x-y}D$ such as $Ga_xAl_yIn_{1-x-y}As$ is formed by three ternaries ABD, BCD, and ACD. Elements A, B, and C share the same sublattice sites as group III atoms, and D is a group V element. The lattice constant of this III–III′–III″–V quaternary follows Vegard's law as

$$a_0(x, y) = a_{\mathrm{AD}}x + a_{\mathrm{BD}}y + a_{\mathrm{CD}}(1-x-y) \tag{4.75}$$

The other quaternary compounds $AB_xC_yD_{1-x-y}$ formed by one group III element and three group V elements also fall into this category. It has a III–V–V′–V″ form

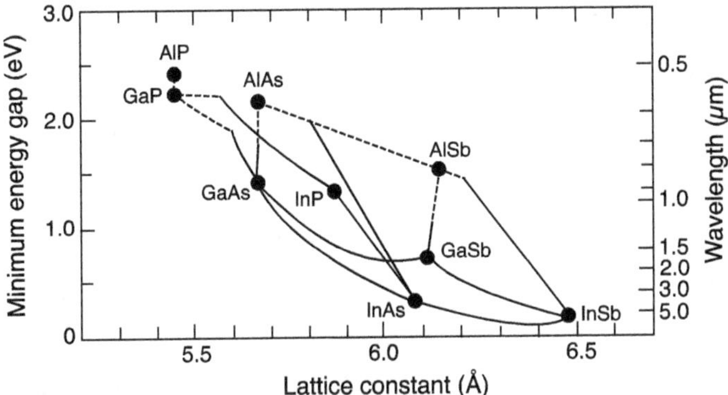

Fig. 4.23 Lattice constant as a function of bandgap energy of III–V compound semiconductors. The bandgap energy of ternaries follows the line connecting the constituent binaries. Solid and dashed lines indicate direct and indirect bandgap, respectively

and follows Vegard's law as

$$a_0(x, y) = a_{AB}x + a_{AC}y + a_{AD}(1 - x - y) \tag{4.76}$$

In the latter case, a type-I quaternary $A_xB_{1-x}C_yD_{1-y}$ is formed by four ternaries ABC, ABD, ACD, and BCD. The group III and group V sublattices are shared by A, B, and C, D, respectively, to form the III–III′–V–V′ quaternary. Its lattice constant is

$$a_0(x, y) = a_{AC}xy + a_{AD}x(1 - y) + a_{BC}(1 - x)y + a_{BD}(1 - x)(1 - y) \tag{4.77}$$

For example, the lattice constant of the $Ga_xIn_{1-x}As_yP_{1-y}$ alloy is given by

$$a(x, y) = 5.8687 - 0.4175x + 0.1896y + 0.0124xy$$

Figure 4.24 shows the linear variation of lattice constant with composition in $Ga_xIn_{1-x}As_yP_{1-y}$ alloy following Vegard's law.

In theory, one can mix three III–III′–V ternaries with the same group III elements but three different group V elements to form a penternary or quinternary alloy III–III′–V–V′–V″. The lattice constant and bandgap coverage do not extend beyond what one can achieve with quaternaries. Therefore, there have been few reported results on quinternary. In addition, there are physical limits that prevent the formation of compounds in certain composition ranges, under equilibrium growth conditions, where a miscibility gap exists. As shown in Fig. 4.8, the nearest-neighbor distance of alloys is not changed significantly. In ternary and quaternary compounds, the distance between group III and group V atoms only changes by <2% as determined by EXAFS technique. For two binaries having very different bond lengths, they

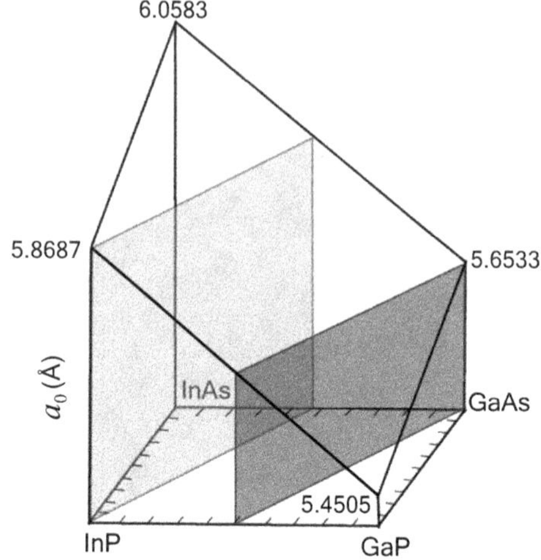

Fig. 4.24 Lattice constant of the $Ga_xIn_{1-x}As_yP_{1-y}$ quaternary alloy. The vertical axes at four corners indicate the lattice constants of binaries. The shaded rectangles indicate the composition range lattice-matched to InP and GaAs, respectively

will not form homogeneous ternary compounds over a certain composition range called the *miscibility gap*. Instead, the mixture contains multiple solid phases (or *phase separation*). However, homogeneous thin films can be achieved grown by non-equilibrium methods such as MBE and MOCVD.

4.5.2 Bandgap Energy

In ternary and quaternary compounds, the crystal structure is seldom completely ordered. The periodic lattice potential is broken due to the randomness in lattice atom distribution. Additionally, the bond length variation induces a local strain. Vegard's law, therefore, does not extend to some physical parameters including E_g, m^*, and μ in ternaries and quaternaries. First consider a ternary semiconductor $A_xB_{1-x}C$ which is an alloy of two semiconductor binaries i (BC) and j (AC). A ternary semiconductor parameter can be expressed in a general form as

$$T_{ij}(x) = (1-x)B_i + xB_j + x(1-x)C_{ij} \tag{4.78}$$

where $T_{ij}(x)$ is the value of the physical property of the ternary alloy composed of binaries i and j, the B_i and B_j are the material property of the given binaries, and C_{ij} is the ternary *bowing parameter* which relates to the deviation from a linear interpolation relationship between two binaries. Thus, the relationship between the bandgap and composition becomes

$$E_g = a + bx + Cx^2 \tag{4.79}$$

where C is the bowing parameter. The bowing parameter varies between 0.2 and 0.8 in III–V compounds.

The other feature in ternary alloys is that, when mixing a direct bandgap binary with an indirect bandgap binary, the lowest conduction bandgap crosses over from direct to indirect at a composition between the two binaries. The energy corresponding to the crossover composition is the highest energy that can be efficiently used for this compound as a light emitter. The bandgap energy as a function of composition in $Al_xGa_{1-x}As$ is shown in Fig. 4.25. This ternary alloy has a direct energy gap from GaAs ($x = 0$) up to $x \approx 0.4$ (40% of Al mole fraction). Beyond the crossover composition ($x > 0.4$), the alloy has a minimum energy gap in the X valley and the bandgap becomes indirect.

For determining the band parameters of quaternary alloy from those of the constituent ternary alloys, a number of approaches with different degree of uncertainties have been developed. A widely adopted method derived using interpolation procedure is shown first. For the type-I quaternary $A_xB_{1-x}C_yD_{1-y}$, the band parameter is expressed as a weighted sum of the related ternary values as

$$Q(x,y) = \frac{x(1-x)[yT_{ABC}(x) + (1-y)T_{ABD}(x)]}{x(1-x) + y(1-y)} + \frac{y(1-y)[xT_{ACD}(y) + (1-x)T_{BCD}(y)]}{x(1-x) + y(1-y)} \tag{4.80}$$

where T_{ij}'s represent the ternary semiconductor parameters following (4.78). For type-II quaternaries such as $AB_xC_yD_{1-x-y}$, the band parameter has the form of

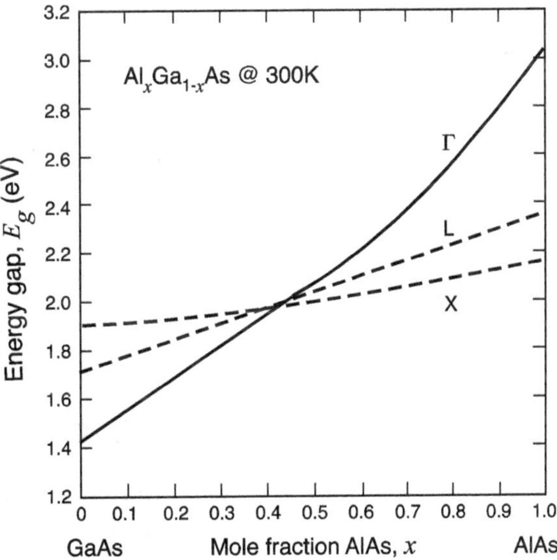

Fig. 4.25 Composition dependence of the direct energy gap Γ and the indirect energy gap X and L for $Al_xGa_{1-x}As$

$$Q(x, y) = xB_{AB} + yB_{AC} + (1 - x - y)B_{AD} - xyC_{ABC} - x(1 - x - y)C_{ABD} - y(1 - x - y)C_{ACD} \quad (4.81)$$

where B_{ij}'s represent the binary semiconductor parameters, and C's are ternary bowing parameters. Nevertheless, fitting calculated quaternary alloy parameters to experiments is necessary to reach a suitable description of a particular quaternary alloy system. For example, the bandgap of the most important quaternary $Ga_xIn_{1-x}As_yP_{1-y}$ as a function of alloy composition is listed as [17]

$$E_g(Ga_xIn_{1-x}As_yP_{1-y}) = 1.35 + 0.668x - 1.068y - 0.069xy + 0.758x^2 + 0.078y^2 - 0.322x^2y + 0.03xy^2 \quad (4.82)$$

The composition dependence of energy gap for a quaternary alloy is constructed from energy bandgap versus composition (E_g–x, y) relations of the four constituent ternaries. As an example, Fig. 4.26 shows a 3D composition–bandgap energy plot for the quaternary alloy $Ga_xIn_{1-x}As_yP_{1-y}$. On each sidewall is the E_g–x, y relation of one of the four ternary alloys. The four ternaries form the boundary of the quaternary energy gap surface. The base of the 3D plot gives the composition in terms of x and y. The intersection of the direct and indirect energy gap surfaces indicates that most of this system is in the direct bandgap region covering 0.36 to ~2 eV. For practical applications, the compositions lattice-matched to InP are of considerable interest because $Ga_xIn_{1-x}As_yP_{1-y}$ has an energy gap that covers from 0.74 to 1.35 eV. This allows the construction of a wide range of $Ga_xIn_{1-x}As_yP_{1-y}$/InP heterostructures for photonic as well as high-speed device applications.

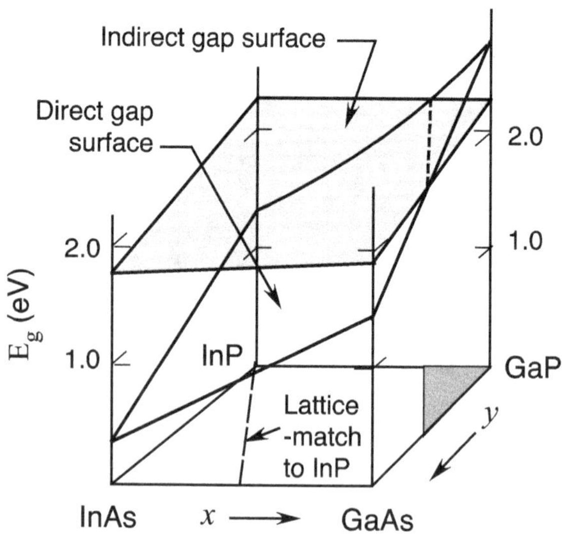

Fig. 4.26 Composition dependence of the bandgap energy surface of the quaternary $Ga_xIn_{1-x}As_yP_{1-y}$ alloy. Each sidewall represents the bandgap energy–composition relationship of a constituent ternary alloy. The square base shows the composition in terms of x and y

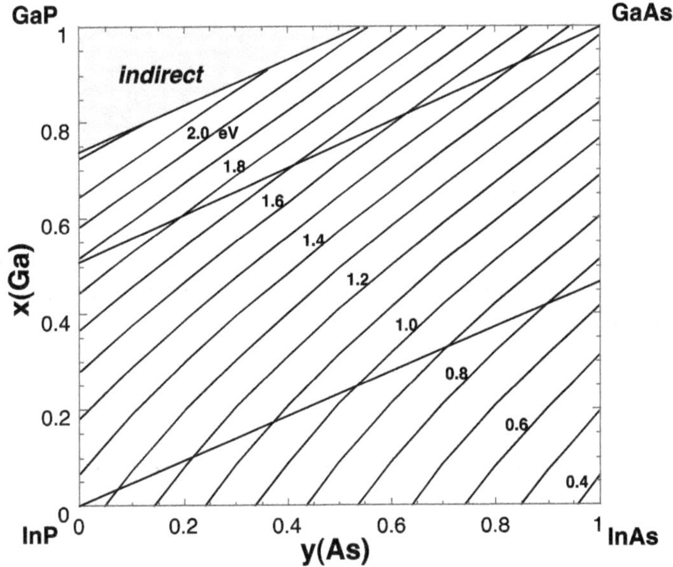

Fig. 4.27 x-y compositional plane for $Ga_xIn_{1-x}As_yP_{1-y}$. The x-y coordinate of any point in the plane gives the composition. The curved lines are constant direct bandgap energy values that were obtained by projection from the direct energy surface in Fig. 4.26. The composition lattices matched to InP and GaAs are shown as straight lines connected to InP and GaAs corners, respectively

From Fig. 4.26 we can construct the energy gap–lattice constant–composition diagram using the projection of the energy gap surface onto the x-y plane. This diagram is given in Fig. 4.27. The constant bandgap energy lines are plotted for a function of composition (x, y). The two straight lines originating from GaAs and InP corners represent the lattice-match compositions to GaAs and InP, respectively.

The interpolations used above by requiring only the boundary binary and ternary values satisfy a necessary but not sufficient condition for describing the quaternary alloy property inside those boundaries. Thus, it is necessary to determine the quaternary bowing parameter for a better description of the quaternary function $Q(x, y)$. A multivariable quadratic interpolation algorithm has been developed to identity the intrinsic quaternary bowing parameter for modeling the compositional dependence of quaternary alloy bandgaps [18].

For a quaternary alloy $A_xB_{1-x}C_yD_{1-y}$, the material parameter function $Q(x, y)$ can be plotted as a surface domain enclosed by four boundaries similar to the composition dependence of the bandgap energy surface shown in Fig. 4.26. The four corners of the surface are defined by four binary alloys, and their material parameters are B_{AC}, B_{AD}, B_{BC}, and B_{BD}. Following (4.78), the ternary parameter expressions are

$$T_{\mathrm{ABV}} = xB_{\mathrm{AV}} + (1-x)B_{\mathrm{BV}} + x(1-x)C_{\mathrm{ABV}},\ \mathrm{V} = \mathrm{C, D} \tag{4.83a}$$

$$T_{\mathrm{IIICD}} = yB_{\mathrm{IIIC}} + (1-y)B_{\mathrm{IIID}} + y(1-y)C_{\mathrm{IIICD}},\ \mathrm{III} = \mathrm{A, B} \tag{4.83b}$$

where $C_{\text{III-V}}$ is the alloy bowing parameter. The quaternary parameter can be expressed in two paths with either x mixing or y mixing:

$$Q_{\text{CD}} = yT_{\text{ABC}}(x) + (1-y)T_{\text{ABD}}(x) + y(1-y)D_{\text{CD}}(x), \tag{4.84a}$$

$$Q_{\text{AB}} = xT_{\text{ACD}}(y) + (1-x)T_{\text{BCD}}(y) + x(1-x)D_{\text{AB}}(y). \tag{4.84b}$$

Requiring the quaternary function $Q(x, y)$ to be unique inside the quaternary surface domain gives $Q_{\text{AB}} = Q_{\text{CD}}$. D is the quaternary surface bowing parameter and can be estimated as

$$D_{\text{CD}} = xC_{\text{ACD}} + (1-x)C_{\text{BCD}} + x(1-x)D_1, \tag{4.85a}$$

$$D_{\text{AB}} = yC_{\text{ABC}} + (1-y)C_{\text{ABD}} + y(1-y)D_2, \text{ and} \tag{4.85b}$$

$$D_1 = D_2 = D.$$

Therefore, the quaternary material parameter function $Q(x, y)$ can be expressed as

$$\begin{aligned} Q(x, y) = {} & y(1-x)B_{\text{BC}} + xyB_{\text{AC}} + (1-y)xB_{\text{AD}} + (1-x)(1-y)B_{\text{BD}} \\ & + x(1-x)(1-y)C_{\text{ABD}} + x(1-x)yC_{\text{ABC}} + (1-x)y(1-y)C_{\text{BCD}} \\ & + xy(1-y)C_{\text{ACD}} + x(1-x)y(1-y)D \end{aligned} \tag{4.86}$$

Letting $B_{\text{BC}} = B_1, B_{\text{AC}} = B_2, B_{\text{AD}} = B_3, B_{\text{BD}} = B_4$, and $C_{\text{ABC}} = C_{12}, C_{\text{BCD}} = C_{14}$, $C_{\text{ACD}} = C_{23}$, $C_{\text{ABD}} = C_{34}$, the above equation can be compacted into the following form:

$$Q(x, y) = \begin{bmatrix} y & y(1-y) & (1-y) \end{bmatrix} \begin{bmatrix} B_1 & C_{12} & B_2 \\ C_{14} & D & C_{23} \\ B_4 & C_{34} & B_3 \end{bmatrix} \begin{bmatrix} (1-x) \\ x(1-x) \\ x \end{bmatrix} \tag{4.87}$$

By matching experimental bandgap values of a quaternary alloy with calculations, the quaternary bowing parameter D is identified and listed in Table 4.6. This equation is valid for both type-I and type-II quaternary alloys. For example, the composition of $\text{Ga}_x\text{In}_{1-x}\text{As}_y\text{P}_{1-y}$ type-I alloy (III–III′–V–V′) can be expressed as $[B_{1y}B_{4(1-y)}]_{(1-x)}[B_{2y}B_{3(1-y)}]_x$ with $B_1 = \text{InAs}$, $B_2 = \text{GaAs}$, $B_3 = \text{GaP}$, and $B_4 = \text{InP}$. For type-II quaternary alloys with the form of III–III′–III″–V, one binary can be assigned twice in the same row or column of the alloy matrix with a zero ternary bowing parameter (i.e., $B_2 = B_3$, $C_{23} = 0$). The composition of $\text{Al}_x(\text{Ga}_y\text{In}_{1-y})_{1-x}\text{P}$ type-II alloy is then given by $[B_{1y}B_{4(1-y)}]_{(1-x)}B_{2x}$. The necessary bandgap and bowing parameters of binary and ternary alloys as well as the quaternary bowing parameters needed in (4.87) for calculating III–V quaternary semiconductor alloy bandgaps are

Table 4.6 Bowing parameters of ternary and quaternary III–V semiconductor alloys

AlP	0.48 (–0.38) 0.19	InP	–0.65 (–0.2) 0	GaP	0 (–0.13) –	AlP	$D_{AlInGaP} = -0.10$ (–0.07)
–0.22 (–0.22) –	0	–0.1 (–0.27) –0.16	–0.11	–0.19 (–0.24) –	0	–0.22 (–0.22) –	
AlAs	–0.7 (0) –0.15	InAs	–0.477 (–1.4) –0.15	GaAs	0.127–1.31x (–0.055) 0	AlAs	$D_{AlInGaAs} = 0.03$
–0.8 (–0.28) –0.15	–2.5	–0.67 (–0.6) –1.2	–0.01	–1.43 (–1.2) –0.6	0	–0.8 (–0.28) –0.15	
AlSb	–0.43 (0) –0.25	InSb	–0.415 (–0.33) –0.1	GaSb	–0.044 (0) –0.3	AlSb	$D_{AlInGaSb} = -2.5$
–2.7 (–2.7) –	0	–1.9 (–0.19) –0.75	0	–2.7 (–1.7) –	0	–2.7 (–2.7) –	
AlP	0.48 (–0.38) 0.19	InP	–0.65 (–0.2) 0	GaP	0 (–0.13) –	AlP	$D_{AlInGaP} = -0.10$ (–0.07)

The numbers between two binary alloys are: First row—the bowing parameter of the direct bandgap ternary alloy (C_{ij}), second row—the numbers in the parentheses are the ternary bowing parameters of the indirect bandgap, and third row—the ternary bowing parameter of spin-orbit splitting Δ_{so}. Numbers in the shaded boxes are bowing parameters of quaternaries (D) associated with four neighboring binary alloys that form type-I quaternary alloys. Bowing parameters of type-II quaternaries associated with three constituent binary alloys in the same row are shown in the last column [18, 19]

listed in Table 4.6. As examples, the composition dependence of bandgap energy diagrams of both type-I and type-II quaternary alloys are calculated and shown in Fig. 4.28.

4.6 III–N and Dilute III–V–N Compound Semiconductors

4.6.1 III–N Compounds

AlN, GaN, InN, and their alloys can crystallize in both wurtzite and zinc-blende lattice structures. Most of the current substrate materials used to grow these compounds have wurtzite structure, which enhances the formation of the same lattice structure. Therefore, all III–N compounds considered here have a wurtzite crystal structure and can be described by a- and c-axis lattice constants. However, under certain conditions, it is possible to grow zinc-blende structure GaN crystals. Due to the lack of suitable substrates, wurtzite III–N epitaxial layers have been grown on sapphires (Al_2O_3) and SiC along the [0001] direction. Usually, the resulting epilayers contain high density of threading dislocations ($\geq 10^9$ cm^{-2}) extending from the hetero-interface toward

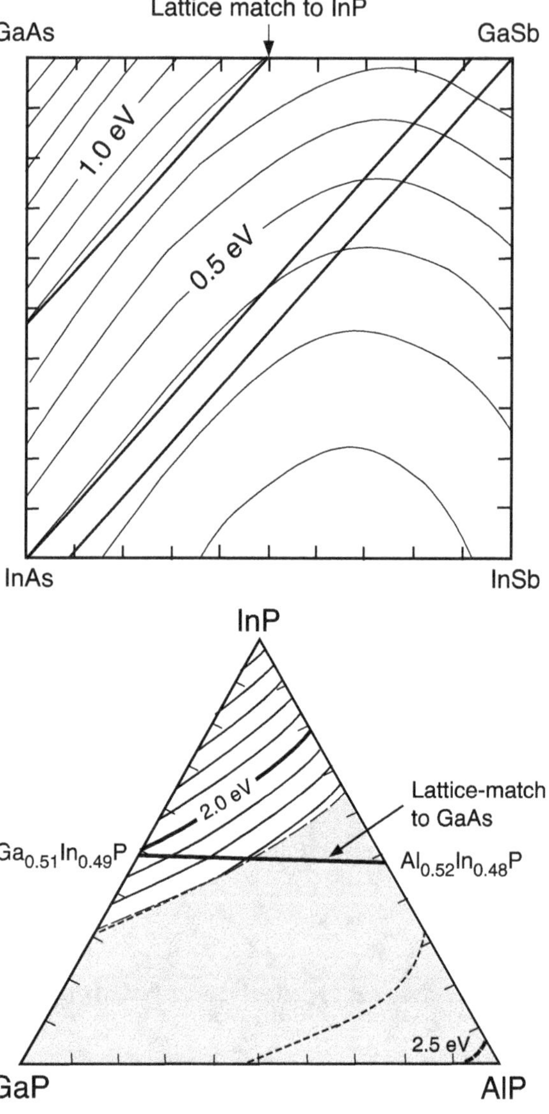

Fig. 4.28 Examples of calculated composition dependence of bandgap energy diagrams of type-I ($Ga_xIn_{1-x}As_ySb_{1-y}$, top) and type-II ($Al_xGa_yIn_{1-x-y}P$, bottom) III–V quaternary alloys. Available lattice-matched substrates are shown in thick straight lines, and the shaded area indicates the region with indirect bandgap. Reprinted with permission from [18], copyright AIP Publishing

the surface. Therefore, there has been a continuous effort to develop large area GaN substrate technology. The wurtzite structure GaN consists of two closely spaced hexagonal lattices as described in Sect. 2.4, one formed by gallium atoms and the other by nitrogen atoms. It has ionicity about 0.5 and shows more ionic properties than covalent properties. Thus, one can simply say that an electron is transferred from an N atom to a Ga atom in the wurtzite GaN crystal to form a Ga^- anion and N^+ cation. Looking along the *c*-axis, as shown in Fig. 4.29, reveals that the (0001) planes

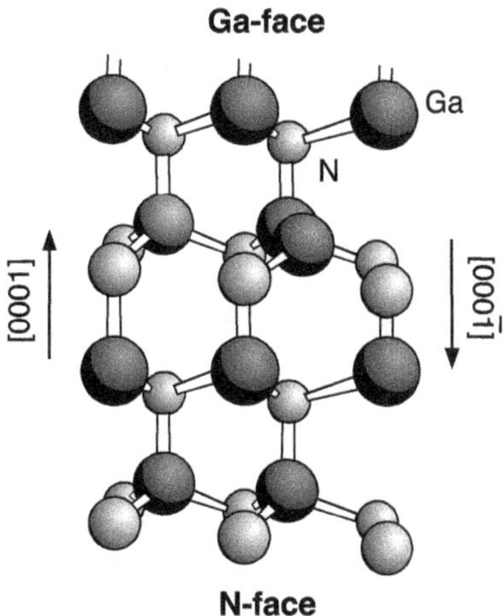

Fig. 4.29 Crystal structure of wurtzite GaN along [0001] direction. Reprinted with permission from [20], copyright AIP Publishing

are alternating anion layers and cation layers with the Ga-plane and N-plane terminating (0001) and $(000\bar{1})$ surfaces, respectively. The ionic bonds linking neighboring (0001) planes with opposing charges generate a spontaneous polarization (P_{SP}) in the direction of $[000\bar{1}]$. In addition, we also notice a large lattice-mismatch between GaN and AlN ($\Delta a/a = 2.4\%$, $\Delta c/c = 3.9\%$). When coupled with the large piezoelectric constants (~1 versus 0.01 in III–V's), a strong strain-induced polarization or piezoelectric polarization (P_{PE}) forms in III-N heterostructures. The direction of the polarization follows dilation or contraction of the *c*-axis. The existence of significant spontaneous and piezoelectric polarizations in III-N heterostructures leads to unique electronic and photonic properties.

III-N layer structures are currently grown using either MBE or MOCVD. These epitaxy techniques will be described in more detail later in Chap. 5. In MBE, molecular beams of Ga and Al are generated from effusion cells using elemental Ga and Al as source materials. Due to the large bond strength of N_2, it is impractical to thermally dissociate N_2 into atomic N. Instead, in plasma-assisted MBE (PAMBE), an RF plasma source is generally used to generate atomic nitrogen and N^+ for III-N growth. Ammonia (NH_3) has also been used as a nitrogen source by direct thermal dissociation on the heated substrate surface at a temperature of ~800 °C. MBE layers are usually grown at ~800 °C directly on (0001) sapphire or *c*-Al_2O_3 under Ga-rich surface conditions to achieve a high layer quality. This particular MBE growth condition yields N-terminated B-face or N-face layers, i.e., the GaN growth direction is along $[000\bar{1}]$ capped with a top nitrogen plane. Polarity reversal into Ga-terminated A-face or Ga-face layer is achievable by inserting a thin AlN layer into the structure.

For III-N layers grown by MOCVD, metalorganics such as trimethylaluminum (TMA) and triethylgallium (TEG) are used to deliver Ga and Al, respectively, and ammonia is used as the nitrogen source. In MOCVD process, III-N materials are grown at much higher temperatures (~1050 °C) than in MBE. To maintain material quality, very high nitrogen flow inside the MOCVD growth chamber is required. In

Table 4.7 Band structure parameters for wurtzite III-N binaries [21, 22]

Parameters	GaN	AlN	InN
a (Å)	3.189	3.112	3.54
c (Å)	5.185	4.982	5.70
E_g^{Γ} (eV) @ 300 K	3.438	6.138	0.64

addition, a low-temperature grown Al(Ga)N nucleation layer on (0001) sapphire is needed before the growth of GaN. The nitrogen-rich growth condition for the nucleation layer yields a Ga-terminated A-face or Ga-face layer and a growth direction along [0001] capped with a top gallium plane. This growth direction is opposite to the MBE grown nitrides and causes a sign change in spontaneous polarization.

The composition dependences of the energy gaps for the ternary alloys AlGaN, GaInN, and AlInN follow the usual quadratic form, similar to other III–V alloys, of (4.79).

$$E_g(A_xB_{1-x}D) = xE_g(AD) + (1-x)E_g(BD) + x(1-x)C \tag{4.88}$$

where C is the bowing parameter. The bowing parameter is always negative for these alloys, which reflects a reduction of the alloy energy gaps. The lattice constants, both in (0001) plane (a) and along c-axis (c), and energy gaps of the III-N binaries are listed in Table 4.7. Using bowing parameters of −0.7, −1.4, and −2.5 in AlGaN, GaInN, and AlInN, respectively, along with bandgap energies of binaries, the composition dependence of the direct energy gap on the lattice constant in (0001) plane (a_0) for these wurtzite structure III-N ternary alloys is plotted in Fig. 4.30.

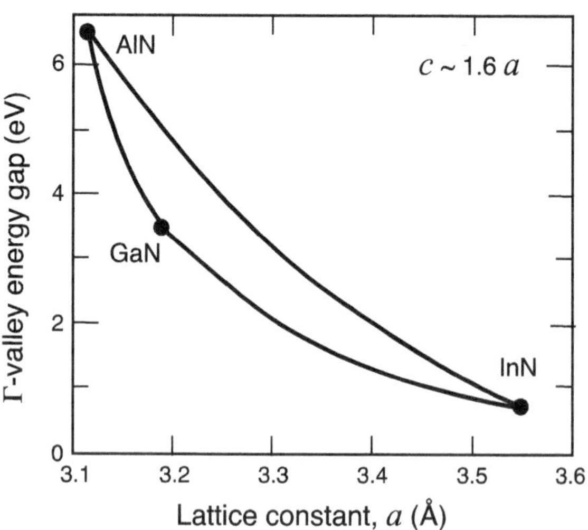

Fig. 4.30 Composition, in terms of lattice constant in the (0001) plane (a), dependence of the direct energy gap Γ for AlGaN, GaInN and AlInN alloys

One interesting property about GaN is the doping behavior. GaN can be doped easily with shallow donors of Si and Ge above 10^{19} cm^{-3}. The Si donor ionization energy is about 12–17 meV. However, doping GaN with acceptors to obtain a high concentration of holes was a difficult problem until the late 1980s. It was discovered that *p*-type doping had been limited by hydrogen passivation of acceptors. To activate hydrogen passivated Mg acceptors, the Mg-doped GaN requires a low-temperature (~300 °C) heat treatment in the form of low-energy electron beam irradiation treatment or thermal annealing in vacuum or in nitrogen atmosphere to dissociate hydrogen atoms which form complexes with Mg atoms. One drawback of Mg doping in GaN is its high acceptor ionization energy of ~150 meV which leads to a low doping efficiency. Currently, hole concentrations of mid 10^{17} cm^{-3} and mid 10^{18} cm^{-3} are achievable in MOCVD and MBE grown layers, respectively.

4.6.2 Dilute III-V-N Compounds

The composition dependence of the energy gap for the III–V ternary alloys usually does not follow Vegard's law due to the randomness in lattice atom distribution and bond length variation. The deviation from a linear interpolation between the two binaries is accounted for by the bowing parameter. When mixing III–V compounds with a small fraction (usually ≤ 2–3%) of nitrogen in dilute nitride compounds such as $GaAs_xN_{1-x}$ and GaP_xN_{1-x}, a very large bowing in bandgap energy exists (Fig. 4.31). This unusual behavior is probably due to the extremely large bond length difference between N and other group V elements (Table 4.1) and/or the unique isoelectronic property of N in III–V alloys. Isoelectronic centers are formed by substituting one atom of the crystal with another atom of the same valence but with large differences in electronegativity and bond length. The isoelectronic center was first observed in GaP:N where a localized potential well is formed around the nitrogen atom. This potential well allows a nitrogen atom to capture an electron, which in turn can bind a hole by Coulomb attraction, thus forming a bound exciton. A more detailed discussion of isoelectronic traps can be found in Chap. 8.

It is the interaction between the spatially localized N isoelectronic level and the conduction band of the underlying non-nitride semiconductor that leads to a splitting of the conduction band into two subbands and a large reduction of the fundamental bandgap. According to the band anti-crossing (BAC) model, if the effect on the valence bands is completely neglected, the dispersion relations for the two coupled subbands are expressed as

$$E_{\pm}(k) = \frac{1}{2}\left\{\left[E^{C}(k) + E^{N}\right] \pm \sqrt{\left[E^{C}(k) - E^{N}\right]^{2} + 4V^{2}x}\right\} \tag{4.89}$$

where $E^{C}(k)$ is the conduction band dispersion of the unperturbed non-nitride semiconductor, E^{N} is the position of the nitrogen isoelectronic impurity level in that

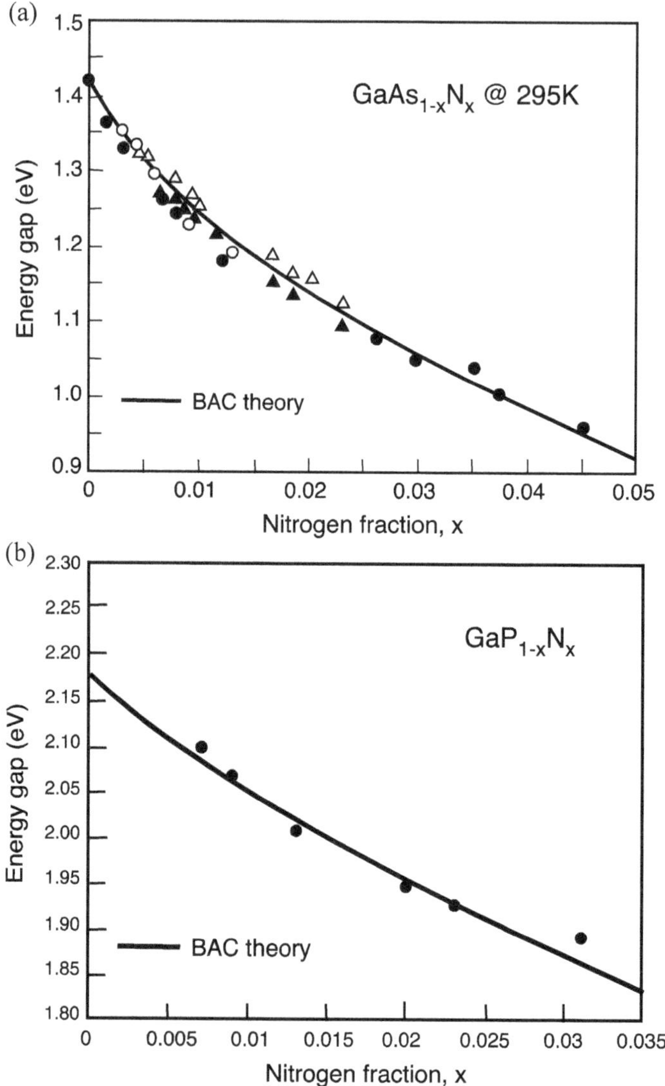

Fig. 4.31 **a** Energy bandgap of $GaAs_{1-x}N_x$ as a function of nitrogen fraction x. **b** BAC model calculated indirect bandgap energy of $GaP_{1-x}N_x$ along with experimental data. Reprinted with permission from [23], copyright IOP

semiconductor, V is the interaction potential between the two bands, and x is the fraction of nitrogen in the alloy. The values of E^N and V for the dilute III–V–N quaternary alloy $Ga_{1-x}In_xAsN$ are [1.65(1–x) + 1.44x – 0.38x(1–x)] and [2.7(1–x) + 2.0x – 3.5x(1–x)], respectively. The values of E^N and V are reduced to 1.65 and 2.7, respectively, for GaAsN. As an example, the anti-crossing characteristics of the

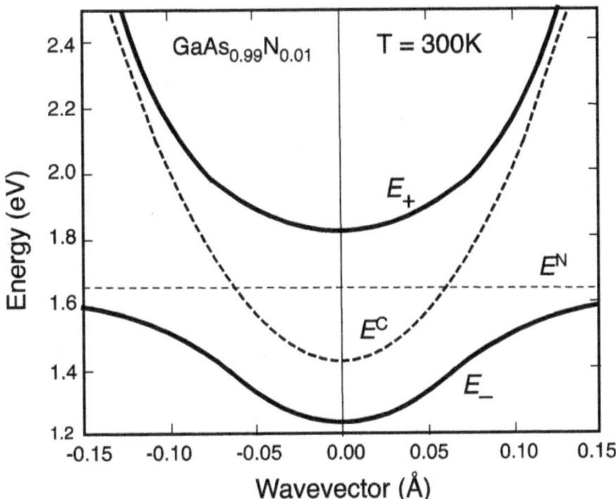

Fig. 4.32 Conduction band dispersion relations for $GaAs_{0.99}N_{0.01}$ at room temperature from the BAC model (solid curves). The unperturbed GaAs conduction band (dashed curve) and the position of the N level (thin dashed line) are also shown. Reprinted with permission from [21], copyright AIP Publishing

dispersion relations for the two coupled conduction bands in $GaAs_{0.99}N_{0.01}$ are shown in Fig. 4.32. The fundamental energy gap is now governed by the transition from E_- to the top of the valence band. One precaution is that the BAC model considers only the interaction between a single, spatially localized nitrogen level and the conduction band of the underlying non-nitride semiconductor. It neglects not only mixing with the *L* and *X* valleys, but also nitrogen pairs and clusters. Therefore, the predictive power of the BAC model for an indirect bandgap material, e.g., GaP_xN_{1-x}, is more limited.

Problems

1. The diatomic chain model considered for phonon (lattice vibration) characteristics have identical springs but different masses. A model with alternating spring constants, α and β, but the same mass, *m*, is appropriate for Si, Ge or diamond crystal. Calculate the ω-k dispersion relation for crystals with a diamond structure.
2. Determine what fraction of holes in Si and InP are heavy holes. Let $m^*_{hh}(\text{Si}) = 0.537$, $m^*_{lh}(\text{Si}) = 0.153m_0$, $m^*_{hh}(\text{InP}) = 0.56m_0$, $m^*_{lh}(\text{InP}) = 0.12m_0$.
3. For Ge: $m^*_L = 1.59m_0$, $m^*_T = 0.0823m_0$, $m^*_{hh} = 0.28m_0$, $m^*_{lh} = 0.043m_0$; and for GaAs: $m^*_e = 0.067m_0$, $m^*_{hh} = 0.50m_0$, $m^*_{lh} = 0.076m_0$.

   (a) Determine the density of states effective mass m_{DOS} of Ge.
   (b) Calculate N_c and N_v of Ge at 300 K.
   (c) Determine what fraction of holes in Ge and GaAs are heavy holes.
   (d) Comment on the room-temperature mobility difference between Ge (μ_e ~ 3900, μ_h ~ 1800 cm^2/Vs) and GaAs (μ_e ~ 9000, μ_h ~ 400 cm^2/Vs).

4. The energy differences $E_c - E_F \geq 3kT$ and $E_F - E_v \geq 3kT$ are defined as thresholds for non-degenerate *n*-type and *p*-type semiconductors, respectively.

(a) Determine the n- and p-type carrier densities for Si and GaAs at the threshold of non-degeneracy. Commenting the differences between n- and p-type materials as well as between Si and GaAs.
(b) Calculate n- and p-type carrier densities of Si and GaAs at 300 K for $E_c - E_F = 0.2kT$ and $E_F - E_v = 0.2kT$, respectively.

5. For a linear extrapolation of the density of states effective mass between GaAs and AlAs, the electron effective mass for the indicated conduction band in $Al_xGa_{1-x}As$ is taken as

$$m^{\Gamma} = (0.067 + 0.083x)m_0$$
$$m^{L} = (0.55 + 0.12x)m_0$$
$$m^{X} = (0.85 - 0.07x)m_0.$$

The energy bandgap variations in the X-, L-, and Γ-conduction bands are also provided as follows:

$$E^{\Gamma}(\text{eV}) = 1.424 + 1.247x, \text{ for } 0 \le x \le 0.4$$
$$E^{L}(\text{eV}) = 1.707 + 0.645x,$$
$$E^{X}(\text{eV}) = 1.899 + 0.21x + 0.055x^2.$$

Determine the fraction of the electrons is in the X-, L-, and Γ-conduction band valleys of non-degenerately doped n-type (a). $Al_{0.3}Ga_{0.7}As$, and (b). $Al_{0.5}Ga_{0.5}As$.

6. Electron–hole pairs (EHPs) are generated by photoexcitation of an undoped $Ga_{0.47}In_{0.53}As$ layer. The increasing EHP concentration causes the quasi-Fermi level of electrons and holes to move away from the equilibrium Fermi level toward the conduction band and valence band, respectively. One special condition, under a strong photoexcitation, which is important for lasers, occurs when the separation in quasi-Fermi levels $(F_c - F_v) = E_g$, a condition known as 'transparency'. At what electron concentration ($n = p$) does the material reach transparency? You have to use Fermi–Dirac integral for both electrons and holes. For $Ga_{0.47}In_{0.53}As$: $m_e^* = 0.041m_0$, $m_{hh}^* = 0.465m_0$, and $m_{lh}^* = 0.0503m_0$. Note, at transparency, $\eta - \zeta = 0$, where

$$\eta = (F_c - E_c)/kT \text{and } \zeta = \chi + \eta = (F_v - E_v)/kT$$

7. For Ge: $m_L^* = 1.59m_0$ and $m_T^* = 0.0823m_0$; for Si: $m_L^* = 0.9163m_0$ and $m_T^* = 0.1905m_0$; and for GaAs: $m_e^* = 0.067m_0$. The measured room-temperature electron mobility of n-type Ge, Si, and GaAs samples doped to 2×10^{17} cm^{-3} are 2750, 600, and 4000 cm^2/V-s, respectively.

(a) Calculate the average relaxation time, τ (scattering probability $= 1/\tau$), for Ge, Si, and GaAs.

(b) Assume, at room temperature, the dominant scattering mechanisms in elemental semiconductors are due to lattice scatterings. The scattering due to optical phonons for Ge can be approximated as

$$\mu_{\text{op}} \cong 900(300/T)\left[\exp\left(\hbar\omega_{\text{op}}/kT\right) - 1\right]$$

The LO phonon frequency, ω_{op}, for Ge is 6 THz. Calculate the lattice scattering mobility, μ_l

(c) Although Ge and GaAs have a similar average relaxation time, the mobility difference is quite large. Comment on the room-temperature mobility difference between Ge and GaAs.

8. The ternary $GaAs_xSb_{1-x}$ is direct bandgap over the full composition range. It has a bandgap energy of 0.775 eV at $x = 50\%$ where it is lattice-matched to InP. The room-temperature bandgap energy and lattice constant of GaAs and GaSb are 1.424 eV and 5.6533 Å, and 0.727 eV and 6.0959 Å, respectively.

   (a) Find the coefficients *a*, *b*, and *c* of the bandgap energy quadratic equation.
   (b) Sketch E_g versus *x*.

9. The ternary $InAs_xSb_{1-x}$ is direct bandgap over the full composition range. It has the smallest bandgap among all III–V compounds at $x = 35\%$. The corresponding emission wavelength at that composition is 11.5 μm. The room-temperature bandgap energy and lattice constant of InAs and InSb are 0.35 eV and 6.0584 Å, and 0.17 eV and 6.4794 Å, respectively.

   (a) Find the coefficients *a*, *b*, and *c* of the bandgap energy quadratic equation.
   (b) Sketch E_g versus *x* and indicate on the sketch the bandgaps when the ternary is lattice-matched to AlSb (6.1355Å) and GaSb (6.09593Å).

10. Derive the quaternary direct bandgap energy as a function of compositions *x* and *y* of the quaternary alloy $Ga_xIn_{1-x}Sb_yAs_{1-y}$ using (4.87). Make a sketch of the E_g versus compositions *x* and *y*. For the iso-energy curves, show only the curve with $E_g = 0.6$ eV, which corresponds to an eye-safe wavelength of ~2.1 μm. Estimate the compositions lattice-matched to InAs and GaSb on this curve.

11. From the E_g versus composition sketch of the $Ga_xIn_{1-x}As_yP_{1-y}$ quaternary alloy (Fig. 4.27):

    (a) Find the range of bandgap energies covered by the lattice-matched alloys on InP and GaAs. (The lattice constant of GaP is 5.45117 Å.)
    (b) Find the *x* and *y* values for the alloy lattice-matched to $Ga_{0.75}In_{0.25}As$ and has bandgap energy of 1.4 eV.
    (c) Derive the quaternary direct bandgap energy as a function of compositions *x* and *y* using (4.87). Compare the results at $x = y = 0.5$ with that derived from (4.82).

12. The alloy $Al_xGa_yIn_{1-x-y}As$ is a type-II quaternary alloy.
    (a) Derive the quaternary direct bandgap energy as a function of compositions x and y using (4.87).
    (b) Make a sketch of the E_g versus composition x and y. Mark clearly the boundary of the direct–indirect crossover and show the compositions lattice-matched to InP, which has a lattice constant of 5.86875 Å. The lattice constants of AlAs, GaAs, and InAs are 5.6605 Å, 5.65325 Å, and 6.0584 Å, respectively.
    (c) What is the range of the bandgap energy covered by this quaternary alloy lattice-matched to InP? At which compositions does it have the bandgaps corresponding to 1.3 and 1.55 μm?

13. In an effort to achieve efficient luminescence from group IV semiconductors, the Ge/α-Sn alloy has been investigated. The lattice constants of α-Sn and Ge are 6.48 Å and 5.6575 Å, respectively. The calculated transition energies along Γ, L, X valleys of the Ge_xSn_{1-x} compound are shown below. Deduce the bandgap energy as a function of composition x for the Γ valley and determine the composition range where this alloy is a semiconductor with direct bandgap energy. Note the material becomes semimetal at $E_g \leq 0$ eV.

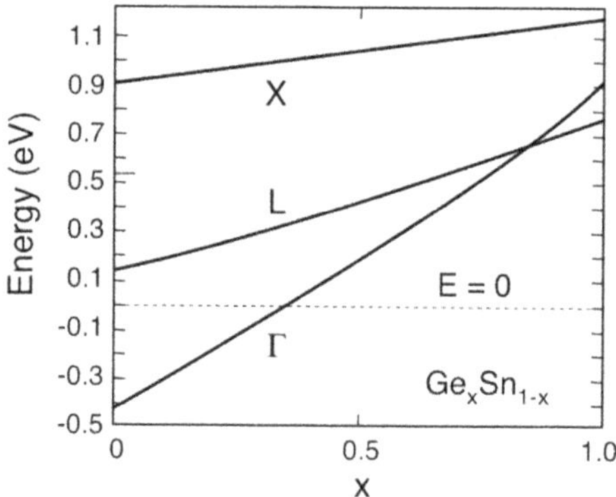

References

1. A.I. Blum, N.P. Mokrovski, A.R. Ragel. Izvest. Akad. Nauk S.S.S.R., Ser. Fiz. **16**, 139 (1952)
2. H. Welker, Z. Naturforsch. **79**, 744 (1952)
3. J.L.T. Waugh, G. Dolling, Phys. Rev. **132**, 2410 (1963)
4. B.N. Brockhouse, Phys. Rev. Lett. **2**, 256 (1959)
5. C.M. Wolfe, G.E. Stillman, W.T. Lindley, J. Appl. Phys. **41**, 3088 (1970)
6. J.C. Mikkelsen Jr., J.B. Boyce, Phys. Rev. **B28**, 7130 (1983)
7. J.D. Oliver Jr., L.F. Eastman, P.D. Kirchner, W.J. Schaff, J. Cryst. Growth **54**, 64 (1981)
8. P.M. Smith, M. Inoue, J. Fre, Appl. Phys. Lett. **37**, 797 (1980)
9. L. W. James. J. P. van Dyke, F. Herman, D. M. Chang, Phys. Rev. B **1**, 3998 (1970)

10. A. Baldareschi, N.O. Lipari, Phys. Rev. **B9**, 1525 (1974)
11. D.J. Chadi, K.J. Chang, Phys. Rev. **B39**, 10063 (1989)
12. D.V. Lang, R.A. Logan, M. Jaros, Phys. Rev. **B19**, 1015 (1979)
13. E.F. Schubert, K. Ploog, Phys. Rev. **B30**, 7021 (1984)
14. D.E. Holmes, R.T. Chen, K.R. Elliot, C.G. Kirkpatrick, Appl. Phys. Lett. **40**, 46 (1982)
15. J. Robertson, B. Falabretti, J. Appl. Phys. **100**, 014111 (2006)
16. W.E. Spicer, I. Lindau, P. Skeath, C.Y. Su, J. Vac. Sci. Technol. **17**, 1019 (1980)
17. E. Kuphal, J. Crystaj Growth **67**, 441 (1984)
18. G.P. Donati, R. Kaspi, K.J. Malloy, J. App. Phys. **94**, 5814 (2003)
19. I. Vurgaftman, J.R. Meyer, L.R. Tam-Mohan, J. App. Phys. **89**, 5815 (2001)
20. O. Ambacher, J. Smart, J.R. Shealy, N.G. Weimann, K. Chu, M. Murphy, W.J. Schaff, L.F. Eastman, R. Dimitrov, L. Wittmer, M. Stutzmann, W. Rieger, J. Hilsenbeck, J. Appl. Phys. **85**, 3222 (1999)
21. I. Vurgaftman, J.R. Meyer, J. App. Phys. **94**, 3675 (2003)
22. J. Wu, W. Walukiewicz, W. Shan, K.M. Yu, J.W. Ager III, S.X. Li, E.E. Haller, H. Lu, W.J. Schaff, J. Appl. Phys. **94**, 4457 (2003)
23. J. Wu, W. Shan, W. Walukiewicz, Semicond. Sci. Technol. **17**, 860 (2002)

Further Reading

1. S. Wang, *Fundamentals of Semiconductor Theory and Device Physics*(Prentice-Hall, 1989)
2. S. Adachi, *Physical Properties of III–V Semiconductor Compounds* (Wiley, 1992)
3. E. F. Schubert, *Doping in III-V Semiconductors* (Cambridge, 1993)

Chapter 5
Material Technologies

Abstract High-quality, single-crystal substrates are essential for the advancement of semiconductor device technologies. The significant development of III-V compound semiconductor electronic and photonic devices in the last several decades has been a result of the progress made in the preparation of bulk single crystals and in the various epitaxial techniques for growing thin layers. Significant improvements have been made in uniformity, reproducibility, dislocation density control, thermal stability, diameter control, and impurity and dopant control. As discussed in this chapter, single crystals of GaAs and InP have been grown from the melt by one of several techniques that include horizontal Bridgman, vertical gradient freeze (VGF), and liquid encapsulated Czochralski (LEC). Nevertheless, due to problems of the very high vapor pressure of nitrogen at the growth temperature, bulk crystals of III-nitrides are still under development. Among epitaxial growth techniques, molecular beam epitaxy (MBE) and metal organic chemical vapor deposition (MOCVD) have evolved from laboratory experiment tools into viable manufacturing technologies. Hybrid epitaxial techniques such as gas source and metalorganic MBE have also been successfully applied for the growth of many GaAs, InP-based heterostructures. In addition, post-growth material processing techniques to modify the as-grown quantum wells and superlattices into new alloys with desired physical properties have been introduced. Using selective impurity diffusion or ion implantation, the impurity-induced layer disordering (IILD) process allows the formation of lateral heterojunctions perpendicular to quantum well and superlattice heterostructures such that devices with three-dimensional carrier and photon confinement are achieved. The other post-growth method is to wet oxidize Al-bearing III-V alloys into insulators, either on the surface or between semiconductor layers, through lateral oxidation from an edge. The insulating layer provides current confinement as well as optical confinement due to a significant refractive index difference created within the heterostructure.

K. Y. Cheng, *III–V Compound Semiconductors and Devices*, Graduate Texts in Physics,
https://doi.org/10.1007/978-3-030-51903-2_5

5.1 Growth of Bulk Crystals

5.1.1 Phase Equilibria

In bulk crystal growth by liquid–solid reactions, an important consideration in the equilibrium is the dependence of melting temperature on atomic composition and pressure. Under one atmosphere (1 atm) pressure, the equilibrium of solid with a liquid of binary compound can be described by the temperature-composition phase diagram as shown in Fig. 5.1. The temperature at which the solid and liquid have the same composition is the congruent melting point (T_F). Ideally, it happens at a composition of 50% A and 50% B atoms. However, the solid at the congruent melting point does not necessarily have the ideal composition of 50%. Rather, it covers a range of composition around 50% which defines the extent of the solid phase and is called the *existence region.* In GaAs, the congruent melting point occurs slightly to the Ga-rich side of the stoichiometric composition. Crystals grown with compositions in the existence region different from the 50% value lead to solids that contain non-stoichiometric defects such as vacancies, interstitials, and antisite defects. Therefore it is necessary to suppress the density of these native defects by carefully controlling the stoichiometry of the melt during crystal growth.

The main difficulty of growing III-V compound semiconductor crystals is that many of them decompose at the melting point, releasing the volatile component. The vapor pressure at the melting point is determined by the vapor pressure of the more volatile group V element rather than the group III element, whose vapor pressure is

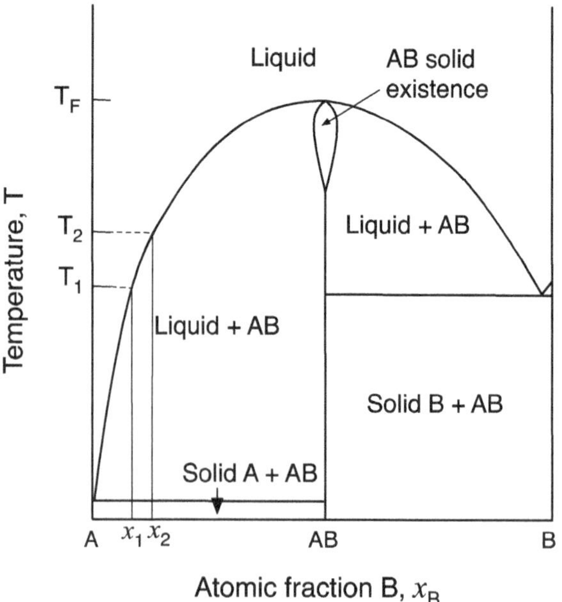

Fig. 5.1 Temperature-composition phase diagram of the III-V binary compound $A^{III}B^{V}$. T_F is the congruent melting point of the compound AB at which the solid and liquid have the same composition. The existence region is greatly exaggerated

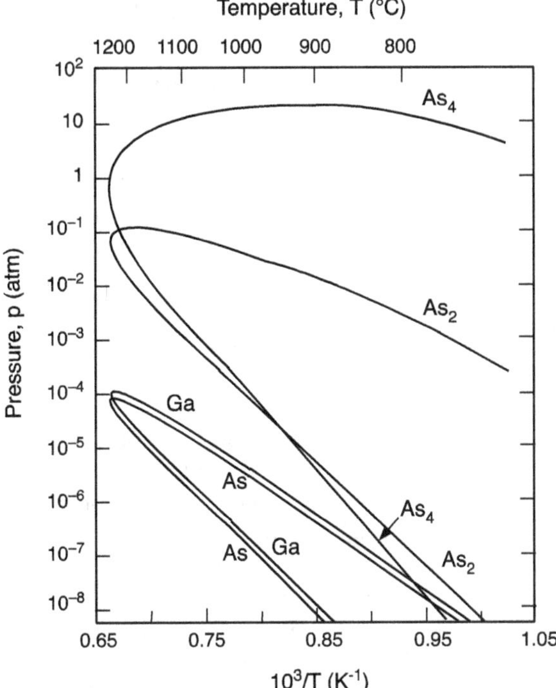

Fig. 5.2 Partial pressures p of As_4, As_2, As, and Ga along the Ga-As liquidus as a function of the reciprocal temperature. Reprinted with permission from [1], copyright Elsevier

lower by several orders of magnitude. Thus growth must take place under a vapor pressure of the volatile group V component to prevent depletion of the group V component from the melt and maintain a stoichiometric composition in the melt during crystal growth. Figure 5.2 shows the partial pressures of As monomers (As), dimers (As_2), tetramers (As_4), and Ga in equilibrium with the Ga-As liquidus and the solid as a function of reciprocal temperature. Near the melting point, the vapor pressure of As (mainly As_4) in equilibrium with GaAs (melting point 1238 °C) is ~0.9 atm, while the vapor pressure of P in equilibrium with InP (melting point 1062 °C) is ~25 atm. These high pressures call for special attentions to prevent As or P depletion from the melt during crystal growth. The growth chamber should also be properly designed to withstand high pressures used during growth. For GaN, the vapor pressure of N at the melting temperature (~2500 °C) is extremely high (P_M ~ 4.5×10^4 atm). For pressure values below P_M, GaN does not melt but decomposes. Thus the standard crystal growth techniques generally used for other III-V alloys (Bridgman, VGF, LEC) cannot be adopted.

5.1.2 Crystal Growth Techniques

(a) Crystal growth by pulling from a crucible

One of the most important techniques for the growth of semiconductor single crystals from the melt is the Czochralski pulling method. Figure 5.3 shows the basic components of a crystal puller according to the Czochralski technique. Here, the crucible contains the melt of the semiconductor material. The seeded single crystal is held in the seed holder which can be moved up and down and rotated. A furnace or an RF coil is used for heating the crucible and the melt. At the beginning of the pulling process, the seed is dipped into the melt, whose temperature is lowered until a small amount of crystalline material is solidified. The seed is then withdrawn from the melt and rotated to maintain thermal geometry and cylindrical geometry. The melt temperature is lowered slowly, and the diameter of the crystal increases. Once the desired diameter is reached, the lowering of the temperature is stopped. Growth at a constant diameter is maintained until the desired length is grown. The usual pull rates range from 1 to 10 mm/h. In this process the growth rate is determined primarily by the rate of the heat been carried away through the crystal and radiation to the surroundings from the solid-liquid interface.

The Czochralski technique may readily be used to grow semiconductor crystals including Si, Ge, and III-V compounds. However, the arsenides and phosphides of indium and gallium have a relatively high vapor pressure at their moderately high melting point. Under normal conditions, these compounds decompose on melting.

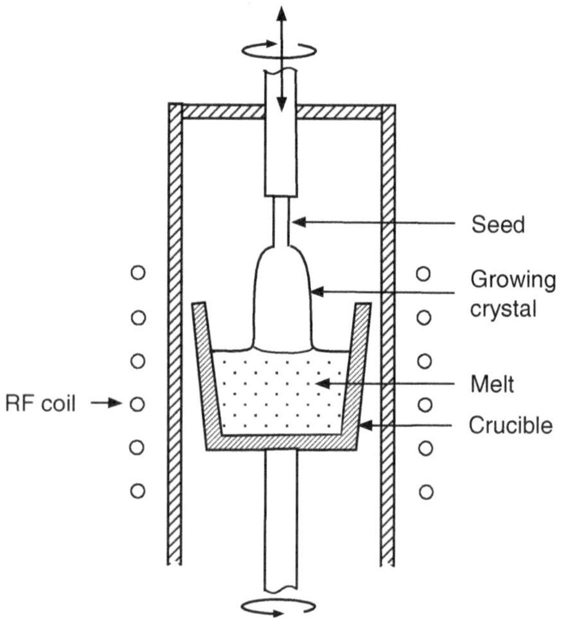

Fig. 5.3 Czochralski crystal growth furnace with rf induction heating

The result is an In- or Ga-rich melt and As- or P-rich vapor phase. A modified Czochralski technique to prevent loss of group V element from the melt is the encapsulation technique where the melt is covered by a thin layer of low vapor pressure molten material that is less dense. B_2O_3 satisfies this requirement and is the commonly used encapsulant. The density of B_2O_3 is 1.5 g-cm^{-3} while those of GaAs and InP melts are 5.71 and 5.1 g-cm^{-3}, respectively. It softens and begins to flow at 450 °C but is immiscible with the melts of III-V compounds. It has a low vapor pressure of only ~0.1 torr at the melting of GaAs. In addition, the growth system is housed inside a pressure chamber such that a counter pressure of an inert gas, which is higher than the partial pressure of the group V element, can be maintained on top of the B_2O_3. This technique is called the liquid encapsulated Czochralski (LEC) technique. LEC furnaces that can be used up to pressures of 100 atm are available commercially.

The LEC growth can be done by preparing the compound ex situ first and carrying out the crystal growth subsequently. Alternatively, the compound is synthesized in situ prior to growth in the same system. The second approach is called the high-pressure (HP) LEC technique due to the high pressure associated with the in situ synthesis process. The direct synthesis of GaAs occurs at about 700 °C under a high-purity nitrogen or argon pressure of 60 atm. The starting materials for the synthesis are stoichiometric quantities of high-purity Ga and As. A pellet of B_2O_3 is placed on top of the charge to encapsulate the melt when molten. High-purity pyrolytic BN (PBN) crucibles are used to grow consistently pure materials. The use of PBN crucibles has become standard in the growth of undoped semi-insulating GaAs crystals. After the compound is synthesized, the temperature is increased above the melting point of GaAs, and the pressure is reduced to 2–4 atm. Single-crystal growth then proceeds as in the conventional Czochralski technique. In the growth of InP single crystals by LEC, the compounding is done ex situ due to the extremely high pressure required.

The III-nitride compounds yield further technology problems. The melting points and vapor pressures of them are extremely high. Therefore, it is difficult to apply the Czochralski technique, or even a modification of this technique such as LEC, to these compounds.

(b) **Horizontal gradient-freeze and Bridgman methods**

The gradient-freeze technique is analogous to the horizontal Bridgman method in that they both produce single crystals by gradual solidification of a molten ingot. The two methods differ only in that solidification by the gradient-freeze technique takes place by the movement of a temperature gradient along the ingot, making the entire system stationary. The horizontal Bridgman method requires the movement of either the furnace or the crucible holding the molten ingot. Figure 5.4 is the schematic of the horizontal gradient furnace used for growing GaAs, InP, and GaSb along with the temperature profile. The growth system consists of a two-zone furnace: the low-temperature zone on the left is used to control the vapor pressure of the volatile component, which is necessary to control the stoichiometry. The high-temperature zone on the right is maintained at a temperature in excess of the melting point of the

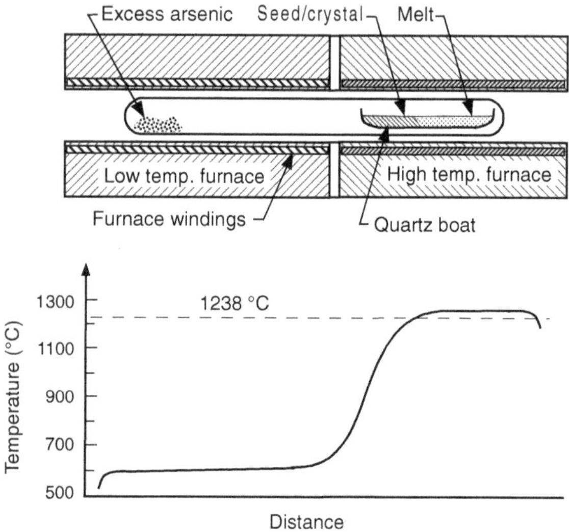

Fig. 5.4 Horizontal gradient-freeze furnace and a typical temperature profile

compound in such a manner as to provide the desired temperature gradient along the length of the melt. In this technique, a sealed quartz tube contains high-purity (6–9's pure) group III metal in a quartz boat at one end and excess high-purity volatile group V element (6–9's pure) at the other end. A porous quartz plug is placed between the two zones and acts as a radiation shield and diffusion barrier. The quartz boat containing group III metal and the seed is located in the high-temperature zone of the furnace with the group III metal above the melting point of the compound. The group V vapor reacts with group III metal to form the III-V melt. The temperature at the high-temperature zone is adjusted to melt a small part of the seed when the contact to the melt is made initially. Solidification is achieved by slowly lowering the temperature of the high-temperature zone from an equilibrium state in which the material is completely molten above the melting point, to a state in which the material is completely solid below the melting point. The temperature gradient along the length of the melt causes the freezing point to traverse from the seed end of the melt to the other. The solidified crystal has to be cooled to room temperature at a slow and even rate to reduce thermal stress induced dislocation generation.

In these horizontal growth techniques, the shape of the crystal is constrained by the geometry of the quartz boat. The cross section of the crystals is typically D-shaped and seeded in the ⟨111⟩ direction. The (100) oriented circular wafers can be sawed from these crystals but with excessive material loss. In general, significantly lower thermal gradients, and hence reduced convective flows, in the melt and extremely low thermal stresses can be realized in the horizontal technique compared to the Czochralski technique. As a result, owing to the reduced thermal stresses and precise control of group V element overpressure, dislocation densities tend to be much less in these crystals. Typical dislocation densities in GaAs crystals grown by the horizontal Bridgman or the horizontal gradient-freeze technique are below 5000 cm^{-2} in 7.5 cm wide D-shaped crystals.

(c) Vertical gradient-freeze (VGF) technique

Today the most common process for large-scale production of round InP and semi-insulating GaAs single crystals has been LEC. But the LEC technique has several inherent problems. One of the most serious drawbacks with the LEC process is the large axial-temperature gradient needed to maintain diameter control. The resulting thermoelastic stresses in the crystal have been shown to exceed the low yield strength of these compound semiconductors and create a high level of dislocations. Typical dislocation densities for undoped InP and GaAs range from 10^4 to 10^5 cm^{-2}. A second drawback of the LEC process is that the melt is coolest at the top surface. This leads to considerable buoyancy-driven convection that creates non-homogeneities in the crystal dopant and incorporates point defects. These drawbacks of the LEC growth of GaAs and InP have been overcome with the vertical gradient-freeze (VGF) technique where growth in low axial- and radial-temperature gradients is combined with the diameter control imposed by a crucible.

The VGF growth apparatus consists of a large pressure vessel that contains the heater assembly, a PBN crucible contains the seed in a well at the bottom and polycrystalline III-V compound above it, and a group V source reservoir located at the coldest point of the growth vessel. The arrangement of the crucible in the vertical furnace and the temperature profile are shown schematically in Fig. 5.5. For growth to occur, the pressure vessel is filled with inert argon gas first to the desired pressure, and the group V heater temperature is raised to increase the group V vapor pressure inside the PBN growth vessel. When the operating group V pressure is established, the polycrystalline III-V compound charge is melted. To begin solidification, the temperature gradient is controlled such that the liquid–solid interface sweeps upward through the melt. Crystallization starts at the seed in the bottom of

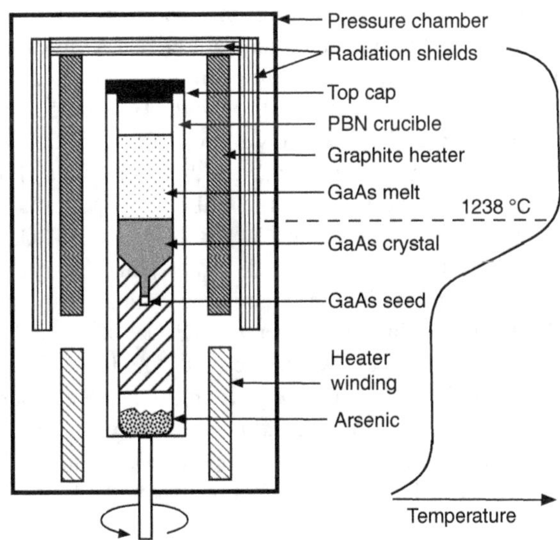

Fig. 5.5 Vertical gradient-freeze (VGF) growth system and a typical temperature profile

the crucible. Because the crucible is cooler at the bottom than at the top, the system is thermally stabilized against convection. During the growth, the group V element from the bottom of the growth vessel sublimates to maintain the appropriate partial pressure and stoichiometry of the melt. After solidification, controlled cooling brings the crystal to room temperature at a rate that minimizes thermal stress and optimizes point defect quenching.

Since the VGF system has axial-thermal symmetry, low axial- and radial-temperature gradients, and slow stable solidification and cooling, the produced crystals contain low levels of thermal stress and high radial uniformity of the electrical properties across wafers. These benefits translate into low dislocation densities of $\leq$100 to 3000 cm^{-2} in crystals that are 7.5 cm in diameter. This dislocation density is considerably lower than in LEC-grown crystals.

(d) **Growth of bulk GaN crystals**

At present, gallium-nitride-based photonic and electronic devices are manufactured mainly by hetero-epitaxial methods on foreign substrates such as sapphire, silicon, and silicon carbide which lead to III-nitride epilayers with high stress, high dislocation density, and mosaic crystal structure. Due to the extremely high vapor pressure of nitrogen at the melting temperature of GaN, the standard bulk crystal growth techniques generally used for other III-V alloys cannot be applied. Currently, the main source of GaN substrates is produced by the hydride vapor-phase epitaxy (HVPE) technique on sapphire substrates. However, this method suffers from all the drawbacks of using heterogeneous substrates on which the growth is initiated: the GaN layer obtained is highly stressed and bowed. Therefore, many promising growth methods have been explored for preparing GaN bulk crystals including the high nitrogen pressure (HNP) solution growth technique, the Na flux method, and the ammonothermal growth technique [2]. In the HNP method, the decomposition of GaN is inhibited by the use of nitrogen under high pressure. The growth of single crystals is performed in molten gallium and requires temperatures of about 1500 °C and very high nitrogen pressures in the order of 1.5×10^4 atm. High quality GaN crystals have been achieved through the HNP process with a low dislocation density (<200 cm^{-2}) and very narrow x-ray diffraction peaks (20–30 arcsec). However, the extreme growth conditions used in this process may limit the future development of a production process on a large scale.

In the process of using Na flux to grow bulk GaN, it consists of the thermal decomposition of sodium azide (NaN_3) and the reaction between created nitrogen with gallium. Sodium is regarded as a catalyst that enables N_2 dissociation. The process takes place at temperatures 300–800 °C and pressure ranging from atmospheric up to 50 atm, which are rather moderate. Thus, it might be possible to use it industrially. However, the sodium used in the process is very reactive and must be handled in controlled conditions. Moreover, the factors like expensive solvents, low yield (one crystal per growth process), and limited scalability may hamper mass production of bulk GaN crystals by this method.

The approach of the ammonothermal technique came from the field of hydrothermal technology of α-quartz mass production. Hydrothermal synthesis is the use of aqueous solvents under high temperature and high pressure to dissolve and recrystallize materials that are relatively insoluble under ordinary conditions. To help increase the solubility of quartz to be synthesized, *mineralizers* (NaOH) are added to the aqueous solvent. Relatively fine particles of α-quartz *nutrient* are placed in the bottom or dissolving zone of the reaction vessel or autoclave, and suitably oriented single-crystal seed plates of α-quartz are suspended in the upper crystallizing zone. A perforated metal disk separates the two zones into two nearly isothermal regions with the lower dissolving zone hotter than the upper growth zone. As the temperature of the autoclave approaches the operation conditions, the nutrient quartz begins to dissolve and saturate the solution. The autoclave is cooler at the top, causing the solution there to become supersaturated. The seed plates begin to grow as the supersaturated solution deposits the solid phase.

The ammonothermal method is an analogue of hydrothermal one, where ammonia instead of water is used as a solvent and nitrides can be grown instead of the oxides. During the crystal growth process, the GaN nutrient is placed in the dissolving zone of the high-pressure autoclave, then transported by convection in the temperature gradient to the crystallizing zone, where GaN is grown on native seeds due to the supersaturation of the solution (Fig. 5.6). In addition, the use of the mineralizers such as sodium or potassium amide (MNH_2, M=Na, K) is necessary in order to enhance

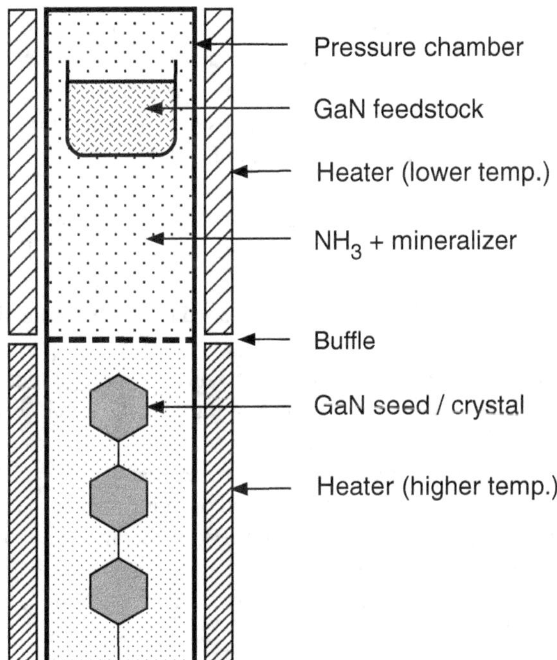

Fig. 5.6 Schematic of the ammonothermal system for GaN bulk crystal growth. The GaN nutrient is located in the (upper) low-temperature dissolving zone, and the crystal growth is taking place in the (lower) high-temperature crystallization zone of the autoclave. Reprinted with permission from [3], copyright Elsevier

the solubility of GaN in ammonia. The typical temperatures and pressures used are 400–600 °C and 1–3 × 10^3 atm, respectively—conditions similar to those used industry in the α-quartz crystal growth. However, there is one significant difference between hydrothermal and ammonothermal methods: under the constant pressure, the solubility of GaN is significantly higher at lower temperatures than at higher temperatures. In hydrothermal method, the solubility of α-quartz is proportional to the temperature. As a consequence, the chemical transport of GaN is directed from the low-temperature dissolving zone to the high-temperature crystallization zone which has to be located below the dissolving zone of the autoclave. Currently, the laboratory-scale ammonothermal crystal growth has allowed the preparation of large (>1.5″ diameter) GaN crystals with narrow x-ray diffraction peak (16 arcsec) and low dislocation density ($<10^3$ cm^{-2}). The success of further development of this promising technique in preparation of GaN crystals with appropriate size and crystalline quality relies on whether a cost-effective production method on an industrial scale can be developed.

5.2 Epitaxy

Epitaxy is a process whereby an oriented crystalline material, usually called the epitaxial layer, is deposited as an extension onto an existing oriented crystal, usually referred to as the substrate. Epitaxial growth can be further divided into two categories: homoepitaxy and hetero-epitaxy. Homoepitaxy is the growth of a layer essentially identical to its substrate as in GaAs on GaAs. In hetero-epitaxy, the epitaxial layer differs chemically from its parent substrate, as in $Al_xGa_{1-x}As$ on GaAs, provided both crystals have the same lattice constant. Lattice-mismatched hetero-epitaxy is also possible in the case where the epitaxial layer is very thin so that defects will not be generated.

Since the early 1960s, there has been strong interest in growing single-crystalline multilayer compound semiconductor structures called heterostructures for device applications. Various deposition techniques such as LPE (Liquid Phase Epitaxy), VPE, and MBE have been developed to fulfill the needs of ever complicated device structures. The process of LPE involves the precipitation of materials from a supersaturated solution onto an underlying substrate under near-equilibrium conditions. The VPE process involves near-equilibrium gas-phase chemical reactions of gaseous sources followed by surface reactions before incorporation into the epitaxial layer. However, in an alternative VPE process using metalorganic sources, called MOCVD, reactions are far from thermodynamic equilibrium. The MBE technique is also a thermodynamically non-equilibrium process. It involves the reaction of thermal beams of constituent atoms or molecules from solid sources with a crystalline substrate under ultra-high-vacuum conditions. MBE using gaseous sources has also been developed. The attributes of these major epitaxy techniques are compared in Table 5.1.

Table 5.1 Summary of the characteristics of epitaxial methods used for the growth of III-V compound semiconductors

Technique	Strength	Weakness
LPE	Simple and low cost apparatus Excellent material quality	Morphology and uniformity problems Small scale Graded interface Difficult to grow Al-In compounds
Trichloride VPE	Simple apparatus High-purity materials	No Al alloys Difficult to grow alloys
Hydride VPE	Large-scale production system	No Al alloys Complex reactor design Use of toxic gases
MOCVD	Highly flexible Abrupt interfaces	Expensive growth system Expensive reactants Use of toxic sources
MBE	Simple process High uniformity Abrupt interfaces In situ monitoring and control	Expensive growth system

5.3 Liquid-Phase Epitaxy

Liquid-phase epitaxy was first demonstrated by H. Nelson in 1963 to fabricate GaAs homojunction diodes from molten metal solutions. Since then, LPE technology has been used successfully to fabricate various types of electronic and optical devices using compound semiconductor materials. Basically, LPE involves the growth of an epitaxial layer from a supersaturated solution on a single-crystal substrate that has a crystal structure and lattice constant similar to those of the growing layer. The supersaturated solution is brought in contact with the substrate for a desired period of time. By regulating the degree of supersaturation in the solution and the contacting time between the solution and the substrate, the amount of material that precipitates onto the substrate can be controlled precisely.

The thermodynamic basis of LPE can be illustrated by the generalized binary (AB) phase diagram shown in Fig. 5.1, where A and B are the group III element and group V element, respectively. In general, the growth solution is rich in one of the major components of the epitaxial layer and dilute in all others. For III-V compound semiconductors, it is usually possible to use the group III metal (i.e., Ga and In) as the solvent for the group V elements such as As, P, and Sb. For example, based on the fact that the solubility of As (element B) in Ga-rich (element A) solutions decreases with decreasing temperature, one can use a Ga-rich solution saturated with As for LPE growth of GaAs. At temperature T_2, the Ga solution is saturated with x_2 atomic percent of As. Cooling the solution to T_1, where x_1 is the corresponding equilibrium atomic percentage of As in the solution, creates a driving force for the precipitation of a congruent compound AB (i.e., GaAs) until the new saturation condition is reached. At the proper growth condition, some of the precipitates may be deposited as an epitaxial layer on a GaAs substrate that is in contact with this solution.

5.3.1 LPE Apparatus

The LPE process requires the use of experimental apparatus that permits growth solutions of desired compositions to be placed in contact with the substrate for periods under controlled temperature cycles. Three different growth techniques are used in LPE: the tipping technique, in which solution-substrate contact is achieved by tipping the furnace; the dipping technique, in which the substrate is dipped vertically into the solution; and the sliding technique, in which the substrate is slid horizontally into contact with the solution. Tipping and dipping systems are comparatively simple and easy to operate. High-quality single epitaxial layers have been produced by both methods. However, such systems become inadequate for the growth of multilayer heterostructures required for many modern devices.

The sliding technique, which allows multilayer growth, has become the principle LPE method. Figure 5.7 illustrates a typical horizontal sliding LPE system. The main components of the apparatus are a graphite multibin-boat with a slider insert, a fused silica tube to provide a protective atmosphere, and a multiple zone resistance heating furnace. The graphite boat has a number of reservoirs, each of which contains a saturated solution corresponding to the epilayer to be grown. The desired electrical conductivity of epilayers can be controlled by adding impurities into the specific growth solutions. Typical *n*- and *p*-type dopants for LPE growth of III-V compounds are from column VI (Te and Se) and column II (Zn) of the periodic table, respectively. The group IV elements (Si, Ge, and Sn) are amphoteric dopants in III-V compounds. Its incorporation and the resulting electrical conductivity depend on the growth conditions. The substrate is placed in the recess of the graphite slider, which can be brought into contact with each solution in turn by sliding under different

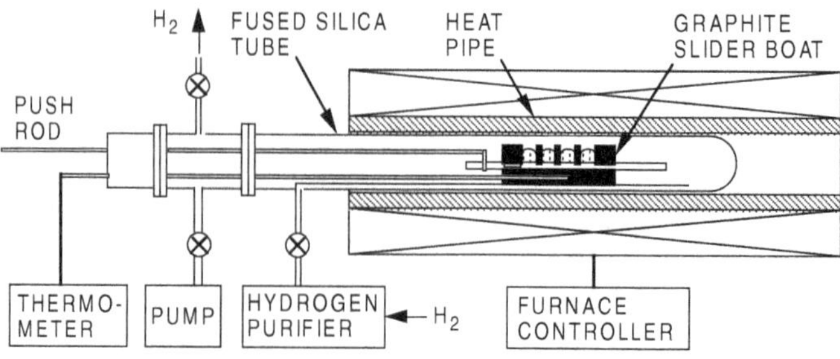

Fig. 5.7 Schematic diagram of a sliding liquid-phase epitaxy system

reservoirs. In this way, multilayer *p-n* junctions and heterostructures with desired compositions and thicknesses can be grown successfully on the substrate, and this operation can be automated easily. In order to ensure a clean wipe-off of the melts and to prevent melt carryover between reservoirs, the clearance between the slider and the bottom of the reservoirs is optimized. The growth process is generally carried out under hydrogen ambient to minimize oxidation of melts.

5.3.2 Phase Diagram

In order to control the alloy compositions and layer thicknesses precisely in LPE processes, it is necessary to determine appropriate phase diagrams that describe relationships between solids and liquids at different temperatures. Phase diagrams for ternary and quaternary III-V compounds have been calculated using simple solution models and solubility data of binary compounds. In III-V alloys, the solution composition-temperature relationship of a binary compound such as GaAs can be described by a single solubility curve as shown in Fig. 5.1. When a ternary compound is formed by combining two binary compounds, a continuous solid solution with different compositions usually develops. This provides the freedom of selecting alloys with a specific lattice constant or bandgap energy. The composition of solutions that are in equilibrium with a solid phase is no longer a single-valued function of temperature as in binary compounds. The ternary liquidus relationships may be conveniently represented by a series of isotherms, each of which gives the concentrations of the two minor constituents in saturated solutions at a particular temperature. Such liquidus isotherms for the Ga-rich saturated solutions of the Al-Ga-As system between 700 and 1000 °C are shown in Fig. 5.8a, where the atomic percentages of Al and As in the solutions are plotted against each other.

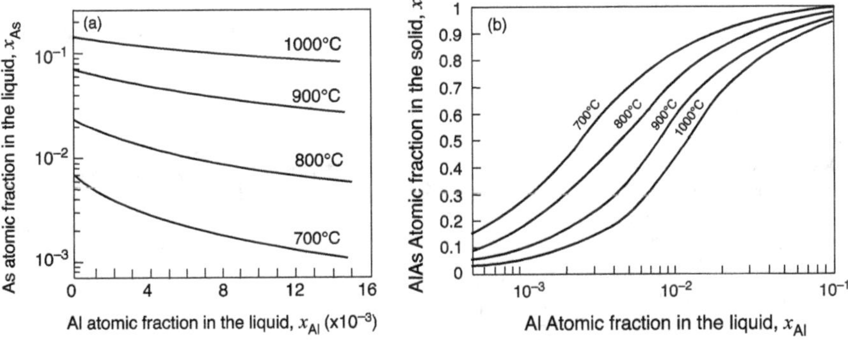

Fig. 5.8 **a** Liquidus isotherms in the Al-Ga-As system between 700 and 1000 °C. The liquidus compositions of the saturated Al-Ga-As solution are related through the mass conservation relationship $x_{Al} + x_{Ga} + x_{As} = 1$. **b** Solidus isotherms for Al in $Al_xGa_{1-x}As$ alloys between 700 and 1000 °C. Reprinted with permission from [4], copyright Elsevier

The ternary phase diagram also provides information about the relationships between liquid and solid phases at different temperatures. Taking advantage of the stoichiometric property of the III-V compounds, the solidus relationships are represented by isotherms giving the solid composition as a function of the concentration of one of the minor constituents of the solution. A set of such isotherms for the Al-Ga-As system between 700 and 1000 °C is shown in Fig. 5.8b, where the mole fraction of AlAs in the epilayer (x in $Al_xGa_{1-x}As$) is plotted against the atomic fraction of Al in the saturated solution, x^l_{Al}. With the knowledge of the solidus relationships of the system, the epilayer compositions can be controlled precisely. For example, in the $Al_xGa_{1-x}As$ system, specifying x^l_{Al} and the temperature completely fixes the composition of the saturated solution through liquidus isotherms like those of Fig. 5.8a.

In extending the thermodynamic treatment to $A^{III}_x B^{III}_{1-x} C^V_y D^V_{1-y}$ type quaternary compounds, such as $Ga_xIn_{1-x}As_yP_{1-y}$, the alloy may be considered as regular mixtures of four ternary compounds. Using quaternary alloys, an added degree of freedom for independent selection of lattice constant and bandgap energy may be obtained. For example, bandgap energies extending from 0.75 to 1.34 eV are readily adjustable in lattice-matched $Ga_xIn_{1-x}As_yP_{1-y}$ layers on InP. However, because of the uncertainties in various interaction parameters, the calculated equilibrium Ga-In-As-P phase diagrams are in poor agreement with experiments. Nevertheless, for $Ga_xIn_{1-x}As_yP_{1-y}$ lattice matched on InP, the solidus and liquidus curves for different compositions at various temperatures have been determined experimentally.

5.3.3 LPE Growth Techniques

After growth parameters are determined through phase diagrams, epitaxial layers can be grown by LPE following the temperature cycle shown in Fig. 5.9. The furnace is initially heated above the saturation temperature T_0 and kept there for sufficient time to enable the melts to attain saturation. After equilibration, the temperature is

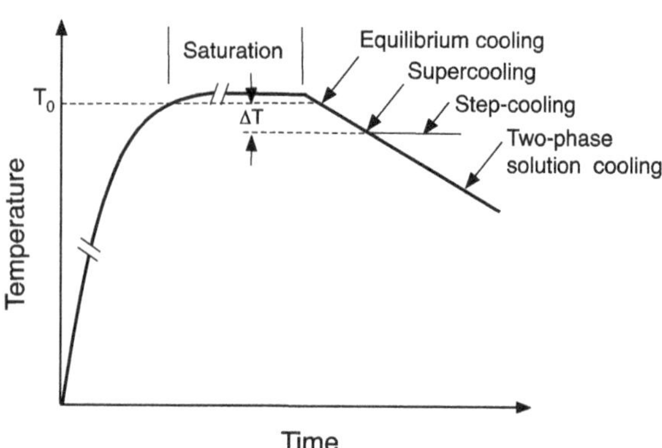

Fig. 5.9 Temperature cycle and solution cooling procedure for liquid-phase epitaxy growth. T_0 is the equilibrium temperature of the saturated solution. The arrows indicate the times at which the growth solution is initially placed in contact with the substrate

ramped down at a linear rate, usually in the range 0.1–0.8 °C/min. Depending on the temperature where the substrate contacts the melt and the cooling procedure, there are four different LPE techniques. They are equilibrium-cooling, step-cooling, supercooling, and the two-phase method. All but the two-phase method provide a typical growth rate of ~1 μm/min. Because of the high growth rate, it is very difficult to grow thin layers (≤100 Å) by LPE.

The equilibrium-cooling technique employs a constant cooling rate throughout the growth cycle. The substrate is brought into contact with the solution at the saturation temperature T_0 to begin the growth. The growth is terminated by sliding the substrate with the epilayer out of the solution. The thickness of the grown layer d is determined by

$$d = \frac{2}{3} K \frac{dT}{dt} t^{3/2} \tag{5.1}$$

where t is the growth time and K is a constant that depends on the diffusivity of each solute and on the solute's mole fraction in the solution at the growth temperature.

In the step-cooling technique, the substrate and solution are cooled at a constant rate to a temperature ΔT below T_0 without spontaneous precipitation and then brought into contact. The constant temperature ($T_0 - \Delta T$) is maintained during the growth period. The thickness d of the grown layer is related to ΔT and the growth time t by the relation

$$d = K \Delta T t^{1/2} \tag{5.2}$$

The supercooling technique is a combination of equilibrium-cooling and step-cooling. The substrate is brought into contact with the solution when both are at a temperature ΔT below the saturation temperature of the solution. Cooling is continued without interruption at the same rate until growth is terminated. The thickness of the grown layer d is given by the sum of (5.1) and (5.2), that is,

$$d = K \left(\Delta T t^{1/2} + \frac{2}{3} \frac{dT}{dt} t^{3/2} \right) \tag{5.3}$$

In the two-phase technique, the temperature is lowered far below T_0 for spontaneous precipitation to occur in the solution. Then the substrate and the solution are brought into contact, and cooling is continued at the same rate without interruption. This technique can be used to grow very thin layers because the presence of the precipitation reduces the growth rate.

5.4 Vapor-Phase Epitaxy

Vapor-phase epitaxy refers to the formation of an epitaxial layer from a gaseous medium of different chemical composition. The VPE process involves vapor-phase transfer of the active species to the VPE reactor, followed by chemical reactions in the gas stream before being brought into contact with the substrate surface. When the appropriate molecule arrives at the surface, there must be adsorption and surface diffusion to a suitable growth site and desorption of the products not needed for growth. Unreacted gaseous reactants and products are thoroughly scrubbed and burned before being exhausted to the atmosphere.

In general, VPE is carried out in an open-tube flow system where transport results from forced convection induced by a rather large flow of a carrier gas. Hydrogen is often used as the carrier gas, and other inert gases such as nitrogen, argon, and helium have also been used. Commonly used reactants are either a gas or a volatile liquid at room temperature. In the latter case, the reactant is carried into the system by passing a carrier gas through the liquid. Reactant partial pressures and resident time in the deposition region are easily controlled externally by changing the gas flow rate using mass flow controllers. The reactor is heated by a multizone furnace (hot wall reactor) or a radio frequency (RF) inductance heater (cold wall reactor) and can be operated at either atmospheric or low pressure. According to their gas flow characteristics, the VPE reactors can be classified in two categories: horizontal and vertical. In a horizontal reactor system, the flow of gases is parallel to the substrate surface. Because of the nature of the laminar flow of gases, a stagnant layer or boundary layer near the substrate surface is developed. The reactant species must diffuse in order to reach the substrate surface through this layer. Therefore, the growth rate varies as a function of the boundary layer thickness. In the vertical reactor design, the gas flow is perpendicular to the substrate surface. This geometry allows the incorporation of a rotating substrate holder design to improve the uniformity. However, thermal convection-induced turbulence alters the laminar flow conditions over the substrate.

Many chemical reactions have been employed for chemical transport of reactants in chemical vapor depositions. Among those useful for VPE growth of semiconductor materials are pyrolysis, reduction, synthesis, and disproportionation. For the VPE growth of III-V compounds, two methods are commonly used: (1) halide transport VPE and (2) MOCVD. In the halide transport VPE system, chloride, bromide, and iodide transport agents have been used with success. However, further development was restricted to chloride VPE because only high-purity chloride species were readily available.

5.4.1 Chloride VPE

In chloride VPE of III-V binary compounds AB, the group III element (A) is transported in the form of metal chloride (ACl) and reacted with group V element (B) according to the reaction

$$4ACl + B_4 + 2H_2 \rightarrow 4AB + 4HCl \quad (5.4)$$

Depending on the group III metal chloride generation method, chloride VPE is further classified into two categories: trichloride VPE and hydride VPE. Both epitaxy techniques are carried out in hot wall reactors composed of two zones set at different temperatures by using multizone furnaces.

(a) **Trichloride VPE**

For the trichloride VPE process, the growth apparatus is shown in Fig. 5.10. In the process of growing GaAs, $AsCl_3$ is used as a source of As and as a reactant with GaAs to transport Ga by the formation of GaCl in the high-temperature source zone according to the following reactions:

$$4AsCl_3 + 6H_2 \rightarrow As_4 + 12HCl \quad (5.5)$$

$$4Ga + As_4 \rightarrow 4GaAs \quad (5.6)$$

$$4GaAs + 4HCl \leftrightarrow 4GaCl + As_4 + 2H_2 \quad (5.7)$$

The $AsCl_3$ vapor is transported by a hydrogen carrier gas to the heated furnace and decomposed into As_4, HCl, and excess H_2. The As_4 dissolves in the liquid Ga source until the solution becomes saturated and a GaAs crust forms over the liquid surface as described by (5.6). By reaction of HCl with the GaAs crust in the high-temperature source zone, GaCl and As_4 are generated and transported to the low-temperature deposition zone for epilayer growth. These reactions are described by (5.7), and the reaction direction is determined by the temperature profile in the reactor. Because the

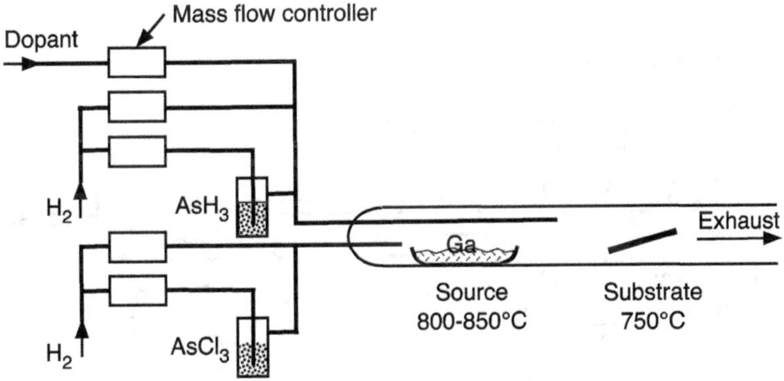

Fig. 5.10 Schematic illustration of a trichloride vapor-phase epitaxy system for GaAs growth

equilibrium of the reaction in (5.7) is established only after the Ga has been saturated with As and a GaAs crust formed, it is vital to maintain a flat temperature profile over the source. A partial dissolution of the GaAs crust and the reaction of the exposed Ga with HCl lead to uncontrolled variations in the reactant concentrations that cause surface morphology problems and loss of growth rate control. This problem can be avoided by the use of a solid GaAs source. Single-epilayer GaInAs and GaInAsP alloys have been prepared by this method.

One unique feature of trichloride VPE is the ability to achieve low background carrier concentration in epitaxial layers. This is due to the fact that $AsCl_3$ and PCl_3 can be distilled into very high-purity liquids. This allows the fabrication of devices consisting of low doping layers. For the growth of intentionally doped layers, both trichloride and hydride VPE techniques use either gaseous or solid dopant sources to achieve the desired electrical conductivity. Doping with *n*- and *p*-type dopants is commonly accomplished with H_2S gas and Zn vapor, respectively.

(b) **Hydride VPE (HVPE)**

For the hydride VPE process, the growth apparatus is very similar to that of the trichloride VPE and is shown in Fig. 5.11. In this process, GaCl is generated directly by passing HCl over the Ga source and As_4 from the pyrolysis of AsH_3 according to

$$2\text{HCl} + 2\text{Ga} \rightarrow 2\text{GaCl} + \text{H}_2 \tag{5.8}$$

$$4\text{AsH}_3 \rightarrow \text{As}_4 + 6\text{H}_2 \tag{5.9}$$

These reactions supply GaCl and As_4 to establish the reaction of (5.4) or (5.7).

The generation of GaCl is accomplished by the complete reaction of HCl with Ga at temperatures above 800 °C, and no critical temperature control over the source is required. In addition, the ability of independent generation of gas phase species allows variation of the Ga-to-As ratio. This is in contrast to the trichloride method where the Ga-to-As ratio is fixed by the reaction of (5.5). These properties make

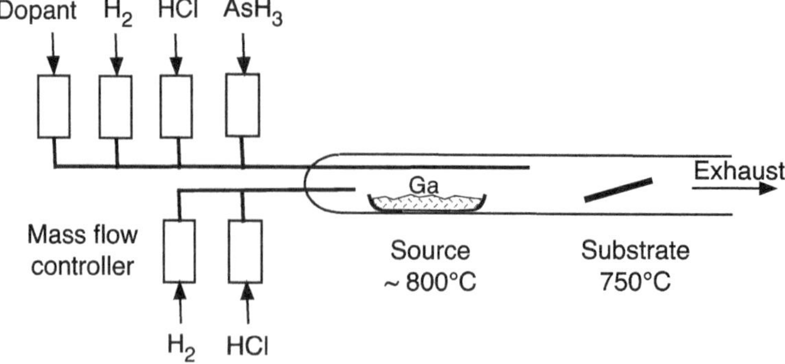

Fig. 5.11 Schematic illustration of a hydride vapor-phase epitaxy system for GaAs growth

the hydride method a preferred growth technique among chloride VPE techniques. For example, it has been used for the mass production of $GaAs_{1-y}P_y$ epitaxial layers on GaAs and GaP substrates for light-emitting diode applications. Most of the VPE growth of GaInAsP alloys has also been done by the hydride method. Nevertheless, because of the many chemical reactions involved in the quaternary alloy growth, modeling is complex and difficult. Predicting alloy composition from vapor flows is further complicated by the formation of deposits on the reactor wall. Therefore, proper gas flows to achieve desired compositions must be determined empirically.

Because the chloride VPE reactions require some time to achieve equilibrium after altering the flow rates, it is impractical to grow mutilayer structures by changing flow rates. One solution is to use a multichamber reactor design where gas reactants for each layer are supplied through separate reaction chambers. After the gas flow equilibrium has been established in each chamber, heterostructures can be grown by introducing the substrate into designated reaction chambers sequentially [5].

An important recent application of the HVPE technique is the epitaxial growth of thick GaN on sapphire as III-nitride substrates since growth rates of tens of microns an hour can be readily achieved. However, the drawback of this method is that even after removal of non-native substrate the freestanding GaN layer obtained in this way is highly stressed and typically bowed. Either gallium monochloride is synthesized upstream in the reactor by reacting HCl gas with liquid Ga metal at 800–900 °C, or presynthesized $GaCl_3$ is used. The gallium chloride is transported to the substrate where reacts with NH_3 at 900–1100 °C to form GaN, via the reaction

$$GaCl + NH_3 \leftrightarrow GaN + HCl + H_2 \tag{5.10}$$

In trichloride and hydride VPE methods, the temperature-dependent growth rates are qualitatively similar and are shown in Fig. 5.12. At low temperatures, the growth

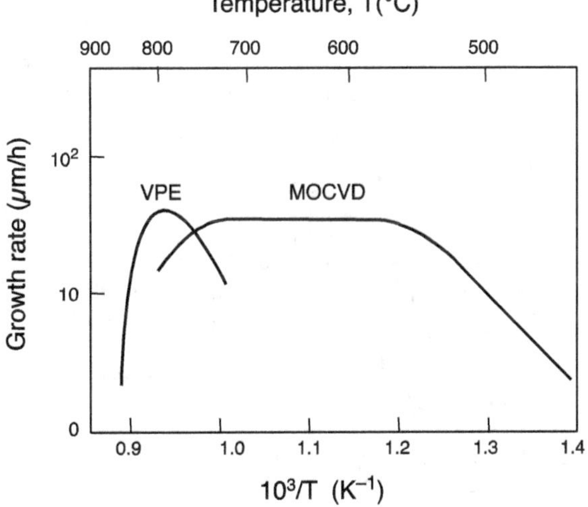

Fig. 5.12 Temperature dependence of GaAs growth rate on (100) GaAs substrates for typical trichloride VPE and MOCVD

rate is kinetically limited by the reduction of chemisorbed As-Ga-Cl complexes to form GaAs. This is evident from the substrate-orientation-dependent growth rates. In the low-temperature regime, the growth rate increases with the temperature. When the temperature is increased, the growth is mass-transfer-limited by the diffusion of reactants to the substrate and causes a decrease of growth rate, which is independent of the substrate orientation.

5.4.2 *Metalorganic Chemical Vapor Deposition*

The MOCVD technique was pioneered by H. M. Manasevit in 1969 to grow GaAs on various substrates using a metalorganic Ga source mixed with arsine. It is also referred to as organometallic VPE (OMVPE) or organometallic CVD (OMCVD). The MOCVD system is simpler than the chloride systems. The growth of GaAs by the pyrolysis of vapor-phase mixture of AsH_3 and $Ga(CH_3)_3$ (trimethylgallium, TMG) or $Ga(C_2H_5)_3$ (triethylgallium, TEG) typifies the process. As shown in Fig. 5.13, TMG or TEG vapor is transported into the cold wall reactor by bubbling H_2 through the liquid source, which is held in a temperature-controlled bubbler. Arsenic is transported in the hydride form (i.e., AsH_3). In addition, dopant sources either in hydride forms or metalorganic forms are injected into the reactor to achieve the desired electrical conductivity. Disilane (Si_2H_6) and diethylzinc (DEZ) are the typical *n*- and *p*-type dopants used in the MOCVD growth of III-V compounds, respectively. The substrate is heated on an inductively heated susceptor to a temperature of 600–800 °C. The metalorganic compound and the hydride diffuse through the boundary layer

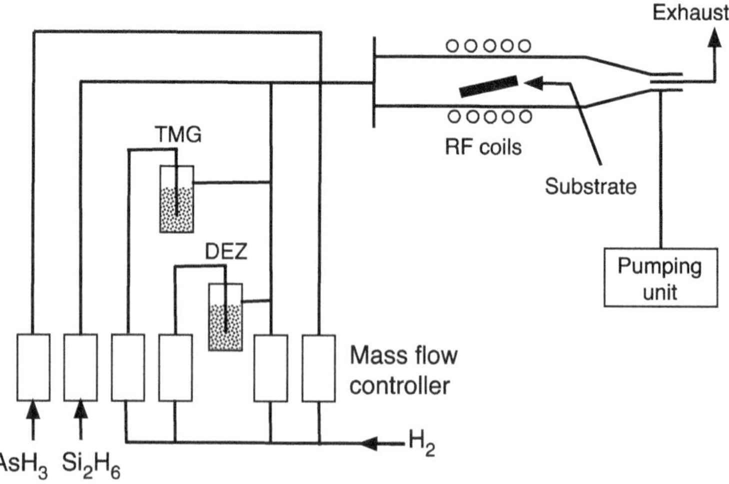

Fig. 5.13 Schematic diagram of an MOCVD system using TMG and AsH_3 sources for the growth of GaAs. Diethylzinc (DEZ) and disilane (Si_2H_6) are *p*- and *n*-type dopant sources, respectively

and decompose on the hot substrate surface in an irreversible reaction to form a GaAs epilayer according to

$$Ga(CH_3)_3 + AsH_3 \rightarrow GaAs + 3CH_4 \quad (5.11a)$$

or

$$Ga(C_2H_5)_3 + AsH_3 \rightarrow GaAs + 3C_2H_6 \quad (5.11b)$$

Because the reaction is irreversible and no thermodynamic equilibrium is involved, a very wide range of growth conditions and abrupt compositional changes of epitaxial structures are possible.

These properties make the MOCVD technique one of the favored mass production methods of III-V heterostructures. Temperature-dependent growth rates have been studied to understand the growth mechanism of MOCVD. Taking GaAs as an example, as seen in Fig. 5.12, for an AsH_3-to-TMG ratio much larger than one, the growth rate is nearly independent of temperature over the range 550–750 °C. This behavior indicates that the growth is determined by diffusion of the TMG species through the boundary layer to the substrate. In this regime, the available metalorganic Ga is the growth-limiting factor, and the growth rate is found to be linearly proportional to the TMG flow rate and independent of arsine partial pressure. In the high-temperature region ($\geq$750 °C), as a result of the desorption of Ga atoms, As molecules, and/or decomposition of the GaAs, the growth rate is under the thermodynamic limitation and begins to decrease. On the other hand, as a result of the inefficient pyrolysis, the growth rate falls off with decreasing temperature below 550 °C.

For the MOCVD growth of III-nitrides, the basic reaction describing the GaN deposition process is expressed as:

$$Ga(CH_3)_3 + NH_3 \rightarrow GaN + 3CH_4 \quad (5.12)$$

The basic growth principles are the same as the growth of III-V compounds except that very high growth temperatures (>1000 °C) are required for the growth of GaN because of high bond-strength of N-H bond in ammonia precursors.

In growing heterostructures by atmospheric pressure MOCVD, changing the layer composition is carried out by switching flows of metalorganic reactant mixtures. The time required to establish a stable flow with minimum turbulence in the reactor prevents the formation of a sharp interface. The growth of indium-containing compounds from trimethylindium (TMI) or triethylindium (TEI) is further complicated by the possible formation of adducts between hydrides and indium-alkyls. To minimize these problems, low-pressure MOCVD (LP-MOCVD) has been developed. LP-MOCVD is usually carried out in a reactor under reduced pressure in the range of 50–100 torr. The large reduction of system pressure and increased flow rate enhance the gas-phase transfer of reactants to, and by-products from, the substrate surface. The practical consequences of these results are very significant. Under low-pressure conditions, the high gas flow rate permits rapid establishment of new gas compositions, which lead to more abrupt changes in composition. Furthermore, a more uniform boundary layer thickness is established under high gas flow rates, leading to good uniformity in layer thickness and composition. Additionally, the reaction rate is slowed, and reaction time is reduced, which, under low-pressure conditions, leads to a minimum adduct formation and a reduced growth rate of about 2–5 μm/h.

For the growth of $A_x^{III}B_{1-x}^{III}C^V$ ternary compounds by MOCVD under a high V/III flow rate ratio, the solid composition x is determined by

$$x = J_A^{III} / \left(J_A^{III} + J_B^{III} \right) \tag{5.13}$$

where J is the flux of the group III element. For the growth of alloys with mixing on a group V sublattice (i.e., $A^{III}C_y^V D_{1-y}^V$), the solid composition y becomes a nonlinear function of vapor composition. This is because of the unequal pyrolysis rates of different group V reactants at the growth temperature. For example, in $GaAs_{1-y}P_y$, a very large ratio of PH_3 to AsH_3 is required to produce alloys with a significant phosphorus content below 750 °C.

5.5 Molecular Beam Epitaxy

Pioneered by A. Y. Cho at Bell Laboratories, molecular beam epitaxy is an ultra-high-vacuum (UHV) deposition technique with several important features. The MBE growth of semiconductor films takes place by the reaction of molecular beams of the constituent elements with a crystalline substrate surface held at a suitable substrate temperature under UHV conditions. The kinetically controlled MBE growth process involves a series of events: adsorption, surface migration and dissociation, and incorporation. In the case of MBE growth of GaAs, in the absence of free surface Ga adatoms, the impinging As_2 molecules will simply re-evaporate from the surface above 500 °C. Dissociation of adsorbed As_2 and subsequent incorporation into the GaAs lattice can occur only when they encounter paired Ga lattice sites while migrating on the surface. Therefore, for the MBE growth of a stoichiometric GaAs epitaxial layer, it is required that only an excess of As species be present while the growth rate is determined by the arrival rate of the Ga flux.

The unique feature of MBE is the ability to prepare epitaxial layers with atomic dimensional precision down to a few angstroms. This ability allows the preparation of novel devices with multilayered epitaxial structures tailored to meet specific needs. Because MBE is done in a UHV environment, many surface analysis techniques may be used during the growth process. This makes the MBE process a highly controllable and reproducible epitaxy method.

5.5.1 MBE Apparatus

The modern MBE system uses a modular configuration that contains a number of building blocks, such as the growth chamber, the sample exchange load-lock, the surface processing chamber, and the surface analysis chamber, which are all interconnected by a UHV transfer tube. A basic MBE system for III-V compounds is shown in Fig. 5.14. The UHV growth chamber is evacuated with a pumping stack that maintains a base pressure of 10^{-11} torr. In addition, a liquid nitrogen-cooled shroud is used to enclose the entire interior surface of the growth chamber in order to minimize contamination from residual water vapor and hydrocarbons during epitaxy. The sample exchange load-lock permits the maintenance of UHV in the growth chamber while changing substrates between successive growth runs.

The substrate is typically mounted on a molybdenum substrate holder attached to a sample manipulator for precise positioning within the growth chamber. The substrate holder can rotate continuously to achieve extremely uniform epitaxial layers. Thermal radiation generated by resistance heating from behind the substrate holder is employed to heat the substrate. On the backside of the manipulator is an

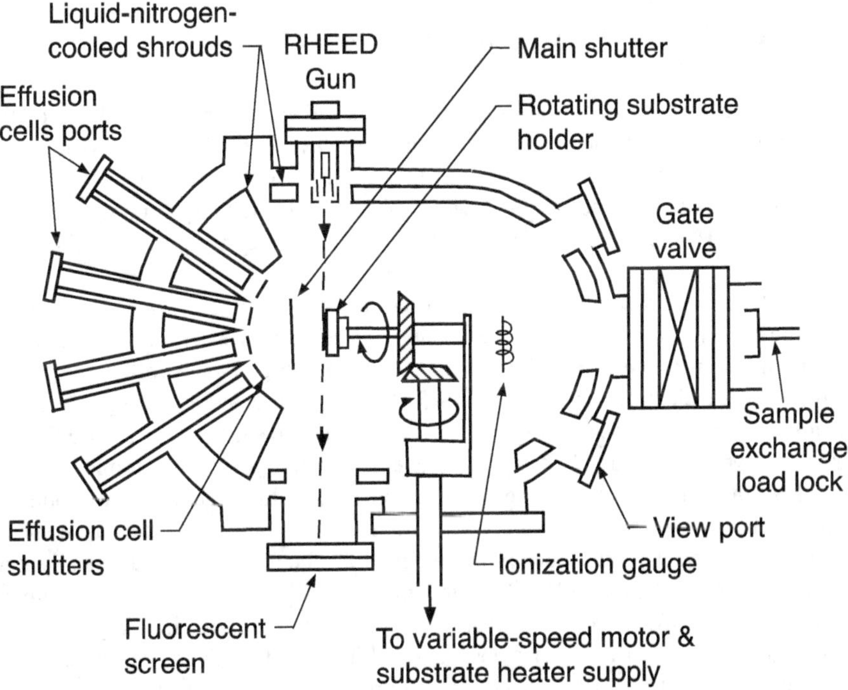

Fig. 5.14 Cutaway top view of a molecular beam epitaxy system. Molecular beams are generated from effusion cells and/or gas injectors mounted on the source flange. Opening and closing of different shutters in front of effusion cell orifices determine the heterostructure grown on the heated substrate. Reprinted with permission from [6], copyright AIP Publishing

ion gauge for beam flux measurements. When rotating the manipulator into position such that the movable ion gauge is facing the effusion cells, the relative flux of each beam can be estimated.

The source flange on the growth chamber contains a viewport and eight or more ports for mounting effusion cells and/or gas injectors. The viewport both facilitates the mounting of an optical pyrometer for substrate temperature measurement and provides a means of directly viewing the substrate during growth. Ultra-high-purity elemental source materials and dopants loaded in PBN effusion cell crucibles are used to generate the desired molecular beams. The effusion cell temperatures are controlled to an accuracy of ± 1 °C to provide the precise amount of beam flux. To initiate or terminate the molecular beam flux, each source is provided with its own externally controlled mechanical shutter. Shutters in front of the orifices can be opened and closed within a tenth of a second, which is much shorter than the typical MBE growth rate of one to two monolayers per second, resulting in abrupt interfaces in the range of one atomic layer. Therefore, the sequence of opening and closing different shutters determines the multilayer heterostructure in terms of both composition and doping profile.

5.5.2 *In Situ Surface Diagnosis Techniques*

In addition to the components used directly for the MBE growth process, the growth chamber may contain various pieces of in situ surface analysis equipment to monitor the surface structure and control the growth conditions. For a modern MBE system to be used for the production of devices, only a reflection high-energy electron diffraction (RHEED) apparatus and a movable ion gauge in the growth chamber are essential.

The RHEED apparatus provides information concerning substrate cleanliness, smoothness, and surface structure before and during growth as a function of growth conditions. To generate RHEED patterns, as shown in Fig. 5.14, a collimated beam of high-energy electrons in the range of 5–40 keV is directed at an angle of 1–2° toward the sample surface orthogonal to the molecular beam paths. Because the de Broglie wavelength of an electron at this energy is a fraction of the atomic spacing on the surface, a diffraction pattern is formed on the fluorescent screen mounted opposite the electron source. In this configuration, the sample surface can be continuously monitored without interrupting the growth procedure. On an atomically flat surface, it shows a streaked RHEED pattern normal to the shadow edge of the sample. Otherwise, the diffraction pattern from a rough surface is formed mainly in transmission through the surface asperities and exhibits a spotty appearance. Figure 5.15 is an example of the (100) GaAs surface morphology evolution during the initial stages of the MBE growth and their corresponding RHEED patterns. The clean but rough starting surface shows a spotty bulk RHEED pattern as seen in Fig. 5.15a. The RHEED pattern changes from spotty to streaky as the surface is smoothed out.

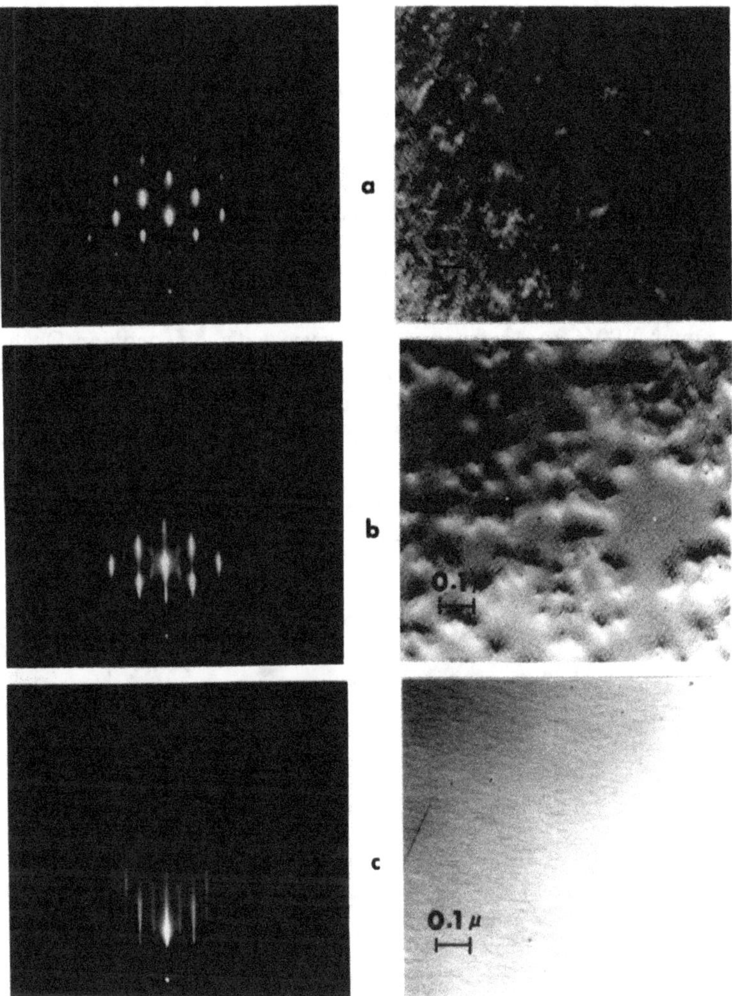

Fig. 5.15 40 keV RHEED patterns of a (100) GaAs surface taken at azimuth $[\bar{1}\bar{1}0]$ and the corresponding photo-micrographs; **a** after chemical etching and heating to 580 °C, **b** after depositing 150 Å of GaAs, and **c** after depositing 1 μm of GaAs. Reprinted with permission from [7], copyright American Vacuum Society

In addition to the spotty-to-streaky transition in the RHEED patterns of a clean crystal surface during growth, additional light streaks appear halfway between the elongated bulk spots along the $[\bar{1}\bar{1}0]$ azimuth as seen in Fig. 5.15b, c. These added features represent the rearrangement of atoms on the surface in order to accommodate the surface dangling bonds and to minimize the surface free energy. Depending on the surface-atom coverage conditions and the electron beam incident directions, the pattern of extra diffraction lines between the bulk streaks assumes different forms.

For example, the RHEED pattern of the As-rich surface (Fig. 5.15c) shows extra ½ integral order streaks along the $[\bar{1}\bar{1}0]$ azimuth and extra ¼ integral order streaks (i.e., three additional light streaks between bulk streaks) along the $[1\bar{1}0]$ azimuth (not shown). The RHEED patterns of Ga-rich (Ga-stabilized) surfaces and As-rich (As-stabilized) surfaces are similar, but are interchanged along $[\bar{1}\bar{1}0]$ and $[1\bar{1}0]$ azimuths. In real space, the two surface structures are related by a simple rotation of 90° about the [001] direction. The relationships between the surface structures and the growth conditions (i.e., the *surface phase diagram*) of (100) GaAs have been established in terms of As_2/Ga flux ratios and substrate temperatures (Fig. 5.16). From the point of view of practical GaAs growth, the As-stabilized structure is preferred. A high-quality smooth (100) GaAs surface can be achieved under this condition. On the other hand, prolonged growth under a Ga-stabilized condition leads to a dull surface caused by the formation of Ga droplets. Overall, because of its simplicity and in situ nature, the RHEED technique is routinely used in MBE to monitor the surface cleaning process prior to epitaxial growth and to optimize growth conditions during growth.

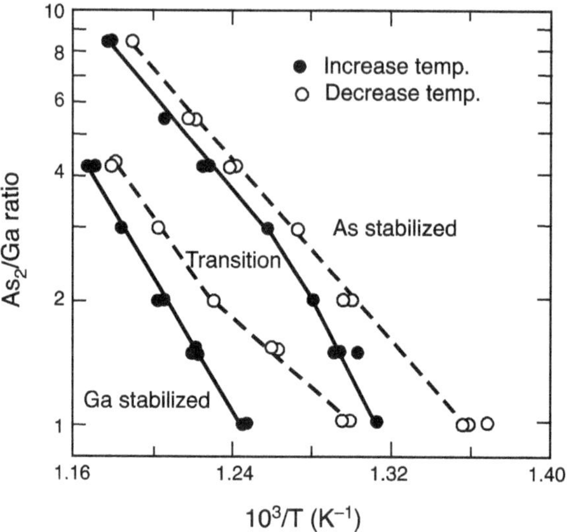

Fig. 5.16 Arsenic-to-gallium ratio in the molecular beam as a function of the substrate temperature when the transition of an As-stabilized (½ order) surface structure and a Ga-stabilized (¼ order) surface structure takes place. The RHEED patterns are observed in the $[\bar{1}\bar{1}0]$ azimuth. During the transition, a 1/3 order is observed. Depending on whether the substrate temperature is increasing or decreasing, two sets of transition boundaries are shown. Reprinted with permission from [7], copyright American Vacuum Society

5.5.3 Flux Control of Molecular Beams

For an ideal Knudson-type effusion cell, the beam flux arriving at the substrate surface positioned at a distance d (cm) from the aperture can be calculated as follows:

$$J = 1.118 \times 10^{22} \frac{pA}{d^2\sqrt{MT}} \cos\theta \ \left(\text{molecules/cm}^2\text{-s}\right) \tag{5.14}$$

where p (torr) is the pressure in the cell, A (cm^2) is the area of the aperture, M is the molecular weight, T (K) is the temperature of the cell, and θ is the angle between the beam and the normal of the surface. However, this equation serves only as a guideline because in practice the ideal pinhole-size cell aperture is enlarged to enhance the growth rate. The beam fluxes emerging from these non-ideal effusion cells are generally determined experimentally.

For in situ calibration of beam fluxes and growth rates, the most convenient and routinely used method is the RHEED intensity oscillation technique. As shown in Fig. 5.17, the equilibrium surface existing before growth is smooth, corresponding to high reflectivity of the specular beam. As growth commences, nucleation islands

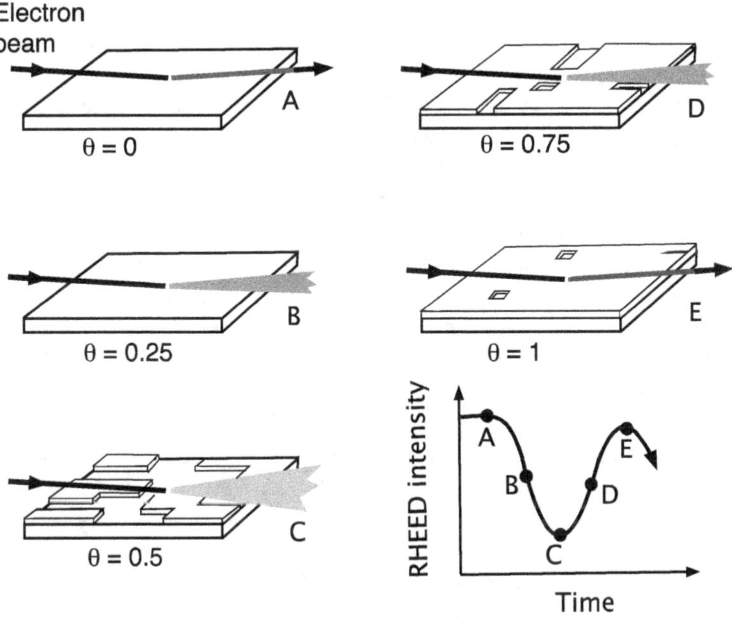

Fig. 5.17 . Real-space representation of the formation of the first complete monolayer of (001) GaAs with respect to RHEED intensity oscillations. The intensity of the diffracted electron beam decreases as surface roughness increases. θ is the fractional layer coverage. The period of the oscillation (e.g., time span between point A and point E) corresponds to the growth rate of one atomic layer. Reprinted with permission from [8], copyright Elsevier

will form at random sites on the surface, leading to a decrease in reflectivity. These islands grow until they coalesce into a smooth surface again. It is expected that the minimum in reflectivity would correspond to 50% coverage by the growing layer. Therefore, the period of the oscillations corresponds precisely to the growth rate of a monolayer. Because the sticking coefficients of the group III elements are unity, once the beam fluxes and the growth rates are calibrated, the alloy composition in the A^{III}-B^{III}-C^{V} alloy is simply determined by the relative group III fluxes reaching the surface. For example, the Al fraction in $Al_xGa_{1-x}As$ can be determined by the relation

$$x = \frac{R(Al_xGa_{1-x}As) - R(GaAs)}{R(Al_xGa_{1-x}As)} \tag{5.15}$$

where R is the growth rate.

5.5.4 Variations of Molecular Beam Sources

Although MBE has long been successful for the growth of arsenic and antimony compounds, the growth of phosphorus compounds by conventional solid source MBE (SSMBE) had been hampered until 1995 by the high vapor pressure and the allotropic property of solid phosphorus. In that year, a new approach for the growth of P-compounds by SSMBE was reported [9]. This technique uses a three-zone valved cracking cell to generate P_2 molecules. The structure of a solid-P valved cracking cell consists of a red P oven, a white P condensing reservoir, and a valved thermal cracking region. During operation, the red P oven is heated to an appropriate temperature in order to generate a sufficient amount of P_4 vapor. The white P reservoir section during this time period is held at a low temperature to condense the vapor into white P while the valve is closed. After the desired quantity of white P has been collected, both the red P oven and the white P reservoir are returned to ambient temperature to finish the distillation process. The P_4 vapor emanating from the white P reservoir is passed through the high-temperature cracking zone and dissociated into P_2 at the desired rate via the adjustable flux control valve. The accumulated white P and its associated high vapor pressure enable the condensing reservoir to be operated at room temperature during growth. This capability strongly inhibits the formation of multiple allotropes and makes accurate P flux control possible. Highly reproducible growth of P-compounds and device structures has been demonstrated.

The gas source molecular beam epitaxy (GSMBE) technique, in which the elemental As and P sources are replaced by gaseous AsH_3 and PH_3, respectively, represents an alternative approach. A further development of the technology replaces the group III elements with gas sources. This is conveniently accomplished using metalorganic group III species in addition to hydride group V sources. This method is generally referred to as metalorganic molecular beam epitaxy (MOMBE) or chemical beam epitaxy (CBE). Both GSMBE and MOMBE use a growth system design similar to SSMBE with modifications in source delivery and UHV pumping methods.

GSMBE and MOMBE use gas-handling systems similar to those used by MOCVD to deliver gas sources into the UHV growth chamber for epitaxy. The major distinction between the GSMBE/MOMBE method and MOCVD centers around the pressure regimes involved in each method. Namely, the latter operate under viscous flow and the former under molecular flow conditions. In molecular flow, the pressure inside the growth chamber is low ($<10^{-4}$ torr), and there exists no effective boundary layer at the growing interface. This leads to a GSMBE/MOMBE beam flux control very similar to SSMBE, and all the analytical equipment ordinarily used to monitor the process can be implemented in the same way.

In GSMBE, after the hydrides have been delivered into the growth chamber, they are thermally cracked at high temperatures. To improve the quality of grown layers and for safety reasons, the hydride gas injector should be able to generate a significantly enhanced dimer-to-tetramer ratio (V_2/V_4) and an improved cracking efficiency ($\geq$99.9%) for both AsH_3 and PH_3 with a minimum switching transient. In the case of MOMBE, the group III metalorganic species, similar to those used in MOCVD, are also delivered from a gas source into the UHV growth chamber through a single gas injector. Unlike the group V hydrides, the mixed group III metalorganics are not predecomposed, and pyrolysis occurs only on the heated substrate surface to prevent any chemical reactions in gas phase. Therefore, the chemical reactions at the growing surface in MOMBE are expected to be considerably more complex than in the case of GSMBE. On the other hand, the lack of a boundary layer on the growth surface in MOMBE distinguishes the surface reaction chemistry from that of MOCVD.

Gallium nitride and related compounds have recently emerged as the leading materials for fabricating blue-green light emitters and high-power electronic devices. For the MBE growth of III-nitride compounds using nitrogen gas source, the stable chemical bonds of molecular nitrogen in a low background pressure environment make thermal cracking unsuccessful in generating abundant atomic nitrogen for epitaxial growth. To supply sufficient energy in breaking chemical bonds of molecular nitrogen, the electron cyclotron resonance (ECR) microwave plasma source and RF plasma source techniques have been developed. In these approaches, known as plasma-assisted molecular beam epitaxy (PAMBE) technique, the microwave or RF energy is coupled into the nitrogen plasma to crack molecular nitrogen efficiently into an atomic source of nitrogen suitable for III-V nitride compound growth.

5.6 Post-growth Modification of Material Structures

5.6.1 *Impurity-Induced Layer Disordering (IILD) of Quantum-Well Heterostructures*

Impurity-induced layer disordering (IILD) is an important post-growth process for III-V photonic device fabrication. It allows the fabrication of heterojunctions perpendicular to the thin epitaxial layers characteristic of quantum well (QW) and superlattice heterostructures. Using standard photomasking procedures of the semiconductor surfaces and selective impurity diffusion or ion implantation, devices with three-dimensional carrier and photon confinement can be fabricated, including buried heterostructure QW lasers and optical waveguides. In addition, IILD has been used to improve the reliability of high-power diode lasers against facet damage by increasing the bandgap of cleaved facets.

(a) **Diffusion mechanisms in semiconductors**

The most practical way of controlling the type and concentration of impurities within specific regions of a semiconductor crystal, besides epitaxial growth, is through solid-state diffusion. Impurity diffusion in semiconductors is a classic subject and has been studied extensively. For an impurity distribution $C(z)$ in the direction normal to the semiconductor surface (in x-y plane), the impurity flux $\varphi(z)$ (per unit time and unit area) can be described by Fick's first law

$$\varphi = -D\frac{\partial C}{\partial z} \tag{5.16}$$

where

$$D \equiv D_0 \exp(-E_a/kT) \tag{5.17}$$

is the diffusion constant and E_a is the activation energy. Also, following the continuity equation, the increase (decrease) of impurity flux in a volume element equals to the net flow of impurities into (out of) the volume element per unit time as given by

$$\frac{\partial C}{\partial t} = -\frac{\partial \varphi}{\partial z} \tag{5.18}$$

These results lead to a second order partial differential equation

$$\frac{\partial C}{\partial t} = D\frac{\partial^2 C}{\partial z^2} \tag{5.19}$$

which is known as Fick's second law. For specific boundary conditions, impurity distributions can be obtained from the solution of Fick's second law.

For example, in the case of impurity diffusion in a semiconductor with a constant impurity concentration C_s at the surface, i.e., the infinite reservoir case, such as a pre-deposited Si film on GaAs, the diffusion profile is a complementary error function,

$$C(z,t) = C_s \operatorname{erfc} \frac{z}{2\sqrt{Dt}} \tag{5.20}$$

where $\sqrt{Dt}$ is the diffusion length.

In the finite reservoir case, a sheet of diffusing impurities with a fixed total number C_{2D} (per cm^2) is deposited on the semiconductor surface for the drive-in diffusion. The diffusion profile at a later time, t, is given by a Gaussian distribution as

$$C(z,t) = \frac{C_{2D}}{\sqrt{4\pi Dt}} \exp\left(-\frac{z^2}{4Dt}\right) \tag{5.21}$$

Diffusion of impurities in semiconductors is achieved through thermally activated random hops of the impurity atoms and vacancies. The diffusion rate depends on a number of factors such as atomic size of the impurity, vacancy concentration, concentration of other defects, and Fermi level position. In the *substitutional diffusion* process, the impurity hops from one substitutional site to a nearby vacant substitutional site. Since one vacancy is required for each diffusion hop, the diffusion coefficient for this process is proportional to the vacancy concentration, [V]. On the other hand, an impurity situated on an interstitial site can hop to a neighboring interstitial site through the *interstitial diffusion* process. However, the common impurities of group II, IV, and VI elements used in III-V semiconductors mainly occupy substitutional sites in the zinc-blende crystal lattice. As a result, there is limited practical relevance of the pure interstitial diffusion process to the doping process. Nevertheless, the mixture of substitutional and interstitial diffusion processes, i.e., the *substitutional–interstitial diffusion* process, is one of the most important impurity diffusion mechanisms in semiconductors. We assume the majority of impurities occupy substitutional sites, and a minor fraction of impurities occupies interstitial sites. But these substitutional impurities have a very insignificant probability of diffusion substantially. Furthermore, these interstitial impurities are less strongly bound to the lattice, which gives them a much higher diffusion constant. Therefore the diffusion process is dominated by the highly mobile interstitials and occurs more rapidly at high doping concentration as compared to lower doping concentration.

Furthermore, impurities frequently diffuse in the form of impurity-defect complexes rather then isolated impurity atoms. The simplest defect is a lattice vacancy near the surface, also known as a *Schottky defect*. It is formed by transferring a lattice atom to the crystal surface. Another defect, called *Frenkel defect*, is a vacancy-interstitial pair in which an atom is relocated from the lattice site to an interstitial position leaving behind a vacancy.

In addition to impurity diffusion, the lattice atoms, e.g., group III and group V elements in III-V compound semiconductors, can also diffuse, and the diffusion process is known as *self-diffusion*. In a homogeneous material, the self-diffusion of lattice atoms must proceed through native defects, e.g., group III vacancies V_{III}, of the crystal. Therefore, the self-diffusion rate will be dependent on the diffusion rate of the native defects. Since the concentration of native defects is much less than the concentration of group III lattice atoms ($\cong 2 \times 10^{22}$ cm^{-3}), the group III self-diffusion rate will also depend on the concentration of the defects. If we consider, for example, self-diffusion of the group III sublattice through group III vacancies, then

$$D_{\mathrm{III}} = f D_{\mathrm{V_{III}}}[V_{\mathrm{III}}] \tag{5.22}$$

where D_{III} is the group III lattice atom diffusion coefficient, $D_{\mathrm{V_{III}}}$ is the group III vacancy diffusion coefficient, $[V_{\mathrm{III}}]$ is the group III vacancy concentration, and the prefactor f contains information about the crystal structure. It has been determined that the activation energy for group III self-diffusion in many undoped III-V alloys is quite high, $E_a \geq 4$ eV. Thus, the self-diffusion is a rather unlikely process in intrinsic III-V compound semiconductors.

(b) **Impurity-induced layer disordering (IILD) process**

Although the probability of self-diffusion of lattice atoms in undoped III-V semiconductors is very low, impurities do have a profound effect on the self-diffusion of lattice atoms in III-V semiconductors. In 1981, W. D. Laidig and N. Holonyak Jr. et al. discovered that, upon Zn diffusion, Al-Ga interdiffusion at the hetero-interfaces of an AlAs-GaAs superlattice and AlGaAs-GaAs quantum wells is drastically increased even at a moderately low diffusion temperature of 575 °C [10]. Specifically, they found that the layers of an AlAs-GaAs superlattice are unstable against Zn diffusion and intermix. The diffusion-induced intermixing, or *disordering*, causes structure change from direct-gap AlAs(15 nm)-GaAs (5 nm) superlattice (E_{g} ~ 1.61 eV) to bulk, homogeneous material of indirect-gap $\mathrm{Al}_x\mathrm{Ga}_{1-x}\mathrm{As}$ (x ~ 0.77, E_{g} ~ 2.08 eV). Several different impurities such as the donors Si, Ge, S, Sn, and Se, as well as the acceptors Be, and Mg, also have been found to cause layer mixing in III-V-based quantum-well heterostructures (Fig. 5.18). So far, the AlGaAs-GaAs material system is the most studied among all III-V alloys. The IILD process has also operated very effectively on the GaAs-GaP system through group V disordering.

The fact that diffusion of impurity (e.g., Zn and Si) greatly enhances the interdiffusion of Al and Ga in the quantum well and superlattice heterostructures indicates that a modification of the substitutional–interstitial diffusion mechanism for impurity diffusion is required. To understand the Zn diffusion-induced layer disordering, Laidig and Holonyak et al. suggested that in the transfer of a Zn atom from a substitutional site to an interstitial site, an intermediate phase occurs which consists of a Zn-vacancy pair, a Frenkel pair, formed by interstitial Zn and a group III vacancy

$$\mathrm{Zn_i^+} + V_{\mathrm{III}} \leftrightarrow (\mathrm{Zn_i}\, V_{\mathrm{III}})^+ \leftrightarrow \mathrm{Zn_{III}^-} + 2h \tag{5.23}$$

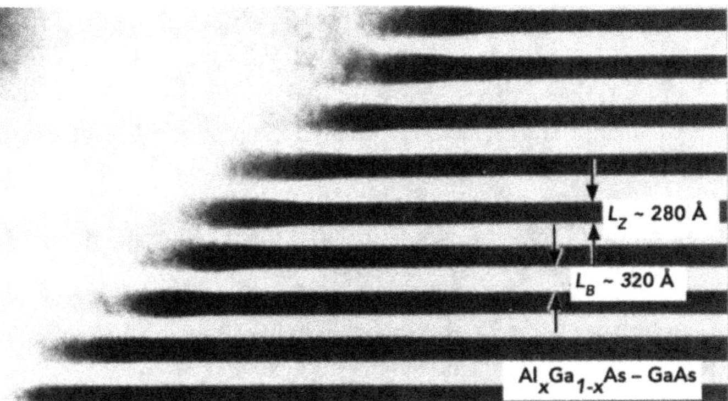

Fig. 5.18 Bright field TEM (transmission electron microscope) micrograph of a 40-period $Al_xGa_{1-x}As$-GaAs superlattice ($x \geq 0.6$, $L_B = 320$ Å, $L_z = 280$Å) that on the left-hand side has been disordered into bulk crystal $Al_{x'}Ga_{1-x'}As$ by Si diffusion. The barrier and well regions in the superlattice are shown as light and dark strips, respectively. Reprinted with permission from [11], copyright AIP Publishing

where h represents a hole, Zn_i^+ and Zn_{III}^- are an interstitial (positive charged) and a substitutional (negative charged) Zn atom, respectively. Neighboring Al and Ga atoms could move into the vacancy of the (Zn_i, V_{III}) pair such that interdiffusion of Al and Ga can be promoted. Consider the reaction between the Zn-vacancy pair and substitutional Zn atom of the above equation: Under equilibrium conditions, the concentrations of these reactants are related by the law of mass action

$$\left[(Zn_i V_{III})^+\right]/\left[Zn_{III}^-\right]p^2 = K \tag{5.24}$$

where p is the hole concentration and K is the equilibrium constant. It indicates that the concentration of the Zn-vacancy pair increases with the square of the Zn doping concentration, i.e., $[V_{III}] \propto p^2$. Since D_{III} is proportional to $[V_{III}]$, it means that the group III self-diffusion coefficient is strongly enhanced by orders of magnitude at high Zn doping concentration. Thus, the formation and doping concentration dependence of native defects form the basis of the lattice atom self-diffusion enhancement, which leads to impurity-induced layer disordering.

Further studies indicate that the impurity-induced layer mixing also depends on the type of doping impurity (n or p) grown into the heterostructure and group V overpressure used in diffusion processes. In other words, it is Fermi level position dependent. During the annealing of AlGaAs-GaAs heterostructures, by adding more or less As to the anneal ampule, the crystal stoichiometry will be changed via generation of native defects. The native defects, which are formed first on the crystal surface by reacting with the As vapor, will diffuse into the bulk until equilibrium is established. When the vapor pressure is on the As-rich side of equilibrium, excess As_{Ga} antisites, As interstitials (As_i), or group III vacancies (V_{III}) are created. Under an As-poor annealing condition, we would expect Ga_{As} antisites, Ga interstitials (Ga_i), or As vacancies (V_{As}) to be created. Of all the point defects possible, only V_{III} and the group III antisite defect, Ga_{As}, are expected to have acceptor-like energy states in the semiconductor. The other native defects are expected to act as donor-type defects. If we assume that the heterostructure is heavily n-type doped, the Fermi level is located very close to the conduction band edge, much above the vacancy acceptor energy level E_a. Thus, most of the group III vacancies will be ionized. The equilibrium concentration of ionized vacancies $[V_{III}^-]$ is related to the number of neutral vacancies $[V_{III}^*]$ following

$$\left[V_{III}^-\right] = \left[V_{III}^*\right] exp[(E_F - E_a)/kT] \tag{5.25}$$

Since $E_F > E_a$, the crystal will have an increased solubility for the acceptor-like V_{III} defects. Therefore, the group III self-diffusion will increase in heavily doped n-type heterostructure when annealed under As-rich condition. However, under As-poor annealing conditions, the n-type heterostructure remains stable with little intermixing.

Similar arguments can be applied to the p-type crystal, where the Fermi level is situated close to the valence band. Under As-poor annealing conditions, the group III self-diffusion proceeds via group III interstitials (Ga_i and Al_i), which are donor-like defects with $E = E_d$. Since $E_d > E_F$, most defects are ionized. Through the same argument of n-type structure, the equilibrium concentration of ionized defects is proportional to $\exp[(E_d - E_F)/kT]$. The solubility of donor-like group III interstitials will be increased to enhance the group III self diffusion. Nevertheless, p-type heterostructure does not intermix under As-rich annealing conditions.

(c) **Application of IILD to QW laser fabrication**

Because IILD allows the fabrication of heterojunctions perpendicular to the thin epitaxial layers, this process can be used to fabricate devices with three-dimensional carrier and photon confinement using standard masking procedures of the crystal surfaces and either selective impurity diffusion or ion implantation. This property makes the IILD well suited to the fabrication of both buried heterostructure quantum-well laser diodes and optical waveguides for optoelectronic integrated circuit applications.

As an example, Fig. 5.19 shows the cross-sectional scanning electron microscope (SEM) image of a double heterostructure QW laser diode after IILD. The starting material is a typical GaAs-AlGaAs double heterostructure wafer which forms 1D current and optical confinements along the layer growth direction in fabricated laser diodes. The active region of the device consists of GaAs multiple QWs separated by $Al_xGa_{1-x}As$ ($x \sim 0.35$) barriers, with barrier and wells sandwiched in a 0.2 μm $Al_xGa_{1-x}As$ waveguide region. An n-type $Al_yGa_{1-y}As$ ($y \sim 0.8$) lower confining layer and a p-type $Al_yGa_{1-y}As$ upper confining layer complete the structure. A thin (~0.1 μm) heavily doped p^+ contact layer is grown on the upper p-type $Al_yGa_{1-y}As$ confining layer in order to aid low-resistance metal contact to the device. To achieve 2D current and optical confinements, Si IILD is then used to process the wafer into

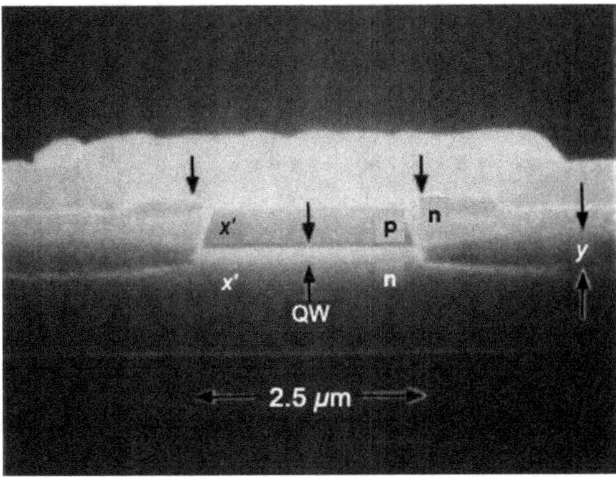

Fig. 5.19 SEM image of the stripe region after Si diffusion (850 °C, 9.5 h) of the QW heterostructure. The original masking stripe width is 5 μm. Lateral diffusion of the Si has narrowed the active region to ~2.5 μm. Arrows show where the Si, through intermixing of the $Al_xGa_{1-x}As$ ($x \sim 0.35$) and $Al_{x'}Ga_{1-x'}As$ ($x' \sim 0.85$) regions, flares the waveguide and forms $Al_yGa_{1-y}As$ ($x < y < x'$). Reprinted with permission from [12], copyright AIP Publishing

buried heterostructure QW laser diodes. As shown in Fig. 5.19, the Si diffusion, which has been masked by a narrow (~5 μm) Si_3N_4 stripe, extends on either side of the active region into the lower *n*-type $Al_yGa_{1-y}As$ confining layer. The Si diffusion converts the crystal to *n*-type and thus forms a lateral current constraining *p*-*n* junction in the high gap upper *p*-type $Al_yGa_{1-y}As$ confining layer. Most important, as the Si diffuses through the QW active region, layer intermixing occurs between the GaAs QWs, the $Al_xGa_{1-x}As$ barriers, and the $Al_xGa_{1-x}As$ waveguide, and also between the waveguide region and the upper and lower $Al_yGa_{1-y}As$ confining layers. The layer intermixing results in an increased bandgap relative to the masked undiffused active region preserved beneath the Si_3N_4 mask, thus giving a built-in lateral refractive index step that provides optical confinement. The lateral *p*-*n* junction formed in the upper confining layer because of the Si diffusion provides current confinement in the laser active region since the high gap junction and the diode active region have different turn-on voltages. This form of IILD buried heterostructure offers the advantage of a simplified fabrication process over that of regrown laser diode structures. An added advantage of the IILD is that by utilizing the lateral diffusion of the Si under the Si_3N_4, it is possible to obtain reduced active region widths (≤0.7 μm) and thus lower lasing thresholds.

5.6.2 Wet Oxidation of Al-Containing III-V Alloys

The predominance of silicon as the semiconductor material of choice was mainly due to, among other things, its ability to grow a native oxide SiO_2 on Si surface. This led directly to the development of the IC industry by providing a convenient patternable mask for impurity diffusion and a high-quality oxide-semiconductor interface for MOSFETs. Conversely, the chief shortcoming of III-V compound semiconductors has been the lack of a high-quality native oxide. Although many techniques have been attempted, including anodic and thermal oxidation, a viable oxide has proven elusive. On the other hand, it has been recognized that $Al_xGa_{1-x}As$ with high Al-composition ($x > 0.7$) is unstable and deteriorates (hydrolyzed) after long exposure to atmospheric condition. During their studies of the reliability issues involved in high Al-composition AlGaAs heterostructures, J. Dallessasse and N. Holonyak Jr. discovered that by oxidizing the semiconductor in water vapor at elevated temperatures, a dense and stable oxide of AlGaAs is formed [13]. This AlGaAs native oxide possesses many desirable properties for integrated semiconductor photonic device fabrication including a low index of refraction (~1.63) and good electrical insulating properties.

(a) **Wet-oxidation reactions**

The wet oxidation of AlGaAs is usually performed by exposing the high Al-composition layers to water vapor, transported in an inert carrier gas, in an elevated temperature environment of ~350 to 600 °C. The oxidation process proceeds either from the plane surface or the exposed edges of the multilayer structure (lateral oxidation). The reactions of wet oxidation of Al-bearing III-V materials are more complex than silicon oxidation, as the interaction of the water molecule with both group III and group V elements needs to be considered. Although the aluminum-water system is well studied in equilibrium and under high-pressure conditions, a systematic examination of reactions between Al-bearing crystalline material, such as AlAs, and water is lacking. Nevertheless, by projecting from the alumina-water system, for the oxidation of $Al_xGa_{1-x}As$, it is expected that $Al(OH)_3$ forms during the room-temperature deterioration process, and at temperatures up to ~100–150 °C, AlO(OH) forms at intermediate temperatures, and amorphous γ-Al_2O_3 is formed in the temperature range greater than ~370 °C. It is also noted that, through secondary-ion mass spectroscopy (SIMS) studies, arsenic is depleted from the high-temperature oxidized layers as well as near the surface of room-temperature hydrolyzed layers, supporting the view that the reaction creates a volatile form of arsenic. It is believed that AsH_3 is formed as the primary product of the direct reaction with the $Al_xGa_{1-x}As$ crystal. Then AsH_3 immediately undergoes a follow-on reaction with the H_2O to form As_2O_3 or elemental arsenic. Thus the likely reactions could be described by

$$2AlAs + 3H_2O \rightarrow Al_2O_3 + 2AsH_3 \qquad \geq 370\,^\circ C \tag{5.26a}$$

$$AlAs + 2H_2O \rightarrow AlO(OH) + AsH_3 \qquad > 150 \sim < 370\,^\circ C \tag{5.26b}$$

$$AlAs + 3H_2O \rightarrow Al(OH)_3 + AsH_3 \qquad \leq 150\,^\circ C \tag{5.26c}$$

(b) **Oxidation process parameters**

The wet-oxidation rate of Al-bearing III-V semiconductors is affected by several process parameters including Al-composition, layer thickness, and oxidation temperature. Following the model of silicon oxidation, under fixed oxidation temperature and constant gas flow containing oxidizing medium (water vapor), the oxidation thickness as a function of time is described by

$$x_0^2 + Ax_0 = Bt \tag{5.27}$$

where x_0 is the thickness of the oxide layer, coefficient B is related to the diffusivity of the reactants in the oxide layer, and coefficient A is determined by the oxide-semiconductor interface-reaction rate constant and the gas-phase mass-transfer coefficient.

In the limit of thin oxide thickness or short oxidation time, one obtains a linear growth of oxide layer thickness with time,

$$x_0 \cong (B/A)t \tag{5.28}$$

where *B/A* is the linear reaction rate constant. This linear behavior is called *reaction-rate limited* case. In the limit of longer oxidation time, a parabolic relationship is approached.

$$x_0^2 \cong Bt \tag{5.29}$$

Then the oxidation rate is *diffusion limited* where the rate is determined by the diffusion of water vapor through the oxide to the reaction front.

For lateral oxidation of $Al_xGa_{1-x}As$, the time dependence of the oxidation thickness has also been observed to vary between linear (reaction-rate limited) and parabolic (diffusion-limited). However, the origin of the shift of oxidation rate from linear to parabolic relationship is different. During the wet-oxidation process, there exists a thin, dense, amorphous region of a few nanometers thick at the oxidation front. Behind this dense mixture of As_2O_3 and Al_2O_3 is a less dense region of amorphous Al_2O_3 that extends to the exposed mesa edge. The time progression of the thickness of this dense region will determine whether linear or parabolic behavior dominates. When the reduction of As_2O_3 to As is sufficiently fast to balance the rate of formation of As_2O_3, a thin layer of dense oxide of relatively constant thickness will be formed near the oxidation front. Under this condition, a linear time-dependent oxidation rate is expected. However if reaction conditions are changed to preferentially increase the formation of As_2O_3 relative to As loss, a steady increasing thickness of the dense, As_2O_3-containing layer will form more quickly. This will lead to a diffusion-limited parabolic time dependence of the oxidation rate. This shift from reaction-rate limited to diffusion-limited time dependence is favored by increasing oxidation temperature or increasing Al content of $Al_xGa_{1-x}As$, both of which increase total oxidation rate.

Next, the temperature-dependent oxidation rate is considered. Experimental results show that the oxidation rate of Al-rich $Al_xGa_{1-x}As$ depends exponentially on oxidation temperature and follows an Arrhenius relationship. It is given by $x = x_0 \exp(-E_a/kT)$, where x_0 is a constant depending on the material, and E_a is the activation energy. As shown in Fig. 5.20, the activation energy maintains a nearly constant value and decreases slowly from 1.90 to 1.85 eV as the AlAs-mole fraction increased from 0.9 to 0.96. However, the activation energy of $Al_xGa_{1-x}As$ drops sharply when the AlAs-mole fraction is increased beyond 0.96. The activation energies reduce to ~1.75 eV and 1.25 eV for $Al_{0.98}Ga_{0.02}As$ and AlAs, respectively. When oxidized at a fixed temperature, this rapid change of activation energy in Al-rich $Al_xGa_{1-x}As$ leads to an extremely strong compositional dependence of the oxidation rates. Thus by making a small change in AlAs-mole fraction near $1 \geq x \geq 0.9$, the oxidation rate can be changed by as much as an order of magnitude, leading to a high degree of oxidation selectivity between AlGaAs layers of different Al concentrations. However, this unique property implies that a stringent composition control is critical to precisely regulate the oxidation rate.

Furthermore, the oxidation rate of AlGaAs is also influenced by its thickness and hetero-interface structures. Specifically, the lateral oxidation rate of AlAs layers departs from a constant rate and decreases sharply for layers thinner than ~60 nm (Fig. 5.21). The drastic reduction of oxidation rate with decreasing AlAs layer thickness is due to the variation of surface energy between AlAs and AlGaAs barriers at the oxide front or change of strain with thickness. This behavior can be used to compensate for the compositional dependence of wet oxidation since thin layers of AlAs may oxidize slower than thick layers of AlGaAs.

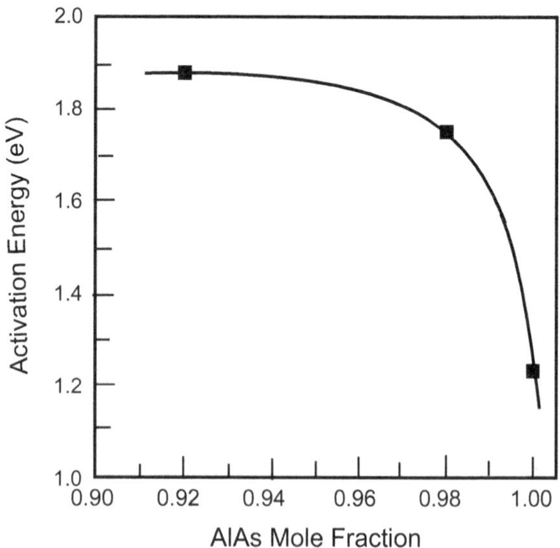

Fig. 5.20 Arrhenius activation energy for the oxidation reaction of $Al_xGa_{1-x}As$ versus Al-composition. Reprinted with permission from [14], copyright IEEE

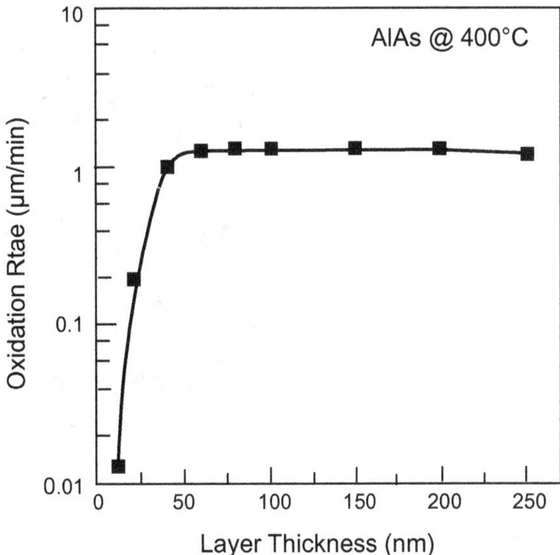

Fig. 5.21 AlAs oxidation rate at 400 °C versus AlAs layer thickness. Reprinted with permission from [14], copyright IEEE

(c) **Application of wet oxidation to device fabrication**

The wet oxidation of Al-bearing III-V alloys converts semiconductor into an insulator that can be used for the design and fabrication of various device structures. The oxide layers can be formed on the surface or through lateral oxidation from an edge to create oxide narrow bands between semiconductor layers. Due to its capability to form insulating layers for current confinement and to provide a significant refractive index difference for optical confinement, a wide variety of devices have consequently been fabricated using the III-V oxidation process.

Edge-emitting lasers with single- and multiple-stripe geometry have been fabricated using the III-V surface oxidation process. The oxide layer is formed in the $Al_{0.8}Ga_{0.2}As$ upper cladding layer of a GaAs-$Al_xGa_{1-x}As$ QW heterostructure laser structure. The high selectivity in oxidation rate between material having different aluminum compositions allows a simple fabrication process that uses the top GaAs contact layer as the oxidation mask. The laser stripe or stripes are defined using photolithography followed by a wet chemical process to etch through the GaAs cap to the $Al_{0.8}Ga_{0.2}As$ layer underneath. Then the photoresist is stripped, and the material is oxidized. For example, an oxide layer ~1000–1500 Å thick is formed in a 3 h oxidation at 400 °C using water vapor provided by a 1.4 scfh N_2 flow through a water bubbler maintained at 95 °C. In Fig. 5.22, a SEM cross section image for a fabricated laser array consisting of ~3 μm stripes on 4 μm centers over a 200 μm aperture is shown. Stable operation to high output powers (>2.5 W) has been achieved in this laser with surface-oxide defined multistripe arrays.

The lateral wet oxidation at the exposed edges of superlattice samples proved to be a more versatile approach for both edge- and surface-emitting photonic devices.

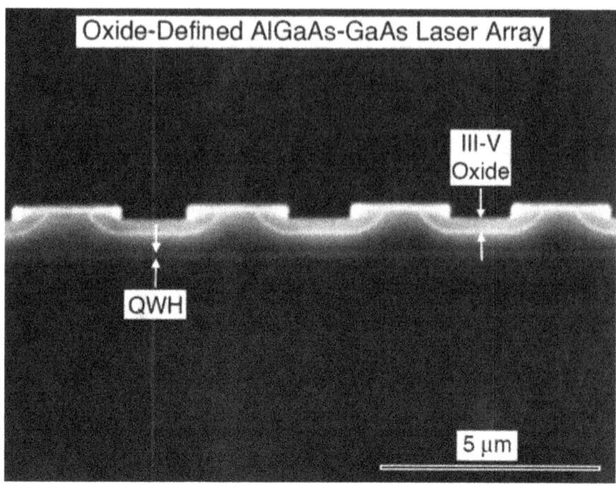

Fig. 5.22 SEM cross section image of a segment of an AlGaAs-GaAs multistripe laser. The total array consists of 50 stripes, with a total aperture width of 200 μm. Reprinted with permission from [13], copyright AIP Publishing

It has been shown that lateral oxidation can be confined to a layer having high Al-composition when kept between two lower composition layers of sufficient thickness. The close proximity of the oxide to the waveguide and quantum-well layers of the laser also demonstrates that the oxide will not degrade device performance even when very close to the device active region. An example of lateral oxidation to form buried apertures for current and optical confinement in a current injection edge-emitting laser is shown in Fig. 5.23. Here, index-guided lasers are formed in the AlGaAs-GaAs-InGaAs material system to produce lasers operating in the 980 nm wavelength range. The high-composition $Al_{\sim 0.85}Ga_{\sim 0.15}As$ layers on the top and bottom of a lower composition QW region are oxidized from the edge of a stripe mesa to form layers that reduce the active device geometry and confine both optically and electrically.

In Fig. 5.23a, the extent of lateral oxidation on the top and bottom confining layers is clearly evident in the cross-sectional SEM. Lower composition layers in the laser waveguide and quantum-well region do not show significant oxidation. The

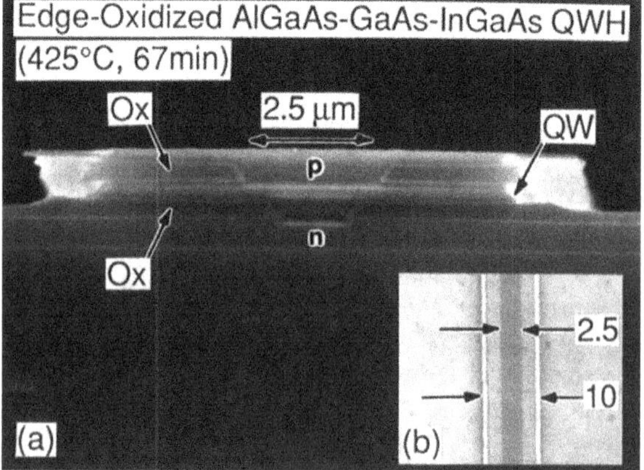

Fig. 5.23 **a** A SEM cross section and **b** a top-down optical photomicrograph of an AlGaAs-GaAs-InGaAs stripe-geometry QW laser formed via lateral oxidation. The slight difference in the rate of lateral oxidation between the upper and lower oxide layers is due to the small composition difference between the layers ($x \sim 0.85$ versus $x \sim 0.87$). Reprinted with permission from [15], copyright AIP Publishing

large difference in oxidation rates, especially between the upper and lower confining layers, highlights the sensitivity of oxidation rate to the aluminum composition of the oxidized layer. The top-down image shown in Fig. 5.23b also demonstrates a clear demarcation between the edge of the mesa (10 μm width) and the lateral extent of the oxidation front (2.5 μm width). Since the lateral oxidation distance can be controlled through Al-composition, and oxidation time and temperature, wet oxidation provides a scalable, manufacturable method for forming small-dimension lasers from larger mesas. Application of oxidation to VCSELs (vertical-cavity surface-emitting lasers), i.e., circular lateral confinement, which employ high-composition $Al_xGa_{1-x}As$ layers to achieve required device performance, is a simple extension and an ideal fit for oxide technology.

References

1. C.D. Thurmond, J. Phys. Chem. Solids **26**, 785 (1965)
2. H. Amano, Jpn. J. Apply. Phys. **52**, 050001 (2013)
3. R. Dwilinski, R. Doradzinski, J. Garczynski, L.P. Sierzputowski, A. Puchalski, Y. Kanbara, K. Yagi, H. Minakuchi, H. Hayashi, J. Crystal Growth **311**, 3015 (2009)
4. M.B. Panish, M. Ilegems, in *Progress in Solid State Chemistry*, vol. 7, ed. by H. Reiss, J.O. McCaldin (Pergamon Press, New York, 1972), pp. 39–83
5. T. Mizatami, M. Yoshida, A. Usai, H. Watanabe, T. Yuasa, I. Hayashi, Jpn. J. Appl. Phys. vo. **19**, L113 (1980)
6. A.Y. Cho, K.Y. Cheng, Appl. Phys. Lett. **38**, 360 (1981)
7. A.Y. Cho, J. Vac. Sci. Technol. **8**, S31 (1971)
8. B.A. Joyce, P.J. Dobson, J.H. Neave, K. Woodbridge, J. Zhang, P.K. Larsen, B. Bolger, Surf. Sci. **168**, 423 (1986)
9. F.F. Briones, *Method of producing white phosphorus for molecular beam epitaxy*. U.S. Patent 5,482,892, January 9, 1996
10. W.D. Laidig, N. Holonyak Jr., M.D. Camras, K. Hess, J.J. Coleman, P.D. Dupkus, J. Bardeen, **38**, 776 (1981)
11. K. Meehan, N. Holonyak Jr., J.M. Brown, M.A. Nixon, P. Gavrilovic, R.D. Burnham, Appl. Phys. Lett. **45**, 549 (1984)
12. D.G. Deppe, K.C. Hsieh, N. Holonyak, R.D. Burnham, R.L. Thornton, J. Appl. Phys. **58**, 4515 (1985)
13. J.M. Dallesasse, N. Holonyak, J. Appl. Phys. **113**, 051101 (2013)
14. K.D. Choquette et al., IEEE Sel. Topic. Quantum Electron. **3**, 916 (1997)
15. S.A. Maranowski, A.R. Sugg, E.I. Chen, N. Holonyak Jr., Appl. Phys. Lett. **63**, 1660 (1993)

Further Reading

1. V. Swaminathan, A.T. Macrander, *Materials Aspects of GaAs and InP Based Structures* (Prentice-Hall, 1991)
2. A. Denis, G. Goglio, G. Demazeau, Mat. Sci. Eng. **R50**, 167 (2006)
3. K. Xu, J.F. Wang, G.Q. Ren, Chin. Phys. B **24**, 066105 (2015)
4. H.C. Casey, M.B. Panish, *Heterostructure Lasers: Part B, Materials and Operating Characteristics* (Academic Press, 1978)

5. G. Beuchet, Halide and chloride transport vapor-phase deposition of InGaAsP and GaAs, in *Semiconductors and semimetals*, ed. by W. T. Tsang, vol. 22, Part A (Orlando, FL: Academic Press, 1985) pp. 261–297
6. A.Y. Cho, *Molecular Beam Epitaxy* (American Institute of Physics, New York, 1994)
7. G. B, Stringfellow, *Organometallic Vapor-Phase Epitaxy: Theory and Practice* (Academic Press, Boston, 1989)
8. M.B. Panish, H. Temkin, *Gas Source Molecular Beam Epitaxy* (Springer-Verlag, Berlin, 1993)
9. D.G. Deppe, N. Holonyak, J. Appl. Phys. **64**, R93 (1988)
10. N. Holonyak, IEEE Sel. Topic. Quantum Electron **4**, 584 (1998)

Chapter 6
Heterostructure Fundamentals

Abstract The development of modern epitaxy methods enables the verification of many core concepts of modern physics and the development of bandgap engineering technology using quantum wells and superlattices. Molecular beam epitaxy (MBE) technology, in particular, not only brings the theory of quantum mechanics from classroom to laboratory through the fabrication of quantum-well structures, but also promotes new discoveries in physics including the fractional quantum Hall effect. In addition, many new physics concepts have been discovered and developed into different physics subdisciplines, including low-dimension (0D and 1D) nanometer-scale structures and technologies. With the ability to form high-quality heterostructures, new breeds of devices including unipolar quantum cascade lasers, high electron-mobility transistors, and quantum-well infrared photodetectors, just to mention a few, have been invented. All these developments rely on the inimitable properties of the heterostructure where the proper energy band alignment provides unique carrier and/or optical confinement. At the inception of III–V heterostructure development, an energy band alignment model based on electron affinity difference between semiconductors was developed. The model provided a simple theoretical basis to promote the development of heterostructure technologies. However, as the epitaxy technique improved, the discrepancy between the simple theoretical predictions and experimental results became undeniable, prompting the development of a number of theoretical band-edge alignment models to include effects of charge redistribution at the hetero-interface. Meanwhile, the advancement of epitaxy techniques allows the formation of strained but otherwise high-quality thin layers in lattice-mismatched heterostructure systems. Thus, to properly describe the equilibrium band diagram of a heterostructure, band-edge alignment models capable of treating strain effects are needed. The development of band-edge alignment models including strain effect constitutes the focus of this chapter.

K. Y. Cheng, *III–V Compound Semiconductors and Devices*, Graduate Texts in Physics,
https://doi.org/10.1007/978-3-030-51903-2_6

6.1 Energy Band Alignment

6.1.1 Anderson's Electron Affinity Model

In an ideal Schottky barrier structure, we use the vacuum as the reference level to determine the metal–semiconductor barrier height ϕ_b. As shown in Fig. 6.1, under the equilibrium condition, the conduction band edge and the Fermi level of the semiconductor below the vacuum level are represented by the electron affinity χ_s and the work function ϕ_s, respectively The work function of the metal is represented by ϕ_m. After forming the metal–semiconductor junction, the Fermi levels on both sides merge into one level to attain an equilibrium condition. This creates an energy barrier in the form of

$$\phi_b = \phi_m - \chi_s \tag{6.1}$$

Anderson's model of heterostructures is based on the same idea [1]. It uses the vacuum as the reference energy level to estimate the bandgap discontinuity at the *p*-*N* heterojunction interface as shown in Fig. 6.2. Note that a lower case *p* and a capital *N* have been used to represent the small and large bandgap materials, respectively, in the heterostructure.

Using the vacuum level as the reference, the conduction band and valence band discontinuities are defined as

$$\begin{cases} \Delta E_c = \chi_1 - \chi_2 \\ \Delta E_v = \left(E_{g1} - E_{g2}\right) - \Delta E_c \end{cases} \tag{6.2}$$

and

$$\Delta E_c + \Delta E_v = E_{g2} - E_{g1} = \Delta E_g \tag{6.3}$$

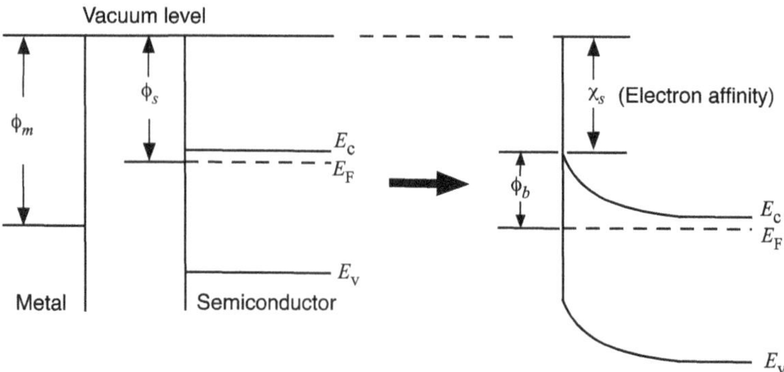

Fig. 6.1 Formation of the Schottky barrier and its energy band structure: Metal and semiconductor are separated in vacuum (left), and metal–semiconductor contact under equilibrium condition (right)

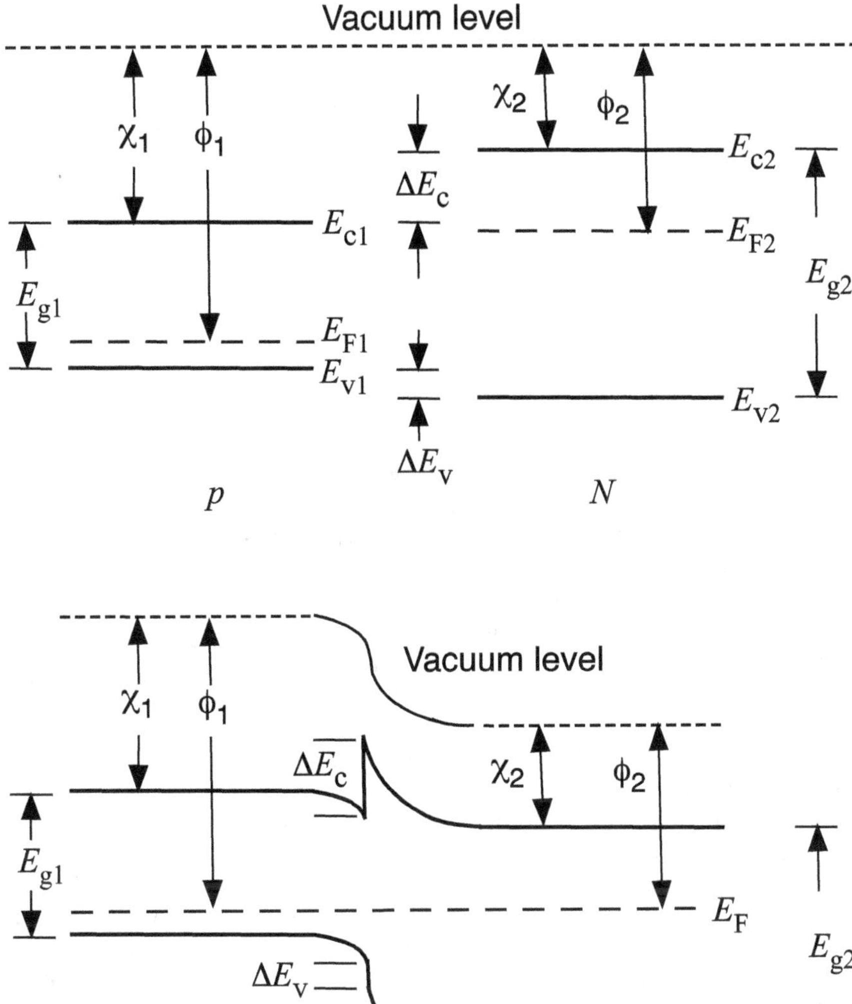

Fig. 6.2 Energy band alignments at the *p-N* heterojunction before and after contacting in Anderson's model

where χ_i and ϕ_i represent the electron affinity and work function of the material, respectively. At equilibrium, after the formation of the heterojunction, the Fermi levels on both sides of the heterojunction line up to the same value. At the same time, the electron affinity and work function in both materials, away from the hetero-interface, have to maintain constant values. These conditions lead to energy band bending at the heterostructure interface. This band bending is similar to the ideal Schottky barrier except that it occurs on both sides of the interface.

This simple model is intuitive and provides a starting point for the understanding of the nature of heterostructures. However, this early theory has a fundamental problem. In the case of an ideal Schottky barrier contact, a flat work function or vacuum level is maintained on the metal side and can be used as a reference level. For an ideal heterojunction inside a bulk solid, there is no vacuum level present to which all energies can be referred. The long-range nature of the Coulomb interaction in a crystal does not allow the zero of energy to be well defined as in an isolated atom. Therefore, there is no unique reference level within each bulk solid with which to compare the potentials for two different solids.

6.1.2 Model-Solid Theory

Numerous theoretical approaches have been pursued to derive an accurate band-edge alignment model. It is important to realize that around the hetero-interface, the redistribution of electronic charge induces dipole moments across the interface and leads to a shift of the bulk bands. Therefore, one has to perform a calculation in which the interface atoms from both types materials are included. The self-consistent interface calculations (SCIC), in which the electrons are allowed to adjust to the specific environment induced by the hetero-interface, have been used to derive the complete solution of this interface problem. Unfortunately, the computational complexity of the early first-principles calculations is very high, which limits their use as a general tool in the exploration and design of novel heterostructures. Therefore a reliable model to predict band offsets for a wide variety of heterostructures, and to do so without heavy calculations, is needed. In the 1980s, several band discontinuity models with varying degrees of success were proposed. These models are W. A. Harrison's 'linear combination atomic orbit' (LCAO) [2], J. Tersoff's 'energy level' [3, 4], M. Cardona and N. E. Christensen's 'linear muffin-tin orbitals' (LMTO)[5], and C. G. van de Walle's 'model solid' calculations [6, 7]. Among these, only the last two allow the incorporation of misfit strain into the calculation, making them usable in band alignment estimation for a wide range of heterostructures. However, the LMTO treatment of strain effects is more complicated, and an exact solution is not available. This leaves the model-solid theory of van de Walle as the most convenient and simple model for band alignment estimation including the strain effect.

The main idea of the model-solid theory is to establish an 'absolute reference energy level' for each semiconductor in a heterostructure such that the band lineups can be calculated. This reference energy level, $E_{v,\,av}$, was derived in the model-solid theory by averaging the three topmost valence bands at Γ, i.e., the heavy-hole, the light-hole, and the spin-orbit split-off bands. The numerical values of $E_{v,\,av}$ for most of the binary compound semiconductors have been derived by van de Walle and are listed in Table 6.1.

Using this reference energy, the position of the topmost valence band for a specific semiconductor can be derived by

Table 6.1 Material parameters for III–V binary semiconductor compounds at room temperature: lattice constant a_0, 300 K bandgap energies E_g^{Γ} (Γ-valley) and E_g^X (X-valley), spin-orbit split-off Δ_{so}, average valence band energy $E_{v,\,av}$, conduction band hydrostatic deformation potential a_c, valence band hydrostatic deformation potential a_v, shear deformation potential b ([001]-direction) and d ([111]-direction), elastic constants C_{11}, C_{12} and C_{44}, and valence band offset (VBO) with respect to InSb

Parameters	GaAs	AlAs	InAs	GaP	AlP	InP	GaSb	AlSb	InSb
a_0 (Å)	5.65325	5.6611	6.0583	5.4505	5.4672	5.8697	6.0959	6.1355	6.4794
E_g^{Γ} (eV)	1.424	3.003	0.354	2.9903	3.5527	1.3529	0.727	2.300	0.1737
E_g^X (eV)	1.8989	2.1641	1.37 1.07 (*L*)	2.2727	2.4878	2.273	1.033 0.753 (*L*)	1.616	0.63 (0 K)
Δ_{so} (eV)	0.341	0.28	0.39	0.08	0.07	0.108	0.760	0.676	0.81
$E_{v,\,av}$ (eV)	−6.92	−7.49	−6.67	−7.40	−8.09	−7.04	−6.25	−6.66	−6.09
a_c (eV)	−7.17	−5.64	−5.08	−8.2	−5.7	−6.0	−7.5	−4.5	−6.94
a_v (eV)	1.16	2.47	1.00	1.7	3.0	0.6	0.8	1.4	0.36
b (eV)	−2.0	−2.3	−1.8	−1.6	−1.5	−2.0	−2.0	−1.35	−2.0
d (eV)	−4.8	−3.4	−3.6	−4.6	−4.6	−5.0	−4.7	−4.3	−4.7
C_{11} (10^{11} dyn/cm^2)	12.21	12.50	8.329	14.05	13.30	10.11	8.842	8.769	6.847
C_{12} (10^{11} dyn/cm^2)	5.66	5.34	4.526	6.203	6.30	5.61	4.026	4.341	3.735
C_{44} (10^{11} dyn/cm^2)	6.00	5.42	3.959	7.033	6.15	4.56	4.322	4.076	3.111
VBO (eV)	−0.8	−1.33	−0.59	−1.27	−1.74	−0.94	−0.03	−0.41	0

(Data from [7] and [8])

$$E_{\mathrm{v}} = E_{\mathrm{v,av}} + \frac{\Delta_{\mathrm{so}}}{3} \tag{6.4}$$

where Δ_{so} is the spin-orbit split-off energy. The value of Δ_{so} is determined experimentally. Once the value of E_{v} is determined, the experimentally determined bandgap value can be used to determine the conduction band edge.

$$E_{\mathrm{c}} = E_{\mathrm{v}} + E_{\mathrm{g}} \tag{6.5}$$

Note, the bandgap energy is a temperature-dependent quantity and often described in the empirical Varshni form

$$E_{\mathrm{g}}(T) = E_{\mathrm{g}}(0) - \frac{\alpha T^2}{\beta + T} \tag{6.6}$$

where α (meV/K) and β (K) are fitting parameters and $E_{\mathrm{g}}(0)$ is the bandgap energy at absolute zero.

In an unstrained heterostructure, once we have determined E_{c} and E_{v} in two semiconductors with similar lattice constant, the band alignment can be calculated using

$$\begin{cases} \Delta E_{\mathrm{v}} = E_{\mathrm{v}}^{B} - E_{\mathrm{v}}^{A} \\ \Delta E_{\mathrm{c}} = E_{\mathrm{c}}^{B} - E_{\mathrm{c}}^{A} = \left(E_{\mathrm{v}}^{B} + E_{\mathrm{g}}^{B}\right) - \left(E_{\mathrm{v}}^{A} + E_{\mathrm{g}}^{A}\right) \end{cases} \tag{6.7}$$

The superscripts A and B correspond to the two semiconductors A and B.

Example: GaAs/AlAs Heterojunction From Table 6.1, we have the following parameters:

$$\begin{aligned} E_{\mathrm{v,av}}(\mathrm{GaAs}) &= -6.92\,(\mathrm{eV}), \\ \Delta_{\mathrm{so}}(\mathrm{GaAs}) &= 0.34\,(\mathrm{eV}), \\ E_{\mathrm{g}}(\mathrm{GaAs}) &= 1.424\,(\mathrm{eV}) \\ E_{\mathrm{v,av}}(\mathrm{AlAs}) &= -7.49\,(\mathrm{eV}), \\ \Delta_{\mathrm{so}}(\mathrm{AlAs}) &= 0.28\,(\mathrm{eV}), \\ E_{\mathrm{g}}(\mathrm{AlAs}) &= 2.164\,(\mathrm{eV}) \end{aligned}$$

Since AlAs has an indirect bandgap, the minimum indirect conduction band value of X-valley, E_{g}^{X}, is used in the calculation.

Then,

$$\begin{aligned} E_{\mathrm{v}}(\mathrm{GaAs}) &= -6.92 + 0.34/3 = -6.8063\,\mathrm{eV} \\ E_{\mathrm{v}}(\mathrm{AlAs}) &= -7.49 + 0.28/3 = -7.3967\,\mathrm{eV} \\ \Delta E_{\mathrm{v}} &= -6.8063 - (-7.3967) = +0.5903\,\mathrm{eV} \end{aligned}$$

This indicates that the valence band in GaAs is above the valence band in AlAs. The valence band discontinuity ΔE_v can also be determined from the valence band offset (VBO) values listed in Table 6.1. Using the empirical VBO values obtained by averaging experimental results, the relative VBO difference between GaAs and AlAs is obtained as

$$\text{VBO(GaAs)}-\text{VBO(AlAs)} = -0.8-(-1.33) = +0.53\,\text{eV}$$

The model-solid theory calculated ΔE_v is overestimated by 60 meV over the relative VBO value. The conduction band discontinuity occurs between the direct conduction band of GaAs and the indirect conduction band of AlAs. Therefore,

$$\begin{aligned}
E_c(\text{GaAs}) &= E_v(\text{GaAs}) + E_g(\text{GaAs}) \\
&= -6.8063 + 1.424 = -5.3823\,\text{eV} \\
E_c(\text{AlAs}) &= E_v(\text{AlAs}) + E_g(\text{AlAs}) \\
&= -7.3967 + 2.1641 = -5.2326\,\text{eV} \\
\Delta E_c &= -5.3823-(-5.2326) \\
&= -0.1497\,\text{eV}
\end{aligned}$$

The minus sign indicates that the direct conduction band in GaAs is below the indirect conduction band in AlAs. The band discontinuity lineup in GaAs/AlAs heterojunction is 'type-I,' meaning that the bandgap of one material (GaAs) is completely inside the bandgap of the other (AlAs). This band alignment is illustrated in Fig. 6.3.

At type-I interface, potential barriers exist for both electrons and holes in going from the smaller gap semiconductor to the larger gap semiconductor. This feature is very useful for double heterostructure lasers where both types of carriers can be confined in a single layer to generate stimulated emission. On the other hand, only one type of 2D carrier gas will form at the interface depending on the doping profile

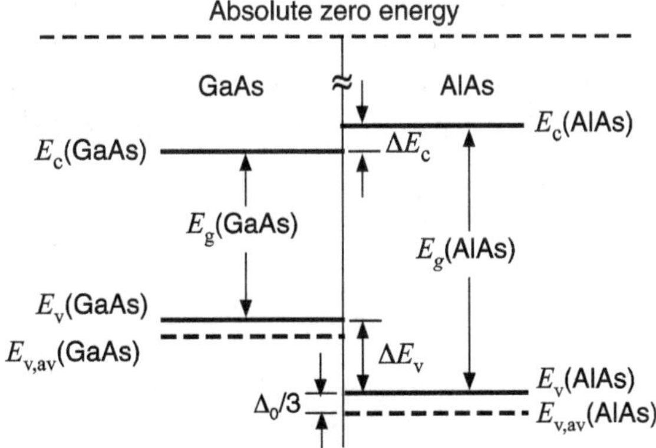

Fig. 6.3 Band lineups at the GaAs/AlAs interface. The average valence band $E_{v,\,av}$ is obtained from the model-solid theory. Spin-orbit splitting and bandgap energies are taken from experiments

across the interface. Interfaces formed by III–V compounds, either with different group III elements such as GaAs-AlAs or with different group V elements such as GaAs-GaP, belong to type-I.

6.1.3 Empirical Band Alignment Models

The experimentally determined band discontinuity data is very useful in designing heterostructure devices for specific characteristics. However, new device designs often require this information for alloy combinations for which no experimental data is available. The first step to derive the band discontinuity data is to establish band parameters of all compound semiconductor binary alloys. The band alignment data derived from the model-solid theory provides a useful guideline but the uncertainty is sometimes too large ($\geq$200 meV) to be useful for device designs. Therefore, it is worthwhile to compare the band discontinuity results of the model-solid theory with other empirically derived results. Using the existing theories coupled with experimental data, the VBOs of binary III–V alloys with respect to InSb (VBO of InSb is set at 0) are listed in Table 6.1. In deriving these VBO values, the properties of transitivity and commutativity are followed while the values of valence band offsets to be independent of temperature. The transitivity property of band offsets means that the sum of the band offsets at the *A–B* and *B–C* interfaces is equal to the *A–C* interface band offset. The commutative property means that the band offset between two semiconductors is independent of the order in which they are grown. Once the valence band discontinuity ΔE_{v} of two closely lattice-matched binary alloys is determined from the VBO difference of the heterojunction, the conduction band discontinuity ΔE_{c} is simply obtained by subtracting ΔE_{v} from the bandgap energy difference ΔE_{g}.

Example: AlSb/InAs/GaSb heterojunctions The lattice mismatches between binary alloys AlSb, GaSb, and InAs are small, and high-quality AlSb and InAs can be grown on GaSb substrates for mid-infrared (mid-IR) device applications. The alignments of conduction and valence bands in this lattice-matched alloy system are derived from the model-solid theory and shown in Fig. 6.4. Using values of $E_{\mathrm{v,\,av}}$, Δ_{so}, and E_{g} from Table 6.1, the maximum valence band edges of AlSb and GaSb are

$$E_{\mathrm{v}}(\mathrm{AlSb}) = -6.66 + 0.676/3 = -6.4347(\mathrm{eV})$$
$$E_{\mathrm{v}}(\mathrm{GaSb}) = -6.25 + 0.760/3 = -5.9967(\mathrm{eV})$$

and

$$\Delta E_{\mathrm{v}} = -6.4347 - (-5.9967) = -0.438(\mathrm{eV})$$

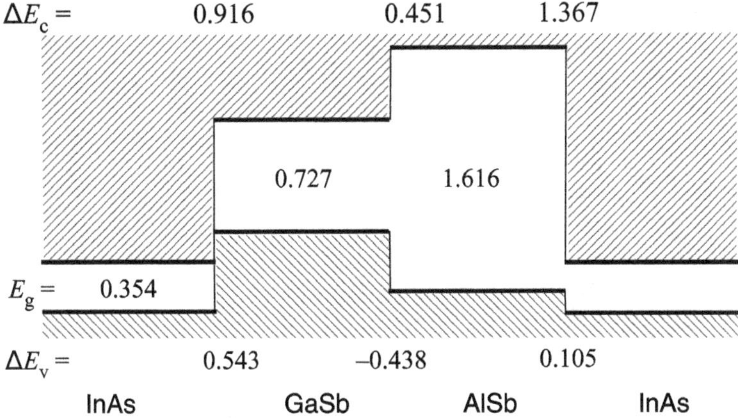

Fig. 6.4 Calculated bandgap energy and conduction and valence band-edge discontinuities in lattice-matched InAs/GaSb/AlSb material systems using model-solid theory

The valence band edge of AlSb is below that of GaSb by 0.438 eV. The measured relative VBO difference between AlSb and GaSb is –0.38 eV. The discrepancy between the theoretical and averaged measurement values is 58 meV. The conduction band discontinuity occurs between the direct conduction band of GaSb and indirect conduction band E_g^X of AlSb.

$$E_c(\text{AlSb}) = E_v(\text{AlSb}) + E_g^X(\text{AlSb}) = -4.8187\,(\text{eV})$$
$$E_c(\text{GaSb}) = E_v(\text{GaSb}) + E_g(\text{GaSb}) = -5.2697\,(\text{eV})$$

and

$$\Delta E_c = -4.8187 - (-5.2697) = 0.4510\,(\text{eV})$$

The indirect conduction band-edge minimum in AlSb is above the direct conduction band edge in GaSb leading to a type-I band alignment in AlSb-GaSb heterostructure.

Next, the band alignments of GaSb and InAs are derived. The maximum valence band edges of GaSb and InAs are

$$E_v(\text{GaSb}) = -5.9967\,(\text{eV})$$
$$E_v(\text{InAs}) = -6.67 + 0.39/3 = -6.540\,(\text{eV})$$

and

$$\Delta E_v = -5.9967 - (-6.540) = 0.5433\,(\text{eV})$$

The valence band edge of GaSb is above that of InAs by 0.5433 eV, which matches with the measured relative VBO difference between GaSb and InAs (0.56 eV) quite well. The minimum conduction band edges of GaSb and InAs are located at

$$E_c(\text{GaSb}) = -5.2697\,(\text{eV})$$
$$E_c(\text{InAs}) = E_v(\text{InAs}) + E_g(\text{InAs}) = -6.1860\,(\text{eV})$$

and

$$\Delta E_c = -5.2697 - (-6.1860) = 0.9163\,(\text{eV})$$

Both the conduction and valence band edges in InAs are below that of GaSb forming a 'type-II' band discontinuity lineup in the GaSb-InAs heterojunction. The GaSb-InAs interface is a typical example of the type-II interface alignment. Due to the large difference in average valence band energies $E_{v,\,av}$ and the spin-orbit splitting, Δ_{so}, between GaSb and InAs, in addition to the small bandgap of InAs, the two bandgaps are completely misaligned in energy. This forms a type-II *broken-gap interface* as shown in Fig. 6.4. The conduction band of InAs is below the valence band of GaSb. The electrons and holes are spatially separated and attracted to the InAs and GaSb sides of the interface, respectively. Type-II interfaces are formed in III-V compound heterostructures with both different group III elements and different group V elements.

If we reduce the difference in average valence band energies and spin-orbit splitting, and increase the bandgap energy of the smaller gap material by using ternary alloys, the type-II *staggered interface* results. AlSb-InAs and $Al_{0.48}In_{0.52}As$-InP systems are two typical examples. The maximum valence band edges of AlSb and InAs are

$$E_v(\text{AlSb}) = -6.4347\,(\text{eV})$$
$$E_v(\text{InAs}) = -6.540\,(\text{eV})$$

and

$$\Delta E_v = -6.4347 - (-6.540) = 0.1053\,(\text{eV})$$

The valence band edge of AlSb is above that of InAs by 0.1053 eV. However, the measured relative VBO difference between AlSb and InAs is 0.18 eV. The minimum conduction band edges of AlSb and InAs are located at

$$E_c(\text{AlSb}) = -4.8187\,(\text{eV})$$
$$E_c(\text{InAs}) = -6.1860\,(\text{eV})$$

and

$$\Delta E_c = -4.8187 - (-6.1860) = 1.3673\,(\text{eV})$$

Both the conduction and valence band edges in InAs are below that of AlSb forming a type-II staggered band discontinuity lineup in AlSb-InAs heterojunction. Again, the electrons and holes are spatially separated by the interface and located in the smaller and larger bandgap sides, respectively. The recombination of electrons and holes in a type-II staggered heterostructure results in emission energy below the gap of either semiconductor. This feature is important for the generation of mid-IR emissions beyond the bandgap energy of the heterostructure.

Next, the valence band alignments for unstrained ternary compounds are considered. When the constituent binary compounds are nearly lattice-matched, using linear extrapolation between two end binary alloys to generate valence band position for ternary compound is appropriate. Otherwise, a strain-induced bandgap bowing parameter should be included. More detailed discussions on strain-induced effects in ternary and quaternary compounds will be developed in Sect. 6.4. In general, the bowing parameter applies only to the conduction band edge. A linear variation is used for the estimation of valence band discontinuity in ternary alloys. This follows the observation in lattice-matched $Al_xGa_{1-x}As/GaAs$ and $(Al_xGa_{1-x})_{0.49}In_{0.51}P/GaAs$ systems that the valence band discontinuity follows the composition linearly.

A useful empirical technique for determining band alignments of unstrained III–V heterostructures was developed by S. Tiwari and D. J. Frank, which had an estimated accuracy of $\approx$100 meV [9]. The empirical approach is a consolidation of band discontinuity data based on two principles: the properties of transitivity and commutativity, and the strong correlation observed between barrier height of *p*-type semiconductor and the valence band discontinuity. The zero point of the energy at each lattice constant has been chosen as the reference level above the *p*-type semiconductor Schottky barrier height. Once the valence band discontinuity is determined for each lattice constant group, the conduction band discontinuity is simply obtained by adding the bandgap value to the *p*-type Schottky barrier height (ϕ_{bv}) listed in Table 4.5. For Si and Ge, ϕ_{bv} equals 0.447 and 0.072 eV, respectively. The band discontinuities for ternary compounds are generated from the binary data incorporating bandgap bowing parameters (Table 4.6). The complete room-temperature conduction band-edge and valence band-edge energies as functions of the lattice constant for unstrained III–V compound semiconductors are compiled in Fig. 6.5. The circles indicate the band edges of the binary semiconductors and the lines show the band edges of the ternary alloys. The portion of the ternary alloy with a direct bandgap is indicated by a solid line.

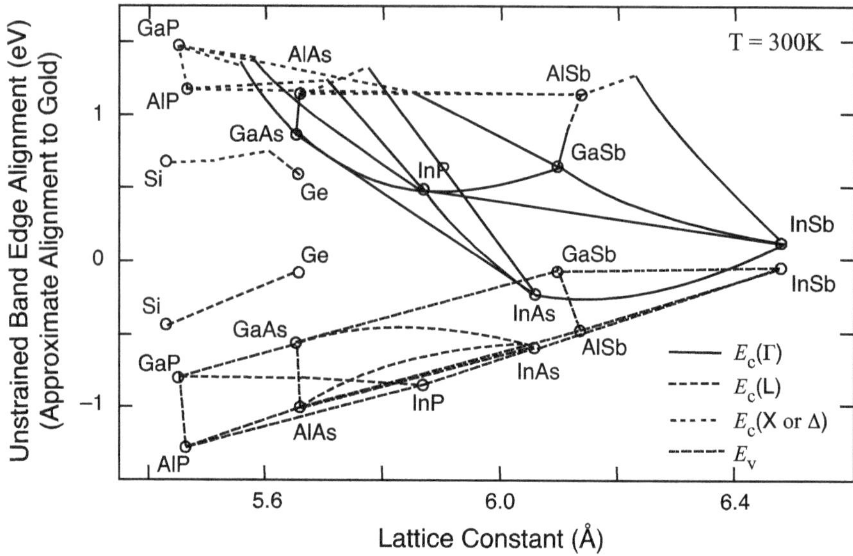

Fig. 6.5 Conduction band-edge and valence band-edge energies plotted as functions of the lattice constant of III–V and elemental semiconductors. The circles indicate the band edges of binaries and the lines show the band edges of ternaries. Reprinted with permission from [9], copyright AIP Publishing

6.2 Strained Layer Structures

When an epitaxial layer is grown on a foreign substrate with different lattice constant, the strain is created in such a heterostructure. The strain substantially alters the physical characteristics (e.g., band structure) of the epilayer and provides another degree of freedom for device structure design. Modern photonic and electronic devices take advantage of this added freedom by incorporating strained quantum-well thin layers into heterostructures with superior characteristics. However, in fabricating heterostructure devices, one of the most critical issues in material preparation is to avoid defect generation due to excessive lattice-mismatch between the substrate and the epitaxial layers. The generated high density of threading dislocation degrades the performance or even causes the device to cease operating. Therefore, it is necessary to examine the strain effect and its influence on bandgap energies closely.

6.2.1 *Critical Layer Thickness (h_c)*

If the lattice mismatch between a bulk substrate and an epitaxial layer is sufficiently small, the first few deposited atomic layers will be strained to match the substrate and its 2D cell structure is shown in Fig. 6.6a. The epilayer is coherently strained,

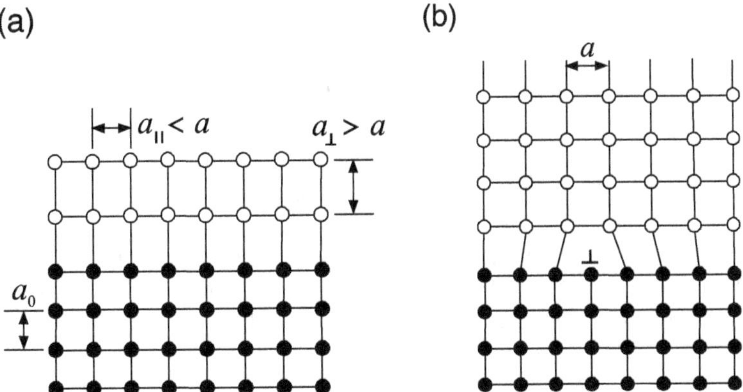

Fig. 6.6 Hetero-epitaxy of cubic crystal with equilibrium lattice constant a on top of a cubic crystal substrate with lattice constant a_0. **a** A coherently strained thin layer assumes an in-plane lattice constant a_0 and a normal lattice constant $a_{\perp} > a$. **b** When the epilayer thickness exceeds the critical layer thickness, h_c, misfit dislocations (⊥) are formed at the interface between the two semiconductors

and it maintains a pseudomorphic interface. As the epilayer thickness increases, the homogeneous strain energy increases rapidly. When the epilayer thickness reaches the critical layer thickness h_c and beyond, conditions become favorable for 'misfit dislocations,' meaning that some atoms within the epilayer are out of registry with the substrate to accommodate a fraction of the misfit as depicted in Fig. 6.6b.

Matthews and Blakeslee developed a mechanical balance model for critical layer thickness calculation [10]. In this model, as shown in Fig. 6.7, a pre-existing segment of threading dislocation bows and elongates under the influence of the misfit stress. At the critical layer thickness h_c, the misfit strain, F_H, equals the interface tension, F_I, and generates a segment of interfacial dislocation line. Beyond this point, the

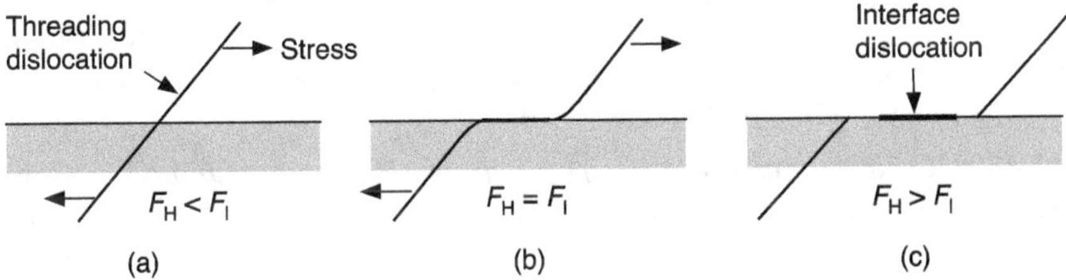

Fig. 6.7 Matthews and Blakeslee's model of misfit dislocation generation at the hetero-epitaxy interface. **a** A segment of threading dislocation exists at the interface of a strained system. **b** When the epilayer reaches the critical layer thickness, $F_H = F_I$, generating a segment of interfacial dislocation line. **c** With further increase of the layer thickness, the interface becomes incoherent and a misfit dislocation is formed [10]

originally coherent interface becomes incoherent and an interface misfit dislocation is generated.

For a single epitaxial layer, the equilibrium critical thickness is determined by

$$h_c = \left(\frac{b}{f_0}\right)\frac{1 - v\cos^2\Theta}{8\pi(1 + v)\cos\lambda}\left[1 + \ln\left(\frac{h_c}{b}\right)\right] \tag{6.8}$$

where $b = a_0/\sqrt{2}$ is Burger's vector, a_0 is the bulk lattice constant of the epilayer, f_0 is the misfit, v is Poisson's ratio, and $\Theta = \lambda = 60°$ and $30°$ in cubic and wurtzite crystal structures, respectively. The misfit and Poisson's ratio are defined as

$$f_0 = \frac{|a_0 - a_s|}{a_s} \tag{6.9}$$

and

$$\nu = \frac{C_{12}}{C_{11} + C_{12}} \tag{6.10}$$

where a_s is the lattice constant of the substrate, and C_{11} and C_{12} are Young's moduli which will be defined in the next section.

The critical thickness of a strained multilayer structure is obtained by the same procedure as the single-layer model except that in the multilayer model, it is now required that $F_H = 2F_I$. The factor 2 arises from the fact that in a multilayer structure such as a quantum well, each interface may support a segment of line dislocation. Therefore,

$$h_c(\text{multilayer}) = 2h_c(\text{single layer}) \tag{6.11}$$

Under normal growth conditions, the Matthews and Blakeslee model gives a fairly good fit to experimental data. However, strain relaxation is a kinetically controlled process. The critical layer thickness is, for example, a function of the growth temperature, which is not counted for in the model. Nevertheless, the Matthews and Blakeslee model is a good starting point to estimate the critical layer thickness.

Example: Critical layer thickness of a $Ga_{0.7}In_{0.3}As$/GaAs quantum well The relevant material parameters from Table 6.1 are listed below.

$$\begin{aligned}
C_{11}(\text{InAs}) &= 8.329 \times 10^{11}, \\
C_{12}(\text{InAs}) &= 4.526 \times 10^{11}, \\
a_0(\text{InAs}) &= 6.0583\,\text{Å} \\
C_{11}(\text{GaAs}) &= 12.21 \times 1011, \\
C_{12}(\text{GaAs}) &= 5.66 \times 10^{11}, \\
a_0(\text{GaAs}) &= 5.6533\,\text{Å}
\end{aligned}$$

The lattice constant, C_{11} and C_{12} of the $Ga_{0.7}In_{0.3}As$ quantum well are calculated using interpolation from binaries as

$$a_0(x) = 0.7 \times 5.6533 + 0.3 \times 6.0583 = 5.7748\ \text{Å}$$
$$C_{11}(x) = (0.7 \times 12.21 + 0.3 \times 8.329) \times 10^{11} = 11.046 \times 10^{11}$$
$$C_{12}(x) = (0.7 \times 5.66 + 0.3 \times 4.526) \times 10^{11} = 5.320 \times 10^{11}$$

Then, we can calculate the Burger's vector, misfit, and Poisson's ratio as follows:

$$b = a_0(x)/\sqrt{2} = 4.083$$
$$f_0 = |(5.7748 - 5.6533)|/5.6533 = 0.0215$$
$$v = 5.320/(11.046 + 5.320) = 0.3251$$

Using $\Theta = \lambda = 60°$, we have

$$h_c = \left(\frac{4.083}{0.0215}\right)\frac{1 - 0.3251\cos^2 60^\circ}{8\pi(1 + 0.3251)\cos 60^\circ}\left[1 + \ln\left(\frac{h_c}{4.083}\right)\right]$$
$$0.0954h_c - 1 = \ln(h_c/4.083)$$
$$h_c = 4.083\exp(0.0954h_c - 1)$$

The numerical solution of h_c is 32.1 Å. For the $Ga_{0.7}In_{0.3}As$/GaAs quantum well, the critical layer thickness is $2 \times 32.1 = 64.2$ Å.

6.2.2 *Cubic Crystal Under Stress*

(a) **Basic definitions**

Since most III–V compound semiconductors have a cubic crystal structure, it is reasonable as well as pragmatic to examine a strained cubic crystal first. Stress σ is the force intensity acting on a material. The stress deforms the crystal and generates strain ϵ. Depending on applied directions, the stress force can be divided into a shear and a normal component as shown in Fig. 6.8. Shear components of stress σ_{ij} $(i \neq j)$ will cause the crystal to rotate. In typical semiconductors, this type of deformation is rare and we will assume that $\sigma_{ij} = 0$ for $i \neq j$. The normal components of stress (σ_{ij}, $i = j$) will cause the crystal to expand or contract along the crystal axes. The deformed crystal remains rectangular. Since we are only considering normal components, the three normal forces σ_{11}, σ_{22}, and σ_{33} can be abbreviated by σ_1, σ_2, and σ_3. The σ_i's are defined as positive for outward directed forces.

In a uniform crystal, within the elastic limit of the material, the induced strain tensors from the applied stress are simply related by Hooke's law, $\epsilon = s\sigma$ or $\sigma = C\epsilon$, where the proportion constant s and C are the *compliance* and *Young's modulus* of the

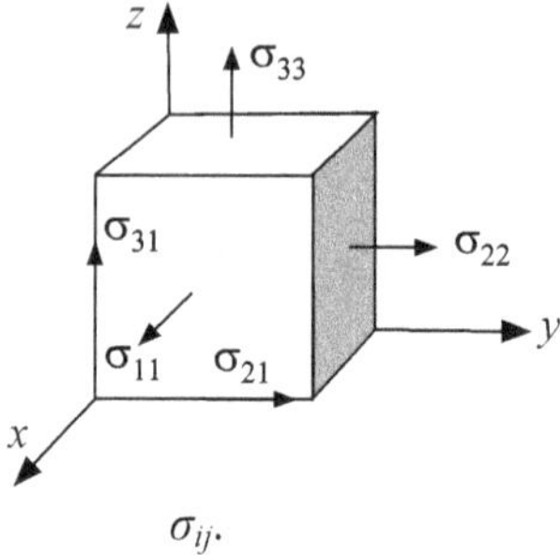

Fig. 6.8 Components of stress tensor, σ_{ij}

crystal, respectively. Under a homogeneous stress σ_{ij}, excluding shear components of stress, the resulting homogeneous strain ϵ_{ij} is a linear combination of all components of the stress, where $i, j = 1, 2$, and 3 corresponding to the crystallographic x-, y-, and z-axis. Thus, for example,

$$\epsilon_{ij} = s_{ij11}\sigma_{11} + s_{ij12}\sigma_{12} + s_{ij13}\sigma_{13} + s_{ij21}\sigma_{21} + s_{ij22}\sigma_{22} + s_{ij23}\sigma_{23} + s_{ij31}\sigma_{31} + s_{ij32}\sigma_{32} + s_{ij33}\sigma_{33} \tag{6.12}$$

and eight similar equations for the other eight components of ϵ_{ij}. Alternatively, the generalized Hooke's law may be also conveniently written as

$$\sigma_{ij} = C_{ijkl}\epsilon_{kl} \tag{6.13}$$

The C_{ijkl} are the 81 elastic stiffness coefficients or the elastic moduli of the crystal, where i, j, k, and l take on values of 1, 2, and 3 corresponding to the crystallographic x-, y-, and z-axis. Under homogeneous stress, a body must experience balanced forces to maintain statistical equilibrium, which implies that some of the strain components must be interrelated to prevent a net rotation of the body. For example, a homogeneous force is applied on the faces along the x-axis and perpendicular to the y- and z-axis of a unit cube. The shear components of the stress have to be balanced, e.g., $\sigma_{23} = \sigma_{32}$, to prevent any rotation or translation of the crystal body. As a consequence, one must have

$$\sigma_{ij} = \sigma_{ji} \text{ and } \epsilon_{kl} = \epsilon_{lk} \tag{6.14}$$

In general,

$$C_{ijkl} = C_{ijlk} \text{ and } C_{ijkl} = C_{jikl} \tag{6.15}$$

Thus the number of independent C_{ijkl} is reduced from 81 to 36 and can be expressed as a 6×6 tensor. We can further simplify the notation of C_{ijkl} by abbreviating the first two and last two suffixes into a pair of single digit number of 1–6 in the following way:

tensor notation	11	22	33	23,32	13,31	12,21
matrix notation	1	2	3	4	5	6

Now we can transform the elastic constant C_{ijkl} into C_{mn} such that

$$C_{ijkl} = C_{mn}(i, j, k, l = 1, 2, 3; m, n = 1, 2, \ldots 6) \tag{6.16}$$

and $\sigma_{ij} = C_{ijkl}\epsilon_{kl}$ becomes $\sigma_m = C_{mn}\epsilon_n$. Thus the array of C_{mn} is expressed in a 6 × 6 matrix as

$$C_{mn} = \begin{bmatrix} C_{11} & C_{12} & C_{13} & C_{14} & C_{15} & C_{16} \\ C_{21} & C_{22} & C_{23} & C_{24} & C_{25} & C_{26} \\ C_{31} & C_{32} & C_{33} & C_{34} & C_{35} & C_{36} \\ C_{41} & C_{42} & C_{43} & C_{44} & C_{45} & C_{46} \\ C_{51} & C_{52} & C_{53} & C_{54} & C_{55} & C_{56} \\ C_{61} & C_{62} & C_{62} & C_{64} & C_{65} & C_{66} \end{bmatrix}. \tag{6.17}$$

By requiring that the symmetry of the crystal be replicated in the elastic coefficient matrix, the matrix can be further reduced. For a cubic crystal, the tensor form of stress and strain is reduced to

$$\begin{bmatrix} \sigma_1 \\ \sigma_2 \\ \sigma_3 \\ \sigma_4 \\ \sigma_5 \\ \sigma_6 \end{bmatrix} = \begin{bmatrix} C_{11} & C_{12} & C_{12} & 0 & 0 & 0 \\ C_{12} & C_{11} & C_{12} & 0 & 0 & 0 \\ C_{12} & C_{12} & C_{11} & 0 & 0 & 0 \\ 0 & 0 & 0 & C_{44} & 0 & 0 \\ 0 & 0 & 0 & 0 & C_{44} & 0 \\ 0 & 0 & 0 & 0 & 0 & C_{44} \end{bmatrix} \begin{bmatrix} \epsilon_1 \\ \epsilon_2 \\ \epsilon_3 \\ \epsilon_4 \\ \epsilon_5 \\ \epsilon_6 \end{bmatrix} \tag{6.18}$$

If only the normal stress is considered in diamond and zinc-blende crystals, the general form of Hooke's law can be simplified as

$$\sigma = \begin{bmatrix} \sigma_1 \\ \sigma_2 \\ \sigma_3 \end{bmatrix} = \begin{bmatrix} C_{11} & C_{12} & C_{12} \\ C_{12} & C_{11} & C_{12} \\ C_{12} & C_{12} & C_{11} \end{bmatrix} \begin{bmatrix} \epsilon_1 \\ \epsilon_2 \\ \epsilon_3 \end{bmatrix} \tag{6.19}$$

where ϵ_n measures the fractional increase ($\epsilon_n > 0$) or decrease ($\epsilon_n < 0$) of the crystal lattice along the nth axis. The above expression assumed that all the off-diagonal elements are equal, and all the diagonal elements are also equal, leaving us with only two elastic moduli, C_{11} and C_{12}, in cubic crystals. In common semiconductors, $C_{11} > C_{12} > 0$ and has a unit of 10^{11} dyn/cm^2. In less symmetric crystals, such as III–V compounds, more C_{mn} components may need to be specified.

For a hexagonal crystal, e.g., wurtzite structure, the relation between stress and strain is expressed as

$$\begin{bmatrix} \sigma_1 \\ \sigma_2 \\ \sigma_3 \\ \sigma_4 \\ \sigma_5 \\ \sigma_6 \end{bmatrix} = \begin{bmatrix} C_{11} & C_{12} & C_{13} & 0 & 0 & 0 \\ C_{12} & C_{11} & C_{13} & 0 & 0 & 0 \\ C_{13} & C_{13} & C_{33} & 0 & 0 & 0 \\ 0 & 0 & 0 & C_{44} & 0 & 0 \\ 0 & 0 & 0 & 0 & C_{44} & 0 \\ 0 & 0 & 0 & 0 & 0 & \frac{1}{2}(C_{11} - C_{12}) \end{bmatrix} \begin{bmatrix} \epsilon_1 \\ \epsilon_2 \\ \epsilon_3 \\ \epsilon_4 \\ \epsilon_5 \\ \epsilon_6 \end{bmatrix} \quad (6.20)$$

(b) **Biaxial strain**

During the epitaxial growth of lattice-mismatched layers, stress is applied equally to all in-plane lattices (*x*-*y* plane) such that $\sigma_1 = \sigma_2$, and no stress is applied along the growth direction (*z*-axis), $\sigma_3 = 0$. For this condition, as shown in Fig. 6.9, the crystal is under biaxial strain. When the lattice constant of the epilayer is smaller than that of the substrate, the stress is directed outward ($\sigma_1 > 0$), and the resulting strain is a biaxial tensile strain. In contrast, the stress is directed inward ($\sigma_1 < 0$) for epilayer with a larger lattice constant than the substrate and gives rise to biaxial compressive strain.

With $\sigma_3 = 0$ and $\epsilon_1 = \epsilon_2$, the strain in the growth direction ϵ_3 can be calculated from Hooke's law (6.19).

$$\begin{aligned} \sigma_3 = 0 &= C_{12}\epsilon_1 + C_{12}\epsilon_2 + C_{11}\epsilon_3 \\ &= 2C_{12}\epsilon_1 + C_{11}\epsilon_3 \end{aligned} \quad (6.21)$$

Therefore,

$$\epsilon_3 = -\frac{2C_{12}}{C_{11}}\epsilon_1 \quad (6.22)$$

Here, the in-plane strains are defined as

$$\epsilon_1 = \epsilon_2 \equiv \frac{a_s - a_e}{a_e} \quad (6.23)$$

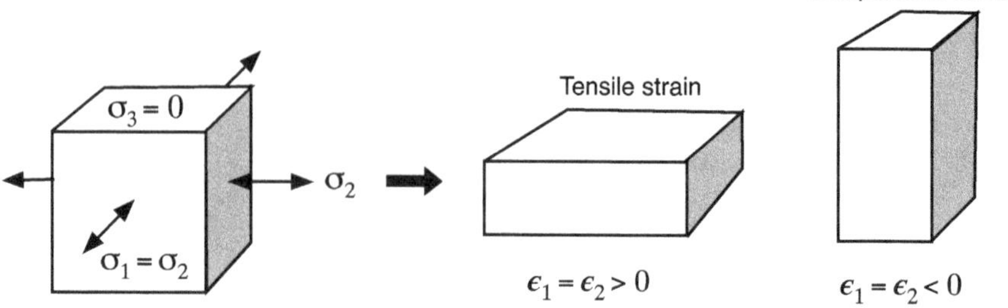

Fig. 6.9 Relationship between stress and strain

where a_s and a_e are the lattice constants of the substrate and the epilayer, respectively. The in-plane stress and strain relationship is also derived from Hooke's law. Using the relationship between the strain components derived above, we can find

$$\begin{aligned}\sigma_1 = \sigma_2 &= C_{11}\epsilon_1 + C_{12}\epsilon_2 + C_{12}\epsilon_3 \\ &= C_{11}\epsilon_1\left[1 + \frac{C_{12}}{C_{11}} - 2\left(\frac{C_{12}}{C_{11}}\right)^2\right]\end{aligned} \tag{6.24}$$

(c) **Hydrostatic stress**

When a uniform pressure is applied equally to all sides of the crystal, the material is under a hydrostatic stress (Fig. 6.10). The inward-directed stress is defined as a differential pressure change, dP, surrounding the crystal.

Applying this relationship to Hooke's law, we have

$$\begin{aligned}-\mathrm{d}P &= \epsilon_1 C_{11} + C_{12}(\epsilon_2 + \epsilon_3) \\ -\mathrm{d}P &= \epsilon_2 C_{11} + C_{12}(\epsilon_1 + \epsilon_3) \\ -\mathrm{d}P &= \epsilon_3 C_{11} + C_{12}(\epsilon_1 + \epsilon_2)\end{aligned} \tag{6.25}$$

The above equations can be simplified by adding both sides together.

$$\mathrm{d}P = -\frac{1}{3}(C_{11} + 2C_{12})(\epsilon_1 + \epsilon_2 + \epsilon_3) \tag{6.26}$$

The strain measures the fractional change of the crystal lattice along a major axis, i.e.,

$$\epsilon_n = \frac{a_s - a_n}{a_n} = \frac{\Delta a_n}{a_n} \tag{6.27}$$

where a_s and a_n are the substrate lattice constant and unstrained lattice constant of the epilayer, respectively. Assume the volume of a cubic crystal is given by $V = a_1a_2a_3$, where a_i defines the length of each side of the cube. Under a uniform pressure, the

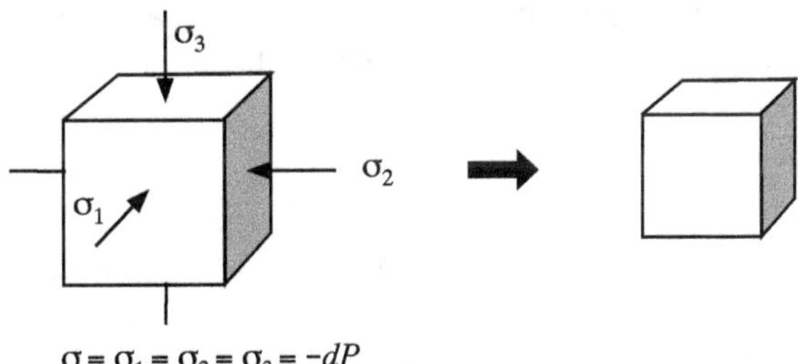

Fig. 6.10 Schematic of hydrostatic stress applied to a crystal

sides of the volume change to $(a_1 + \Delta a_1)$, $(a_2 + \Delta a_2)$, and $(a_3 + \Delta a_3)$. Then the volume change becomes

$$\begin{aligned}\Delta V &= (a_1 + \Delta a_1)(a_2 + \Delta a_2)\Delta a_3 \\ &\quad + (a_1 + \Delta a_1)a_3\Delta a_2 + a_2a_3\Delta a_1 \\ &\cong a_1a_2\Delta a_3 + a_1a_3\Delta a_2 + a_2a_3\Delta a_1\end{aligned} \tag{6.28}$$

In the final form of the volume change, the higher-order terms of Δa_i are neglected for $\Delta a_i \ll a_i$. Since the strain $\epsilon_n \equiv \Delta a_n/a_n$, the total strain of the crystal under the uniform pressure is

$$\begin{aligned}\epsilon_1 + \epsilon_2 + \epsilon_3 &= \frac{\Delta a_1}{a_1} + \frac{\Delta a_2}{a_2} + \frac{\Delta a_3}{a_3} \\ &= \frac{a_1a_2\Delta a_3 + a_1a_3\Delta a_2 + a_2a_3\Delta a_1}{a_1a_2a_3}\end{aligned} \tag{6.29}$$

The total strain is, therefore, directly related to the induced fractional volume change.

$$\epsilon_1 + \epsilon_2 + \epsilon_3 \cong \Delta V/V = \mathrm{d}(\ln V) \tag{6.30}$$

The strain-induced volume change also produces a bandgap variation in a semiconductor that is expressed as

$$\mathrm{d}E = a_\mathrm{d}\frac{\Delta V}{V} = a_\mathrm{d}(\epsilon_1 + \epsilon_2 + \epsilon_3) \tag{6.31}$$

where the proportional constant a_d (eV) is the *hydrostatic deformation potential*. Combining dE and dP, we can derive the expression of a_d as

$$a_\mathrm{d} = -\frac{1}{3}(C_{11} + 2C_{12})\frac{\mathrm{d}E}{\mathrm{d}P}. \tag{6.32}$$

Because $\mathrm{d}E/\mathrm{d}P > 0$ and C_{mn} are positive values, $a_\mathrm{d} < 0$. This relationship allows one to determine a_d from the pressure dependence of the energy gap. The hydrostatic deformation potential a_d has a conduction band component (a_c) and a valence band component (a_v). They are related by the following equations:

$$a_\mathrm{d} = a_\mathrm{c} - a_\mathrm{v} \tag{6.33a}$$

$$\Delta E_\mathrm{c}^\mathrm{hy} = a_\mathrm{c}\frac{\Delta V}{V} = a_\mathrm{c}(\epsilon_1 + \epsilon_2 + \epsilon_3) \tag{6.33b}$$

$$\Delta E_\mathrm{v}^\mathrm{hy} = a_\mathrm{v}\frac{\Delta V}{V} = a_\mathrm{v}(\epsilon_1 + \epsilon_2 + \epsilon_3) \tag{6.33c}$$

where ΔE_c^{hy} and ΔE_v^{hy} are strain-induced shift of the conduction and valence band edges, respectively.

6.2.3 Strain Effect on Band-Edge Energies

In a strained lattice structure, due to elastic deformation energy, there is a corresponding bandgap energy change from its equilibrium value. The elastic deformation energy consists of two parts: the hydrostatic term and the shear term. The hydrostatic term is proportional to the change in volume of the crystal lattice created by the strain. The shear term represents the energy shift due to the asymmetry in the strain parallel and perpendicular to the stress plane. By adding these two energy terms into the model-solid approach, the variation of the band edges in a strained heterojunction can be derived. Again, we will determine the absolute reference energy level first. This energy level in the valence band is modified by the hydrostatic component of the strain and expressed as

$$E_{v,av}^{s} = E_{v,av} + \Delta E_{v,av} = E_{v,av} + a_v \frac{\Delta V}{V} \tag{6.34}$$

where $E_{v,\,av}$ is the reference energy level of the unstrained material. The position of the topmost valence band in a biaxial-strained semiconductor becomes

$$\begin{aligned} E_v &= E_{v,av}^{s} + \frac{\Delta_{so}}{3} \\ &= E_{v,av} + a_v(\epsilon_1 + \epsilon_2 + \epsilon_3) + \frac{\Delta_{so}}{3} \end{aligned} \tag{6.35}$$

where

$$\epsilon_1 = \epsilon_2 = \frac{a_s - a_e}{a_e} \text{ and } \epsilon_3 = -2\epsilon_1 \frac{C_{12}}{C_{11}} \tag{6.36}$$

Therefore, the topmost valence band can be written as

$$E_v = E_{v,av} + \frac{\Delta_{so}}{3} + 2a_v\epsilon_1\left(\frac{C_{11} - C_{12}}{C_{11}}\right) = E_v^0 + \Delta E_v^{hy} \tag{6.37}$$

where $E_v^0 \equiv E_{v,av} + \Delta_{so}/3$ is the unstrained valence band edge. ΔE_v^{hy} represents the strain contribution to the valence band-edge shift and is defined as

$$\Delta E_v^{hy} \equiv a_v(\epsilon_1 + \epsilon_2 + \epsilon_3) = 2a_v\epsilon_1\left(\frac{C_{11} - C_{12}}{C_{11}}\right) \tag{6.38}$$

When growing epilayers with a larger lattice constant than the substrate, the in-plane strain ($\epsilon_1 = \epsilon_2$) is negative in the compressive strained system. Since $a_v > 0$, and $C_{11} > C_{12} > 0$, the resulting strain-induced valence band shift has a negative value ($\Delta E_v^{hy} < 0$). The topmost valence band edge is shifted downward by the compressive strain. For the epilayer with a smaller natural lattice constant than the substrate, the in-plane strain becomes positive in the tensile strained system. This leads to a positive valence band-edge shift, where the topmost valence band edge moves upward.

Under biaxial strains, the conduction band edge also gets modified by the hydrostatic component of the strain from the unstrained value by an amount of

$$\Delta E_c^{hy} \equiv a_c \frac{\Delta V}{V} = a_c(\epsilon_1 + \epsilon_2 + \epsilon_3) = 2a_c\epsilon_1 \left(\frac{C_{11} - C_{12}}{C_{11}} \right) \tag{6.39}$$

The new position of the biaxial-strained conduction band edge becomes

$$E_c = E_v^0 + E_g + \Delta E_c^{hy} \tag{6.40}$$

Since a_c is negative in direct bandgap materials, the strain effect on the conduction band-edge shift is opposite to that of the valence band.

In compressive strained systems, $\epsilon_1 = \epsilon_2 < 0$, $a_c < 0$, and $C_{11} > C_{12} > 0$, the resulting strain-induced conduction band shift has a positive value ($\Delta E_v^{hy} > 0$). The conduction band edge is shifted upward by the compressive strain. Under tensile strains, $\epsilon_1 = \epsilon_2 > 0$ and $\Delta E_v^{hy} < 0$. The conduction band edge is now shifted downward by the tensile strain.

The topmost valence band edge of Si, Ge, and most III–V compound semiconductors, such as GaAs, InP, and InAs, is twofold degenerate. For most III–V semiconductors, the split-off bands are several hundred milli-electron volts below the heavy-hole and light-hole bands. Therefore, the coupling between these bands can be ignored to simplify discussions. Under lattice-mismatch-induced biaxial strain, the heavy- and light-hole states at $k = 0$ are split by the shear component of the strain, ΔE_v^{sh}. The band-edge energy of the heavy-hole band and light-hole band is further modified by ΔE_v^{sh} as defined below.

$$\Delta E_v^{sh} \equiv b\left(\frac{\epsilon_1 + \epsilon_2}{2} - \epsilon_3 \right) = b\epsilon_1 \left(\frac{C_{11} + 2C_{12}}{C_{11}} \right) \tag{6.41}$$

where $b < 0$ is the shear deformation potential. At the zone center ($k = 0$), the shifts of band-edge energies of heavy-hole and light-hole bands due to shear strain are

$$\Delta E_{hh}^{sh} = +\Delta E_v^{sh} = b\epsilon_1 \left(\frac{C_{11} + 2C_{12}}{C_{11}} \right) \tag{6.42}$$

and

$$\Delta E_{\rm lh}^{\rm sh} = -\Delta E_{\rm v}^{\rm sh} = -b\epsilon_1\left(\frac{C_{11} + 2C_{12}}{C_{11}}\right) \tag{6.43}$$

The general expressions for the biaxial-strained band edges are

$$E_{\rm v} = E_{\rm v,av} + \frac{\Delta_{\rm so}}{3} + \Delta E_{\rm v}^{\rm hy} + \max\left(\Delta E_{\rm hh}^{\rm sh}, \Delta E_{\rm lh}^{\rm sh}\right) \tag{6.44}$$

$$E_{\rm c} = E_{\rm v,av} + \frac{\Delta_{\rm so}}{3} + E_{\rm g} + \Delta E_{\rm c}^{\rm hy} \tag{6.45}$$

Under compressive strains, $\epsilon_1 < 0$ and $b < 0$, the resulting shear strain has a positive value ($\Delta E_{\rm v}^{\rm sh} > 0$). The heavy-hole band edge is shifted upward and the light-hole band edge moves downward. In tensile strain systems, $\epsilon_1 > 0$. This leads to a negative shear strain component ($\Delta E_{\rm v}^{\rm sh} < 0$). The split valence band edges move downward and upward for heavy-hole and light-hole bands, respectively.

The complete bandgap diagram for cubic semiconductors under different stresses is shown in Fig. 6.11. Combining (6.38) through (6.43), the bandgap energies for conduction band to heavy-hole band (C–HH) transition and conduction band to light-hole band (C–LH) transition can be summarized as follows:

$$E_{\rm c-hh} = E_{\rm g} + \Delta E_{\rm c}^{\rm hy} - \Delta E_{\rm v}^{\rm hy} - \Delta E_{\rm v}^{\rm sh} \tag{6.46}$$

$$E_{\rm c-lh} = E_{\rm g} + \Delta E_{\rm c}^{\rm hy} - \Delta E_{\rm v}^{\rm hy} + \Delta E_{\rm v}^{\rm sh} \tag{6.47}$$

Remember, $\Delta E_{\rm c}^{\rm hy}$ and $\Delta E_{\rm v}^{\rm sh}$ are positive and $\Delta E_{\rm v}^{\rm hy}$ is negative under compressive strains. Under tensile strains, these parameters change signs.

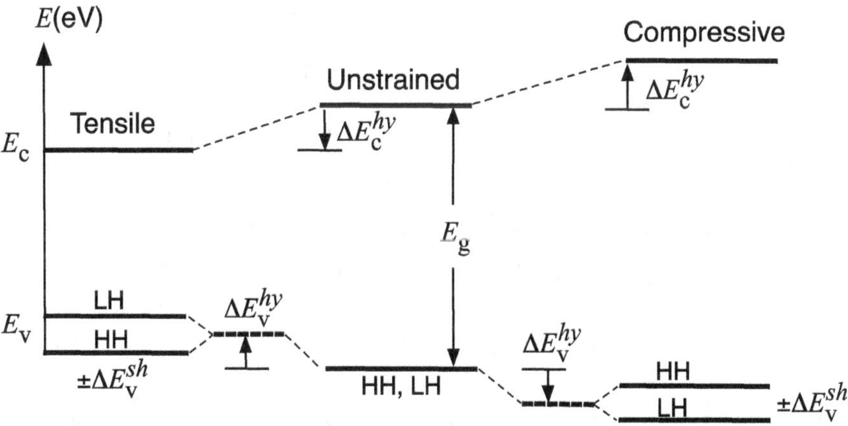

Fig. 6.11 Qualitative conduction and valence band-edge shifts for biaxial tensile and compressive strain

Since $a_d = a_c - a_v$, it is convenient to group the two hydrostatic strain terms into a single term and redefine as

$$\Delta E^{hy} = \Delta E_c^{hy} - \Delta E_v^{hy} = a_d(\epsilon_1 + \epsilon_2 + \epsilon_3) \equiv -\delta E^{hy} \tag{6.48}$$

and

$$\Delta E_v^{sh} = b\left[\frac{1}{2}(\epsilon_1 + \epsilon_2) - \epsilon_3\right] \equiv -\frac{1}{2}\delta E^{sh} \tag{6.49}$$

The effective bandgap energies are given by

$$E_{c-hh} = E_g + \Delta E^{hy} - \Delta E_v^{sh} = E_g - \delta E^{hy} + \frac{1}{2}\delta E^{sh} \tag{6.50}$$

$$E_{c-lh} = E_g + \Delta E^{hy} + \Delta E_v^{sh} = E_g - \delta E^{hy} - \frac{1}{2}\delta E^{sh} \tag{6.51}$$

Example: Strained Ge/Si heterojunction As an example of the application, we consider the pseudomorphic Si/Ge interface on (001)Si. The relevant material parameters of Si and Ge are listed in Table 6.2.

From the table, we found the following unstrained parameters:

$$E_{v,av}(\text{Si}) = -7.03\,(\text{eV}),\ \Delta_{so}(\text{Si}) = 0.044(\text{eV})$$
$$E_{v,av}(\text{Ge}) = -6.35\,(\text{eV}),\ \Delta_{so}(\text{Ge}) = 0.30\,(\text{eV})$$

The position of the topmost valence band in Si is

$$E_v(\text{Si}) = -7.03 + 0.044/3 = -7.015\,(\text{eV}).$$

Since the Ge layer is compressively strained, the valence band is also modified. For the strained Ge, $\epsilon_1 = \epsilon_2 = (5.4309 - 5.6575)/5.6575 = -0.04005$, $\epsilon_3 = 0.03009$, so that using $a_v = 1.24$, $b = -2.9$, and $\Delta V/V = -0.05002$, we get the topmost absolute reference energy level as

$$\begin{aligned} E_v(\text{Ge}) &= E_{v,av}(\text{Ge}) + \Delta_{so}/3 + \Delta E_v^{hy} + \Delta E_v^{sh} \\ &= -6.35 + (0.30/3) + 1.24(-0.05) \\ &\quad + (-2.9)(-0.04005)[(12.40 + 8.26)/12.40] \\ &= -6.1185\,(\text{eV}) \end{aligned}$$

The valence band offset $\Delta E_v = E_v(\text{Ge}) - E_v(\text{Si}) = -6.1185 - (-7.015) = 0.91(\text{eV})$. This value agrees well with the experimentally determined value for ΔE_v of 0.81 eV.

Table 6.2 Material parameters for semiconductor Si and Ge at room temperature

	a_0 (Å)	Δ_{so} (eV)	$E_{v,av}$ (eV)	a_c(dir) (eV)	a_c(ind) (eV)	a_v (eV)	b (eV)	E_g (eV)	C_{11} (10^{11} dyn/cm^2)	C_{12} (10^{11} dyn/cm^2)
Si	5.4309	0.044	−7.03	1.98	4.18	2.64	−2.1	1.124	16.577	6.393
Ge	5.6575	0.30	−6.35	−8.24	−1.54	1.24	−2.9	0.664	12.40	4.13

6.3 Band-Edge Energies in Strained Ternary and Quaternary Alloys

To derive valence band positions for an alloy, linear interpolation between the nearly lattice-matched pure materials (binary compounds) is appropriate. When the constituent materials are not lattice-matched, a strain contribution should be included in the calculation. This is because, in the alloy, one material is effectively expanded, whereas the other is compressed. For a ternary alloy $A_xB_{1-x}C$, the topmost valence band position has the following expression [11]:

$$E_v(x) = xE_v(\mathrm{AC}) + (1-x)E_v(\mathrm{BC}) + x(1-x)\left[-3\Delta a_v\left(\frac{\Delta a}{a_0}\right)\right] \quad (6.52)$$

where $a_0 = xa_0(\mathrm{AC}) + (1-x)a_0(\mathrm{BC})$ and $\Delta a = a_0(\mathrm{AC}) - a_0(\mathrm{BC})$. The last term in the equation shown above includes a bowing parameter C due to hydrostatic strain contribution, where

$$C \equiv 3\Delta a_v\left(\frac{\Delta a}{a_0}\right) = 3[a_v(\mathrm{AC}) - a_v(\mathrm{BC})]\left(\frac{\Delta a}{a_0}\right) \quad (6.53)$$

Example: $Ga_{0.47}In_{0.53}As$/InP lattice-matched heterojunction From Table 6.1, we have the following parameters:

$$\begin{aligned}
E_{v,av}(\mathrm{GaAs}) &= -6.92\ (\mathrm{eV}),\ \Delta_{so}(\mathrm{GaAs}) = 0.34\ (\mathrm{eV}),\\
E_g(\mathrm{GaAs}) &= 1.424\ (\mathrm{eV})\\
E_{v,av}(\mathrm{InAs}) &= -6.67\ (\mathrm{eV}),\ \Delta_{so}(\mathrm{InAs}) = 0.39\ (\mathrm{eV}),\\
E_g(\mathrm{InAs}) &= 0.354\ (\mathrm{eV})\\
E_{v,av}(\mathrm{InP}) &= -7.04\ (\mathrm{eV}),\ \Delta_{so}(\mathrm{InP}) = 0.108\ (\mathrm{eV}),\\
E_g(\mathrm{InP}) &= 1.353\ (\mathrm{eV})
\end{aligned}$$

Then,

$$\begin{aligned}
E_v(\mathrm{GaAs}) &= -6.92 + 0.34/3 = -6.806\,\mathrm{eV}\\
E_v(\mathrm{InAs}) &= -6.67 + 0.38/3 = -6.540\,\mathrm{eV}
\end{aligned}$$

Although $Ga_{0.47}In_{0.53}As$/InP heterojunction is a lattice-matched system, both GaAs and InAs are lattice-mismatched to InP. Therefore, it is necessary to use (6.52) to calculate the topmost valence band edge of $Ga_{0.47}In_{0.53}As$ as follows:

$$\begin{aligned}
E_v(\mathrm{Ga_{0.47}In_{0.53}As}) &= 0.47 \times E_v(\mathrm{GaAs}) + 0.53 \times E_v(\mathrm{InAs})\\
&\quad + 3 \times 0.47 \times 0.53 \times (-1.16 + 1.0)(5.653\text{–}6.058)/5.87\\
&= -6.6568\,\mathrm{eV}
\end{aligned}$$

$$E_v(\text{InP}) = -7.04 + 0.108/3 = -7.004\,\text{eV}$$
$$\Delta E_v = -7.004 - (-6.6568) = -0.3472\,\text{eV}$$

This indicates that the valence band in InP is below the valence band in $Ga_{0.47}In_{0.53}As$. This calculated value is very close to the experiment results (between 0.325 and 0.38 eV) determined by various methods. Next, we will determine the conduction band discontinuity, ΔE_c.

$$\begin{aligned} E_c(\text{InP}) &= E_v(\text{InP}) + E_g(\text{InP}) \\ &= -7.004 + 1.353 = -5.651\,\text{eV} \\ E_g(\text{Ga}_{0.47}\text{In}_{0.53}\text{As}) &= xE_g(\text{GaAs}) + (1-x)E_g(\text{InAs}) + x(1-x)C \\ &= 0.47 \times 1.424 + 0.53 \times 0.354 - 0.47 \times 0.53 \times 0.477 \\ &= 0.7381\,\text{eV} \\ E_c(\text{Ga}_{0.47}\text{In}_{0.53}\text{As}) &= E_v(\text{Ga}_{0.47}\text{In}_{0.53}\text{As}) + E_g(\text{Ga}_{0.47}\text{In}_{0.53}\text{As}) \\ &= -6.6568 + 0.7381 = -5.9187\,\text{eV} \\ \Delta E_c &= -5.651 - (-5.9187) = 0.2677\,\text{eV} \end{aligned}$$

The conduction band edge of InP is 0.2677 eV above $Ga_{0.47}In_{0.53}As$. Therefore, the $Ga_{0.47}In_{0.53}As$/InP heterojunction has a type-I band alignment.

Example: $Ga_{0.7}In_{0.3}As$/GaAs strained quantum well for 0.98 μm lasers In this problem, we need to know E_v, C_{11} and C_{12} in order to calculate ΔE_c and ΔE_v of the $Ga_{0.7}In_{0.3}As$/GaAs system. From the previous examples, we found

$$\begin{aligned} E_v(\text{GaAs}) &= -6.806\,\text{eV}, \\ E_c(\text{GaAs}) &= -5.382\,\text{eV}, \text{ and} \\ E_v(\text{InAs}) &= -6.54\,\text{eV} \end{aligned}$$

Since this is a strained system, except for C_{11} and C_{12}, we cannot use linear interpolation between GaAs and InAs to calculate E_v. Instead, we use (6.52) to include the strain effect.

$$\begin{aligned} E_v(\text{Ga}_{0.7}\text{In}_{0.3}\text{As}) &= 0.3E_v(\text{InAs}) + 0.7E_v(\text{GaAs}) \\ &\quad + 3x(1-x)[-a_v(\text{InAs}) + a_v(\text{GaAs})] \\ &\quad \{(a_{\text{InAs}} - a_{\text{GaAs}})/[x\,a_{\text{InAs}} + (1-x)a_{\text{GaAs}}]\} \\ &= -6.720\,\text{eV} \end{aligned}$$

Using linear interpolation, we found $C_{11}(\text{GaInAs}) = 10.815 \times 10^{11}$ and $C_{12}(\text{GaInAs}) = 5.124 \times 10^{11}$. Then

$$E_c(\mathrm{Ga_{0.7}In_{0.3}As}) = E_v(\mathrm{Ga_{0.7}In_{0.3}As}) + E_g(\mathrm{Ga_{0.7}In_{0.3}As}) + 2a_c\epsilon_1(C_{11}-C_{12})/C_{11}$$

where

$$\begin{aligned}
\epsilon_1 &= \frac{a_s - a_e}{a_e} = \frac{5.653 - 5.775}{5.775} = -0.0211 \\
a_c &= -6.543 \\
E_g(\mathrm{Ga_{0.7}In_{0.3}As}) &= 0.7 \times 1.424 \\
&\quad + 0.3 \times 0.354 - 0.7 \times 0.3 \times 0.477 = 1.003\,\mathrm{eV} \\
E_c(\mathrm{Ga_{0.7}In_{0.3}As}) &= -6.730 + 1.003 + 0.145 = -5.582\,\mathrm{eV} \\
\Delta E_c &= -5.382 + 5.582 = 0.20\,\mathrm{eV}
\end{aligned}$$

The heavy-hole energy level in $\mathrm{Ga_{0.7}In_{0.3}As}$ can be calculated as follows (6.44)

$$E_{hh}(\mathrm{Ga_{0.7}In_{0.3}As}) = E_v(\mathrm{Ga_{0.7}In_{0.3}As}) + 2a_v\epsilon_1\left(\frac{C_{11} - C_{12}}{C_{11}}\right) + b\epsilon_1\left(\frac{C_{11} + 2C_{12}}{C_{11}}\right)$$

Using linear interpolation, we found $a_v = 1.112$ and $b = -1.94$. Then,

$$\begin{aligned}
E_{hh}(\mathrm{Ga_{0.7}In_{0.3}As}) &= -6.720 - 0.0145 + 0.0469 \\
&= -6.6877\,\mathrm{eV} \\
\Delta E_v &= -6.806 + 6.6877 = -0.1183\,\mathrm{eV}
\end{aligned}$$

Therefore, the $\mathrm{Ga_{0.7}In_{0.3}As/GaAs}$ system has a type-I heterostructure.

The methods of deriving quaternary semiconductor parameters from those of the underlying binary and ternary alloys have been discussed in Sect. 4.5.2. Both quaternary bandgap energy E_g and spin-orbit splitting Δ_{so} can be derived from (4.87) using bowing parameters listed in Table 4.6. To calculate $E_{v,\,av}$ using (4.87), the necessary ternary bowing parameters C are obtained according to (6.53). For strained quaternary alloys, additional relevant parameters including ΔE_v^{hy} and δE^{sh} are needed to derive the valence band-edge energy E_v that can be obtained from binary data using Vegard's law.

Example: $(\mathrm{Al_{0.48}In_{0.52}As})_{0.6}(\mathrm{Ga_{0.47}In_{0.53}As})_{0.4}$/InP lattice-matched heterojunction
For this unstrained III–V quaternary compound, only a one-dimensional subset of the vast two-dimensional space of compositions (x, y) that is lattice-matched to a common binary substrate material (i.e., InP) is considered. Then the band parameters of a quaternary alloy may be represented as a combination of two lattice-matched constituents of ternary alloys. Letting α (e.g., $\mathrm{Al_{0.48}In_{0.52}As}$) and β (e.g.,

$Ga_{0.47}In_{0.53}As$) represent two lattice-matched ternary end point constituents, the combination of α and β with arbitrary composition z forms the lattice-matched quaternary $\alpha_{1-z}\beta_z$. Although the two ternary constituents have identical lattice constants, their binary constituents are not lattice-matched. Therefore, the expression of the valence band alignment of the quaternary has to include a bowing parameter due to hydrostatic strain contributions. In this example, the composition of the quaternary ($z = 0.4$) has the form of $Al_x(Ga_yIn_{1-y})_{1-x}As$ with $x(AlAs) = 0.288$ and $y(GaAs) = 0.188$. To apply (4.87) to this type-II quaternary alloy, one binary can be assigned twice in the same row or column of the matrix with a zero ternary bowing factor. Let B_1, $B_2 = B_3$, and B_4 represent binary parameters of GaAs, AlAs, and InAs, respectively, and $C_{23} = 0$. To calculate $E_{v,av}$ it is necessary to get ternary bowing factors of deformation potentials first. Following (6.53),

$$C_{12} = 3\left(a_v^{GaAs} - a_v^{AlAs}\right)(a_{GaAs} - a_{AlAs})/a_{InP} = 0.0053$$
$$C_{14} = 3\left(a_v^{GaAs} - a_v^{InAs}\right)(a_{GaAs} - a_{InAs})/a_{InP} = -0.0331$$
$$C_{34} = 3\left(a_v^{AlAs} - a_v^{InAs}\right)(a_{AlAs} - a_{InAs})/a_{InP} = -0.2984$$

Together with reference energy parameters ($E_{v,\,av}$) $B_1 = -6.92$, $B_2 = -7.49$, $B_4 = -6.67$ (eV), and set $D = 0.03$ in (4.87), $E_{v,av}$ reaches a value of $E_{v,av}(Q) = -6.9917$ (eV). Next we need to calculate Δ_{so} of the quaternary alloy using (4.87) with the following spin-orbit split-off parameters: $B_1 = 0.341$, $B_2 = B_3 = 0.28$, $B_4 = 0.39$ (eV), $C_{12} = 0$, $C_{14} = -0.15$, $C_{34} = -0.15$, and $D = 0$. The result of $\Delta_{so}(Q) = 0.310$ (eV). Thus the valence band edge of the quaternary alloy becomes

$$E_v(Q) = E_{v,av}(Q) + \frac{\Delta_{so}(Q)}{3} = -6.8884\text{eV}$$

From the previous example, we found $E_{v,av}(\text{InP}) = -7.004$ eV. Thus

$$\Delta E_v = -6.8884 + 7.004 = 0.1156\text{eV}.$$

The valence band edge of the quaternary is located at a higher energy than InP by 0.1156 eV.

6.4 Strained Nitrides with Wurtzite Crystal Structure

6.4.1 *Band-Edge Energies of Strained Nitrides*

The energy band edges of wurtzite crystals are modified under stress similarly to those of the zinc-blende crystal. For example, for the commonly encountered case of growing a strained wurtzite nitride layer pseudomorphically on the (0001) surface of a sapphire substrate, the epitaxial layer experiences in-plane biaxial strain but zero stress along the z-direction. For simplicity, the shear components of stress are neglected (i.e., $\epsilon_4 = \epsilon_5 = \epsilon_6 = 0$). Therefore, the lattice-mismatch induced strains are expressed as

$$\epsilon_1 = \epsilon_2 = \frac{a_s - a_e}{a_e} \text{ and } \epsilon_3 = -\frac{2C_{13}}{C_{33}}\epsilon_1 \tag{6.54}$$

where a_s and a_e are the lattice constants of the substrate and the epitaxial layer, respectively. The second relation is obtained by solving $\sigma_3 = 0$ in (6.20). The induced strains for other stress conditions can also be derived using proper boundary conditions.

The strain-induced elastic deformation energy causes a bandgap energy change from its equilibrium value as in the zinc-blende crystal structure. Again, the elastic deformation energy consists of two parts: the hydrostatic term and the shear term. To describe the valence band structure of GaN under strain, six valence band deformation potentials (D_i) are necessary. The parameters D_1 and D_2 are associated with the valence band hydrostatic deformation potential, a_v, and D_3 and D_4 are related to the shear deformation potential, b.

The valence band-edge energies at $k = 0$ in the strained crystal are modified as

$$\begin{cases} E_1 = \Delta_1 + \Delta_2 + \lambda_\epsilon + \theta_\epsilon \\ E_2 = \dfrac{\Delta_1 - \Delta_2 + \theta_\epsilon}{2} + \lambda_\epsilon + \sqrt{\left(\dfrac{\Delta_1 - \Delta_2 + \theta_\epsilon}{2}\right)^2 + 2\Delta_3^2} \\ E_3 = \dfrac{\Delta_1 - \Delta_2 + \theta_\epsilon}{2} + \lambda_\epsilon - \sqrt{\left(\dfrac{\Delta_1 - \Delta_2 + \theta_\epsilon}{2}\right)^2 + 2\Delta_3^2} \end{cases} \tag{6.55}$$

where $\Delta_1 = \Delta_{cr}$, $\Delta_2 = \Delta_3 = \Delta_{so}$, and the strain-induced energy shifts are

$$\lambda_\epsilon = D_1\epsilon_3 + D_2(\epsilon_1 + \epsilon_2) \tag{6.56}$$

$$\theta_\epsilon = D_3\epsilon_3 + D_4(\epsilon_1 + \epsilon_2) \tag{6.57}$$

The conduction band edge at $k = 0$ is modified and expressed as

$$E_c = E_g + \Delta_1 + \Delta_2 + P_{c\epsilon} \tag{6.58}$$

$$P_{c\epsilon} = a_{cz}\epsilon_3 + a_{ct}(\epsilon_1 + \epsilon_2) \tag{6.59}$$

where $P_{c\epsilon}$ is the conduction band hydrostatic energy shift and related to the longitudinal and transverse interband deformation potentials a_1 and a_2, respectively, as

$$a_1 = a_{cz} - D_1 \quad a_2 = a_{ct} - D_2 \tag{6.60}$$

Table 6.3 Band structure parameters for wurtzite nitride binaries: in-plane lattice constant a, longitudinal (c-axis) lattice constant, bandgap energy at 0 K $E_g(0)$, bandgap fitting parameters α and β, spin-orbit split-off energy Δ_{so}, crystal-field splitting energy Δ_{cr}, longitudinal hydrostatic deformation potential a_1, in-plane hydrostatic deformation potential a_2, valence band hydrostatic deformation potential D_1 and D_2, shear deformation potential D_3 and D_4, elastic constants $C_{11}, C_{12}, C_{13}, C_{33}$, and C_{44}, piezoelectric coefficients e_{13} and e_{33}, and spontaneous polarization P_{sp} [12]

Parameters	GaN	AlN	InN
a (Å) @300 K	3.189	3.112	3.545
c (Å) @300 K	5.185	4.982	5.703
$E_g(0)$ (eV) @0 K	3.510	6.25	0.69[a]
α (meV/K)	0.909	1.799	0.41[a]
β (K)	830	1462	454[a]
Δ_{cr} (eV)	0.010	−0.169	0.040
Δ_{so} (eV)	0.017	0.019	0.005
a_1 (eV)	−4.9	−3.4	−3.5
a_2 (eV)	−11.3	−11.8	−3.5
D_1 (eV)	−3.7	−17.1	−3.7
D_2 (eV)	4.5	7.9	4.5
D_3 (eV)	8.2	8.8	8.2
D_4 (eV)	−4.1	−3.9	−4.1
C_{11} (10^{11} dyn/cm^2)	39.0	39.6	22.3
C_{12} (10^{11} dyn/cm^2)	14.5	13.7	11.5
C_{13} (10^{11} dyn/cm^2)	10.6	10.8	9.2
C_{33} (10^{11} dyn/cm^2)	39.8	37.3	22.4
C_{44} (10^{11} dyn/cm^2)	10.5	11.6	4.8
e_{31} (C/m^2)[b]	−0.34	−0.53	−0.41
e_{33} (C/m^2)[b]	0.67	1.50	0.81
P_{sp} (C/m^2)	−0.034	−0.090	−0.042

[a][13]
[b][14]

The relevant band structure parameters for wurtzite nitride binary alloys are shown in Table 6.3.

Thus, the interband transition energy of the system is

$$\begin{aligned} E_c - E_1 &= E_g + P_{c\epsilon} - \lambda_\epsilon - \theta_\epsilon \\ &= (a_{cz} - D_1)\epsilon_3 + (a_{ct} - D_2)(\epsilon_1 - \epsilon_2) \\ &\quad - D_3\epsilon_3 - D_4(\epsilon_1 + \epsilon_2) \end{aligned} \tag{6.61}$$

Figure 6.12 shows the energy bands of a strained GaN crystal as functions of the in-plane strain using E_v^0 of an unstrained GaN as a reference energy at 0 eV.

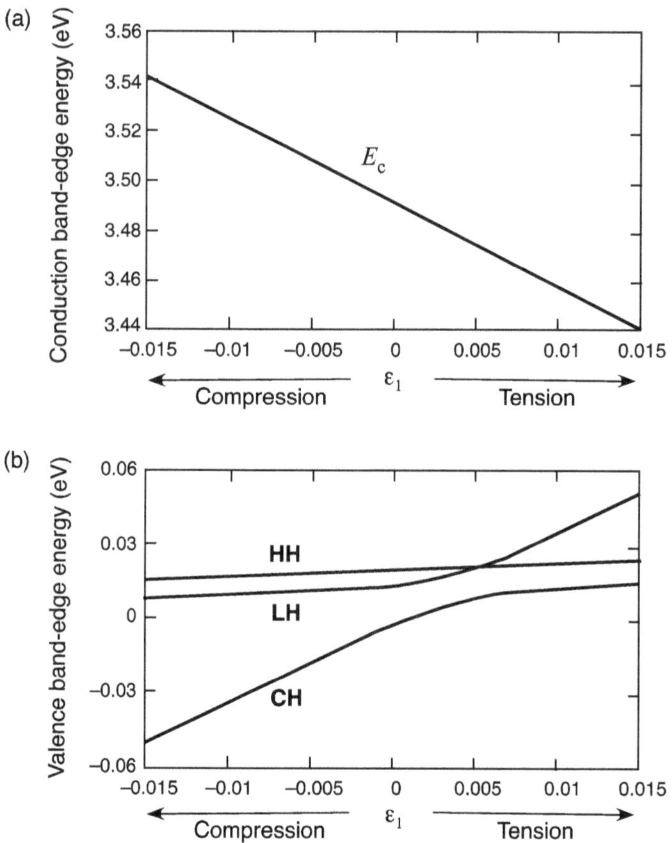

Fig. 6.12 Band energies of **a** the conduction band and **b** the HH, LH, and CH bands of a strained GaN crystal as functions of the in-plane strain ϵ_1. The reference energy is set at E_v^0 of an unstrained GaN crystal. Reprinted with permission from [15], copyright IEEE

6.4.2 Natural Band Alignments of III-Nitrides

The band discontinuities at the heterojunction of III-N systems are complicated due to the large electrostatic fields originating from the polar interface and the ionic nature of the wurtzite crystal. Because of this and other strain-induced effects, the natural band alignment could not be accurately determined. The measured band alignments of binary III-nitride heterojunctions do not even follow the commutative property such that, e.g., GaN/AlN and AlN/GaN structures are not equivalent.

Recently, the 'natural' band alignments of wurtzite III-nitride semiconductor heterojunctions have been determined directly on the non-polar side-facet of a vertically aligned nanorod array grown along the *c*-axis. The nanorod structure warrants a near fully relaxed lattice structure without the influence of spontaneous and piezoelectric polarization fields. Using microscopic-area photoelectron spectroscopy, the

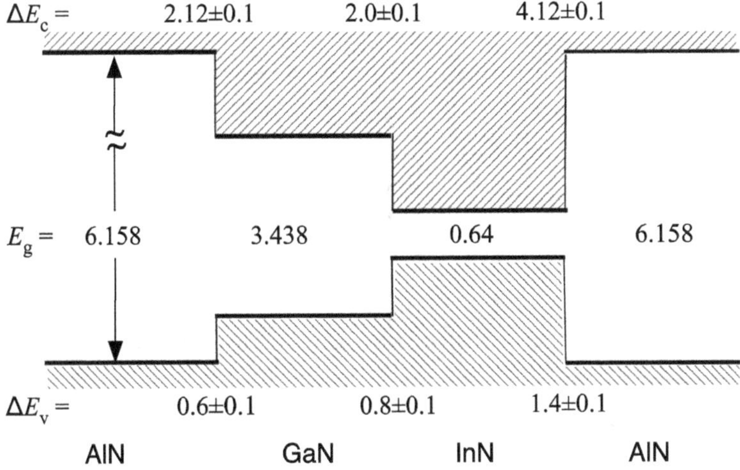

Fig. 6.13 Schematic of 'natural' band alignments for InN, GaN, and AlN. The corresponding values of E_g, ΔE_v, and ΔE_c are shown in the graph. The determined VBOs for GaN/AlN, InN/GaN, and InN/AlN heterojunctions are 0.6 ± 0.1 eV, 0.8 ± 0.1 eV, and 1.4 ± 0.1 eV, respectively [16]

VBOs of InN/GaN, GaN/AlN, and InN/AlN are measured to be 0.8 ± 0.1, 0.6 ± 0.1, and 1.4 ± 0.1 eV, respectively. The measured ΔE_v along with E_g and deduced ΔE_c are summarized in Fig. 6.13. Type-I heterostructures are formed between all wurtzite III-nitrides.

6.5 Construction of Heterostructure Band Diagrams

Once the conduction band and valence band discontinuities are determined, we can proceed to construct the approximate heterostructure band diagrams without detailed calculation. However, an accurate treatment requires a self-consistent numerical solution of Poisson's equation for the electrostatic potential as shown in the next chapter.

6.5.1 Anisotype N-p *Heterojunctions*

The band diagram construction method for a heterojunction is very similar to that for a doped homostructure, but we need to take account of the discontinuities ΔE_c and ΔE_v. Using an *N-p* heterojunction as an example, as shown in Fig. 6.14, one starts with flat bands in each material as determined by Anderson's rule and the Fermi levels set by the doping on each side. The capital and lower case letters of the *N-p*

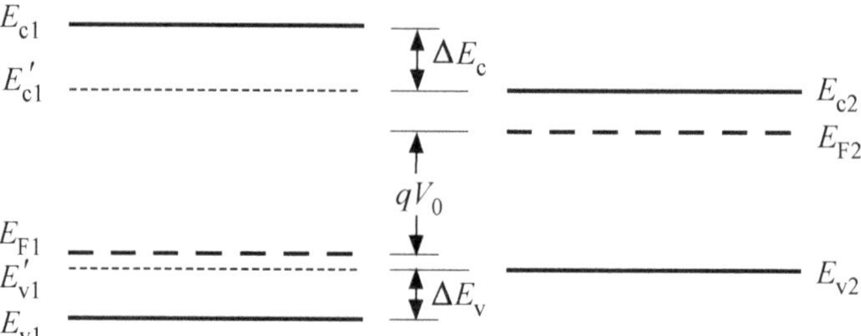

Fig. 6.14 Equilibrium band diagrams (flat band) of a *P-n* heterojunction before contacting each other. The relative position of the conduction band edge is determined by ΔE_c and ΔE_v

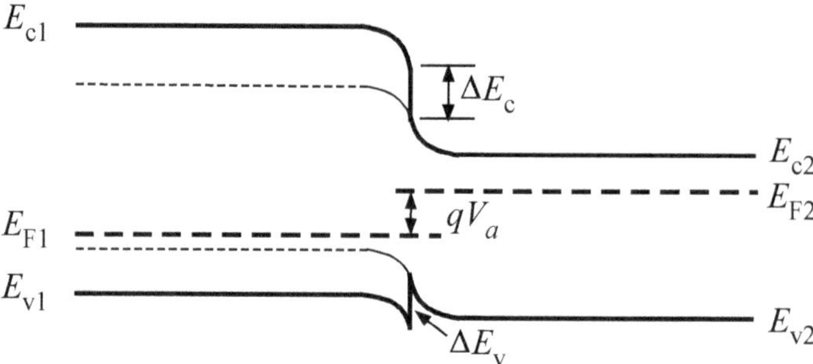

Fig. 6.15 Completed energy band diagram of an isotype *P-n* heterostructure under a bias voltage of V_a

heterojunction represent that the larger and smaller bandgap semiconductors are *n*-type and *p*-type, respectively. To temporarily cancel out the effect of discontinuities, draw lines $E'_{c1} = E_{c1} - \Delta E_c$ and $E'_{v1} = E_{v1} + \Delta E_v$. Now the 'effective' bandgap is the same on both sides.

Next, as shown in Fig. 6.15, align the Fermi levels. The difference in Fermi levels far from the junction is set with the applied voltage, V_a. So, $E_{F2} - E_{F1} = qV_a$. Then join E'_{c1} to E_{c2}, and E'_{v1} to E_{v2}, with parallel S-shaped curves due to the electrostatic potential. The slope of the curve is set by the charge density distribution.

Now restore E_{c1} on side 1 as a line at $E'_{c1} + \Delta E_c$ and E_{v1} at $E'_{v1} - \Delta E_v$, including the discontinuities in ΔE_c and ΔE_v at the junction. The band discontinuities are inserted at the mid-points of S-shaped curves. This completes the sketch of the band diagram shown in Fig. 6.15.

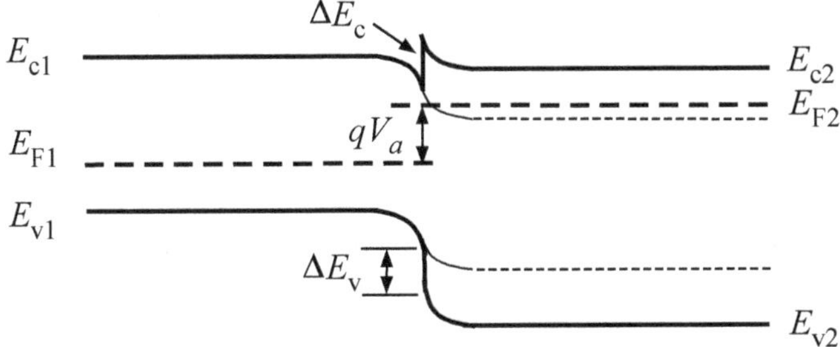

Fig. 6.16 Energy band diagram of an isotype *p-N* heterujunction under a bias of V_a

6.5.2 *Anisotype* p-N *Heterojunctions*

The procedure for constructing the band diagram for a *p-N* heterojunction is similar to that for the *P-n* heterojunction shown above. The finished band diagram is shown in Fig. 6.16.

6.5.3 *Isotype* N-n *and* P-p *Heterojunctions*

Unlike in homostructures, a potential barrier can be formed in heterostructures even when both sides are doped similarly. This type of junction is called the isotype heterojunction as shown in Fig. 6.17. In *N-n* heterojunctions, the band discontinuity is mostly located in the valence band and acts as a hole blocker. In a properly designed N^+-n^- heterojunction, electrons trapped in the potential energy notch in the conduction band form a two-dimensional electron gas (2DEG). Field-effect transistors utilizing these electrons conducting parallel to the interface show very interesting properties. On the other hand, the band discontinuity in a *P-p* heterojunction is mostly across the conduction band and will prevent electrons from flowing through.

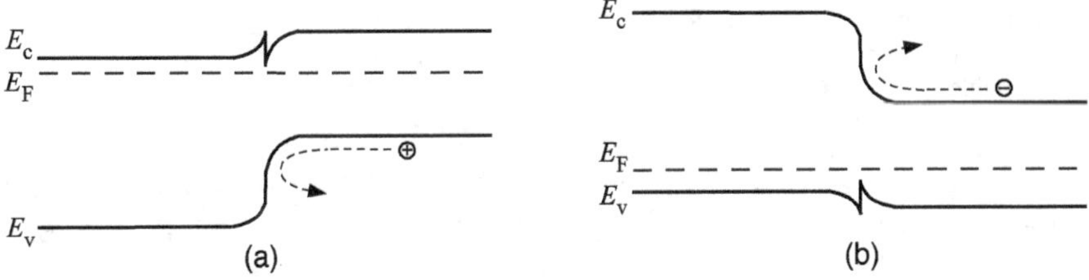

Fig. 6.17 Two types of isotype heterostructures: **a** *N-n* heterojunction showing the hole blocking due to ΔE_V. Electrons may be trapped in the energy notch formed in the conduction band. **b** *P-p* heterojunction showing the electron blocking due to ΔE_c

The combination of isotype and anisotype heterostructures can further confine carriers for new transport and photonic device applications as illustrated in the following sections.

6.5.4 Current Confinement in Double Heterostructures

The energy band diagrams for *P-p-N* or *P-n-N* double heterostructures (DH) are combinations of the anisotype and isotype heterojunctions previously described. In the zero-bias *P–n–N* DH shown in Fig. 6.18, for example, the conduction band discontinuity ΔE_c provides a barrier to electrons at the *P-n* heterojunction and thus confines the injected electrons to the small bandgap *n*-layer. The valence band discontinuity ΔE_v at the *n-N* heterojunction provides a potential barrier to holes and thus prevents hole injection into the *N*-layer. These barriers result in both minority and majority confinement by the *P-n-N* double heterostructure.

When a forward bias is applied, the distribution of V_a between the *P-n* and *n-N* heterojunctions is determined by the requirement of current continuity through the device. The band diagram of a *P-n-N* DH under forward bias is shown in Fig. 6.19. The carrier distributions as functions of energy are also shown.

Although the DH provides majority and minority carrier confinement at zero bias, the band bending associated with applied bias V_a lowers the barrier heights. The electron and hole distribution in the conduction band and valence bands may extend beyond the band discontinuities ΔE_c and ΔE_v, respectively. The electrons and holes above ΔE_c and ΔE_v, respectively, can leak out of the *n*-region. In a DH laser structure, these leakage currents will not contribute to stimulated emission. Another interesting situation in a DH is the carrier accumulation in the active region under a strong forward bias. Due to the carrier confinement by both *P-n* and *n-N* heterojunctions, the injected carrier concentration exceeds the majority carrier concentration. This is labeled ‘superinjection’ by Zh. I. Alferov [17].

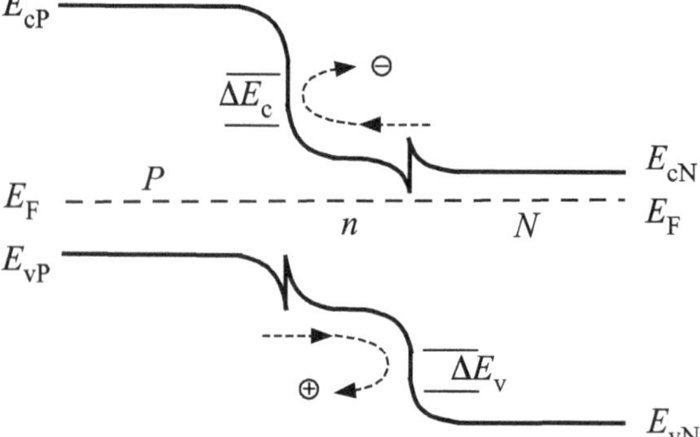

Fig. 6.18 Equilibrium energy band diagram of a *P-n-N* double heterostructure. Electrons and holes are blocked by *P-n* and *n-N* heterojunctions, respectively, and trapped inside the *n*-type middle layer

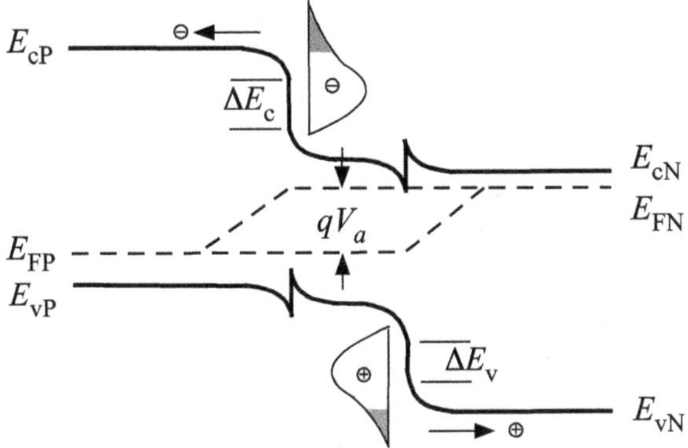

Fig. 6.19 The energy band diagram of a *P-n-N* double heterostructure under a forward bias V_a. Also shown are the electron and hole distributions as functions of energy. Most carriers are trapped inside the smaller bandgap *n*-layer. Only those carriers with energy higher than the barrier, shown as the shaded areas, can escape

Problems

1. $InAs_xSb_{1-x}$ ternary alloy has the highest electron mobility among all semiconductors. Recently, it has been used as the channel material of the psuedomorphic high-electron-mobility transistor (pHEMT). It has also been considered as a channel material for future III–V MOSFET applications. An $InAs_xSb_{1-x}/Al_yIn_{1-y}Sb$ pHEMT relies on the conduction band discontinuity at the heterojunction to form a high mobility two-dimensional electron gas (2DEG). Assume the elastic modulii, C_{11} and C_{12}, and the deformation potentials, a and b, vary linearly between the two binaries and are listed below (in 10^{11} dyn/cm^2):
   $C_{11}(AlSb) = 8.769$; $C_{11}(InSb) = 6.669$; $C_{11}(InAs) = 8.329$
   $C_{12}(AlSb) = 4.341$; $C_{12}(InSb) = 3.645$; $C_{12}(InAs) = 4.526$

   (a) Determine the composition x of $InAs_xSb_{1-x}$ which is lattice-matched to GaSb.
   (b) Calculate the conduction band discontinuity ΔE_c and valence band discontinuity ΔE_v between the lattice-matched $InAs_xSb_{1-x}$ and GaSb. Please use both model-solid theory and the empirical method by Tiwari and Frank. Is this a type-I or type-II heterostructure?
   (c) If we wish to grow an $InAs_xSb_{1-x}$ on GaSb with a compressive strain of 1.8%, find the critical layer thicknesses of $InAs_xSb_{1-x}$.
   (d) Find the critical layer thicknesses of $Al_{0.8}In_{0.2}Sb$ grown on GaSb,
   (e) Calculate the conduction band discontinuity ΔE_c between $InAs_xSb_{1-x}$ (lattice-matched to GaSb) and strained $Al_{0.8}In_{0.2}Sb$. This value is very important for the pHEMT application.

2. At the 2008 International Electron Device Meeting (https://doi.org/10.1109/iedm.2008.4796798), Intel demonstrated a *p*-channel, pseudomorphic, high hole mobility transistor by using compressively strained InSb QW as the channel material. Although the heavy-hole band is lifted above the light-hole band in the growth direction under compressive strain, the hole effective mass becomes lighter in the direction parallel to the growth surface. (See *Phys. Rev.* B, vol. 43, pp. 9649, (1991)). The barrier material is $Al_{0.35}In_{0.65}Sb$.

(a) Calculate the critical thickness of the InSb QW grown on $Al_{0.35}In_{0.65}Sb$. The $Al_{0.35}In_{0.65}Sb$ barrier material is very thick and unstrained.
(b) For a 5 nm InSb QW, determine the strain (ϵ), ΔE_c and ΔE_v. Assume no defects are generated in this heterostructure.

3. Consider the binary alloys AlSb, GaSb, and InAs. The lattice mismatches between these compounds are small, and high-quality AlSb/InAs heterostructures can be grown on GaSb substrates.

   (a) Use the model-solid theory and the data provided in Table 6.1 to find the band discontinuities, ΔE_c and ΔE_v, at AlSb/InAs and GaSb/InAs interfaces. Compare your results with experimental values shown in Fig. 6.4.
   (b) Find the critical layer thickness of InAs and AlSb grown on GaSb.
   (c) Draw the band diagram (conduction band and valence band) of the following heterostructure grown on GaSb:
   AlSb/InAs (30 Å)/GaSb (20 Å)/InAs (30 Å)/AlSb
   The three center layers form the active region of a W-type laser. Discuss the type of heterostructures formed.
   (d) Calculate the emission energy of the laser structure of part (c) using the infinite quantum-well approach and the following effective mass data. The emission energy corresponds to the difference between the electron state in InAs QWs and the heavy-hole state in the GaSb QW.

$$m_e^*(\text{InAs}) = 0.024m_0, \quad m_{hh}^*(\text{InAs}) = 0.37m_0,$$
$$m_e^*(\text{GaSb}) = 0.044m_0, \quad m_{hh}^*(\text{GaSb}) = 0.34m_0,$$
$$m_e^*(\text{AlSb}) = 0.92m_0, \quad m_{hh}^*(\text{AlSb}) = 0.94m_0,$$

4. Both $Ga_xIn_{1-x}As$ and $GaAs_ySb_{1-y}$ can be grown lattice-matched on InP substrates. Assume the valence band energy of ternary alloys varies linearly between the two binaries. (A small error will be introduced, but is negligible for lattice-matched cases like this.)

   (a) Determine compositions x and y in $Ga_xIn_{1-x}As$ and $GaAs_ySb_{1-y}$, respectively, that are lattice-matched to InP.
   (b) Use the model-solid theory to find the band discontinuities, ΔE_c and ΔE_v, at $Ga_xIn_{1-x}As$/InP and $GaAs_ySb_{1-y}$/InP interfaces.

5. Strained $Ga_xIn_{1-x}As$ grown on InP substrate is an important heterostructure for both photonic and electronic device applications. The bandgap energy variation of $Ga_xIn_{1-x}As$ is expressed as
   $E_g(\Gamma) = 0.358 + 0.589x + 0.477x^2$
   Assume the elastic modulii, C_{11} and C_{12}, and the deformation potentials, a and b, vary linearly between the two binaries.

   (a) Find the critical layer thickness of $Ga_{0.3}In_{0.7}As$ grown on InP.
   (b) Find the conduction band discontinuity, ΔE_c, and the valence band discontinuity, ΔE_v of strained $Ga_{0.3}In_{0.7}As$ thin layer grown on InP.

(c) Determine the band discontinuities, ΔE_c and ΔE_v, between $Al_{0.48}In_{0.52}As$ and InP
(d) From the results of parts (b) and (c), calculate the band discontinuities ΔE_c and ΔE_v between $Ga_{0.3}In_{0.7}As$ and $Al_{0.48}In_{0.52}As$
(e) For a pHEMT structure consisting of (i) semi-insulating InP substrate, (ii) 12 nm-thick undoped $Ga_{0.3}In_{0.7}As$, and (iii) 20 nm-thick Si-doped $Al_{0.48}In_{0.52}As$ ($n \sim 2\times10^{18}$ cm^{-3}), plot the 'flat' band diagram; i.e., the Fermi levels are not aligned, for simplicity. However, in the $Al_{0.48}In_{0.52}As$ layer, the exact location of E_F should be calculated.
(f) Plot a realistic qualitative band diagram of the structure of part (e). Assume the contact potential on the $Al_{0.48}In_{0.52}As$ side of the $Al_{0.48}In_{0.52}As$/$Ga_{0.3}In_{0.7}As$ heterojunction is ~0.12 eV.

6. The bandgap energy of the quaternary alloy $Ga_xIn_{1-x}As_yP_{1-y}$ lattice-matched to InP can be expressed by $E_g(y) = 1.35 - 0.775y + 0.149y^2$ and $x = 0.47y$.

(a) Determine the composition x and y such that the quaternary has a bandgap corresponding to 1.3 μm.
(b) Determine ΔE_c and ΔE_v between the quaternary and InP.

7. In an $Al_{0.25}Ga_{0.75}N$/GaN high-electron mobility transistor, the undoped $Al_{0.25}Ga_{0.75}N$ barrier is 200 Å in thickness and the GaN buffer layer grown on a sapphire substrate is fully relaxed. Calculate the conduction band discontinuity at the heterojunction.

References

1. R.L. Anderson, Solid-State Electron. **5**, 341 (1962)
2. W.A. Harrison, *Electronic Structure and the Properties of Solids* (Freeman, San Francisco, 1980)
3. J. Tersoff, Phys. Rev. **B30**, 4874 (1984); Phys. Rev. **B32**, 6968 (1985)
4. J. Tersoff, *Heterojunction*, vol. Discontinuities (North Holland, Amsterdam, 1987). Chapter 1
5. M. Cardona, N.E. Christensen, Phys. Rev. **B35**, 6182 (1987)
6. C.G. Van de Walle, R.M. Martin, Phys. Rev. **B35**, 8154 (1987)
7. C.G. Van de Walle, Phys. Rev. **B39**, 1871 (1989)
8. I. Vurgaftman, J.R. Meyer, L.R. Ram-Mohan, J. Appl. Phys. **89**, 5815 (2001)
9. S. Tiwari, D.J. Frank, Appl. Phys. Lett. **60**, 630 (1992)
10. J.W. Mathews, A.E. Blakeslee, J. Crystal Growth. **27**, 118 (1974)
11. Cardona and Christensen, Phys. Rev. **B37**, 1011 (1988)
12. I. Vurgaftman, J.R. Meyer, J. Appl. Phys. **94**, 3675 (2003)
13. J. Wu, W. Walukiewicz, W. Shan, K.M. Yu, J.W. Ager III, S.X. Li, E.E. Haller, H. Lu, W.J. Schaff, J. Appl. Phys. **94**, 4457 (2003)
14. A. Zoroddu, F. Bernardini, P. Ruggerone, V. Fiorentini, Phys. Rev. **B64**, 045208 (2001)
15. S.L. Chuang, IEEE J. Quantum Electron. **32**, 1791 (1996)
16. C.-T. Kuo, K.-K. Chang, H.-W. Shiu, C.-R. Liu, L.-Y. Chang, C.-H. Chen, S. Gwo, Appl. Phys. Lett. **99**, 122101 (2011)
17. Z.I. Alferov, V.B. Khalfin, R.F. Kazarinov, Sov. Phys. Solid State. **8**, 2313 (1967) [Translated from Fiz. Tverd. Tela. **8**, 3102 (1966)]

Further Reading

1. H.C. Casey, M.B. Panish, *Heterostructure Lasers: Part A* (Academic Press, New York, Fundamental Principles, 1978)
2. J.F. Nye, *Physical Properties of Crystals* (Oxford University Press, New York, 1985)
3. S.L. Chuang, Phys. Photon. Dev., 2nd edn. (John Wiley & Sons, 2009)

Chapter 7
Electrical Properties of Compound Semiconductor Heterostructures

Abstract The *p-n* junction and metal–semiconductor junction are the basic building blocks of modern electronic devices such as diodes and transistors. Under the equilibrium condition, a potential barrier is formed at these junctions such that no net charge carriers flow across the junction. An externally applied forward bias is needed to cause the flow of charged carriers and, thus, the current. In semiconductor heterostructures, as discussed in this chapter, due to the formation of energy band discontinuities in both conduction and valence bands at the heterojunction, extra energy barriers exist without resort to *p-n* junctions. For example, under the equilibrium condition, type-I *n-N* and *p-P*-type heterostructures can block flows of holes and electrons, respectively. Furthermore, a triangular potential well is formed at the heterojunction to generate 2D electron gas in *n-N*-type heterostructures and 2D hole gas in *p-P*-type heterostructures. One important feature of heterostructures is the formation of the quantum well that has a far-reaching impact on modern electronic and photonic device advancements. In this chapter, the electrical properties of heterojunctions will be developed first, followed by the analysis of quantum-well heterostructures and superlattices.

7.1 Abrupt Heterojunction Under Equilibrium

7.1.1 Qualitative Analysis of a p-N *Heterojunction in Equilibrium*

(a) **General solutions of the potential distributions**

The general procedure of deriving the potential distribution in a *p-N* or *n-P* heterojunction closely resembles that of a homojunction. The mathematics is also based on Poisson's equation and the charge neutrality condition across the junction and within the depletion region. To illustrate the calculation procedures, a *p*-GaAs/*N*-AlGaAs type-I heterojunction is used as an example. The equilibrium energy band diagram of this anisotype *p-N* heterojunction, as shown in Fig. 7.1, is developed in the following.

K. Y. Cheng, *III–V Compound Semiconductors and Devices*, Graduate Texts in Physics,
https://doi.org/10.1007/978-3-030-51903-2_7

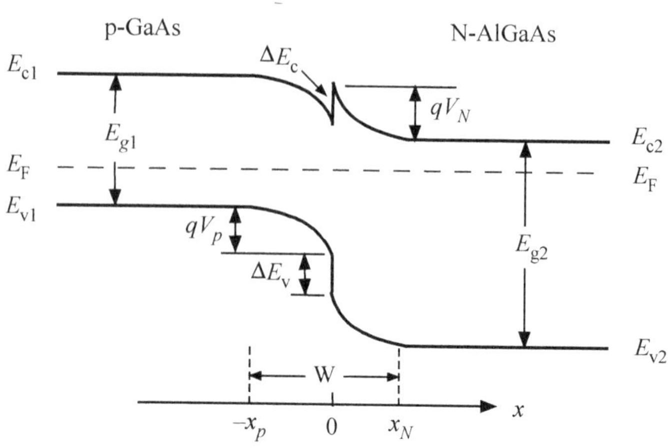

Fig. 7.1 Equilibrium energy band diagram of an anisotype *p*-GaAs/*N*-AlGaAs heterojunction

For a *p-N* heterojunction, the charge neutrality condition requires that

$$qAN_a x_p = qAN_D x_N \tag{7.1}$$

where A is the cross section of the junction, N_a and N_D are the doping concentrations, and x_p and x_N are the depletion widths of the *p*- and *N*-type materials, respectively. The distributions of the electric field intensity $\boldsymbol{F}$ and the potential distribution V are derived from Poisson's equation.

$$\frac{\mathrm{d}^2 V(x)}{\mathrm{d}x^2} = -\frac{\rho}{\epsilon} = -\frac{q}{\epsilon}\left(p - n + N_D^+ - N_a^-\right) = -\frac{\mathrm{d}F}{\mathrm{d}x} \tag{7.2}$$

Neglecting the free carriers p and n in the depletion region, Poisson's equation on either side of the heterojunction becomes

$$-\frac{\mathrm{d}F}{\mathrm{d}x} = \begin{cases} -qN_D/\epsilon_N, & 0 < x < x_N \\ qN_a/\epsilon_p, & -x_p < x < 0 \end{cases} \tag{7.3}$$

where ϵ_p and ϵ_N are the permittivity in the *p*- and *N*-regions, respectively. The electric field intensities on both sides of the junction are obtained by integration.

$$F = \begin{cases} qN_D x/\epsilon_N + A, & 0 < x < x_N \\ -qN_a x/\epsilon_p + B, & -x_p < x < 0 \end{cases} \tag{7.4}$$

where A and B are constants and can be solved using the boundary condition. Similar to a homojunction, the electric field intensity exists only within the depletion region of a heterostructure, i.e., $F = 0$ for $x \geq x_N$ and $x \leq -x_p$. This leads to

$$F = \begin{cases} qN_D(x - x_N)/\epsilon_N, & 0 < x < x_N \\ -qN_a(x + x_p)/\epsilon_p, & -x_p < x < 0 \end{cases} \tag{7.5}$$

These equations indicate that the electric field intensity is a linear function of position x. F has the following forms on either side of the junction:

$$\begin{cases} F(+) = -qN_D x_N/\epsilon_N, & x = 0^+ \\ F(-) = -qN_a x_p/\epsilon_p, & x = 0^- \end{cases} \tag{7.6}$$

Due to the difference in permittivity ($\epsilon_p \neq \epsilon_N$), unlike homojunctions, $F(+)$ and $F(-)$ are not equal at the junction. However, the normal displacement vector $\boldsymbol{D} = \epsilon \boldsymbol{F}$ is continuous at $x = 0$, which gives $\epsilon_N F(+) = \epsilon_p F(-)$. Since $F = -dV/dx$, further integration of F gives

$$\begin{cases} V_N(x) = -(qN_D/2\epsilon_N)x^2 + (qN_D/\epsilon_N)xx_N + C, & 0 < x < x_N \\ V_p(x) = (qN_a/2\epsilon_p)x^2 + (qN_a/\epsilon_p)xx_p + D, & -x_p < x < 0 \end{cases} \tag{7.7}$$

The constants C and D are determined by setting $V_p(x) = 0$ at $x = -x_p$ and $V_N(x) = V_0$ at $x = x_N$.

$$C = V_0 - qN_D x_N^2/2\epsilon_N \quad \text{and} \quad D = qN_a x_p^2/2\epsilon_p \tag{7.8}$$

The potential distribution across the junction has the following final form:

$$V(x) = \begin{cases} V_0 & x_N < x \\ V_0 - (qN_D/2\,\epsilon_N)(x_N - x)^2 & 0 < x < x_N \\ (qN_a/2\,\epsilon_p)(x + x_p)^2 & -x_p < x < 0 \\ 0 & x < -x_p \end{cases} \tag{7.9}$$

Here, V_0 is the total built-in potential and is the sum of potential drop in the p- and N-regions, i.e. $V_0 = V_p + V_N$. At $x = 0$, $V_p(0) = V_p$. Therefore,

$$V_p = V_0 - (qN_D/2\epsilon_N)x_N^2 = (qN_a/2\epsilon_p)x_p^2 \tag{7.10a}$$

$$V_0 - (qN_D/2\epsilon_N)x_N^2 + (qN_a/2\epsilon_p)x_p^2 = V_N + V_p \tag{7.10b}$$

and

$$V_N = (qN_D/2\epsilon_N)x_N^2 \tag{7.10c}$$

However, at $x = 0$, the potential distribution curve has a kink at the junction to reflect the unequal electric field intensities. By combining (7.10b) into (7.9), the potential distribution across the depletion region on N-side becomes

$$V(x) = qN_a x_p^2/2\epsilon_p + (qN_D/2\epsilon_N)(2xx_N - x^2) \quad 0 < x < x_N \tag{7.11}$$

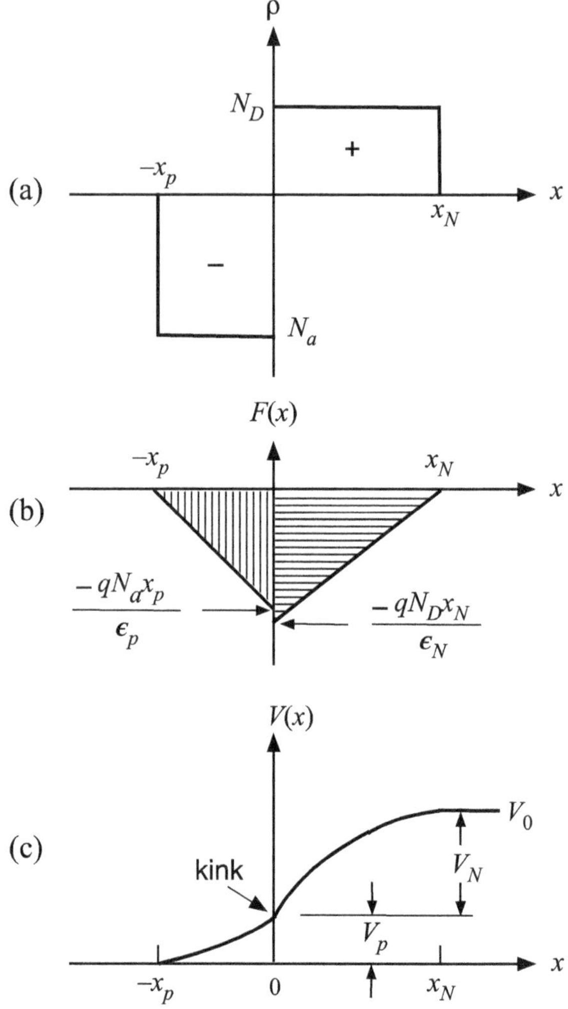

Fig. 7.2 **a** Charge carrier, ρ, **b** electric field intensity, F, and **c** potential, V, distributions of the p-GaAs/N-AlGaAs isotype heterojunction

Now, we can plot the charge distribution, electric field intensity, and potential distribution across the p-N heterojunction as shown in Fig. 7.2.

(b) **Depletion width** (W)

The total depletion width across the p-N junction, W, is the sum of x_N and x_p.

$$W = x_N + x_p \tag{7.12}$$

x_N and x_p are related through the charge neutrality condition (7.1). By replacing x_p or x_N in (7.10b) by x_N or x_p, respectively, the contact potential is related to x_N and x_p as

$$\begin{cases} x_N = \sqrt{\dfrac{2V_0\epsilon_N\epsilon_p}{qN_D}\dfrac{N_a}{(\epsilon_p N_a + \epsilon_N N_D)}} \\ x_p = \sqrt{\dfrac{2V_0\epsilon_N\epsilon_p}{qN_a}\dfrac{N_D}{(\epsilon_p N_a + \epsilon_N N_D)}} \end{cases} \tag{7.13}$$

(c) **Band-edge profile**

Once we know the potential distribution across the *p-N* heterojunction, the band-edge profile can also be derived. Under the equilibrium condition, there is no current flow across the junction. The electron current density is expressed as

$$J_N = q\mu_n nF + qD_n\frac{\mathrm{d}n}{\mathrm{d}x} = 0 \tag{7.14}$$

where q is the charge, μ_n is the electron mobility, n is the electron density, and D_n is the electron diffusion coefficient. The diffusion coefficient and mobility are related through Einstein's relation, $D_n = \mu_n(kT/q)$. Therefore, the electric field intensity becomes

$$F = -\frac{kT}{q}\frac{(\mathrm{d}n/\mathrm{d}x)}{n} \tag{7.15}$$

The equilibrium electron density is determined by the Fermi energy, E_{FN}, relative to the conduction band edge, $E_{FN} - E_c$, and is shown below.

$$\begin{aligned} n &= N_c \exp\left(\frac{E_{FN} - E_c}{kT}\right) \text{and} \\ \frac{\mathrm{d}n}{\mathrm{d}x} &= n\frac{\mathrm{d}}{\mathrm{d}x}\left(\frac{E_{FN} - E_c}{kT}\right) = -\frac{n}{kT}\frac{\mathrm{d}E_c}{\mathrm{d}x} \end{aligned} \tag{7.16}$$

We found

$$F(x) = \frac{1}{q}\frac{\mathrm{d}E_c(x)}{\mathrm{d}x} = -\frac{\mathrm{d}V(x)}{\mathrm{d}x} \tag{7.17}$$

Therefore, $E_c(x) = -qV(x)$. The band edges are parallel to the negative potential distribution. The valence and conduction band edges of the *p-N* heterojunction are plotted in Fig. 7.1 and given by

$$E_v(x) = \begin{cases} -qV_0 - \Delta E_v & x_N < x \\ -qV_0 + \frac{q^2N_D}{2\epsilon_N}(x_N - x)^2 - \Delta E_v & 0 < x < x_N \\ -\frac{q^2N_a}{2\epsilon_p}(x + x_p)^2 & -x_p < x < 0 \\ 0 & x < -x_p \end{cases} \tag{7.18}$$

$$E_c(x) = \begin{cases} E_v(x) + E_{g1}, \ x > 0 \\ E_v(x) + E_{g2}, \ x < 0 \end{cases} \tag{7.19}$$

7.1.2 *Equilibrium* n-N *Heterojunction*

In an isotype *n-N* heterojunction, the charge redistribution near the junction will also produce a space-charge region just like in an anisotype heterostructure. Figure 7.3 shows the qualitative equilibrium band diagram of an *n*-GaAs/*N*-AlGaAs heterostructure. Electrons will spill over from the larger bandgap *N*-region to the small bandgap *n*-region as the Fermi levels line up. The charge distribution around the junction (Fig. 7.4) is

$$\rho(x) = \begin{cases} qN_D & 0 < x < x_N \\ -q[n(x) - N_d] & x < 0 \end{cases} \tag{7.20}$$

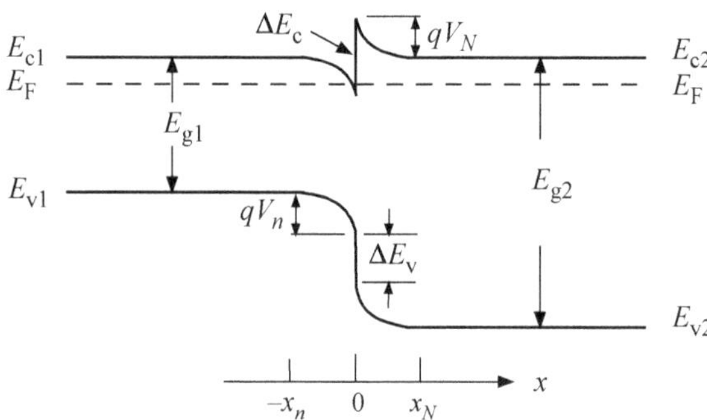

Fig. 7.3 Qualitative equilibrium energy band diagram of an isotype *n*-GaAs/*N*-AlGaAs heterojunction

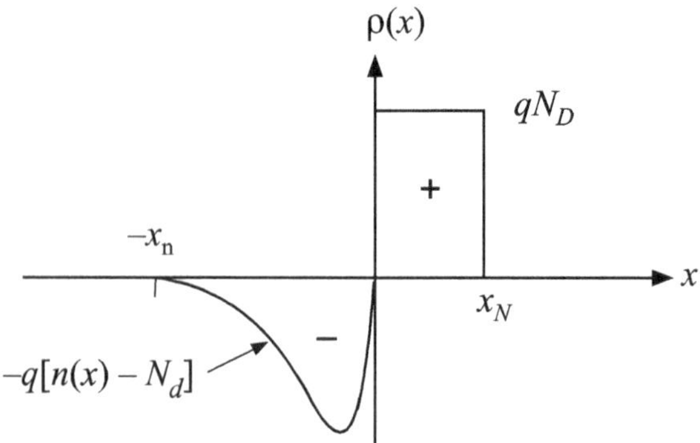

Fig. 7.4 Charge distribution around an *n-N* isotype heterojunction

The charge density in the n-region includes both the ionized impurities, N_d, and electrons injected from the N-region, $n(x)$. Note that $n(x)$ is determined by the allowed wave function in the triangular quantum well.

According to Poisson's equation, the electric field intensity follows

$$\frac{\mathrm{d}F}{\mathrm{d}x} = \begin{cases} qN_D/\epsilon_N & 0 < x < x_N \\ -q[n(x) - N_d]/\epsilon_n & x < 0 \end{cases} \tag{7.21}$$

We can determine the exact solution for $F(x \geq 0)$ in the N-region by using the boundary condition of $F = 0$ at $x = x_N$.

$$F(x) = qN_D(x - x_N)/\epsilon_N \quad 0 \leq x \leq x_N \tag{7.22}$$

At $x = 0^+$, the maximum electric field intensity is

$$F(0^+) = -qN_D x_N/\epsilon_N \tag{7.23}$$

However, the electric field in the n-region is complicated by the non-uniform distribution of the two-dimensional electron gas (2DEG), n_s, in the triangular quantum well. It has a form of

$$F(x) = (-q/\epsilon_n) \int_{-\infty}^{x} [n(x') - N_d] \mathrm{d}x' \quad x \leq 0 \tag{7.24}$$

At $x = 0^-$ the electric field intensity has a maximum value of

$$F(0^-) = \left(-\frac{q}{\epsilon_n}\right) \int_{-\infty}^{0} [n(x') - N_d] \mathrm{d}x' = -\frac{q}{\epsilon_n}(n_s - N_d x_n) \tag{7.25}$$

where $n_s \equiv \int_{-\infty}^{0} n(x)\mathrm{d}x$ is the concentration of the 2DEG.

For a perfect heterojunction without any interface states, these two maximums are related by $\epsilon_n F(0^-) \approx \epsilon_N F(0^+)$. Therefore, $n_s = N_D x_N + N_d x_n$. The electric field intensity distribution in the depletion region of the n-N heterojunction is shown in Fig. 7.5.

Since the band-edge profile is a function of the potential distribution, which in turn relates to the electric field intensity, we need to derive $V(x)$ first, and $F(x) = -dV(x)/dx$. In the N-region,

$$V(x) = -\int F(x)\mathrm{d}x = -\frac{qN_D}{\epsilon_N}\left(\frac{x^2}{2} - x x_N\right) + C \tag{7.26}$$

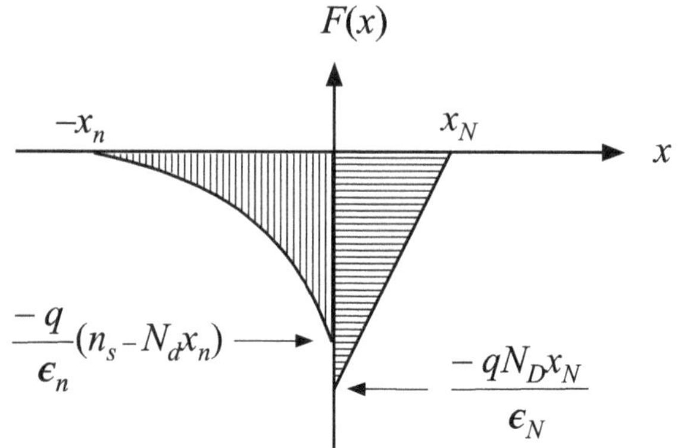

Fig. 7.5 Electric field intensity distribution in an *n-N* isotype heterostructure. Note the maximum intensities of the two sides do not meet at $x = 0$

Setting $V = V_0$ at $x = x_N$, we can solve the constant C. The potential distribution in the N-region is shown in Fig. 7.6 and expressed as

$$V(x) = V_0 - \frac{qN_D}{2\epsilon_N}(x - x_N)^2 \quad 0 \leq x \leq x_N \tag{7.27a}$$

$$V(0) = V_n = V_0 - \frac{qN_D}{2\epsilon_N}x_N^2 \tag{7.27b}$$

and

$$V_N = V_0 - V_n = \frac{qN_D}{2\epsilon_N}x_N^2 \tag{7.27c}$$

The energy band edges on the N-side are parallel to the negative potential distribution, $E(x) = -qV(x)$, and similar to the N-side in a *p-N* heterostructure.

Here, V_0 is determined by the Fermi level difference in the heterostructure.

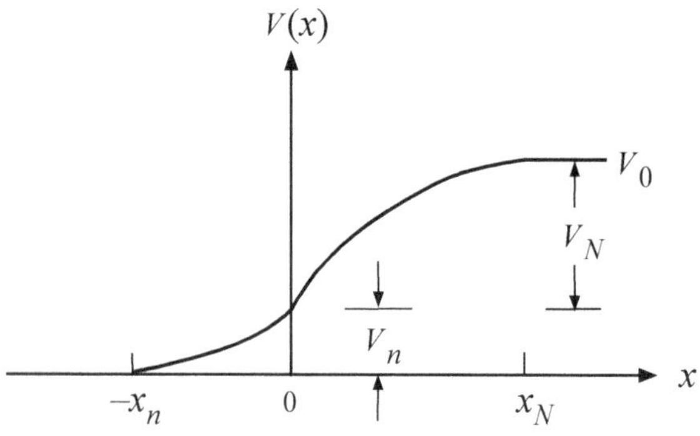

Fig. 7.6 Potential function in the depletion region of an isotype *n-N* heterostructure

$$V_0 = \frac{E_{FN} - E_{Fn}}{q} = \frac{1}{q}[(E_{cn} - E_{Fn}) + \Delta E_c - (E_{cN} - E_{FN})] \tag{7.28}$$

where E_{FN} and E_{Fn} are the bulk Fermi levels of the N- and n-regions, respectively. The Fermi level positions in the bulk materials are determined by

$$(E_{cn} - E_{Fn}) = kT\ \ln(N_d/N_{cn}); \quad (E_{cN} - E_{FN}) = kT\ \ln(N_D/N_{cN}) \tag{7.29}$$

where N_{cN} and N_{cn} are the effective density of states in N- and n-regions, respectively.

The calculation of potential profile on the n-side, however, is complicated by the non-uniform distribution of the 2DEG. Nevertheless, the general feature of the energy band edge on the n-side is similar to that of the p-side in a p-N heterostructure.

7.2 *p-N* Heterojunction Under Bias

7.2.1 *Potential Distribution Profile*

Under forward and reverse biases (V_a), the contact potential (V_0) of a p-N heterojunction is modified by an applied bias of (V_f) and ($-V_r$), respectively. This leads to a modification of the potential distribution profile as well as the depletion widths. The applied bias is distributed between the p- and N-sides, as illustrated in Fig. 7.7:

$$V_a = V_1 + V_2 \tag{7.30}$$

where V_1 is the portion of V_a on the p-side, while V_2 is the portion of V_a on the N-side.

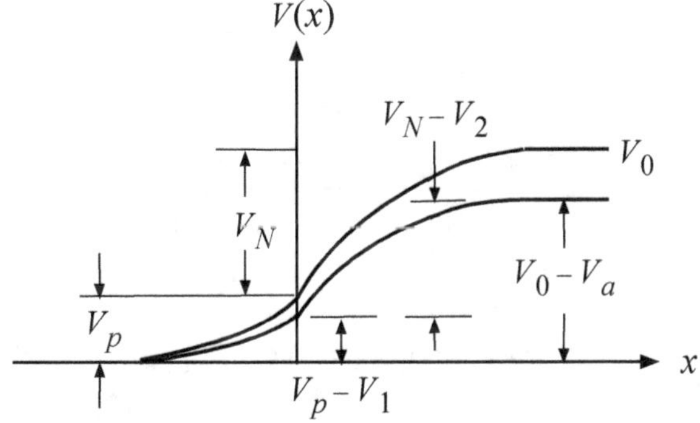

Fig. 7.7 Variation of the potential function under bias in the depletion region of a p-N heterojunction

The applied bias modifies the depletion widths as

$$\begin{cases} x_N = \sqrt{\dfrac{2(V_0 - V_a)\epsilon_N\epsilon_p}{qN_D} \dfrac{N_a}{(\epsilon_p N_a + \epsilon_N N_D)}} \\ x_p = \sqrt{\dfrac{2(V_0 - V_a)\epsilon_N\epsilon_p}{qN_a} \dfrac{N_D}{(\epsilon_p N_a + \epsilon_N N_D)}} \end{cases} \tag{7.31}$$

The previous expressions for the potential distribution and energy band diagram of a *p–N* heterojunction are modified accordingly by the applied bias.

7.2.2 *Carrier Injection Across the* p-N *Heterojunction*

(a) **Carrier densities in the depletion region**

Under applied forward bias V_a, the barrier heights are lowered to allow carriers to flow across the heterojunction and drift through the depletion region. Once passing the depletion region, carrier motion in the opposite side of the junction is governed by diffusion, as depicted in Fig. 7.8. We will follow the same procedures used for homojunctions, with some modifications, to derive the carrier distributions and current flows.

For an ideal current–voltage model, there are four basic assumptions: (a) *Non-degenerate materials*: The doping levels are low so that carrier densities may be represented by the exponential approximation for the Fermi–Dirac function. (b) *Low-level injection*: The injected minority carrier densities are small compared to the

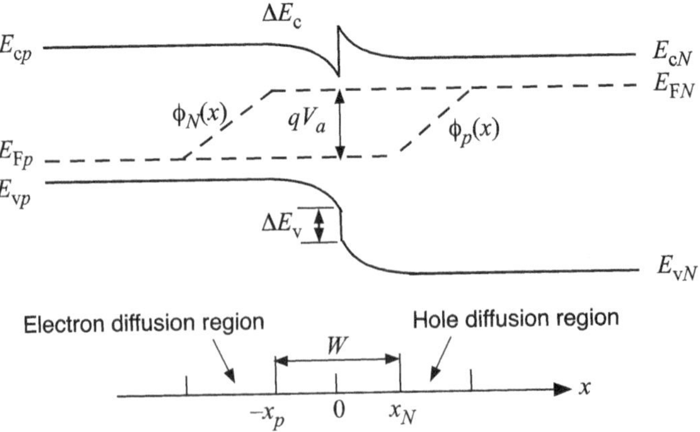

Fig. 7.8 Qualitative energy band diagram of an isotype *p-N* type-I heterostructure under a forward bias V_a

majority carrier densities. (c) *Abrupt depletion region*: The space-charge depletion layer has an abrupt boundary, and the built-in electric field only exists inside this region. Outside the depletion region, the semiconductor is assumed to be neutral. (d) *Ideal junction*: There is no generation and recombination current in the depletion region so that the electron and hole currents are constant through the depletion region.

Under thermal equilibrium conditions, the majority carrier concentration at the edges of the depletion region follows the Boltzmann distribution and is given as

$$n_{N0} = N_{cN} \exp[(E_{FN} - E_{cN})/kT] \tag{7.32a}$$

and

$$p_{p0} = N_{vp} \exp[(E_{vp} - E_{Fp})/kT] \tag{7.32b}$$

where N_{cN} and N_{vp} are the effective densities of states for the N- and p-regions, respectively. The minority carrier densities at thermal equilibrium are given by

$$p_{N0} = n_{iN}^2/n_{N0} \tag{7.33a}$$

and

$$n_{p0} = n_{ip}^2/p_{p0} \tag{7.33b}$$

where n_{iN} and n_{ip} are the intrinsic carrier concentrations of the N- and p materials, respectively. The equilibrium majority carrier densities can also be expressed as

$$n_{N0} = n_{iN} \exp[(E_{FN} - E_{iN})/kT] \tag{7.34a}$$

and

$$p_{p0} = n_{ip} \exp[(E_{ip} - E_{Fp})/kT] \tag{7.34b}$$

where E_{iN} and E_{ip} are the intrinsic levels in the N and p regions, respectively. The product of carrier densities is simply

$$n_{N0}p_{N0} = n_{iN}^2 \exp[(E_{FN} - E_{iN})/kT] \exp[(E_{iN} - E_{FN})/kT] = n_{iN}^2 \tag{7.35}$$

Under low-level injections, the majority carrier concentrations at the edges of the depletion region are

$$\begin{cases} n_N = n_{N0} \cong N_D & \text{at} \quad x = x_N \\ p_p = p_{p0} \cong N_a & \text{at} \quad x = -x_p \end{cases} \tag{7.36}$$

The applied bias V_a is directly related to the Fermi energy difference and is expressed as

$$qV_a = E_{FN} - E_{Fp}. \tag{7.37}$$

Since there is no generation and recombination inside the depletion region, the quasi-Fermi levels, $\phi_N(x)$ and $\phi_p(x)$, maintain a constant value and are equal to the bulk Fermi levels, E_{FN} and E_{Fp}, respectively. Under low-level injections, the carrier densities at the junction are now given by

$$n_N = n_{iN} \exp[(q\phi_{n0} - E_{iN})/kT] \tag{7.38a}$$

and

$$p_N = n_{iN} \exp\left[\left(E_{iN} - q\phi_{p0}\right)/kT\right] \tag{7.38b}$$

where ϕ_{n0} and ϕ_{p0} are the quasi-Fermi levels for electrons and holes, respectively, at the junction. The product of the majority and minority carrier densities at the junction is

$$\begin{aligned} n_N p_N &= n_{iN}^2 \exp\left[\left(q\phi_{n0} - E_{iN} + E_{iN} - q\phi_{p0}\right)/kT\right] \\ &= n_{iN}^2 \exp\left[\left(q\phi_{n0} - q\phi_{p0}\right)/kT\right] \end{aligned} \tag{7.39a}$$

$$n_N p_N = n_{iN}^2 \exp(qV_a/kT) \tag{7.39b}$$

At the edge of the depletion layer, $x = x_N$, the minority carrier density is

$$p_N = \left(n_{iN}^2/n_N\right) \exp(qV_a/kT) = p_{N0} \exp(qV_a/kT) \tag{7.40a}$$

By the same procedure, the minority electron density at the edge of the depletion layer in the p-region, $x = -x_p$, may also be written as

$$n_p = \left(n_{ip}^2/p_p\right) \exp(qV_a/kT) = n_{p0} \exp(qV_a/kT) \tag{7.40b}$$

It is clear that the minority carrier concentrations at the edges of the depletion region increase exponentially with the applied positive bias, V_a.

(b) **Injected current densities across the *p-N* heterojunction**

The carrier transport is governed by the continuity equation. In one dimension, the movement of electrons and holes is described by

$$\begin{cases} \dfrac{\partial n}{\partial t} = \dfrac{1}{q}\dfrac{\partial j_n}{\partial x} + g(x) - \dfrac{n - n_0}{\tau_n} \\ \dfrac{\partial p}{\partial t} = -\dfrac{1}{q}\dfrac{\partial j_p}{\partial x} + g(x) - \dfrac{p - p_0}{\tau_p} \end{cases} \tag{7.41}$$

where j_n and j_p are electron and hole current densities, $g(x)$ is the generation rate, n_0 and p_0 are the thermal equilibrium values of n and p, and τ_n and τ_p are the electron and hole lifetimes, respectively.

At steady state, $\dfrac{\partial n}{\partial t}$ and $\dfrac{\partial p}{\partial t}$ are zero, and without the external excitation, $g(x)$ is also zero. The electron and hole currents including both drift and diffusion current densities are given by

$$\begin{cases} j_n = q\mu_n nF + qD_n\dfrac{dn}{dx} \\ j_p = q\mu_p pF - qD_p\dfrac{dp}{dx} \end{cases} \tag{7.42}$$

In the neutral region, $F = 0$, into which the carriers are injected, only the diffusive component will be considered. The continuity equations are given as

$$\begin{cases} D_n\dfrac{d^2 n}{\partial x^2} - \dfrac{n - n_0}{\tau_n} = 0 \\ D_p\dfrac{d^2 p}{\partial x^2} - \dfrac{p - p_0}{\tau_p} = 0 \end{cases} \tag{7.43}$$

In p-type neutral region, the minority carrier density is $n_0 = n_{p0}$ and the solution of the continuity equation is

$$n(x) = C_1 \exp\left(-x/\sqrt{D_n\tau_n}\right) + C_2 \exp\left(x/\sqrt{D_n\tau_n}\right) + n_{p0} \tag{7.44}$$

Using the boundary condition of $n(x = -\infty) = n_{p0}$, and $n(-x_p) = n_{p0}\exp(qV_a/kT)$, we found $C_1 = 0$ and $C_2 = n_{p0}\left[\exp(-qV_a/kT) - 1\right]\exp\left(x_p/\sqrt{D_n\tau_n}\right)$.

The injected electron distribution in the p-region has the form

$$n(x) - n_{p0} = n_{p0}\left[\exp(-qV_a/kT) - 1\right]\exp\left[\left(x_p + x\right)/L_n\right] \tag{7.45}$$

and $L_n \equiv \sqrt{D_n\tau_n}$ is the diffusion length. Since there is no electric field in the neutral region, the injected current from minority carriers is mainly due to diffusion.

$$j_n\left(-x_p\right) = qD_n\left.\frac{dn}{dx}\right|_{-x_p} = qD_n n_{p0}\left[\exp(-qV_a/kT) - 1\right]\frac{1}{L_n}\exp\left[\left(x_p + x\right)/L_n\right]\Big|_{-x_p} \tag{7.46}$$

Therefore, the minority carrier-induced injection currents on either side of the depletion region are given by

$$\begin{cases} j_n(-x_p) = \dfrac{qD_n}{L_n} n_{p0}\left[\exp(-qV_a/kT) - 1\right] \\ j_p(x_N) = \dfrac{qD_p}{L_p} p_{N0}\left[\exp(-qV_a/kT) - 1\right] \end{cases} \tag{7.47}$$

The total current density in a *p-N* diode becomes

$$\begin{aligned} J &= \left(\frac{qD_n}{L_n} n_{p0} + \frac{qD_p}{L_p} p_{N0}\right)\left[\exp(-qV_a/kT) - 1\right] \\ &= J_0\left[\exp(-qV_a/kT) - 1\right] \end{aligned} \tag{7.48}$$

This is similar to the *I-V* characteristic of a *p-n* homojunction diode. Nevertheless, the influence of the heterojunction is revealed from the ratio of the electron and hole current components.

$$\left|\frac{j_n}{j_p}\right| = \frac{D_n L_p}{D_p L_n} \frac{n_{p0}}{p_{N0}} = \frac{D_n L_p}{D_p L_n} \frac{n_{ip}^2}{p_{p0}} \frac{n_{N0}}{n_{iN}^2} \tag{7.49}$$

Since $n_{iN}^2 = N_{cN} N_{vN} \exp(-E_{gN}/kT)$ and $n_{ip}^2 = N_{cp} N_{vp} \exp(-E_{gp}/kT)$, the current ratio is given by

$$\left|\frac{j_n}{j_p}\right| = \frac{D_n L_p}{D_p L_n} \frac{n_{N0}}{p_{p0}} \frac{N_{cN} N_{vN}}{N_{cp} N_{vp}} \exp\left(\frac{E_{gN} - E_{gp}}{kT}\right) \tag{7.50}$$

The exponential term dominates the current ratio. For $E_{gN} > E_{gp}$, $j_n \gg j_p$, i.e., the current will be injected from the wide-bandgap material into the narrow bandgap material. In *p-n* homojunction diodes, $E_{gN} = E_{gn} = E_{gp}$ and the exponential term in the current ratio is reduced to one.

7.3 Quantum-Well Heterostructures

7.3.1 *Infinitely Deep Square Potential Well*

(a) **One-dimensional (1D) potential well**

In general, it is of particular interest to study the movement of a particle in a static square potential energy distribution $V(\boldsymbol{r})$ of size a. To keep matters simple, we shall resort to a 1D time-independent treatment. As shown in Fig. 7.9, the infinitely deep well can be described by

$$\begin{cases} V(z) = 0, \quad 0 < z < a \\ V(z) = \infty, \; z < 0; \; z > a \end{cases} \tag{7.51}$$

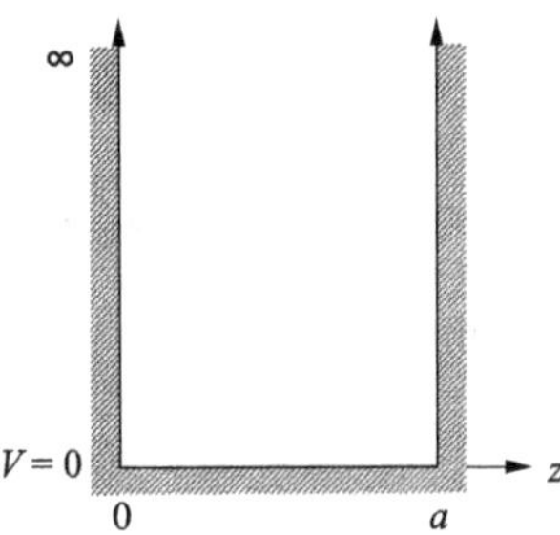

Fig. 7.9 Infinitely deep potential well

Thus we can write the 1D time-independent Schrödinger's wave equation as

$$-\frac{\hbar^2}{2m}\frac{\partial^2\psi(z)}{\partial z^2} + V(z)\psi(z) = E\psi(z) \tag{7.52}$$

where $\psi(z)$, $\hbar$, and E are the wave function, Planck constant ($\hbar = h/2\pi$), and energy, respectively. For the well width a less than the Bohr radius $a_B(= 4\pi\epsilon\hbar^2/m^*q^2)$ of the well material, the potential well is called a *quantum well* (QW). In the region ($0 < z < a$) where $V = 0$, the time-independent wave equation can be written as

$$\frac{\partial^2\psi}{\partial z^2} + k^2\psi = 0 \tag{7.53}$$

where $k = \sqrt{2mE/\hbar^2}$

The wave function does not exist outside the infinitely deep potential well. The allowed wave functions inside the potential well have the form

$$\psi(z) = A\sin kz + B\cos kz \tag{7.54}$$

where A and B are constants. To determine the constants A and B, we need to use boundary conditions in (7.54). At $z = 0$, $\psi(0) = 0$ and $B = 0$. At $z = a$, $\psi(a) = A \sin ka = 0$. This condition can be satisfied only when

$$k_n a = n\pi \quad n = 1, 2, 3, \ldots. \tag{7.55}$$

k_n represents a discrete sequence of values of k, where

$$k_n = \frac{n\pi}{a} = \sqrt{\frac{2mE_n}{\hbar^2}} \tag{7.56}$$

Since k is related to the energy E, this condition defines a discrete set of allowed energies

$$E_n = \frac{\hbar^2 k_n^2}{2m} = \frac{1}{2m}\left(\frac{n\pi\hbar}{a}\right)^2 \quad n = 1, 2, 3, \ldots \tag{7.57}$$

In contrast to the classical result, the allowed energy states have become discrete levels and their values are proportional to the quantum number n^2. The corresponding wave functions of the system within the potential well, by incorporating k_n (7.56), must be of the form

$$\psi_n(z) = A_n \sin\left(\frac{n\pi z}{a}\right) \quad 0 < z < a \tag{7.58}$$

By normalization of the wave function, we can determine the constant A_n.

$$\int_{-\infty}^{\infty} \psi_n^* \psi_n dz = A_n^2 \int_0^a \sin^2(n\pi z/a) dz = 1 \tag{7.59}$$

given $A_n = \sqrt{2/a}$. Then the time-independent wave function becomes

$$\psi_n(z) = \sqrt{\frac{2}{a}} \sin\left(\frac{n\pi z}{a}\right) \tag{7.60}$$

The first three allowed wave functions and the probability amplitudes of a bounded particle are shown in Fig. 7.10. The lowest state has energy $E_1 > 0$. This contrasts with classical mechanics where the state of the lowest energy has no kinetic energy

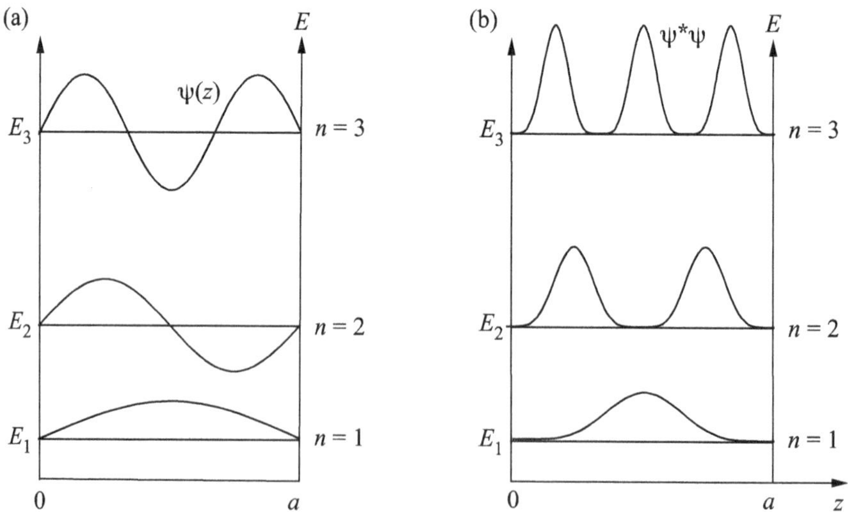

Fig. 7.10 **a** Wave functions and **b** probability amplitudes for the first three allowed energy in an infinite deep potential well

and $E_1 = 0$. The wave functions in the potential well are not constant but show a symmetry property. The probability of finding the particle inside the infinitely deep potential well over a region Δz is

$$\psi\psi^*\Delta z = \frac{2\Delta z}{a}\sin^2\left(\frac{n\pi z}{a}\right) = \frac{\Delta z}{a}\left[1 - \cos\left(\frac{2n\pi z}{a}\right)\right] \tag{7.61}$$

which is not a constant. Again, this contrasts with the result of classical mechanics.

(b) **Three-dimension (3D) potential well**

In a real QW structure, electrons are moving three dimensionally (3D) within the QW. The allowed energy states of these 3D electrons in an infinitely deep QW can be calculated by solving the 3D time-independent Schrödinger wave equation.

$$\left[-\frac{\hbar^2}{2m_w^*}\nabla^2 + V(r)\right]\psi(r) = E\psi(r) \tag{7.62}$$

where m_w^* is the effective mass of the QW. In this QW, the potential function $V(r)$ depends only on the coordinate z normal to the layer. Therefore, we can assume $V(r) = V(z)$, and the wave equation becomes

$$\left[-\frac{\hbar^2}{2m_w^*}\left(\frac{\partial^2}{\partial x^2} + \frac{\partial^2}{\partial y^2} + \frac{\partial^2}{\partial z^2}\right) + V(z)\right]\psi(x, y, z) = E\psi(x, y, z) \tag{7.63}$$

Since there is no barrier in the x- and y- directions, the solutions of the wave equation in these two directions are plane waves. We can assume the solution has the form

$$\psi(x, y, z) = \exp(ik_x x)\exp(ik_y y)\phi(z) \tag{7.64}$$

We substitute this into the Schrödinger equation and find it is a function of z only.

$$\left[\frac{\hbar^2 k_x^2}{2m_w^*} + \frac{\hbar^2 k_y^2}{2m_w^*} - \frac{\hbar^2}{2m_w^*}\frac{d^2}{dz^2} + V(z)\right]\phi(z) = E\phi(z) \tag{7.65}$$

The energy-related terms of the plane waves can be moved to the right-hand side, giving

$$\left[-\frac{\hbar^2}{2m_w^*}\frac{d^2}{dz^2} + V(z)\right]\phi(z) = \left(E - \frac{\hbar^2 k_x^2}{2m_w^*} - \frac{\hbar^2 k_y^2}{2m_w^*}\right)\phi(z) \tag{7.66}$$

This is a one-dimensional Schrödinger equation in z. The solutions of the wave function within the well are

$$\psi_n(x, y, z) = \exp(ik_x x)\exp(ik_y y)\sqrt{\frac{2}{L}}\sin\left(\frac{n\pi}{L}z\right) \quad 0 \le z \le L \tag{7.67}$$

where L is the well width and n is an integer. The corresponding energy eigenvalues are

$$E_n = \frac{\hbar^2 k_x^2}{2m_w^*} + \frac{\hbar^2 k_y^2}{2m_w^*} + \frac{\hbar^2}{2m_w^*}\left(\frac{n\pi}{L}\right)^2 = \frac{\hbar^2 k_x^2}{2m_w^*} + \frac{\hbar^2 k_y^2}{2m_w^*} + \lambda_n \tag{7.68}$$

and

$$\lambda_n = \frac{\hbar^2}{2m_w^*}\left(\frac{n\pi}{L}\right)^2. \tag{7.69}$$

The energy is quantized in the z-direction only. Therefore, electrons still can move freely in the x-y plane of a QW. The E-k diagram of the infinitely deep QW is shown in Fig. 7.11. For a fixed n, the allowed energy is shifted to λ_n. The energy function for each n gives a parabola, called an *electronic subband*, starting at the energy $E_n = \lambda_1 > 0$ when plotted against k. In between λ_1 and λ_2, there are only states in the lowest subband. For $\lambda_2 < E < \lambda_3$, there are states in the two lowest subbands, $n = 1$ and 2. For each subband, the 2D electrons contribute to the DOS (cm^{-2} eV^{-1}) in a step function of height $(m_w^* \pi/\hbar^2)$, starting at λ_1. The total DOS looks like a staircase with jumps at the energies of the subbands.

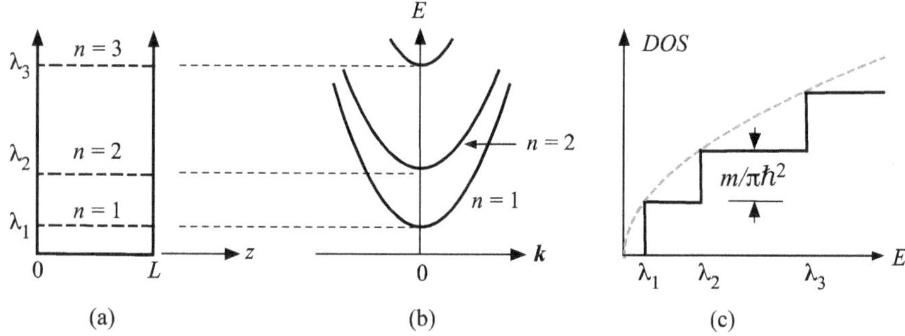

Fig. 7.11 **a** Energy levels in an infinite potential well with confinement in z only, **b** total energy including the transverse kinetic energy for each subband, and **c** step-like density of states (DOS) of a two-dimensional system. The light dashed curve in (**c**) is the parabolic DOS for unconfined three-dimensional electrons

7.3.2 Realistic QW with Finite Depth Barriers

In a realistic QW shown in Fig. 7.12 with a well width L and a finite barrier height V_0, separate wave equations may be used to describe the wave functions and energy eigenvalues in the well and barrier regions. In the well region, $|z| < L/2$, the Schrödinger equation is given by

$$\frac{d^2\psi_0}{dz^2} + \alpha^2\psi_0 = 0 \tag{7.70a}$$

and

$$\alpha = \sqrt{\frac{2m_w^* E}{\hbar^2}} \tag{7.70b}$$

The solutions of the wave function in the QW are

$$\psi_0 = \begin{cases} A\cos(\alpha z) & \text{(even function)} \\ A\sin(\alpha z) & \text{(odd function)} \end{cases} \tag{7.71}$$

In the barrier regions, the wave functions are ψ_+ and ψ_- in the regions of $z > L/2$ and $z < -L/2$, respectively. The associated Schrödinger wave equation in the barrier is

$$\frac{d^2\psi_\pm}{dz^2} - \beta^2\psi_\pm = 0 \tag{7.72a}$$

and

$$\beta = \sqrt{\frac{2m_b^*(V_0 - E)}{\hbar^2}} \tag{7.72b}$$

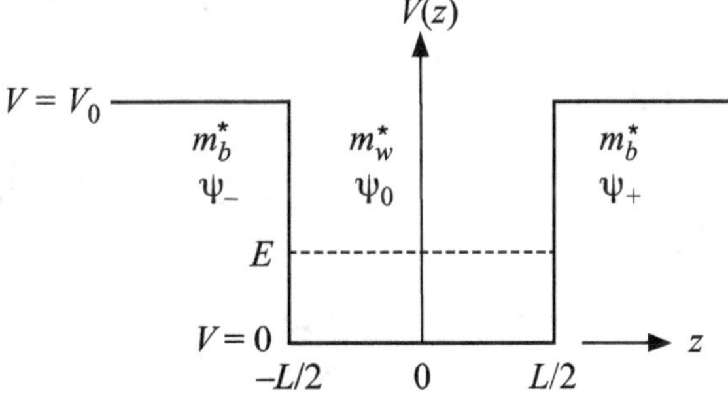

Fig. 7.12 Finite quantum well of depth V_0 and width L

where m_b^* is the effective mass of the barrier material. The solutions of the wave function in the barrier region have the form

$$\psi_{\pm} = \begin{cases} B\exp[-\beta(z - L/2)] & z > L/2 \\ B\exp[\beta(z + L/2)] & z < -L/2 \end{cases} \tag{7.73}$$

To solve the constants A and B, we have to apply the connection conditions at the well-barrier interface, $|z| = L/2$.

$$\psi_0 = \psi_{\pm} \quad \text{and} \quad \frac{1}{m_w^*}\frac{d\psi_0}{dz} = \frac{1}{m_b^*}\frac{d\psi_{\pm}}{dz} \tag{7.74}$$

The second boundary condition for the derivative of wave functions has been modified to include the current flow continuity at the hetero-interface.

(a) **Even function**

For even functions, at $z = L/2$, the boundary conditions lead to

$$A\cos(\alpha L/2) = B$$

and

$$-\alpha\frac{A}{m_w^*}\sin(\alpha L/2) = -\beta\frac{B}{m_b^*}\exp[-\beta(z - L/2)]_{z=L/2} = -\beta\frac{B}{m_b^*}$$

Eliminating A and B, we reach the following solution:

$$\frac{\alpha}{m_w^*}\tan(\alpha L/2) = \frac{\beta}{m_b^*} \tag{7.75a}$$

or

$$\beta = \left(\frac{m_b^*}{m_w^*}\right)\alpha\tan(\alpha L/2) \quad \text{(even function)} \tag{7.75b}$$

(b) **Odd function**

For odd functions, we use the same method to find solutions. At $z = L/2$, the solutions are given by

$$\beta = -\left(\frac{m_b^*}{m_w^*}\right)\alpha\cot(\alpha L/2) \quad \text{(odd function)} \tag{7.76}$$

The above two equations have no analytical solutions. Therefore, the graphic or numerical method will be used. First, we multiply $(L/2)$ to the above two equations.

$$\begin{cases} \left(\beta \frac{L}{2}\right) = \left(\alpha \frac{L}{2}\right)\left(\dfrac{m_b^*}{m_w^*}\right) \tan\left(\alpha \frac{L}{2}\right) & \text{(even)} \\ \left(\beta \frac{L}{2}\right) = -\left(\alpha \frac{L}{2}\right)\left(\dfrac{m_b^*}{m_w^*}\right) \cot\left(\alpha \frac{L}{2}\right) & \text{(odd)} \end{cases} \tag{7.77}$$

Replacing β' for $\beta\sqrt{m_w^*/m_b^*}$, the above equations become

$$\begin{cases} \left(\beta' \frac{L}{2}\right) = \sqrt{\dfrac{m_b^*}{m_w^*}} \left(\alpha \frac{L}{2}\right) \tan\left(\alpha \frac{L}{2}\right) & \text{(even)} \\ \left(\beta' \frac{L}{2}\right) = -\sqrt{\dfrac{m_b^*}{m_w^*}} \left(\alpha \frac{L}{2}\right) \cot\left(\alpha \frac{L}{2}\right) & \text{(odd)} \end{cases} \tag{7.78}$$

These equations are the functions of $(\beta' L/2)$ and $(\alpha L/2)$. Also

$$\alpha^2 + \beta'^2 = \frac{2m_w^* E}{\hbar^2} + \frac{m_w^*}{m_b^*} \frac{2m_b^*(V_0 - E)}{\hbar^2} = \frac{2m_w^* V_0}{\hbar^2} \tag{7.79a}$$

or

$$\left(\alpha \frac{L}{2}\right)^2 + \left(\beta' \frac{L}{2}\right)^2 = \frac{2m_w^* V_0}{\hbar^2} \left(\frac{L}{2}\right)^2 \equiv R^2 \tag{7.79b}$$

This is an equation of a circle defined by $(\beta' L/2)$ and $(\alpha L/2)$. The above equation can be solved either graphically or numerically. Since the well width, well effective mass, and barrier height are known values, R can be determined. Replacing $(\beta' L/2)$ by R and $(\alpha L/2)$ in (7.78), the wave functions can be solved numerically through the following equations of $(\alpha L/2)$. Thus, the values of α are determined and can be used for energy states calculation.

$$\begin{cases} \sqrt{R^2 - \left(\alpha \frac{L}{2}\right)^2} = \sqrt{\dfrac{m_b^*}{m_w^*}} \left(\alpha \frac{L}{2}\right) \tan\left(\alpha \frac{L}{2}\right) & \text{(even)} \\ \sqrt{R^2 - \left(\alpha \frac{L}{2}\right)^2} = -\sqrt{\dfrac{m_b^*}{m_w^*}} \left(\alpha \frac{L}{2}\right) \cot\left(\alpha \frac{L}{2}\right) & \text{(odd)} \end{cases} \tag{7.80}$$

In the graphic approach, the intersections of the circle of radius R and the tangent and cotangent functions determine the solution of the allowed wave function in the QW as shown in Fig. 7.13. From the allowed α's, we can obtain the energy eigenvalues according to (7.70b).

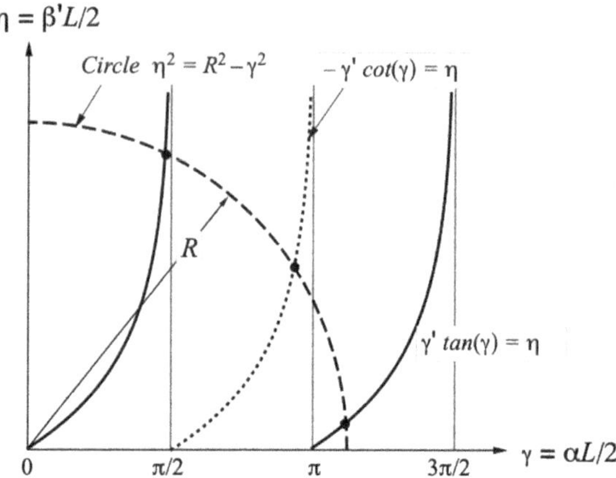

Fig. 7.13 Solutions of the finite depth 1D potential well using graphical method. The solid and dotted curves represent the even and odd transcendental equations, respectively, where $\gamma' = \gamma\sqrt{m_b^*/m_w^*}$. The intersections of R and the transcendental equations are the solutions, $\alpha_1 L/2$, $\alpha_2 L/2$, and $\alpha_3 L/2$

7.3.3 Hole Energy Levels in the QW

The valence band structure of *bulk* semiconductors is degenerated near the zone center. The dispersion of the heavy and light-hole bands is very roughly described by

$$E(k) = E_v - \frac{\hbar^2 k^2}{2m_h^*} \tag{7.81}$$

where m_h^* equals the effective mass m_{hh}^* and m_{lh}^* for heavy and light holes, respectively. The parabolic and isotropic nature of this simple model does not hold for real semiconductors. The shape of the constant energy surface depends on the direction of $\boldsymbol{k}$ and has the shape of the *warped sphere*. The complicated situation gets worse if the cubic symmetry of the crystal is broken. In general, despite its shortcomings, we will still use the simple parabolic model for the E-k relation since a precise description is cumbersome.

In a quantum well, the potential step along the z-direction further destroys the isotropic nature of the crystal and modifies the band shape as illustrated in Fig. 7.14. Two important features are observed: First, the heavy and light holes are no longer degenerated at $k = 0$. The wave functions of heavy and light holes have the symmetry of p orbitals (Sect. 3.6). The orientations of p orbitals for heavy holes are oriented

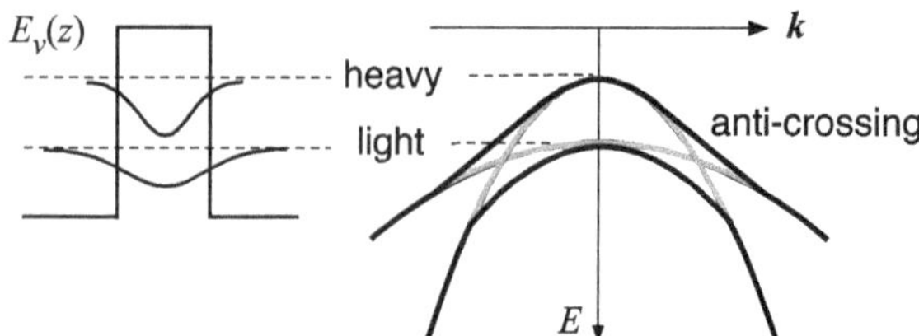

Fig. 7.14 Energy band model of holes in a quantum well. The 'heavy' and 'light' hole bands cross each other, leading to 'anti-crossing' behavior shown by the dark lines

Table 7.1 Luttinger parameters of III–V binary compounds [1]

III–V	GaAs	AlAs	InAs	GaP	AlP	InP	GaSb	AlSb	InSb
γ_1	6.98	3.76	20.0	4.05	3.35	5.08	13.4	5.18	34.8
γ_2	2.06	0.82	8.5	0.49	0.71	1.60	4.7	1.19	15.5
γ_3	2.93	1.42	9.2	2.93	1.23	2.10	6.0	1.97	16.5

normal to z, while the p orbital orientations for light holes are oriented along z. The heavy mass along z for heavy holes means that the bound state is deep in the QW and thus decoupled from the light-hole energy state. Second, the 'heavy' and 'light' characters are strongly mixed. For heavy holes, the orientations of p orbitals are in the xy-plane and have a lighter in-plane mass. This means that the kinetic energy rises rapidly as a function of k in the plane of the QW. On the other hand, the orientation of p orbitals for light holes is perpendicular to that of heavy holes. The light-hole mass is lighter along z and heavy in the xy-plane. This means a low binding energy (high kinetic energy) in the QW. The heavy holes are actually lighter, and the light holes become heavier for transverse motion. Therefore, the light-hole and heavy-hole bands cross over at some k.

The heavy-hole and light-hole effective masses along the [100] (z-direction) in a QW are modified in terms of Luttinger parameters (γ_i) as

$$m^*_\mathrm{hh} = m_0/(\gamma_1 - 2\gamma_2),\ m^*_\mathrm{lh} = m_0/(\gamma_1 + 2\gamma_2) \tag{7.82}$$

The Luttinger parameters for III-V alloys are listed in Table 7.1.

7.3.4 *Strained Quantum Wells*

For a strained quantum-well structure, we can assume that the band edges of the barrier material are unaltered. However, the allowed energy states of the well are modified to include strain-induced band discontinuity changes in the well material. The band-edge energy diagrams of strained QWs are compared with an unstrained QW in Fig. 7.15. The band-edge energy of the QW under strain can be calculated using the procedure outlined in Sect. 6.3. Once the band-edge positions of the QW material are determined, the allowed energy states can be calculated using equations from the previous section to include the effective masses of the well and barriers.

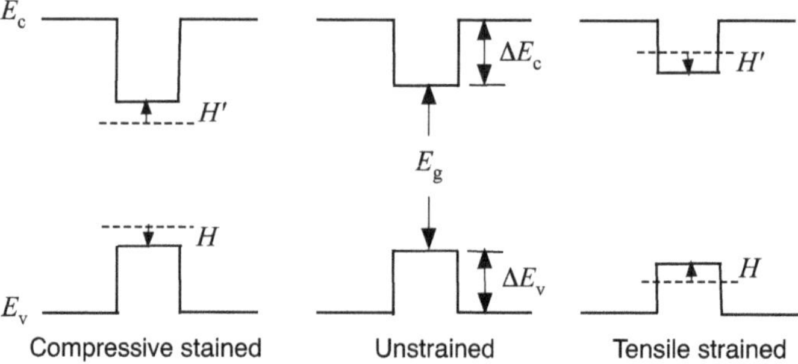

Fig. 7.15 Energy band diagrams for QWs under various stress conditions. The dashed lines of strained structures indicate the location of unstrained band edges. The composition of barriers is the same in all three cases while the composition of the QW is changed to generate strain

7.3.5 Infinite Deep Triangular Quantum Well

The triangular potential well is the most common potential well formed in a semiconductor quantum well under an applied bias and at the semiconductor hetero-interface. When a constant electric field intensity F is applied to a quantum well, as shown in Fig. 7.16a, the bottom of the potential well tilts up forming a triangular well. To simplify the discussion, let us consider a potential well wide enough so that only the vertical potential barrier at $z = 0$ and the tilted barrier are considered (Fig. 7.16b). The potential energy distribution of such a structure can be described as

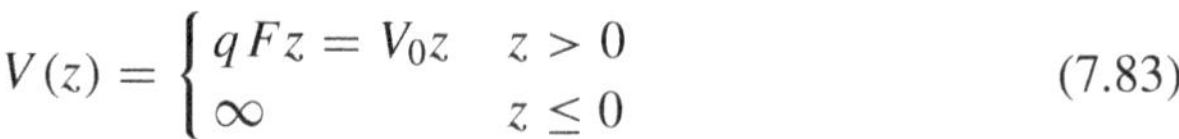

$$V(z) = \begin{cases} qFz = V_0 z & z > 0 \\ \infty & z \leq 0 \end{cases} \tag{7.83}$$

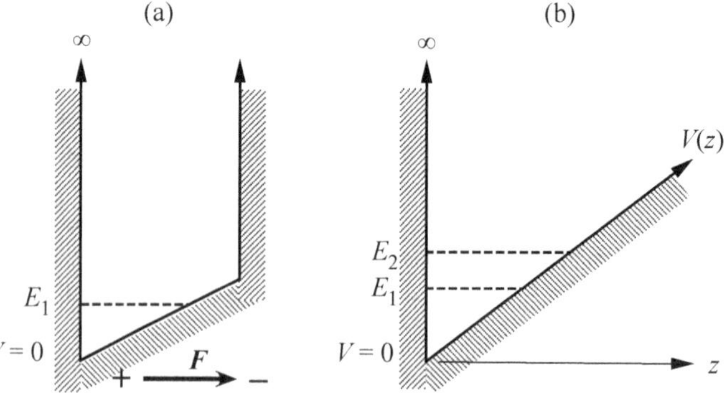

Fig. 7.16 **a** Infinitely deep quantum well under an applied electric field, F. **b** A triangular potential well $V(z) = qFz$. The dashed lines are the allowed energy states

where q is the electron charge and $V_0 = qF$, which is a constant for a specific F. The potential energy increases with z in the tilted region. The wave equation for such a system in the triangular well is

$$-\frac{\hbar^2}{2m}\frac{\partial^2\psi}{\partial z^2} + qFz\psi = E\psi \quad (z > 0) \tag{7.84}$$

(a) **Airy function**

Equation (7.84) can be solved rigorously in closed form through the following procedures. The first step to solve this equation is to get rid of the physical quantities E and z and replace them with pure numbers. The physical problem of (7.84) is thereby reduced to a pure mathematical problem. We do this by defining a 'length unit' L, which can be done by rearranging the equation into the form

$$-\frac{\hbar^2}{2mqF}\frac{\partial^2\psi}{\partial z^2} + (z - b)\psi = 0 \tag{7.85}$$

where $b \equiv (E/qF)$ has a unit of length. The second term has a dimension of $(\text{length})^{+1}$, and the first term has the unit of $(\text{length})^{-2}$ from $1/z^2$, so the factor in front of it must have dimensions of $(\text{length})^{+3}$. This suggests that we should define the length unit L to eliminate this constant by setting

$$\bar{z} \equiv \frac{z - b}{L},\ L^3 \equiv \frac{\hbar^2}{2mqF} \tag{7.86}$$

Similarly, we can define an 'energy unit' E_0 following (7.83) as

$$E_0 = qFL = \left[\frac{(qF\hbar)^2}{2m}\right]^{1/3} \tag{7.87}$$

Replacing these in (7.85), the wave equation becomes

$$-L^3\frac{\mathrm{d}^2\psi}{\mathrm{d}z^2} + L\bar{z}\psi = 0 \quad \text{and} \quad \mathrm{d}z = L\mathrm{d}\bar{z} \tag{7.88}$$

It further reduces to a simple form

$$\frac{\mathrm{d}^2\psi}{\mathrm{d}\bar{z}^2} - \bar{z}\psi = 0 \tag{7.89}$$

The solutions of this equation are a linear superposition of modified Bessel functions of the order $\pm 1/3$, or *Airy functions*. These functions have the following asymptotic properties for large $\bar{z}$

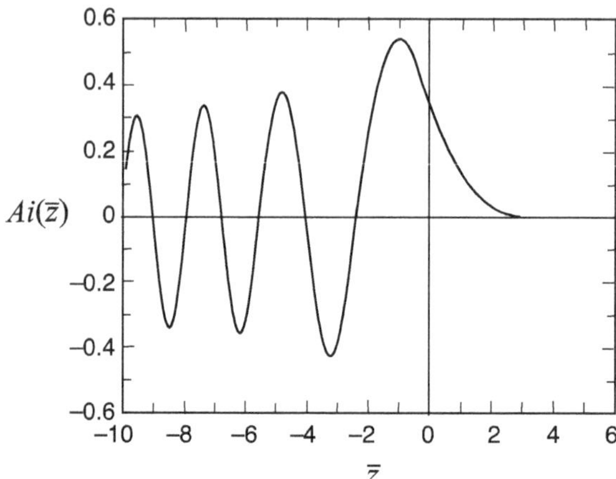

Fig. 7.17 The Airy function $A_i(\bar{z})$

$$\begin{cases} A_i(\bar{z}) \sim \bar{z}^{-1/4} \exp\left(-2\bar{z}^{2/3}/3\right) \\ B_i(\bar{z}) \sim \bar{z}^{-1/4} \exp\left(+2\bar{z}^{2/3}/3\right) \end{cases} \tag{7.90}$$

Clearly we must reject the B_i solutions as they expand rapidly at $\bar{z} \to +\infty$. Thus

$$\psi(\bar{z}) \propto A_i(\bar{z}) = A_i[(z-b)/L] \tag{7.91}$$

The A_i solutions are plotted in Fig. 7.17. $A_i(\bar{z})$ decays exponentially in positive (inside the sloped barrier), and there is an infinite number of negative values of where $A_i(\bar{z}) = 0$ denoted by a_n. These values of a_n are the nth roots of the Airy function. The boundary condition $\psi = 0$ at $z = 0$ then requires

$$\bar{z} = a_n = \frac{b}{L} = \frac{E_n}{qFL} \tag{7.92}$$

for negative $\bar{z}$ and $E_0 = qFL$. This leads to the energy eigenvalues of

$$E_n = a_n E_0 = a_n \left[\frac{(qF\hbar)^2}{2m}\right]^{1/3} \quad n = 1, 2, 3, \ldots . \tag{7.93}$$

Once again, numerical techniques must be used to obtain a_n with

$$a_n \approx \left[\frac{3\pi}{8}(4n-1)\right]^{2/3} \tag{7.94}$$

So that

$$E_n = a_n E_0 \approx \left(\frac{\hbar^2}{2m}\right)^{1/3} \left[\frac{3\pi qF}{8}(4n-1)\right]^{2/3} n = 1, 2, 3, \ldots \tag{7.95}$$

The allowed energy levels and the corresponding wave functions of a triangular well are shown in Fig. 7.18. Since the well broadens as the energy is raised, the energy levels get closer together as n increases. The wave functions vanish at $z = 0$ where the potential barrier is infinitely high. Due to the finite barrier height at $z > 0$, wave functions penetrate into the barrier and decay exponentially.

(b) Fang-Howard wave function

In most discussions involving triangular quantum wells in semiconductors, we are most interested in the energy and waveform of the ground state ($n = 1$). The ground state of the triangular well can be conveniently obtained by a calculation using the following trial function proposed by Fang and Howard [2]:

$$\psi(z) = Az\exp(-\alpha z) \tag{7.96}$$

where α is the trial parameter and A is a normalization parameter. This wave function reflects the feature of the waveform derived through rigorous mathematical calculations discussed earlier. The wave function vanishes at $z = 0$, i.e., at the vertical interface, and decays exponentially in the slope barrier. To evaluate these two parameters, the normalization procedure is applied where

$$\int_0^\infty \psi^*\psi dz = A^2\int_0^\infty z^2\exp(-2\alpha z)dz = 1 \tag{7.97}$$

Equation (7.97) can be solved using the result of $\int_0^\infty z^n\exp(-az)dz = \Gamma(n+1)/a^{n+1}$. Thus,

$$A = 2\alpha^{3/2} \tag{7.98}$$

The normalized wave function of the ground state in a triangular potential well becomes

$$\psi(z) = 2\alpha^{3/2}z\exp(-\alpha z) \tag{7.99}$$

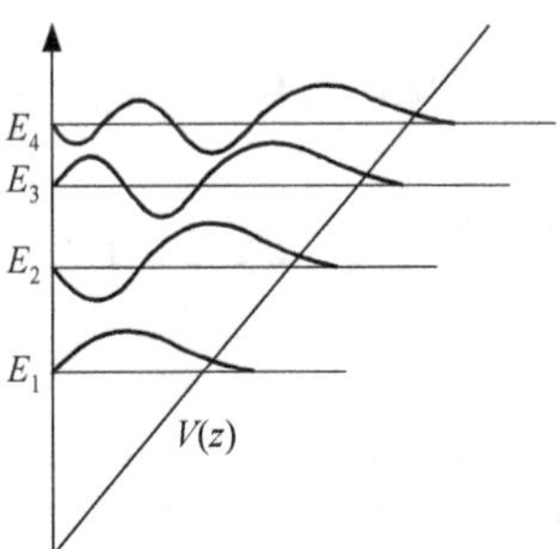

Fig. 7.18 Energy levels and wave functions in an infinitely deep triangular potential well $V(z) = qFz$

which is the *Fang–Howard wave function*.

Then the energy of the ground state E_1 can be obtained through the calculation of the energy expectation value of the wave function.

$$\langle E\rangle = \int_0^\infty \psi^* H\psi dz = \int_0^\infty \psi^*\left(-\frac{\hbar^2}{2m}\frac{\partial^2}{\partial z^2} + qFz\right)\psi dz$$
$$= \int_0^\infty \left[\left(-\frac{\hbar^2\alpha^{3/2}ze^{-\alpha z}}{m}\right)\frac{\partial^2}{\partial z^2}(2\alpha^{3/2}ze^{-\alpha z}) + 4qFz^3\alpha^3e^{-2\alpha z}\right]dz \quad (7.100)$$

$$\langle E\rangle = E_1 = \frac{(\alpha\hbar)^2}{2m} + \frac{3qF}{2\alpha} \quad (7.101)$$

The energy expectation value is a function of the triangular well potential qF and α. For a fixed qF, the trial parameter α and the energy of the ground state, corresponding to an energy minimum of the potential well, can be derived by taking $\partial\langle E\rangle/\partial\alpha = 0$.

$$\frac{\partial\langle E\rangle}{\partial\alpha} = \frac{\alpha\hbar^2}{m} - \frac{3qF}{2\alpha^2} = 0 \quad (7.102)$$

Thus,

$$\alpha = \left(\frac{3qFm}{2\hbar^2}\right)^{1/3} \quad (7.103)$$

Using this calculated trial parameter the energy of the ground state is obtained as

$$E_1 = \frac{9qF}{4\alpha} = \frac{3}{2}\left(\frac{3qF\hbar}{2\sqrt{m}}\right)^{2/3} \quad (7.104)$$

The ground state energy obtained using this result is approximately equal to that calculated from the Airy function approach within an error of ~6%.

7.4 Superlattices and Minibands

7.4.1 Square Barrier

A one-dimensional potential barrier is located with $V(z) = V_0$ for $0 \leq z \leq a$ and $V(z) = 0$ elsewhere. In region I, a wave with energy E is incident from the left onto the barrier as shown in Fig. 7.19. We are interested in the fraction of the incident

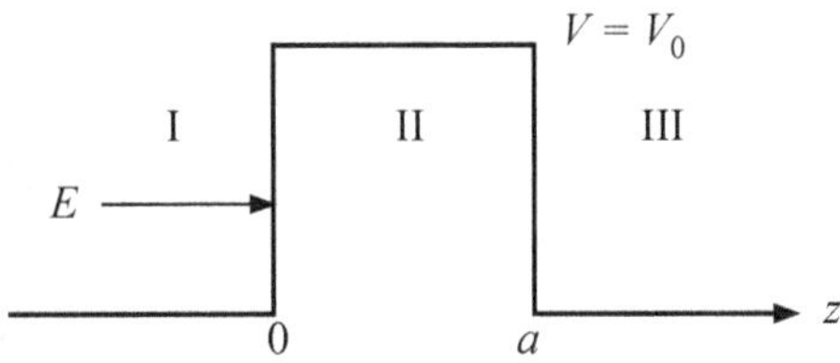

Fig. 7.19 Energy diagram of a square potential barrier with a barrier width a and height V_0

wave that tunnels through the barrier (region II) into region III and the fraction of the incident wave that gets reflected.

The general solutions of the Schrödinger equation in these three regions are

$$\begin{cases} \psi_1(z) = A\exp(ik_1z) + B\exp(-ik_1z) \text{ (Region I)} \\ \psi_2(z) = C\exp(ik_2z) + D\exp(-ik_2z) \text{ (Region II)} \\ \psi_3(z) = F\exp(ik_1z) + G\exp(-ik_1z) \text{ (Region III)} \end{cases} \tag{7.105}$$

$$k_1 = \sqrt{\frac{2mE}{\hbar^2}},\, k_2 = \sqrt{\frac{2m(E-V_0)}{\hbar^2}}$$

A, C, and F are amplitudes of forward propagating waves, and B, D, and G are amplitudes of reflection waves. Since there is no reflection wave in region III, $G = 0$.

Using the fact that the wave function and its derivative are continuous at boundaries, we arrive the following conditions:

At $x = 0$:

$$A + B = C + D \text{ and } k_1(A - B) = k_2(C - D) \tag{7.106a}$$

At $x = a$:

$$C\exp(ik_2a) + D\exp(-ik_2a) = F\exp(ik_1a) \tag{7.106b}$$

$$k_2\left[C\exp(ik_2a) - D\exp(-ik_2a)\right] = k_1F\exp(ik_1a) \tag{7.106c}$$

Solving the above equations by eliminating C and D, we obtain the amplitude ratios of reflection and transmission waves as

$$\frac{B}{A} = \frac{\left(k_1^2 - k_2^2\right)\left[1 - \exp(i2k_2a)\right]}{(k_1 + k_2)^2 - (k_1 - k_2)^2\exp(i2k_2a)} \tag{7.107a}$$

$$\frac{F}{A} = \frac{4k_1k_2\exp[i(k_2 - k_1)a]}{(k_1 + k_2)^2 - (k_1 - k_2)^2\exp(i2k_2a)} \tag{7.107b}$$

These lead to the probabilities of transmission (T) and reflection (R) of the system:

$$T = \left|\frac{F}{A}\right|^2 = \frac{4E(E - V_0)}{4E(E - V_0) + V_0^2 \sin^2 k_2 a} \tag{7.108a}$$

$$R = \left|\frac{B}{A}\right|^2 = \frac{V_0^2 \sin^2 k_2 a}{4E(E - V_0) + V_0^2 \sin^2 k_2 a} \tag{7.108b}$$

For $E > V_0$, the transmitted flux is not a constant as expected from the classical theory. It becomes unity only when $\sin k_2 a = 0$, i.e., $k_2 a = n\pi$ and $n = 1, 2, 3 \ldots$, and has multiple peaks.

For $E < V_0$, k_2 becomes imaginary. Let $\kappa = ik_2$ and $\sin k_2 a = i \sinh \kappa a$. The transmission and reflection coefficients become

$$T = \frac{4E(E - V_0)}{4E(E - V_0) + V_0^2 \sinh^2 \kappa a} \tag{7.109a}$$

$$R = \frac{V_0^2 \sinh^2 \kappa a}{4E(E - V_0) + V_0^2 \sinh^2 \kappa a} \tag{7.109b}$$

Again, in contrast to the classical theory, the tunneling probability does not reduce to zero when E becomes less than V_0.

7.4.2 *Resonant Tunneling Through Double Barriers*

In a single QW, the energy eigenvalues are determined precisely by the well width and barrier height. The allowed state in a QW is called a *bound state*. When the thickness of the two barriers on each side of the QW is reduced as depicted in Fig. 7.20, the confined electron can tunnel out of the QW. In a single square barrier case shown above (Fig. 7.19), the tunneling probability of a wave with an energy of $E < V_0$ is very small. However, when the wave is confined between two identically thin barriers instead, new physics appears. Since the electron can leak out of the well, the electron is not confined in bound states. Rather, the electron resides in a resonant state. The

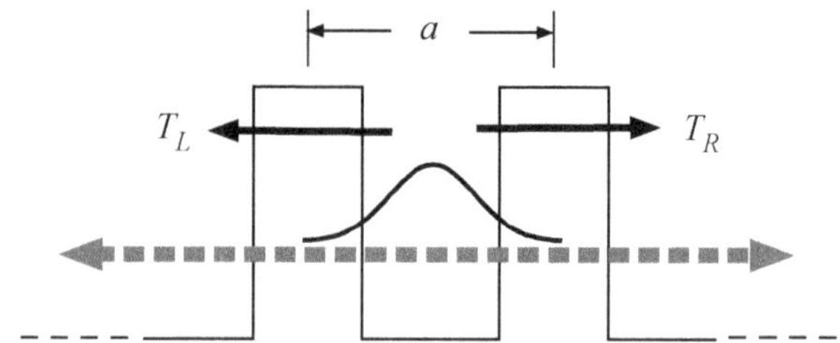

Fig. 7.20 Schematic of a double-barrier system where the two barriers have the same thickness

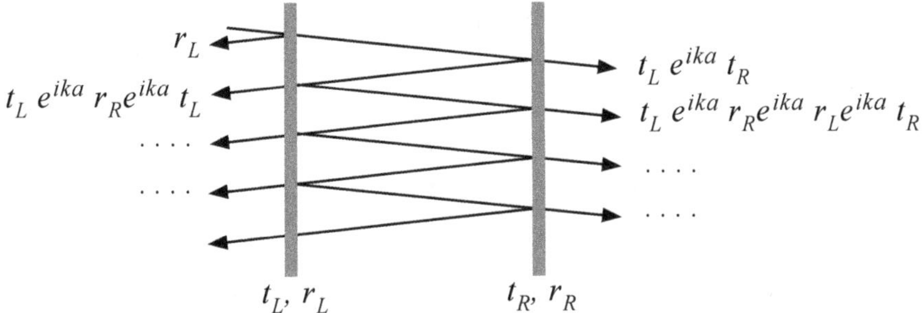

Fig. 7.21 Transmission and reflection amplitudes in a Fabry–Perot cavity. t_R and t_L, and r_R and r_L are the transmission and reflection amplitudes for the right and left mirrors, respectively

energy of this state cannot be precisely determined but is spread into a range $\hbar/\tau$, where τ is the lifetime of an electron in the well before it tunnels away.

In general, the transmission probability T of two barriers is approximately equal to the product of the values for the two individual barriers, $T_L T_R$. In a symmetric structure, where the two barriers are identical, the transmission probability at the resonant state rises dramatically above the product $T_L T_R$ and reaches its maximum value of unity. This process is called the *resonant tunneling*.

The electron oscillation in and transmission through the barriers of a symmetric double-barrier system can be treated as an analogous optical Fabry–Perot cavity (Fig. 7.21). An incident plane wave is bouncing back and forth between the two partially transparent mirrors. At each bounce, the wave loses some of its amplitude by transmission through the mirror. Use t_R and t_L, and r_R and r_L, as the transmission and reflection amplitudes for the right and left mirrors, respectively. The total wave transmitted through the mirror on the right is the sum of individual transmitted waves at each bounce.

$$\begin{aligned} t &= t_L \exp(ika) t_R + t_L \exp(ika) r_R \exp(ika) r_L \exp(ika) t_R \\ &\quad + t_L \exp(ika) r_R \exp(ika) r_L \exp(ika) r_R \exp(ika) r_L \exp(ika) t_R \\ &\quad + \ldots = t_L \exp(ika) t_R \left[1 + r_R r_L \exp(i2ka) + r_R^2 r_L^2 \exp(i4ka) + \ldots\right] \end{aligned} \tag{7.110}$$

This is a geometric series of $x = r_R r_L \exp\,(i2ka)$, and $|x| < 1$. The sum of such a series is

$$1 + x + x^2 + x^3 + x^4 + \ldots = 1/(1 - x)$$

The total transmission amplitude to the right, including the phase factor $\exp(ika)$, is given by

$$t = \frac{t_L t_R}{1 - r_L r_R \exp(i2ka)} \exp(ika) \tag{7.111}$$

If we take the complex reflection amplitudes r_L and r_R in polar forms such that $r_L = |r_L|\exp(i\phi_L)$ and $r_R = |r_R|\exp(i\phi_R)$, the flux transmission coefficient $T = |t|^2$ becomes

$$T = \frac{T_L T_R}{\left(1 - \sqrt{R_L R_R}\right)^2 + 4\sqrt{R_L R_R}\sin^2(\phi/2)} \tag{7.112}$$

where $\phi = \phi_L + \phi_R + 2ka$ is the total phase angle and $R_i = 1 - T_i$. T peaks at $\sin(\phi/2) = 0$ or $\phi = 2n\pi$. For $T_L \ll 1$ and $T_R \ll 1$, and expanding $\sqrt{R_L}$ and $\sqrt{R_R}$ with binomial series, the approximate peak transmission coefficient T_{pk} is given by

$$T_{pk} = \frac{T_L T_R}{\left(1 - \sqrt{R_L R_R}\right)^2} \approx \frac{4 T_L T_R}{(T_L + T_R)^2} \tag{7.113}$$

T_{pk} equals one when $T_L = T_R$ despite that the individual transmission coefficient for each barrier is very small.

7.4.3 Superlattice and Miniband

When stacking thin layers of lattice-matched GaAs and AlGaAs alternatively, with fixed thickness and composition, a series of identical QW/barrier structures is formed. The periodic band-edge energy variation of the structure creates an artificial 1D lattice constant in the growth direction in addition to the natural 3D lattice of GaAs and AlGaAs. This structure is called a *superlattice*. An electron can tunnel from one well to its neighbors in a superlattice similar to that of the double-barrier structure. Therefore, resonant states with finite energy distribution are replacing the sharp bound states associated with the well. The spreading of the allowed state in a superlattice can also be understood from the general theory of a 1D crystal. Since the period of the superlattice is much larger and the periodic potential is weaker than the crystal lattice, the resultant energy bands and gaps are much smaller. Therefore, the allowed energy bands are called *minibands*.

Since a superlattice is a 1D structure, the Kronig–Penney model is perfect for the understanding of the relationships between the allowed energy, layer thickness, and layer composition in superlattices. The model used to treat the 1D periodic crystal lattice can be directly applied to a superlattice with some modifications. As shown in Fig. 7.22, the conduction band discontinuity ΔE_c between the well and the barrier materials becomes the potential barrier height V_0. The free electron mass used in the original model is now replaced by the effective masses m_w^* and m_b^* in the well and barrier, respectively. Furthermore, the connection rule for the derivative of the wave functions is modified to include the effect of mass change when electrons move from one material into the other; i.e.,

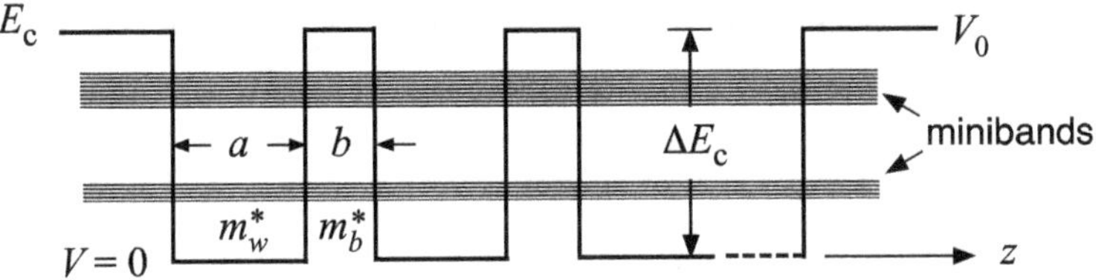

Fig. 7.22 Schematic of the conduction band edge in a superlattice heterostructure with well width a and barrier thickness b

$$\frac{1}{m_w^*}\frac{\mathrm{d}\psi_w}{\mathrm{d}z} = \frac{1}{m_b^*}\frac{\mathrm{d}\psi_b}{\mathrm{d}z} \tag{7.114}$$

The final solution for the $E < \Delta E_c$ case is readily obtained as

$$\frac{1}{2}(\eta - 1/\eta)\sin(\alpha a)\sinh(\gamma b) + \cos(\alpha a)\cosh(\gamma b) = \cos\lambda(a+b) \tag{7.115}$$

where $\eta = \frac{\gamma m_w^*}{\alpha m_b^*}$; $\alpha = \sqrt{\frac{2m_w^* E}{\hbar^2}}$; $\gamma = \sqrt{\frac{2m_b^*(\Delta E_c - E)}{\hbar^2}}$; $d = a + b$

and λ $(-\pi/d < \lambda < \pi/d)$ is the wave number in the superlattice. The interpretation of this result is similar to the model developed in Chap. 3 with modified masses. The allowed energy states are bunched into discrete minibands separated by energy gaps.

7.4.4 Density of States in Superlattice

The dispersion of N states in each miniband of an N-period superlattice destroys the abruptness of the two-dimensional DOS in each well. The DOS of a superlattice can be estimated by considering it as the combination of the one-dimensional confinement of a quantum well along the growth direction and a new periodic lattice along the same direction. Set the density of states of the 1D superlattice as D_{1D}. The density of states of the QW has the form $D_{2D} = m^*/(\pi\hbar^2)$. Then the three-dimensional density of states is given by the integral

$$D_{3D} = \frac{1}{2}\int_{-\infty}^{\infty} D_{1D}(E')D_{2D}(E')\mathrm{d}E' = \frac{m^*}{2\pi\hbar^2}\int_{-\infty}^{\infty} D_{1D}(E')dE' \tag{7.116}$$

The factor of 1/2 avoids double counting the spin. The value of D_{1D} in a 1D free-electron system was derived in Chap. 3 and has the form

$$D_{1D} = \frac{1}{\pi\hbar}\sqrt{\frac{2m^*}{E'}} \tag{7.117}$$

To incorporate the band-edge information of the 1D system, it is replaced by

$$D_{\mathrm{1D}} = \frac{2}{\pi\hbar v(E')} \tag{7.118}$$

where $v(E')$ is the velocity. Here we used $p = \hbar k$ and $k^2 = 2m^* E'/\hbar^2$ to get $v^2(E') = 2E'/m^*$. For a one-dimensional periodic lattice with a periodicity c, the allowed energy band in k-space is approximated by a cosine function with a bandwidth Δ.

$$E'(k) = (\Delta/2)(1 - \cos kc) \tag{7.119}$$

and

$$v(E') = \frac{1}{\hbar}\frac{\mathrm{d}E'}{\mathrm{d}k} = \frac{c\Delta}{2\hbar}\sin kc. \tag{7.120}$$

The sine function is reduced through the cosine function.

$$\cos kc = 1 - \frac{E'}{\Delta/2} = \frac{(\Delta/2) - E'}{\Delta/2}, \tag{7.121a}$$

and

$$\sin kc = \sqrt{(\Delta/2)^2 - [(\Delta/2) - E']^2}/(\Delta/2). \tag{7.121b}$$

Therefore,

$$v(E') = (c/\hbar)\sqrt{(\Delta/2)^2 - [(\Delta/2) - E']^2} \tag{7.122a}$$

and

$$D_{\mathrm{1D}}(E') = (2/c\pi)/\sqrt{(\Delta/2)^2 - [(\Delta/2) - E']^2}. \tag{7.122b}$$

The integration of the density of states for a superlattice becomes

$$\begin{aligned} D_{\mathrm{3D}} &= \frac{m^*}{2\pi\hbar^2}\int_{-\infty}^{E}\frac{2}{c\pi}\frac{\mathrm{d}E'}{\sqrt{(\Delta/2)^2 - [(\Delta/2) - E']^2}} \\ &= \frac{m^*}{c\pi^2\hbar^2}\sin^{-1}\left[\frac{E' - (\Delta/2)}{(\Delta/2)}\right]_{-\infty}^{E} = \frac{m^*}{c\pi^2\hbar^2}\left\{\sin^{-1}\left[\frac{E - (\Delta/2)}{(\Delta/2)}\right] - \sin^{-1}(-\infty)\right\} \end{aligned} \tag{7.123}$$

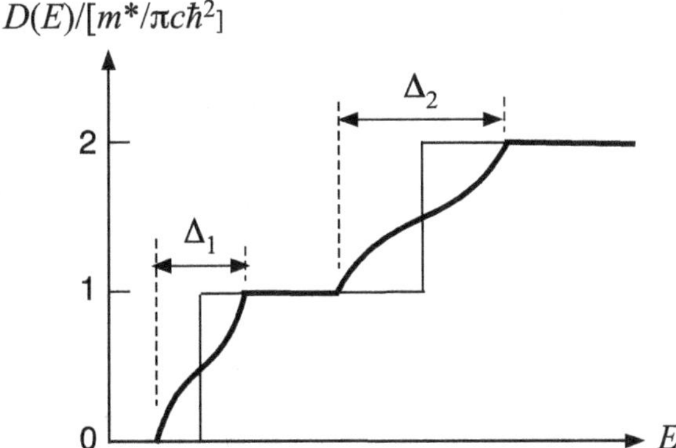

Fig. 7.23 Normalized density of states as a function of energy for a superlattice and a multiple quantum well (thin line) scaled to the same value in the plateau. The two lowest bands have widths Δ_1 and Δ_2

For $\sin^{-1}(-\infty) = -\pi/2$, the density of states of a superlattice is shown in Fig. 7.23 and given by

$$D_{3D} = \frac{m^*}{c\pi\hbar^2}\left\{\frac{1}{2} + \frac{1}{\pi}\sin^{-1}\left[\frac{E-(\Delta/2)}{(\Delta/2)}\right]\right\} \quad 0 \leq E \leq \Delta \tag{7.124}$$

The density of states of a superlattice consists of two terms: a stepwise function due to 1D confinement in QWs and an arcsine function originated from the 1D superlattice. When the barriers are thick, the superlattice can be seen as a multiple QW structure. It maintains a step-like density of states. The communication of electrons between wells increases with decreasing barrier thickness and broadens the sharp step of the density of states into an arcsine of width Δ.

7.5 Heterostructures in Electric Fields

7.5.1 Uniform Electric Field on a Bulk Semiconductor—Franz–Keldysh Effect

When a semiconductor is placed in an electric field, the absorption coefficient changes with the applied field. The phenomenon observed independently by W. Franz and by L.V. Keldysh is known as the *Franz–Keldysh effect* [3, 4]. The physical origin of this change can be understood by solving the Schrödinger wave equation of the bulk semiconductor in a uniform electric field.

Consider an electron with charge q moving in a uniform electric field F along the z-direction. As shown in Fig. 7.24, the potential energy of the system is $V(z) = qFz$, which increases with z for $F > 0$. The one-dimensional Schrödinger wave equation has the following form:

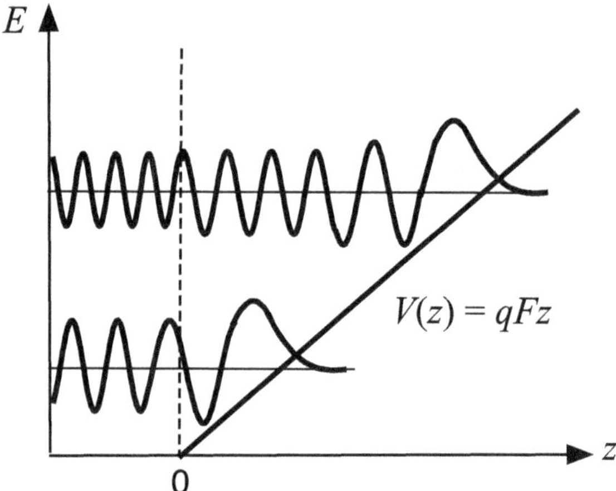

Fig. 7.24 Two wave functions for electrons of different energies in a semiconductor under a uniform electric field F

$$\left(-\frac{\hbar^2}{2m}\frac{\partial^2}{\partial z^2}+qFz\right)\psi(z)=E\psi(z) \tag{7.125}$$

This is the same wave equation in a triangular potential well as has been discussed in Sect. 7.3. The unnormalized solutions of wave functions are

$$\psi(z)=A_i\left[\frac{z-(E/qF)}{L}\right]=A_i\left(\frac{qFz-E}{E_0}\right) \tag{7.126}$$

where $L=\left(\dfrac{\hbar^2}{2mqF}\right)^{1/3}$ and $E_0=\left[\dfrac{(qF\hbar)^2}{2m}\right]^{1/3}=qFL.$

Unlike in triangular quantum wells, there is no confinement for electrons, and they can travel all space. Therefore, we can take any energy value in the wave function solution. The wave functions oscillate for $z < E/qF$. The undulation becomes more rapid as $\bar{z}=(z-E/qF)/L$ gets more negative. For $z > E/qF$, the wave tunnels into the potential and decay gradually. Although there is no obvious confinement in the system, the nature of these waves is standing rather than propagating. This is due to the interference between the incoming and the reflection waves, generated by collision with the potential at $z = E/qF$.

The Franz–Keldysh effect of the electric field dependence of the fundamental absorption edge can be thought of as *photon-assisted tunneling* through the energy gap. In a semiconductor, an optical absorption occurs only if the incident wave with the associated energy, $\hbar\omega$, is larger than the bandgap energy. For $\hbar\omega < E_g$, absorption in a semiconductor is impossible because there is no state available within the forbidden gap. In an electric field, due to the tilted energy band, there are states in both conduction and valence bands at all energies. The electron wave functions in the conduction and valence bands have exponentially decaying amplitude in the energy gap. When there is no photon absorption, a valence band electron must tunnel through

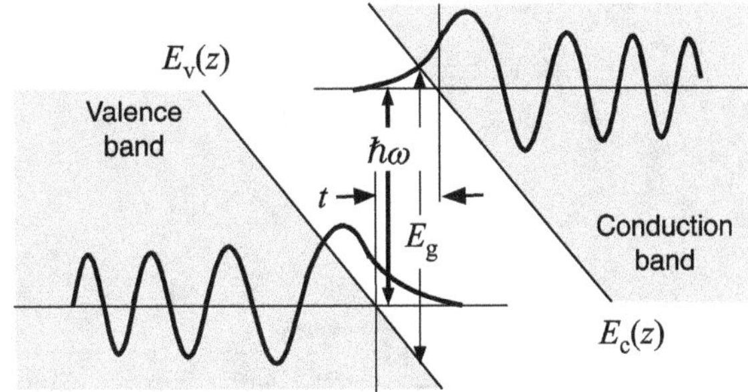

Fig. 7.25 Frenz–Keldysh effect on inter-band absorption. The energy band diagram showing the wavefunction overlap with absorption of a photon of an energy $\hbar\omega < E_g$

a large triangular barrier with a barrier height E_g and a barrier thickness E_g/qF to reach the conduction band. However, with the absorption of a photon with energy $\hbar\omega < E_g$, as shown in Fig. 7.25, the tunneling barrier height is reduced to $(E_g - \hbar\omega)$ and the barrier thickness becomes $t = \left(E_g - \hbar\omega\right)/qF$. It is clear that the tunneling probability is considerably enhanced with photon absorption and depends on the electric field intensity as well as the photon energy.

7.5.2 *Quantum Well in an Electric Field—Quantum-Confined Stark Effect (QCSE)*

The absorption property of quantum wells is also modified by the applied uniform electric field F and is closely related to the Franz–Keldysh effect in bulk semiconductors. It is called the quantum-confined Stark effect (QCSE). The *Stark effect* generally refers to a shift in the atomic energy upon the application of an electric field. It is well known in atomic physics that, in an electric field F, the 2*s* and 2*p* energy states in a hydrogen atom split into $E_2 + 3qaF$, E_2, and E_2–$3qaF$ levels, where $E_2 = -13.6/n^2$ ($n = 2$) and a is the Bohr radius. The splitting of the energy level proportions to the first power of the field strength is called the linear Stark effect.

Under the electric field, as shown in Fig. 7.26, the electron and hole wave functions in the well are pulled toward the opposite side of the quantum well. The reduced overlap of the two wave functions results in a corresponding reduction in absorption and in luminescence. The electron and hole distributions peak at different portions of the quantum well with an average separation $\langle x \rangle$. The spatial distribution of electrons and holes creates an electric dipole moment

$$p = -q\langle x \rangle = \epsilon_0 \alpha F \tag{7.127}$$

where α is the polarizability. This induced dipole moment lowers the state energy by an amount $\Delta E = -\boldsymbol{p} \cdot \boldsymbol{F}/2 = \epsilon_0 \alpha F^2/2$. Therefore, the energy separation between the state of the conduction band and the state of the valence band in a quantum well is reduced by the electric field. In addition, the probability of tunneling for low barrier

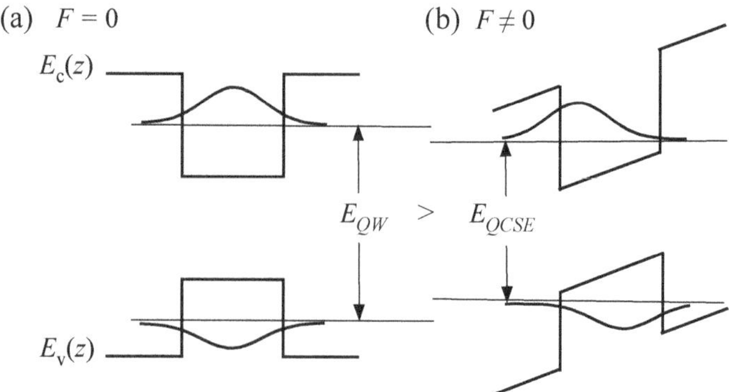

Fig. 7.26 **a** Absorption energy E_{QW} between the bound states in a quantum well with flat bands. **b** Under an applied electric field ($F \neq 0$), the tilted bands lower the energy of both bound states and reduce the absorption energy to E_{QCSE}

holes increases rapidly with increasing field F. Finally, the QCSE merges with the Franz–Keldysh effect as the quantum-well width is increased.

7.6 Polarization Fields in Wurtzite Heterostructures

As shown in Fig. 7.27, the wurtzite structure GaN has a bilayer structure when viewed along the c-axis. The (0001) planes are alternating cation layers and anion layers with

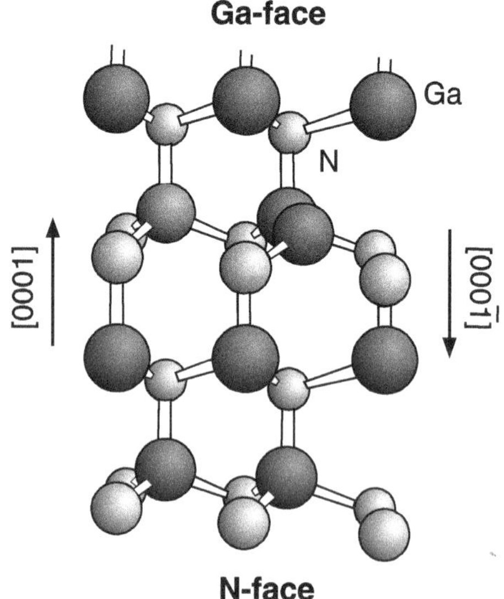

Fig. 7.27 Schematic diagram of the crystal structure of a wurtzite Ga-face GaN. Reprint with permission from [5], copyright AIP Publishing

the Ga-plane and N-plane terminating (0001) and $(000\bar{1})$ surfaces, respectively. The corresponding (0001) and $(000\bar{1})$ faces are the A-face and B-face, respectively. The ionic bonds linking neighboring (0001) planes with opposing charges generate a spontaneous polarization (P_{SP}) in the direction of $[000\bar{1}]$. Thus, the spontaneous polarization in A-face and B-face samples will have opposite signs. In addition, we also notice a large lattice-mismatch between GaN and AlN ($\Delta a/a = 2.4\%$, $\Delta c/c = 3.9\%$). When coupled with the large piezoelectric constants (~1 versus 0.01 in other III-V's), a strong strain-induced polarization or piezoelectric polarization (P_{PE}) forms in III-N heterostructures. The direction of the polarization follows dilation or contraction of the c-axis. The existence of significant spontaneous and piezoelectric polarizations in III-N heterostructures results in extra charge distribution at the heterojunction and modified energy band profile, which lead to unique electronic and photonic properties.

In the absence of external electric fields, the total polarization of a III-nitride heterostructure is the sum of the equilibrium lattice P_{SP} and the strain-induced P_{PE}. In this discussion, we consider polarizations along the $\langle 0001 \rangle$ axis, since this is the direction along which epitaxial layers and heterostructures are usually grown.

Since (0001) and $(000\bar{1})$ surfaces of GaN are non-equivalent, the spontaneous polarization for GaN and AlN depends on whether the surface is Ga/Al-face or N-face. For III-N alloys grown on c-plane sapphire substrates, the MOCVD and MBE growth methods produce Ga/Al-face (A-face) and N-face (B-face) layers, respectively. If we consider the [0001] axis (z-axis) as the epilayer growth direction, the spontaneous polarization for both GaN and AlN is found to be negative. This is because the spontaneous polarization of wurtzite structure GaN crystal is pointing in $[000\bar{1}]$ direction, as shown in Fig. 7.27. The spontaneous polarization of ternary III-nitride alloys is given as a function of composition in the following [6]:

$$
\begin{aligned}
P_{SP}(\mathrm{Al}_x\mathrm{Ga}_{1-x}\mathrm{N/GaN}) &= -0.090x - 0.034(1-x) + 0.021x(1-x)\ C/m^2 \\
P_{SP}(\mathrm{In}_x\mathrm{Ga}_{1-x}\mathrm{N/GaN}) &= -0.042x - 0.034(1-x) + 0.037x(1-x)\ C/m^2 \\
P_{SP}(\mathrm{Al}_x\mathrm{In}_{1-x}\mathrm{N/GaN}) &= -0.090x - 0.042(1-x) + 0.070x(1-x)\ C/m^2
\end{aligned}
\tag{7.128}
$$

The net values of the spontaneous polarization for all ternary III-nitride alloys are negative relative to [0001] growth direction.

Again, if only the stress along the [0001] growth direction is considered, the strain-induced piezoelectric polarization P_i^{PE} and the strain components ϵ_i are related as

$$
\begin{bmatrix} P_1^{PE} \\ P_2^{PE} \\ P_3^{PE} \end{bmatrix} = \begin{bmatrix} 0 & 0 & 0 \\ 0 & 0 & 0 \\ e_{31} & e_{31} & e_{33} \end{bmatrix} \begin{bmatrix} \epsilon_1 \\ \epsilon_2 \\ \epsilon_3 \end{bmatrix} \tag{7.129}
$$

where e_{3i} are the piezoelectric coefficients. The strain components are

$$\epsilon_1 = \epsilon_2 = \frac{a_s - a_e}{a_e} \quad \epsilon_3 = -\frac{2C_{13}}{C_{33}}\epsilon_1$$

where a_s and a_e are the lattice constants of the substrate and the epitaxial layer, respectively. Thus, the induced piezoelectric field polarization becomes

$$P_1^{\mathrm{PE}} = P_2^{\mathrm{PE}} = 0 \tag{7.130a}$$

$$P_3^{\mathrm{PE}} = P_{\mathrm{PE}} = e_{31}(\epsilon_1 + \epsilon_2) + e_{33}\epsilon_3 = 2\left(e_{31} - \frac{C_{13}}{C_{33}}e_{33}\right)\epsilon_1 \tag{7.130b}$$

Here we replaced P_i^{PE} by P_{PE} for simplicity. Since e_{31} is negative and e_{33}, C_{13}, and C_{33} are positive, the value in the parenthesis of (7.130b) is always negative for all compositions of ternary III-nitrides. The piezoelectric polarization P_{PE} is negative (along the $[000\bar{1}]$ direction) for tensile strain and positive for compressive strain. Using the linear extrapolation of the elastic and piezoelectric constants listed in Table 6.3, the nonlinear dependence of the piezoelectric polarization on the alloy composition for III-nitride ternary alloys grown on GaN substrate can be derived as follows [6]:

$$\begin{aligned}
&P_{\mathrm{PE}}(\mathrm{Al}_x\mathrm{Ga}_{1-x}\mathrm{N/GaN}) = -0.0525x + 0.0282x(1-x)\ C/m^2\\
&P_{\mathrm{PE}}(\mathrm{In}_x\mathrm{Ga}_{1-x}\mathrm{N/GaN}) = 0.148x - 0.0424x(1-x)\ C/m^2\\
&P_{\mathrm{PE}}(\mathrm{Al}_x\mathrm{In}_{1-x}\mathrm{N/GaN}) = -0.0525x + 0.148(1-x) + 0.0938x(1-x)\ C/m^2
\end{aligned} \tag{7.131}$$

The piezoelectric polarizations induced in tensile-strained $\mathrm{Al}_x\mathrm{Ga}_{1-x}\mathrm{N}$/GaN and compressively strained $\mathrm{In}_x\mathrm{Ga}_{1-x}\mathrm{N}$/GaN heterostructures have negative and positive values, respectively, over the whole composition range. However, for the $\mathrm{Al}_x\mathrm{In}_{1-x}\mathrm{N}$/GaN heterostructure, the piezoelectric polarization becomes zero at $x \approx 0.81$, where the heterostructure becomes lattice -matched.

The total polarization P of the wurtzite structure III-nitrides is the sum of the spontaneous and piezoelectric polarization,

$$P = P_{\mathrm{SP}} + P_{\mathrm{PE}} \tag{7.132}$$

Due to the net polarization field in dielectric materials, e.g., GaN and $\mathrm{Al}_x\mathrm{Ga}_{1-x}\mathrm{N}$, a polarization-induced sheet charge density may exist at the heterostructure interface. For example, at the surface of an $\mathrm{A}_x\mathrm{B}_{1-x}\mathrm{N}$ layer as well as at the interface of the $\mathrm{A}_x\mathrm{B}_{1-x}\mathrm{N}$/GaN heterostructure, where A and B are group III elements, the total polarization changes abruptly, causing a fixed two-dimensional polarization sheet charge density σ (C/cm^2) given by

$$\sigma_{\mathrm{ABN}} = P_{\mathrm{ABN}} = P_{\mathrm{SP}}(\mathrm{ABN}) + P_{\mathrm{PE}}(\mathrm{ABN}) \quad \text{for surfaces} \tag{7.133}$$

$$\sigma_{\text{ABN/GaN}} = P(\text{bottom}) - P(\text{top}) = P_{\text{GaN}} - P_{\text{ABN}}$$
$$= P_{\text{SP}}(\text{GaN}) - [P_{\text{SP}}(\text{ABN}) + P_{\text{PE}}(\text{ABN})] \quad \text{for interfaces} \tag{7.134}$$

respectively. Here, the relaxed GaN and tensile-strained $A_xB_{1-x}N$ are used in the heterostructure. The charges at the top of the $A_xB_{1-x}N$ layer are usually compensated for by charged surface states. If there is no additional surface charge, the induced electric field F in $A_xB_{1-x}N$ can be determined from the boundary condition at the surface:

$$D = \epsilon F + P_{\text{ABN}} = 0 \tag{7.135}$$

$$F = -P_{\text{ABN}}/\epsilon \tag{7.136}$$

The induced electric field in $A_xB_{1-x}N$ causes the energy band edges to tilt with a constant slope even under equilibrium conditions. The equilibrium energy band diagram and charge density in an $Al_xGa_{1-x}N$/GaN single heterostructure are shown in Fig. 7.28. The fixed interface sheet charge $+\sigma_{\text{AlGaN/GaN}} = +\sigma_{\text{pE}}$ can attract free electrons to the triangular QW and forms a sheet charge of two-dimensional electron gas (2DEG), $-\sigma_{\text{2DEG}}$. More discussions of 2DEG at the heterostructure interface will be given in Chap. 9.

Next, we turn our attention to a multiple heterojunction structure, such as multiple quantum wells (MQW). The continuity of the displacement flux D normal to the surface in the barrier and well, in the absence of free surface carriers, gives

$$D = \epsilon_w F_w + P_w = \epsilon_b F_b + P_b \tag{7.137}$$

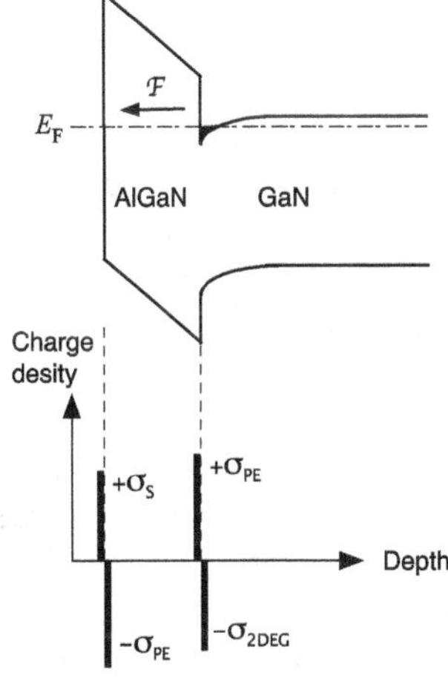

Fig. 7.28 Band diagram of a strained AlGaN/GaN single heterostructure grown on A-face GaN. The sheet charge density distribution (σ) in the structure is also shown

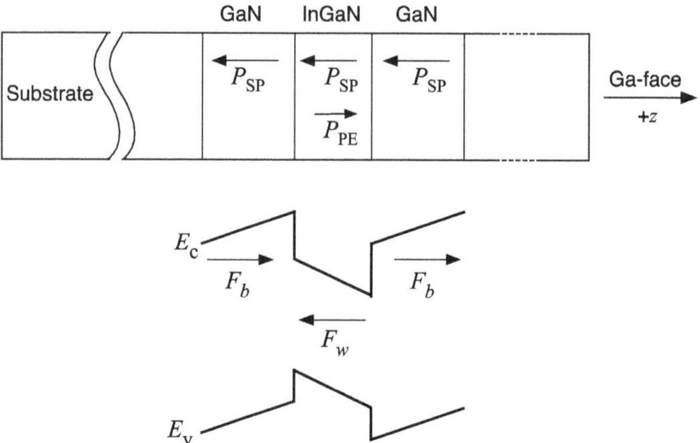

Fig. 7.29 Layer structure of an InGaN/GaN multiple quantum-well heterostructure grown on A-face GaN and the QW energy band profile. Only one QW structure is displayed

where the subscripts w and b represent the parameter in the well and barrier layers, respectively. For a MQW structure, the net voltage drop over one period of barrier (L_b) and well (L_w) is zero.

$$F_w L_w + F_b L_b = 0 \tag{7.138}$$

The electric fields in the well and barrier regions are solved using (7.137) and (7.138).

$$F_w = \frac{L_b}{\epsilon_b L_w + \epsilon_w L_b}(P_b - P_w) \tag{7.139a}$$

$$F_b = \frac{-L_w}{\epsilon_b L_w + \epsilon_w L_b}(P_b - P_w) \tag{7.139b}$$

We notice that the fields in the well and barrier are opposite in direction. Figure 7.29 shows the layer structure of an $In_xGa_{1-x}N$/GaN MQW grown on A-face GaN surface and its energy band-edge profile. In the QW, the electrons and holes are separated spatially even under equilibrium conditions.

Problems

1. The GaAs/$Al_{0.25}Ga_{0.75}As$ high-electron-mobility transistor (HEMT) is one of the major high-speed devices used in wireless communications. The $Al_{0.25}Ga_{0.75}As$ layer is doped with silicon at a doping density of ~10^{18} cm^{-3}. The undoped GaAs has a p-type background near ~10^{14} cm^{-3}. Other relevant material parameters are listed below.

	Relative dielectric constant (ϵ_r)	m_e^*/m_0	m_{hh}^*/m_0	m_{lh}^*/m_0
GaAs	12.85	0.067	0.5	0.076
AlAs	10.06	0.71	0.409	0.153

Draw the band diagram of this *p*–*N* heterojunction to scale including the following information:

(a) The conduction band discontinuity, ΔE_c, and the valence band discontinuity, ΔE_v;
(b) Fermi level positions;
(c) Contact potential on each side; and
(d) Depletion width.
(e) Estimate the ground state energy in the triangular quantum well formed at the *p*–*N* GaAs/$Al_{0.25}Ga_{0.75}As$ heterojunction. An infinitely high barrier at the heterojunction is assumed in this problem.

2. Derive the depletion width x_N and x_p of a *p-N* heterojunction as a function of the contact potential V_0 and doping levels, N_a and N_D.
3. The doping levels in a *p*-GaAs/*N*-$Al_{0.3}Ga_{0.7}As$ heterojunction are $N_a = 5\times 10^{16}$ cm^{-3} and $N_D = 5\times10^{17}$ cm^{-3}. The dielectric constant of $Al_xGa_{1-x}As$ is expressed as $\epsilon_s = \epsilon_0(13.1 - 3.0x)$.

   (a) Calculate the contact potential.
   (b) Calculate the depletion width of the heterojunction.
   (c) Plot the detail equilibrium band diagram of the *p-N* junction.

4. (a) Determine the allowed energy levels in the conduction band of a 60 Å GaAs/ $Al_{0.3}Ga_{0.7}As$ quantum well using realistic material parameters. The electron effective masses of GaAs and AlAs are $0.067m_0$ and $0.71m_0$, respectively. The conduction band discontinuity between GaAs and $Al_{0.3}Ga_{0.7}As$ is $\Delta E_c = 0.2$ eV.
   (b) Repeat part (a) without considering the mass difference between barriers and well. Use the GaAs effective mass of the well in your calculation. Comment on whether this approach is appropriate.
5. At the interface of an undoped AlGaAs/GaAs heterojunction, a triangular quantum well in the conduction band is formed on the GaAs side. Assume the potential barrier is infinitely high at the heterojunction interface and the potential variation on the GaAs side under equilibrium condition can be approximated by a straight line with a value of 5×10^4 V/cm.

   (a) Estimate the ground state energy of this quantum well.
   (b) Compare the results obtained using the exact solution of the Airy function and the Fang-Howard approximation.
   (c) What is the corresponding well width at the ground state energy? The effective mass of GaAs is $m_e^* = 0.067\ m_0$ for electrons.

6. In a rectangular AlGaAs/GaAs quantum well, the well is 0.3 eV deep for electrons. Calculate the ground state energy of an electron in the rectangular GaAs quantum well using the same well width as calculated from the last problem. Which quantum well, rectangular or triangular, gives higher ground state energy, and why?
7. Derive the transmission amplitude (t) of a square barrier structure including the mass difference between the barrier (m_b^*) and its surroundings (m_s^*). The barrier has a thickness a and barrier height V_0.
8. A superlattice structure formed on InP substrate is made of stacking lattice-matched $Ga_{0.47}In_{0.53}As$ and $Al_{0.48}In_{0.52}As$ layers alternatively. The conduction band discontinuity between these two ternary alloys is $\Delta E_c = 0.47$ eV. The electron effective masses of $Ga_{0.47}In_{0.53}As$ and $Al_{0.48}In_{0.52}As$ are $0.041m_0$ and $0.075m_0$, respectively. Estimate the bound states in such a superlattice structure where both barriers and wells are equal to 30 Å.
9. Assume the conduction band discontinuity (ΔE_c) at the $Al_{0.3}Ga_{0.7}As$/GaAs interface is 0.2 eV.

    (a) Calculate the allowed energy states of a 40 Å GaAs/$Al_{0.3}Ga_{0.7}As$ quantum well.
    (b) A single 40 Å $Al_{0.3}Ga_{0.7}As$/GaAs barrier structure is formed. Calculate the transmission coefficient (T) at the lowest energy level obtained in part (a).
    (c) An $Al_{0.3}Ga_{0.7}As$/GaAs superlattice has barriers and wells with the same thickness of 40 Å. Estimate the miniband width of the lowest allowed energy band.

10. A triangular quantum well in the conduction band is formed in the InGaN/GaN heterostructure where the undoped InGaN layer has an In composition of 15% and a thickness of 20 nm.

    (a) Determine the type of the sheet charge created in the quantum well.
    (b) Calculate the 2D sheet charge density.

11. The purpose of this assignment is to learn how to use *SimWindows* simulation software by David Winston for calculating the energy band diagram of heterostructures. Other commercial softwares, such as Silvaco Atlas (www.silvaco.com), can also be used. You can download *SimWindows* simulation program through the following link: http://simwindows.wixsite.com/simwindows.

    The active region of an n-$Al_{0.3}Ga_{0.7}As$/GaAs (undoped)/p-$Al_{0.3}Ga_{0.7}As$ double-heterostucture (DH) diode laser has a 0.1 μm thick undoped GaAs active region. Under a forward bias, electrons and holes are injected into the GaAs active region from n- and p-sides of the DH structure, respectively. Plot the equilibrium energy band diagram of the DH structure using *SimWindows* as a function of thickness. The thickness range of interest is 0.1 μm (p-AlGaAs) −0.1 μm GaAs −0.1 μm (n-AlGaAs). The doping concentrations of the n- and p-doped AlGaAs are at the same level of 2×10^{17} cm^{-3}. Plot the energy band diagram again for the same structure, except with the GaAs layer heavily doped to $p = 5 \times 10^{18}$ cm^{-3}.

References

1. I. Vurgaftman, J.R. Meyer, L.R. Ram-Mohan, J. Appl. Phys. **89**, 5815 (2001)
2. F.F. Fang, W.E. Howard, Phys. Rev. Lett. **16**, 797 (1966)
3. W. Franz, Z. Naturforsch. **13a**, 484 (1958)
4. L.V. Keldysh, Soviet Physics, JETP **7**, 788 (1958)
5. O. Ambacher et al., J. Appl. Phys. **85**, 3222 (1999)
6. O. Ambacher et al., J. Phys. Condens. Matter **14**, 3399 (2002)

Further Reading

1. H.C. Casey, M.B. Panish, *Heterostructure Lasers: Part A, Fundamental Principles* (Academic, New York, 1978)
2. S. Wang, *Fundamentals of Semiconductor Theory and Device Physics* (Prentice-Hall, Englewood Cliffs, 1989)
3. J.H. Davies, *The Physics of Low-Dimensional Semiconductors* (Cambridge University Press, Cambridge, United Kingdom, 1998)
4. S.L. Chuang, *Physics of Photonic Devices*, 2nd edn. (Wiley, New York, 2009)

Chapter 8
Optical Properties of Compound Semiconductor Heterostructures

Abstract Due to the nature of the indirect energy band structure, efficient light emission from silicon is impossible. On the other hand, direct energy band III–V compound semiconductors, such as GaAs, InP, and GaN, are efficient light-emitting materials. By mixing two or three III–V binaries, one can prepare a series ternary or quaternary compounds with added freedoms in the selection of the bandgap energy. Bandgap engineering in heterostructures further allows the creation of new functional devices as will be discussed in Chaps. 9 and 10. Following the discussions of electrical properties of heterojunctions in Chap. 7, we will study their optical properties in this chapter including two major optical processes: the optical absorption and radiative optical transition. The absorption coefficient, spontaneous emission rate, and stimulated emission rate are interrelated through Einstein relations. Finally, the transparency condition of stimulated emission in a semiconductor is discussed.

8.1 Basic Optical Properties of Dielectric Medium

8.1.1 *Maxwell's Equations*

The time-varying electric and magnetic field intensities are expressed as $\boldsymbol{E}(x, y, z, t) = \boldsymbol{E}(x, y, z)\exp(i\omega t)$ and $\boldsymbol{H}(x, y, z, t) = \boldsymbol{H}(x, y, z)\exp(i\omega t)$. The propagation of electromagnetic waves is governed by the Maxwell's equations shown below.

$$\begin{cases} \nabla \times \boldsymbol{E} = -\dfrac{\partial \boldsymbol{B}}{\partial t} & \text{Faraday's law} \\ \nabla \times \boldsymbol{H} = \boldsymbol{J} + \dfrac{\partial \boldsymbol{D}}{\partial t} & \text{Ampere's law} \\ \nabla \cdot \boldsymbol{D} = \rho & \text{Gauss' law} \\ \nabla \cdot \boldsymbol{B} = 0 & \text{Gauss' law} \end{cases} \tag{8.1}$$

Here, $\boldsymbol{B}$ is the magnetic flux density, $\boldsymbol{D}$ is the displacement flux density, $\boldsymbol{J}$ is the current density, σ is the conductivity, and ρ is the charge density. The relations between $\boldsymbol{E}$, $\boldsymbol{H}$, $\boldsymbol{D}$, $\boldsymbol{J}$, and $\boldsymbol{B}$ are

K. Y. Cheng, *III–V Compound Semiconductors and Devices*, Graduate Texts in Physics,
https://doi.org/10.1007/978-3-030-51903-2_8

$$\begin{cases} \boldsymbol{D} = \epsilon \boldsymbol{E} = \epsilon_0 \boldsymbol{E} + \boldsymbol{P} \\ \boldsymbol{J} = \sigma \boldsymbol{E} \\ \boldsymbol{B} = \mu \boldsymbol{H} \\ \nabla \cdot \boldsymbol{J} = -\partial \rho / \partial t \end{cases} \tag{8.2}$$

$\boldsymbol{P}$, ϵ, and μ are the polarization vector, permittivity (or dielectric constant), and permeability of the wave propagation medium, respectively. The permittivity of the free space is ϵ_0. Also, for a linear, homogeneous, and isotropic medium, the polarization vector is related to the applied electric field by

$$\boldsymbol{P} = \epsilon_0 \chi \boldsymbol{E} \tag{8.3}$$

Thus,

$$\boldsymbol{D} = \epsilon_0 (1 + \chi) \boldsymbol{E} = \epsilon \boldsymbol{E} \tag{8.4}$$

where χ is the electric susceptibility. When the wave is propagating across the boundary between two materials, the boundary conditions are

$$\begin{cases} \hat{\boldsymbol{n}} \cdot (\boldsymbol{D}_1 - \boldsymbol{D}_2) = \rho_s \\ \hat{\boldsymbol{n}} \cdot (\boldsymbol{B}_1 - \boldsymbol{B}_2) = 0 \\ \hat{\boldsymbol{n}} \times (\boldsymbol{E}_1 - \boldsymbol{E}_2) = 0 \\ \hat{\boldsymbol{n}} \times (\boldsymbol{H}_1 - \boldsymbol{H}_2) = \boldsymbol{J}_s \end{cases} \tag{8.5}$$

The unit vector $\hat{\boldsymbol{n}}$ represents the normal of the interface and is oriented from region 2 to region 1.

8.1.2 Electrical and Optical Constants in a Lossless (σ = 0) Dielectric

For a plane wave propagating in the z-direction, the electric and magnetic fields have the following characteristics:

$$E(x, y, z, t) = E_x(z, t),\ E_y = E_z = 0,\ \partial/\partial x = \partial/\partial y = 0; \text{ and}$$
$$H(x, y, z, t) = H_y(z, t),\ H_x = H_z = 0$$

The wave equation of such a plane wave is given by

$$\frac{\partial^2 E_x}{\partial z^2} = \mu_0 \epsilon \frac{\partial^2 E_x}{\partial t^2} \tag{8.6}$$

The solution of $E_x(z, t)$ is generally written as

$$E_x(z, t) = E_x(z) \exp(i\omega t) \tag{8.7}$$

Substituting it in the wave equation gives

$$\frac{\partial^2 E_x}{\partial z^2} = -\mu_0 \epsilon \omega^2 E_x = -k^2 E_x \tag{8.8}$$

where $k \equiv \omega\sqrt{\mu_0 \epsilon}$ is the *propagation constant*. The wave equation has the solution for the wave propagating in the $+z$-direction:

$$E_x = A \exp[i(\omega t - kz)] = A \cos(\omega t - kz) \tag{8.9}$$

From Faraday's law, H_y is readily found to be

$$H_y = \frac{\epsilon \omega A}{k} \cos(\omega t - kz) \tag{8.10}$$

The propagation of the plane wave in a lossless medium is illustrated in Fig. 8.1. By analyzing the electric field intensity at a fixed time, say $t = 0$ or $t = \pi/2\omega$, we find the wavelength of the field to be

$$\lambda = 2\pi/k \quad \text{or} \quad k = 2\pi/\lambda$$

Here, k has the value of the wave number and is called the phase or propagation constant. The wave propagation velocity is deduced from $\upsilon = z/t = \omega/k$.

$$\upsilon = \frac{\omega}{k} = \frac{1}{\sqrt{\mu_0 \epsilon}} \tag{8.11}$$

In free space, the propagation velocity has the speed of light, c, and

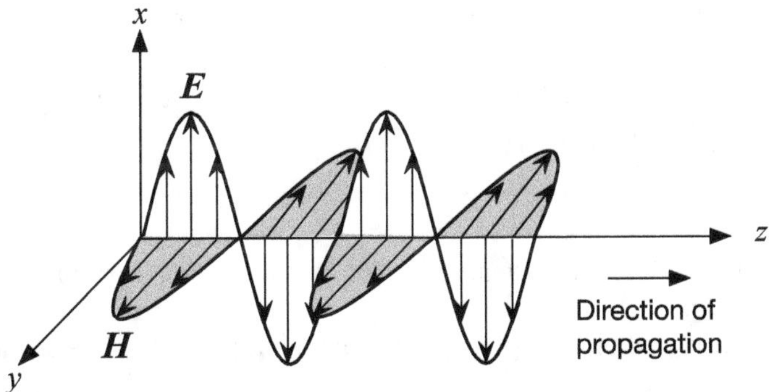

Fig. 8.1 Wave propagation in a lossless medium along the z-direction at $t = \pi/2\omega$. The electric field intensity crosses the z-axis at π/k, $2\pi/k$, $3\pi/k$, etc.

$$c = \frac{1}{\sqrt{\mu_0 \epsilon_0}} \tag{8.12}$$

The ratio of the free space velocity to the velocity in a dielectric medium is defined as the *refractive index* $\overline{n}$:

$$\overline{n} = \sqrt{\frac{\epsilon}{\epsilon_0}} = \sqrt{1+\chi} \tag{8.13}$$

The frequency, ν, of the wave in free space and dielectric medium is the same:

$$\nu = \frac{\upsilon}{\lambda} = \frac{c}{\lambda_0} \tag{8.14}$$

We obtain the wavelength in the dielectric medium as $\lambda = \lambda_0/\overline{n}$ and $k = \overline{n}k_0$.

8.1.3 Electrical and Optical Constants in a Lossy ($\sigma \neq 0$) Medium

Since $\sigma \neq 0$, the magnetic field intensity has the form

$$-\frac{\partial H_y}{\partial z} = \sigma E_x + \epsilon\left(\frac{\partial E_x}{\partial t}\right) \tag{8.15}$$

The wave equation becomes

$$\frac{\partial^2 E_x}{\partial z^2} = i\omega\mu_0(\sigma + i\epsilon\omega)E_x = K^2 E_x. \tag{8.16}$$

Here $K \equiv \sqrt{i\omega\mu_0(\sigma + i\epsilon\omega)} = (\alpha/2) + i\beta$. is the *complex propagation constant*. The solutions of the wave equation for the $+z$ propagating plane wave are

$$\begin{cases} E_x(t) = A\exp[-(\alpha z/2) + i(\omega t - \beta z)] = A\exp(-\alpha z/2)\cos(\omega t - \beta z) \\ H_y(t) = \dfrac{(\sigma + i\omega t)}{K} E_x(t) \end{cases} \tag{8.17}$$

The electric field intensity attenuates in the direction of wave propagation in z, as illustrated in Fig. 8.2.

In the lossless medium, the propagation constant term in the wave equation is

$$k^2 = \epsilon\mu_0\omega^2 = \overline{n}^2 k_0^2 \tag{8.18}$$

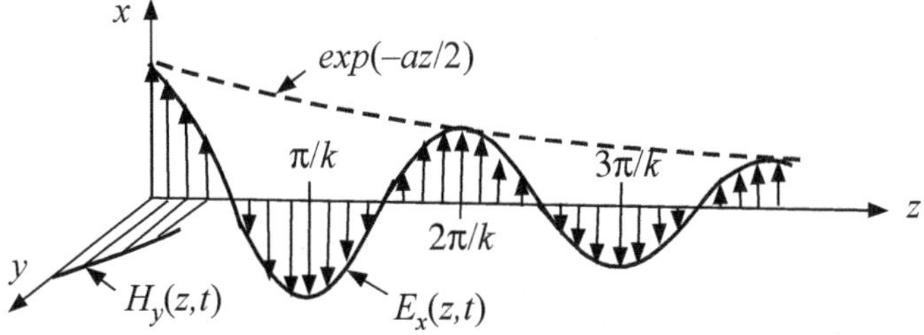

Fig. 8.2 Traveling wave in a lossy medium as represented by (8.17) at $t = 0$ and $\alpha = k/2\pi$

In the same manner for the lossy case,

$$K^2 = i\omega\mu_0(\sigma + i\epsilon\omega) = -\epsilon\mu_0\omega^2 + i\sigma\mu_0\omega = -\overline{N}^2 k_0^2 \tag{8.19}$$

where $\overline{N}$ is the *complex refractive index* with the real part $\overline{n}$ and imaginary part $\overline{\kappa}$:

$$\overline{N} = \overline{n} - i\overline{\kappa} \tag{8.20}$$

Combining the above two equations with $k_0 = 2\pi/\lambda_0$ gives the *complex dielectric constant* $\overline{\epsilon}$:

$$\begin{aligned}\overline{N}^2 &= \overline{n}^2 - \overline{\kappa}^2 - i2\overline{n\kappa} = \left(\epsilon\mu_0\omega^2 - i\sigma\mu_0\omega\right)/k_0^2 \\ &= \frac{\epsilon}{\epsilon_0} - i\frac{\sigma}{\epsilon_0\omega} = \epsilon_1 + i\epsilon_2 = \overline{\epsilon}\end{aligned} \tag{8.21}$$

Equating real and imaginary parts gives

$$\begin{cases}\overline{n}^2 - \overline{\kappa}^2 = \dfrac{\epsilon}{\epsilon_0} = \epsilon_1 \\ 2\overline{n\kappa} = \dfrac{\sigma}{\epsilon_0\omega} = \epsilon_2\end{cases} \tag{8.22}$$

and

$$\begin{cases}\overline{n} = \frac{1}{\sqrt{2}}\left(\epsilon_1 + \sqrt{\epsilon_1^2 + \epsilon_2^2}\right)^{1/2} \\ \overline{\kappa} = \frac{1}{\sqrt{2}}\left(-\epsilon_1 + \sqrt{\epsilon_1^2 + \epsilon_2^2}\right)^{1/2}\end{cases} \tag{8.23}$$

Thus, $\overline{N}$ and $\bar{\epsilon}$ are dependent variables. Furthermore, the complex propagation constant

$$K = (\alpha/2) + i\beta = \sqrt{-\overline{N}^2 k_0^2} = i(\overline{n} - i\overline{\kappa})k_0 \tag{8.24}$$

which gives

$$\begin{cases} \alpha = 4\pi\overline{\kappa}/\lambda_0 \propto \sigma \\ \beta = \overline{n}k_0 = k \end{cases} \tag{8.25}$$

The attenuation term α is related to the conductivity and is the *absorption coefficient*. β is identical to the *propagation constant* k of the lossless medium. The quantity $\overline{\kappa}$ is called the *extinction coefficient*. Therefore, the electrical properties $\epsilon(\omega)$ and $\sigma(\omega)$ are related to the optical properties $\overline{n}(\omega)$ and $\alpha(\omega)$ through equations listed above.

8.1.4 Dielectric Constant

In electromagnetic theory, the dielectric constant is related to the applied electric field intensity and polarization. The interaction of a light wave and a bounded electron can be modeled using a dipole oscillator model developed by Hendrick Lorentz in 1878. The electric field of the light wave induces forced oscillations of the atomic dipole through the driving forces exerted on the bound electrons. As shown in the model of Fig. 8.3, the displacement of the atomic dipoles can be seen as damped harmonic oscillators since the oscillating dipoles can lose their energy through an interaction with phonons. We assume that the mass of the nucleus of the host atom is much larger than the electron mass, so the displacement x of the electron is governed by the following equation of motion:

$$m\frac{\mathrm{d}^2x}{\mathrm{d}t^2} + m\gamma\frac{\mathrm{d}x}{\mathrm{d}t} + \beta x = -qE(t). \tag{8.26}$$

where β, γ, and $E(t)$ are the spring constant, damping rate, and the time-varying electric field, respectively. The three terms on the LHS of (8.25) represent the acceleration, damping, and restoring force, respectively.

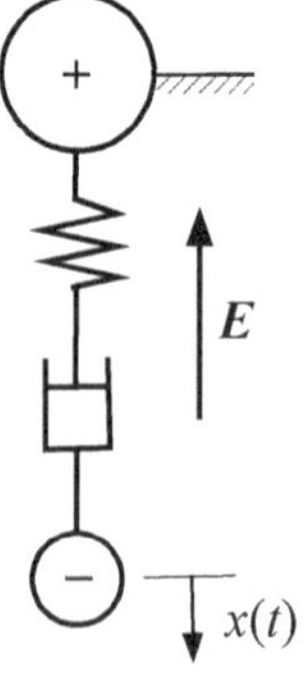

Fig. 8.3 A simple model of a dipole. When interacting with an applied electric field, *E*, the bound electron vibrates about is equilibrium position while the atom is stationary

The interaction of the atom with a monochromatic light wave of angular frequency ω is considered. For the time-dependent electric field with the form $E(t) = E_0\exp(-i\omega t)$, we look for solutions of the form $x(t) = x_0\exp(-i\omega t)$. The solution of the equation of motion for a displacement x is

$$x(t) = -\frac{qE_0/m}{\omega_0^2 - \omega^2 - i\omega\gamma}\exp(-i\omega t) \tag{8.27}$$

where $\omega_0 \equiv \sqrt{\beta/m}$ is the natural oscillation frequency. The displacement of electrons from their equilibrium position produces a time-varying dipole moment $p(t)$. The resonant polarization $P_R = Np(t) = -Nqx(t)$, where N is the atomic density. Therefore, P_R is proportional to $x(t)$ according to (8.27). However, the magnitude of P_R is small unless the frequency is close to ω_0.

The induced time-varying polarization under the influence of the electric field produces an electric displacement flux $\boldsymbol{D} = \epsilon_0\boldsymbol{E} + \boldsymbol{P} = \epsilon\boldsymbol{E}$. We are interested in the optical response near ω_0, so we include the resonant polarization part into the expression.

$$D = \epsilon_0 E + P + P_R = \epsilon_0 E + \epsilon_0\chi E + P_R = \epsilon_0\epsilon_r E \tag{8.28}$$

$$P_R = -Nqx = \frac{Nq^2}{m}\left(\frac{E}{\omega_0^2 - \omega^2 - i\omega\gamma}\right) \tag{8.29}$$

The complex relative dielectric constant is expressed as

$$\epsilon_r(\omega) = \epsilon_1 + i\epsilon_2 = 1 + \chi + \frac{Nq^2}{\epsilon_0 m}\left(\frac{1}{\omega_0^2 - \omega^2 - i\omega\gamma}\right) \tag{8.30}$$

The real and imaginary parts of the dielectric constant are related to the refractive index and absorption coefficient of the medium and are expressed as

$$\begin{cases} \epsilon_1(\omega) = 1 + \chi + \dfrac{Nq^2}{\epsilon_0 m}\dfrac{\omega_0^2 - \omega^2}{\left(\omega_0^2 - \omega^2\right)^2 + (\gamma\omega)^2} \\ \epsilon_2(\omega) = \dfrac{Nq^2}{\epsilon_0 m}\dfrac{\gamma\omega}{\left(\omega_0^2 - \omega^2\right)^2 + (\gamma\omega)^2} \end{cases} \tag{8.31}$$

The low- and high-frequency limits of $\epsilon_r(\omega)$ are expressed as

$$\begin{cases} \epsilon_r(0) \equiv \epsilon_{st} = 1 + \chi + \dfrac{Nq^2}{\epsilon_0 m\omega_0^2} \\ \epsilon_r(\infty) \equiv \epsilon_\infty = 1 + \chi \end{cases} \tag{8.32}$$

where ϵ_{st} and ϵ_∞ are static and optical relative dielectric constants, respectively. The difference of these two constants is

$$\epsilon_{st} - \epsilon_\infty = Nq^2/\epsilon_0 m\omega_0^2$$

Near the resonance where $\omega \approx \omega_0 \gg \gamma$, the dielectric constant can be rewritten, using the approximation of $\delta\omega = (\omega_0 - \omega)$ and $\omega_0^2 - \omega^2 = (\omega_0 + \omega)(\omega_0 - \omega) \approx 2\omega_0\delta\omega$ as

$$\begin{cases} \epsilon_1(\delta\omega) = \epsilon_\infty - (\epsilon_{st} - \epsilon_\infty)\dfrac{2\omega_0\delta\omega}{4(\delta\omega)^2 + \gamma^2} \\ \epsilon_2(\delta\omega) = (\epsilon_{st} - \epsilon_\infty)\dfrac{\gamma\omega_0}{4(\delta\omega)^2 + \gamma^2} \end{cases} \tag{8.33}$$

Figure 8.4 shows the frequency dependence of ϵ_1 and ϵ_2 predicted by (8.33) for a given set of ϵ_{st}, ϵ_∞, γ (s^{-1}), and ω_0 (rad/s). The frequency response of ϵ_1 has a complicated shape with two extremes at ($\omega_0 \pm \gamma/2$). For ϵ_2, it is a strong peak function of ω with a maximum value at ω_0 and a full-width at half-maximum of γ.

Figure 8.4 also shows the refractive index and extinction (absorption) coefficient calculated from ϵ_1 and ϵ_2 using (8.23). We see that $\overline{n}$ approximately follows the frequency dependence of $\sqrt{\epsilon_1(\omega)}$, while $\overline{\kappa}$ or α is a strongly peaked function of ω and more or less follows $\epsilon_2(\omega)$. The correspondence between $\overline{n}$ and $\sqrt{\epsilon_1(\omega)}$, and between $\overline{\kappa}$ and $\epsilon_2(\omega)$ would be exact if $\overline{\kappa}$ or α were much smaller than $\overline{n}$ as in weakly absorbing media. In semiconductors, this corresponds to the absorption in the transparent spectral region where $E < E_g$.

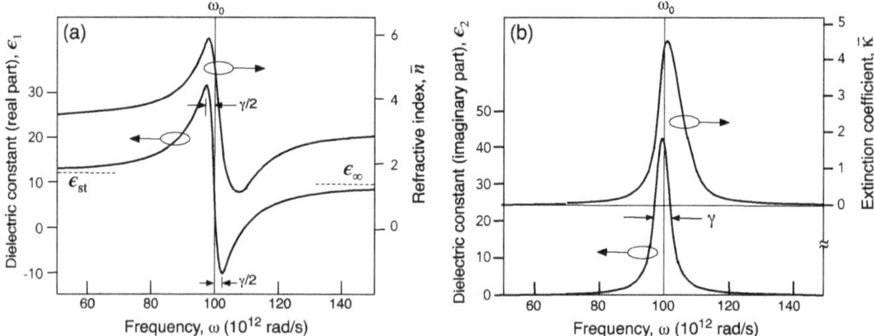

Fig. 8.4 Frequency dependence of the **a** real and **b** imaginary parts of the complex dielectric constant of a dipole oscillator at frequencies close to resonance calculated with $\omega_0 = 10^{14}$ rad/s, $\gamma = 5 \times 10^{12}$ s^{-1}, $\epsilon_{st} = 12.1$, and $\epsilon_\infty = 10$. Also shown are the **a** refractive index and **b** extinction coefficient derived from the dielectric constant. Reprinted with permission from [1] copyright Oxford University Press

8.1.4.1 Refractive Index

The refractive index of a semiconductor is related to its complex dielectric constant $\epsilon(\omega) = \epsilon_1(\omega) + i\epsilon_2(\omega)$. For most photonic applications, such as lasers, we are interested in the optical transitions with energy smaller than the bandgap energy. In the region below the bandgap, $\epsilon_2(\omega)$ has no contribution to the optical dispersion and can be neglected. Therefore, we can assume that

$$\overline{n} \cong \sqrt{\epsilon_1(\omega)} = \sqrt{\epsilon/\epsilon_0} \tag{8.34}$$

In III–V alloys, $\overline{n}$ in the transparency region is given by [2, 3]

$$\overline{n}(\omega) = \left\{ A^* \left[f(x) + \frac{1}{2}\left(\frac{E_g}{E_g + \Delta_{so}}\right)^{3/2} f(y) \right] + B^* \right\}^{1/2} \tag{8.35}$$

with $f(z) = \frac{1}{z^2}\left(2 - \sqrt{1+z} - \sqrt{1-z}\right)$ where $x = \hbar\omega/E_g$ and $y = \hbar\omega/\left(E_g + \Delta_{so}\right)$.

The two terms in the brackets correspond to the free electron–hole pair contributions arising from the E_g and $E_g + \Delta_{so}$ gaps, respectively. $A*$ and $B*$ are the strength parameters for the fundamental and higher-lying band transitions, respectively. The numerical values of $A*$ and $B*$ required to calculate the refractive index are listed below for several important III–V alloys. For $Ga_xIn_{1-x}As_yP_{1-y}$ lattice-matched to an InP system, by fitting the experimental results, we get:

$$A^*(y) = 8.40{-}3.40y \quad \text{and}$$
$$B^*(y) = 0.60 + 3.40y$$

Figure 8.5 shows the fitted calculation results of refractive indices up to the bandgap energy for each As composition y between 0 and 1.

For the $Al_xGa_{1-x}As$/GaAs system, the fitted parameters are:

$$A^*(x) = 6.64 + 16.92x \quad \text{and}$$
$$B^*(x) = 9.20{-}9.22x$$

The measured refractive indices of the $Al_xGa_{1-x}As$ system are plotted for x up to 0.38 in Fig. 8.6a. The room-temperature refractive indices at energy of 1.38 eV over the whole composition range are shown in Fig. 8.6b.

For the $Al_xGa_{1-x}N$ system, only the fundamental transition E_g is considered, i.e., the $f(y)$ term in (8.35) is set to be zero. In the transparent spectral region, for $x_{Al} < 0.4$, the fitting parameters are [4]:

$$A^*(x) = 9.82661{-}8.21608x{-}31.5902x^2 \quad \text{and}$$

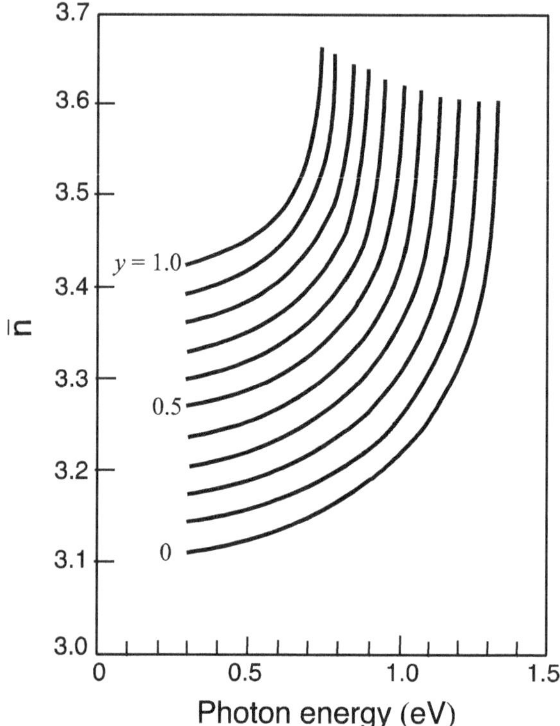

Fig. 8.5 Refractive index of $Ga_xIn_{1-x}As_yP_{1-y}$ lattice-matched to InP up to the bandgap energy for each y-composition. Reprinted with permission from [2] copyright AIP Publishing

$$B^*(x) = 2.73591 + 0.84249x - 6.28321x^2$$

For alloys with higher Al composition ($x_{Al} > 0.4$) the fitting parameters are [5]:

$$A^*(x) = (3.17 \pm 0.39)x^{1/2} + (9.98 \pm 0.27) \quad \text{and}$$
$$B^*(x) = -(2.2 \pm 0.2)x + (2.66 \pm 0.12).$$

Alternatively, the refractive indices of the $Al_xGa_{1-x}N$ system in the transparent spectral region have also been determined using the Kramers–Kronig relations coupled with high-accuracy experimental results of spectroscopic ellipsometry. The results near room temperature are plotted for x up to 0.64 in Fig. 8.7.

8.2 Absorptions in Semiconductors

Absorption in semiconductors is expressed in terms of the absorption coefficient $\alpha(\omega)$. The absorption coefficient is commonly defined as the fraction of the power loss dI/I per incremental length dz:

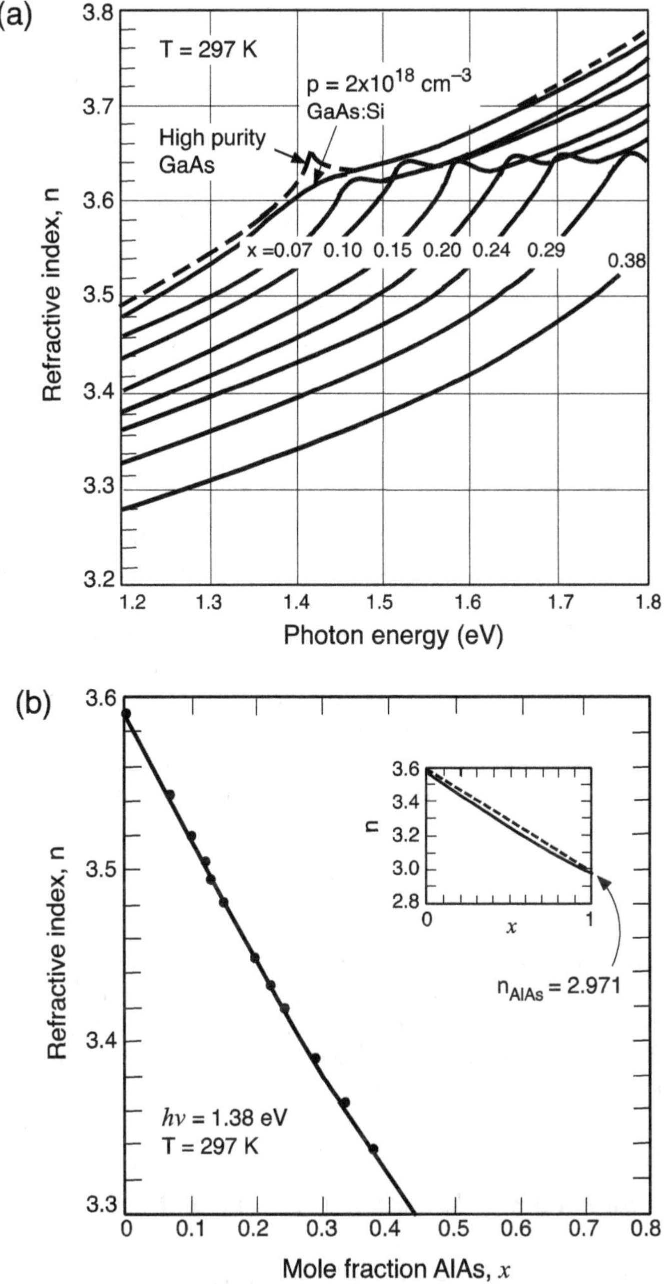

Fig. 8.6 **a** Refractive index as a function of photon energy for $Al_xGa_{1-x}As$. **b** Refractive index of $Al_xGa_{1-x}As$ at an emission wavelength of 1.38 eV. Reprinted with permission from [3] copyright AIP Publishing

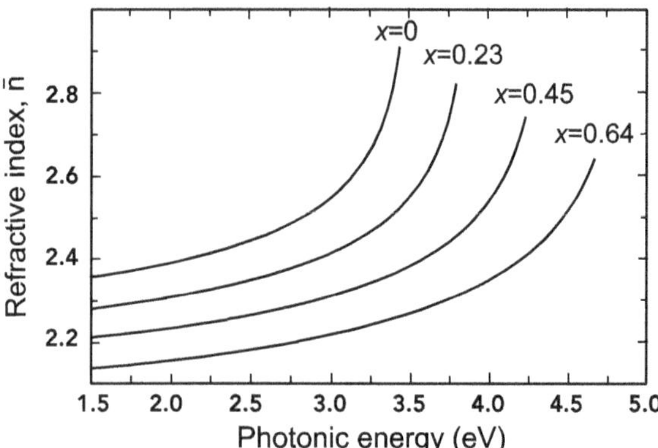

Fig. 8.7 Refractive index as a function of photon energy for $Al_xGa_{1-x}N$ at 290 K with $x_{Al} = 0, 0.23, 0.45$, and 0.64. Reprinted with permission from [6], copyright AIP Publishing

$$\alpha \equiv -\frac{dI/I}{dz} \quad \left(\text{cm}^{-1}\right) \tag{8.36}$$

The absorption coefficient associated with the optical process depends on the probability per unit time that an electron will make a transition from the valence band to the conduction band under the influence of radiation of intensity $I(\omega)$ at the frequency ω. The problem of calculating this transition probability is, however, strictly quantum mechanical, since the exact energy band structures involved cannot be determined using classical physics. In general, the absorption coefficient for a given photon energy $h\nu$ is proportional to the transition probability W_{if} from the initial state with an electron density of n_i to the final state with n_f of available empty state density. This process must be summed for all possible transitions between states separated by an energy difference equal to $h\nu$.

$$\alpha(h\nu) = A \sum W_{if} n_i n_f \tag{8.37}$$

In semiconductors, as illustrated in Fig. 8.8, various optical absorption processes generated by the incident electromagnetic radiation include high-lying band transitions, excitons, band-edge absorption, imperfection absorption, free carrier absorption, and Reststrahlen absorption. In the following, only those near-band-edge transitions will be considered in more detail.

8.2.1 Allowed Direct Transition Between Bands

For optical transitions between the valence band and conduction band, both momentum conservation and energy conservation conditions have to be fulfilled. Because the momentum of a photon, *h/λ* (λ ~μm), is very small compared to the crystal momentum, *h/a* (*a* ~Å), the photon absorption process should conserve the

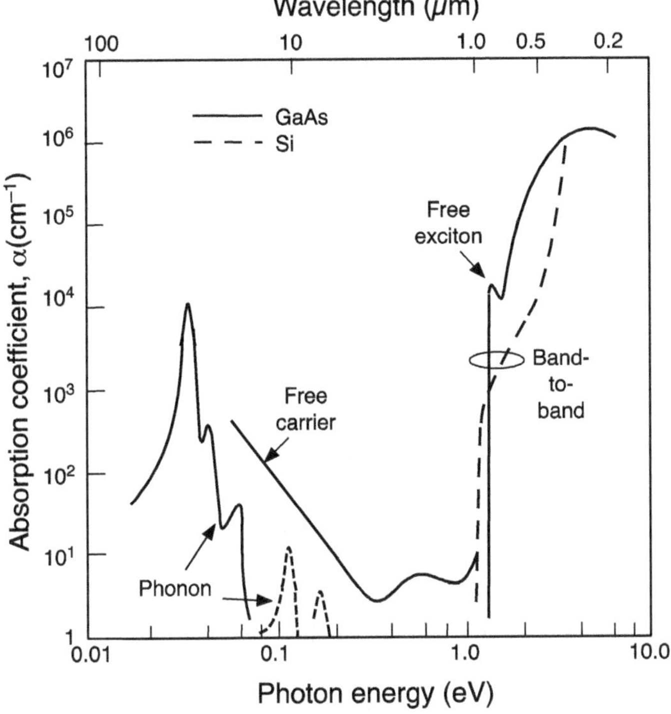

Fig. 8.8 Optical absorption coefficient as a function of photon energy for Si and GaAs at room temperature. The major absorption mechanisms are shown

momentum of the electron. Therefore, for transitions in direct bandgap semiconductors, all the momentum-conserving transitions are allowed. As shown in Fig. 8.9, every initial state at E_i is associated with a final state at E_f such that

$$\hbar\omega = h\nu = E_f - E_i \tag{8.38}$$

In a parabolic band,

$$E_f = E_g + \frac{\hbar^2 k^2}{2m_e^*} \tag{8.39a}$$

and

$$E_i = -\frac{\hbar^2 k^2}{2m_h^*} \tag{8.39b}$$

Therefore,

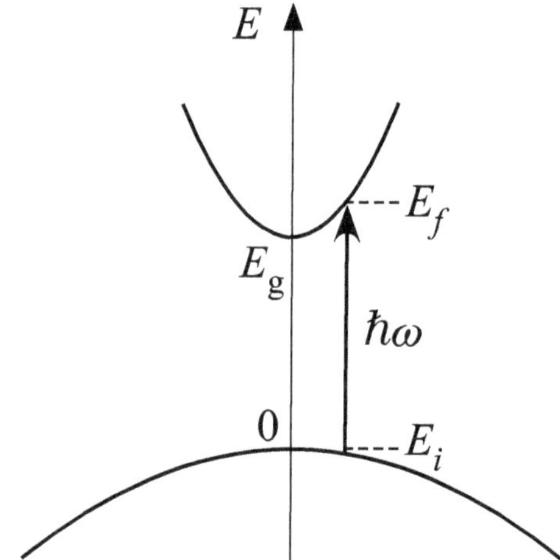

Fig. 8.9 Energy band diagram of a direct bandgap semiconductor. The top of the valence band at the zone center is set to have zero energy

$$h\nu = E_f - E_i = E_g + \frac{\hbar^2 k^2}{2}\left(\frac{1}{m_e^*} + \frac{1}{m_h^*}\right) \tag{8.40}$$

Following (3.29), the joint DOS is given by

$$D(h\nu) = \frac{\left(2m_r^*\right)^{3/2}\sqrt{h\nu - E_g}}{2\pi^2\hbar^3} \tag{8.41}$$

where m_r^* is the reduced mass given by

$$\frac{1}{m_r^*} = \frac{1}{m_e^*} + \frac{1}{m_h^*} \tag{8.42}$$

The transition probability from an initial state to a group of final states is derived from the first-order, time-dependent perturbation theory as

$$W_i = \frac{2\pi}{\hbar}\left|M_{if}\right|^2 D\left(E_f\right) \tag{8.43}$$

Here M_{if} is the matrix element of the perturbation which connects the states i and f of the system, and $D(E_f)$ is the density of final states. Hence, the absorption coefficient for allowed direct transition is

$$\alpha(h\nu) = A^*\sqrt{h\nu - E_g} \tag{8.44}$$

for which M_{if} is independent of k. The constant A^* has a value of ~2×10^4 for GaAs. The absorption coefficients as functions of energy for a number of elemental and III–V compound semiconductors are shown in Fig. 8.10. The direct bandgap material

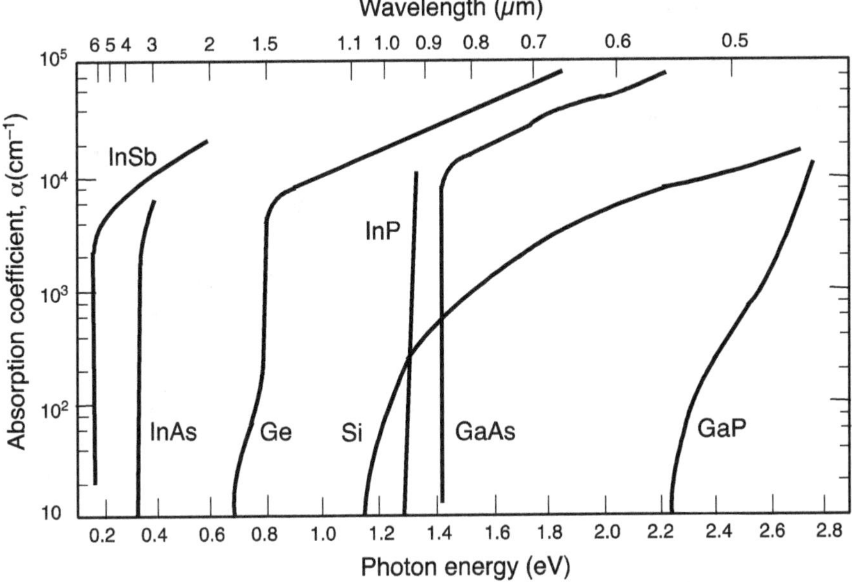

Fig. 8.10 Absorption coefficient versus photon energy for selected element and III–V compound semiconductors at room temperature. Reprinted with permission from [7] copyright IEEE

has a rather sharp increase in absorption coefficient at energies slightly larger than the bandgap.

8.2.2 Indirect (Non-vertical) Transitions Between Indirect Valleys

The mechanism for indirect optical transitions generally involves the simultaneous emission or absorption of a phonon with the absorption of the photon (Fig. 8.11). The electron, initially at E_i in the valence band at $k \approx 0$, is raised in a vertical transition to the intermediate state, also at $k \approx 0$, in the conduction band. It then absorbs, or emits, a phonon of wave vector $\pm k_\mathbf{p}$ and makes the transition to the conduction band minimum at E_f.

The photon energy is related to the phonon absorption ($-\hbar\omega_\mathrm{p}$) or phonon emission ($+\hbar\omega_\mathrm{p}$) by

$$h\nu_a = E_f - E_i - \hbar\omega_\mathrm{p} \tag{8.45a}$$

$$h\nu_e = E_f - E_i + \hbar\omega_\mathrm{p} \tag{8.45b}$$

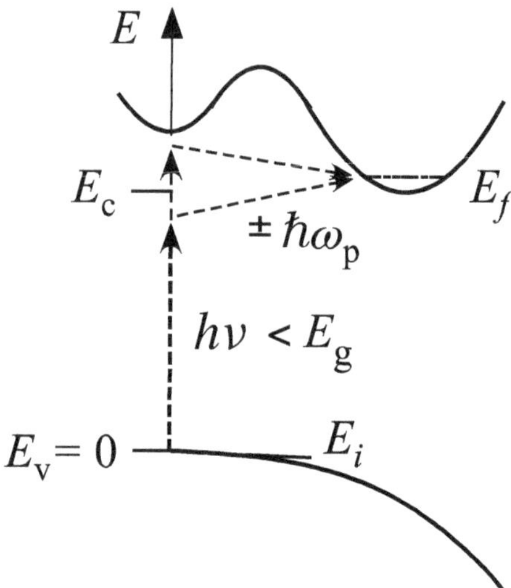

Fig. 8.11 Energy band diagram of an indirect bandgap semiconductor. The phonon absorption-assisted transition between E_i and E_f is shown

The transition probability for an indirect transition depends on the total matrix element and the effective density of states. The density of final conduction band states is

$$D_c(E_f) = A_c\sqrt{E_f - E_c} = A_c\sqrt{\delta - E_i} \tag{8.46}$$

where δ is the total energy of states above E_c and below E_v, and is given by

$$\delta = h\nu - E_g \pm \hbar\omega_p \tag{8.47}$$

The density of states in the initial valence band is

$$D_v(E_i) = A_v\sqrt{E_v - (-E_i)} = A_v\sqrt{E_i} \tag{8.48}$$

The effective density of states for the transition is

$$\delta D(h\nu) \propto \int_0^\delta D_c(E)D_v(E)\mathrm{d}E_i = A_cA_v\int_0^\delta \sqrt{\delta - E_i}\sqrt{E_i}\mathrm{d}E_i \tag{8.49}$$

Therefore,

$$D(h\nu) = A_a\delta^2 = A_a\left(h\nu - E_g \pm \hbar\omega_p\right)^2 \tag{8.50}$$

Since this is the phonon-assisted transition, we have to include the probability of phonon absorption or emission in the calculation. The density of phonons with frequency ω_p is given by Planck's law and has a Bose–Einstein distribution.

$$N(\hbar\omega_{\mathrm{p}}) = 1/\left[\exp(\hbar\omega_{\mathrm{p}}/kT) - 1\right] \tag{8.51}$$

The absorption coefficient due to indirect transition involving phonon absorption becomes

$$\alpha_a = \frac{A_a\left(h\nu - E_{\mathrm{g}} + \hbar\omega_{\mathrm{p}}\right)^2}{\exp(\hbar\omega_{\mathrm{p}}/kT) - 1} \tag{8.52}$$

For phonon emission process, the probability is proportional to $N(\hbar\omega_p) + 1$. The absorption coefficient due to indirect transition involving phonon emission is given by

$$\alpha_e = \frac{A_e\left(h\nu - E_{\mathrm{g}} - \hbar\omega_{\mathrm{p}}\right)^2}{1 - \exp(-\hbar\omega_{\mathrm{p}}/kT)} \tag{8.53}$$

The total absorption coefficient for optically allowed indirect transition is then

$$\alpha = \alpha_a + \alpha_e \tag{8.54}$$

As illustrated in Fig. 8.12a, the plot of $\sqrt{\alpha}$ versus photon energy should yield a curve consisting of two straight-line segments. The extrapolation of the straight lines to $\alpha = 0$ gives the values of ($E_{\mathrm{g}} \pm \hbar\omega_{\mathrm{p}}$). At very low temperatures, the phonon density is very small, and the segment associated with α_a diminishes. Indeed, in a high-purity Si sample measured between 195 and 415 K (Fig. 8.12b), the phonon absorption segment decreases with the reduction of measurement temperature.

8.2.3 *Transitions in Highly Doped Semiconductors*

Absorption in semiconductors depends on the detailed energy band structure of the material including the bandgap energy, the density of states in the vicinity of the band edge, and the type of optical transition allowed. In general, both conduction band and valence band near the band edge can be approximated by parabolic functions for undoped and lightly doped semiconductors as described in the previous sections. However, at high doping concentrations, many characteristics of semiconductors change, such as the density of states distribution near the band edge, the energy of the fundamental gap, and the impurity ionization energy. If the semiconductor is heavily doped, the conduction band becomes considerably filled due to the finite density of states, and the Fermi level is inside the band. Absorption transitions from the top of valence band to the bottom of the conduction band cannot occur due to the band filling effect. Hence, the absorption transition energy shifts from ~E_{g} for undoped semiconductor to >E_{g} in heavily doped *n*-type semiconductors. The shift of the absorption edge to energy higher than E_{g} due to band filling is called *Burstein–Moss*

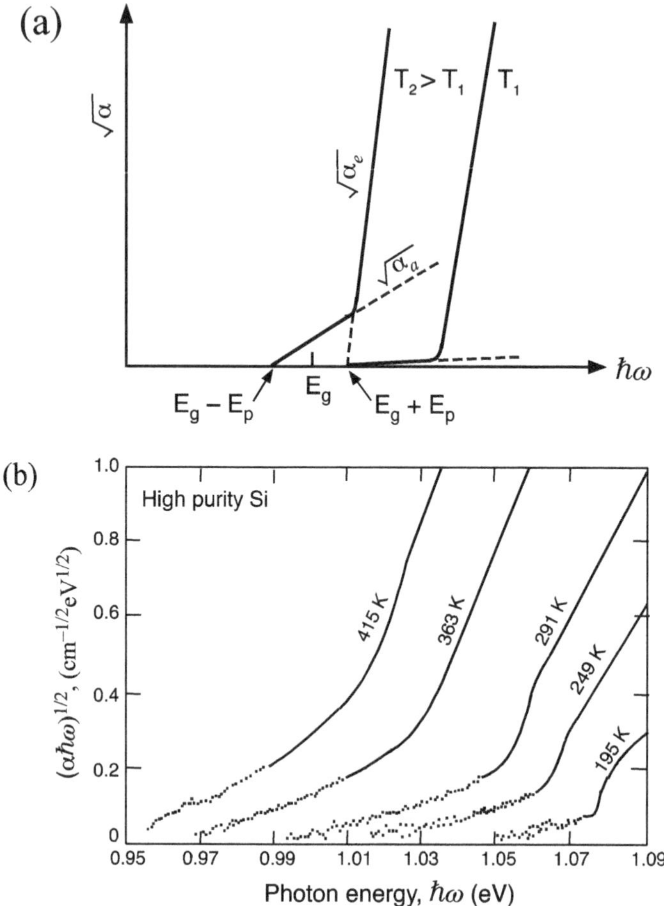

Fig. 8.12 **a** Temperature dependence of α_e and α_a. **b** Square root absorption coefficient as a function of photon energy for a high-purity Si measured at different temperatures. Reprinted with permission from [8] copyright American Physical Society)

shift (Fig. 8.13a). The measured absorption coefficients in *n*-type GaAs for different doping concentrations are reproduced in Fig. 8.14. The Burstein–Moss shift starts to affect the absorption edge as the carrier density exceeds 6×10^{17} cm^{-3}. For *p*-type semiconductors, however, the Burstein–Moss shift is less important. This is due to the high density of states of the valence band originating from the much heavier mass of holes as compared to electrons.

Impurities in semiconductors are generally represented as a localized level at a fixed energy with respect to the conduction or valence bands. However, it was observed that the ionization energy for both donors and acceptors decreased as the impurity concentration was increased. The ionization energy of *n*- and *p*-type GaAs

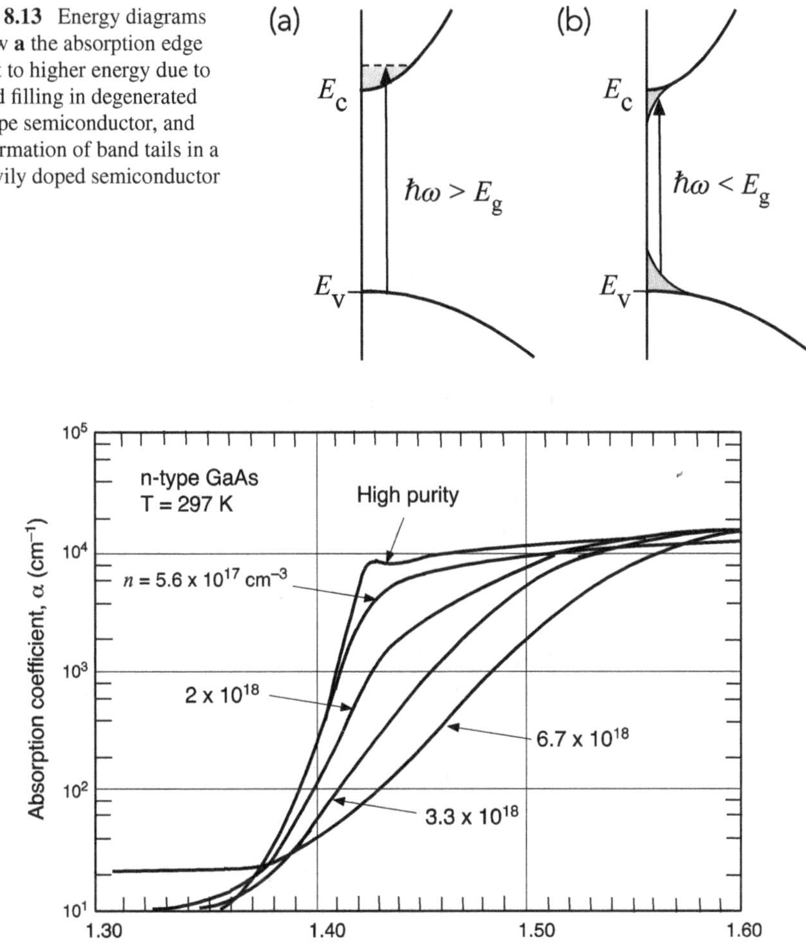

Fig. 8.13 Energy diagrams show **a** the absorption edge shift to higher energy due to band filling in degenerated n-type semiconductor, and **b** formation of band tails in a heavily doped semiconductor

Fig. 8.14 Measured absorption coefficient in n-type GaAs at room temperature for different doping concentrations. Reprinted with permission from [9] copyright AIP Publishing

goes to zero for free carrier densities near 2×10^{16} cm^{-3} and 5×10^{18} cm^{-3}, respectively. These results strongly suggest that impurities in semiconductors can be treated by the localized level concepts only in undoped and lightly doped semiconductors. Thus, the random distribution of impurities in semiconductors is a more realistic assumption. The random distribution of charged impurities in the crystal with high impurity concentration results in potential fluctuations and causes the band edges to vary spatially. This leads to the formation of *band tails* which contain states with energy below the unperturbed conduction band edge or above the unperturbed valence band edge. Thus, the band tail states extend into the forbidden gap. (Figure 8.13b) These band tail states affect the density of states significantly but only in the vicinity of the band edge. However, the spatially averaged position of the band edge does not change.

At high doping concentrations, the magnitude of the bandgap energy reduces with doping concentration—a phenomenon referred to as *bandgap narrowing*, *bandgap shrinkage*, or *bandgap renormalization*. The main reasons for bandgap narrowing are many-body effects of free carriers. At high free carrier concentrations, the small carrier-to-carrier distance allows electrons to interact with each other through the long-range Coulomb potential. The carrier–carrier interaction causes other electrons in the vicinity to spatially redistribute in order to reduce the total energy of the electron system. Therefore, bandgap narrowing occurs in heavily doped semiconductors, and the change in energy gap (ΔE_g) follows a one-third power of the doping concentration.

The Burstein–Moss shift and bandgap narrowing are two competing effects that cause the Fermi level to change in opposite directions. In heavily doped *n*-type semiconductors, the Burstein–Moss shift causes the Fermi level to increase while the Fermi level moves downward by bandgap narrowing. As shown in Fig. 8.14, the Burstein–Moss shift prevails in the *n*-type GaAs due to the light electron effective mass, resulting in a blue-shift of the absorption edge. For heavily doped *p*-type GaAs, a completely opposite behavior is observed as shown in Fig. 8.15. The bandgap narrowing and the formation of band tail states—rather than the Burstein–Moss shift—dominate the near-band-edge optical absorption, as is evident from the red-shift of the optical absorption edge with increasing doping concentration.

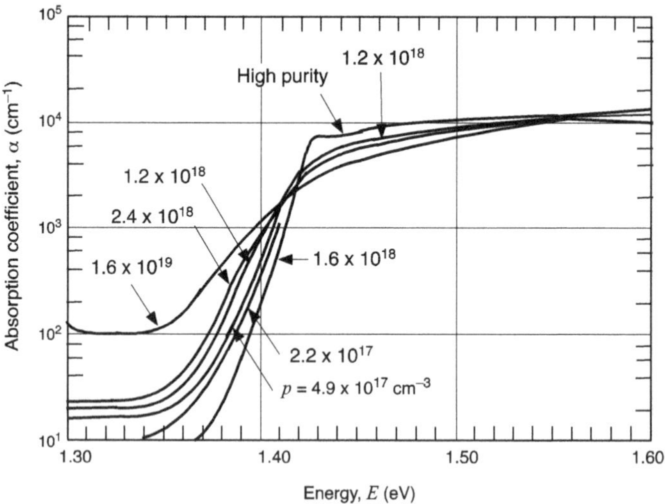

Fig. 8.15 Measured absorption coefficient in *p*-type GaAs at room temperature for different doping concentrations. Reprinted with permission from [9] copyright AIP Publishing

8.2.4 Exciton Absorption

Exciton represents a system of electron and hole bound to one another by their mutual Coulomb attraction. A bound electron–hole pair or exciton can be created by the absorption of a photon. The free electron can orbit about the free hole as if this were a hydrogen-like atom with a reduced mass m_r.

Since the exciton exist in the crystal and not in free space, the binding energy is reduced by the reduced mass and the dielectric constant.

$$E_x = -\left(\frac{m_r q^4}{2h^2\epsilon^2}\right)\frac{1}{n^2} = -\frac{(m_r/m_0)}{\epsilon_r^2}\frac{R_y}{n^2} \tag{8.55}$$

where $R_y = 13.6$ eV and n is an integer ≥ 1. Using $m_r = 0.059m_0$ and $\epsilon_r = 12.85$ for GaAs, the lowest exciton energy level ($n = 1$) has a value of -4.9 meV. Since the exciton can move through the crystal as a unit, it does not have a well-defined potential in the semiconductor energy diagram. The calculated binding energy is customarily referenced to the conduction band minimum. At $n = \infty$, it merges with the conduction band edge. For the ground state ($n = 1$) of the exciton, the binding energy is quite small, and the exciton can be easily dissociated at room temperature.

Since the electron and the hole of an exciton must move together through the crystal, their translational velocities must be identical. This condition places a restriction on the regions in E-k space where excitons can be found, such that both carriers move at the same velocity ($\upsilon \propto \mathrm{d}E/\mathrm{d}k$). Thus, excitons can be created by photon absorption at any critical point, where

$$\frac{\mathrm{d}E_c}{\mathrm{d}k} - \frac{\mathrm{d}E_v}{\mathrm{d}k} = 0 \tag{8.56}$$

In direct bandgap materials, the ground-state level ($n = 1$) of an exciton is easily observed as an absorption peak on the lower energy side of the absorption edge. The free exciton occurs when the photon energy is $h\nu = E_g - E_x$, where E_x is the binding energy of the exciton. At low temperatures, the exciton absorption has a very pronounced transition, which broadens with temperature. Figure 8.16 shows that the exciton feature diminishes in GaAs as the temperature is increased toward room temperature.

If the exciton is formed with the simultaneous absorption or emission of a phonon, as seen in indirect transitions, the increase in absorption coefficient occurs at

$$h\nu = E_g - E_x \pm \hbar\omega_p \tag{8.57}$$

for the transition with phonon emission or absorption. With the participation of the phonon in the transition, the exciton may acquire an additional momentum from the phonon, modifying the associated kinetic energy. The addition of the kinetic energy broadens the exciton level into the band. Although no absorption peak is expected, the presence of indirect exciton absorption should still be noticeable as a

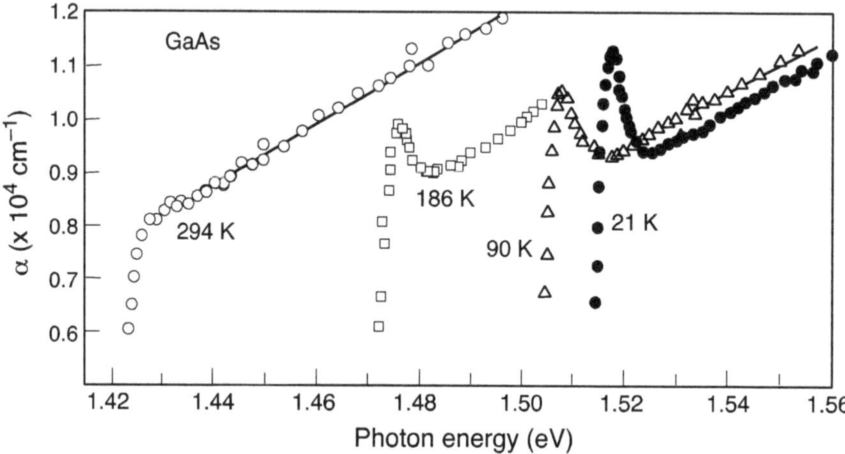

Fig. 8.16 Exciton absorption in GaAs at various temperatures. The exciton absorption peak becomes more pronounced at low temperatures. Reprinted with permission from [10] copyright American Physical Society

sharp rise in the absorption coefficient. In addition, since there are two transverse and one longitudinal phonons in each of the acoustic and optical branches of the phonon spectrum, more than one phonon can participate in the transition with various combinations. Therefore, a large number of steps in the absorption edge can be detected. The rich absorption spectra of the indirect bandgap GaP are shown in Fig. 8.17 as an example.

In addition to free excitons, a free hole can combine with a neutral donor to form a positively charged excitonic ion complex called a *donor bound exciton*. As shown in Fig. 8.18, in such a complex, the electron bound to the donor travels in a wide orbit about the donor while the associated hole moves in the electrostatic field of the dipole determined by the instantaneous position of the electron. An electron associated with a neutral acceptor also forms an 'acceptor bound exciton.'

Due to the two-dimensional nature of quantum wells, the exciton binding energy in a quantum well is enhanced due to a large overlap of the electron and hole wave functions. As in the 3D case, the binding energy of the exciton in a quantum well can be derived by solving a 2D hydrogen model. It has the value

$$E_x^{2D} = -\frac{m_r q^4}{2h^2\epsilon^2}\frac{1}{(n-1/2)^2} = -\left[\frac{(m_r/m_0)R_y}{\epsilon_r^2}\right]\frac{1}{(n-1/2)^2} \quad n = 1, 2, 3, \ldots \tag{8.58}$$

The diameter of the exciton orbit can also be determined as

$$a_x^{2D} = \frac{4\pi\epsilon\hbar^2}{m_r q^2} = 0.53\left(\frac{\epsilon_r}{m_r/m_0}\right) \tag{8.59}$$

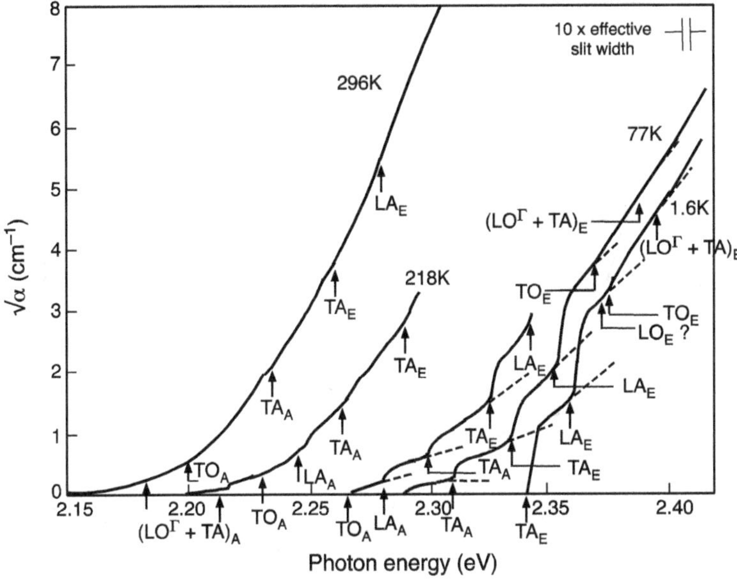

Fig. 8.17 Square root absorption coefficient of GaP measured at different temperatures. Reprinted with permission from [11] copyright American Physical Society

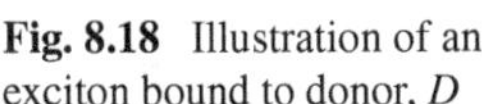
Fig. 8.18 Illustration of an exciton bound to donor, D

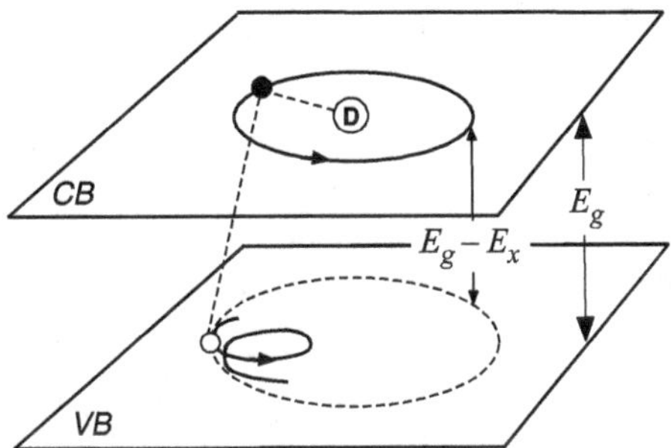

For $n = 1$, the binding energy and orbit radius of an exciton in a GaAs quantum well are $4E_x$ and ~100 Å, respectively. Using MBE growth technique, GaAs/$Al_{0.2}Ga_{0.8}As$ QWs with well widths 140 and 210 Å were demonstrated at Bell Laboratories for the first time. When measured at a very low temperature of 2 K, shown in Fig. 8.19, the QW absorption steps are clearly observed with associated exciton peaks. Due to the increased binding energy, in some materials such as InGaAs/InAlAs QW, the excitonic effect could be observed even at room temperature.

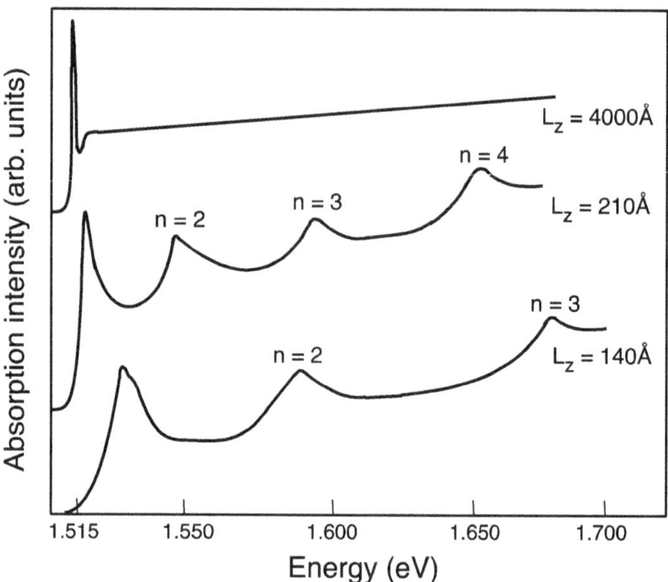

Fig. 8.19 Optical absorption spectra of bulk and quantum-well heterostructures. Reprinted with permission from [12] copyright American Physical Society

8.2.5 Absorption Due to Isoelectronic Traps

While the probability of radiative recombination taking place in an indirect bandgap semiconductor is extremely low, the situation was changed by the discovery of N isoelectronic traps in GaP in 1965. Such a trap is called *isoelectronic* because it has the same number of valence electrons as the atom it replaces when it sits substitutionally on the lattice site. The substitution of nitrogen for a group V atom in GaP establishes a short-range potential well resulting from the combination of the difference in electronegativity between the N atom and the group V atom that it replaces, and the hydrostatic deformation of lattice around the N site that arises from the large covalent radius difference. This potential well allows the N atom to capture an electron, which in turn can bind a hole by Coulomb attraction, thus forming a bound exciton at about 10 meV below the conduction bandedge. Since the electron is bound to the immediate vicinity of the N impurity (real space), the wave function of the electron is diffused in k-space. The wave function of the electron at the isoelectronic trap is much more extensive in k-space. Thus, there is an enhancement near $k = 0\ (\Gamma)$ which results in increased electron–hole recombination in the Γ-band region. At $k = 0$, the shallow N isoelectronic trap depth with a favorable conduction band structure leads to an increase of the electron distribution probability by almost three orders of magnitude and enables the generation of relatively efficient green (near bandgap) electroluminescence in GaP:N LEDs.

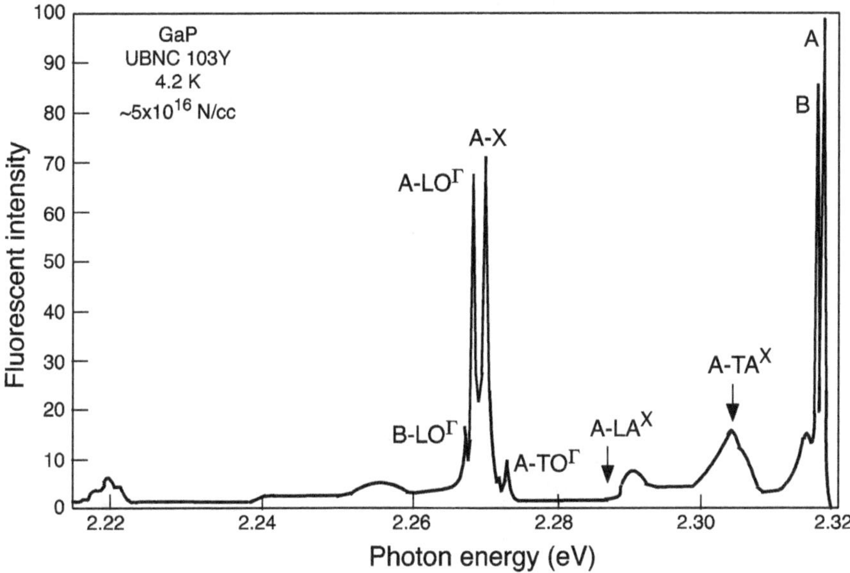

Fig. 8.20 Low-temperature photoluminescence spectrum from a lightly N-doped GaP crystal. The bound exciton emissions A and B lines are prominent transitions. Reprinted with permission from [13] copyright American Physical Society

The features of the absorption spectrum due to isoelectronic traps in a nitrogen-doped GaP have different origins. In a lightly nitrogen-doped (~5 $\times$ 10^{16} cm^{-3}) sample, the dominant transitions are the near bandgap, zero-phonon exciton transition lines A and B, and associated various acoustic and optical phonon replicas (Fig. 8.20). When the doping concentration is increased, neighboring nitrogen traps may interact with each other in pairs and form a new set of energy levels for their own bound excitons. At high nitrogen concentrations, as shown in Fig. 8.21, the interacting isoelectronic centers form a new set of energy levels such as NN_1, NN_2, and NN_3. At even higher concentrations, the NN lines dominate the absorption spectrum, and the A and B lines are absent. Since the binding energy of an exciton bound to paired isoelectronic centers depends on the distance between the two atoms of a pair, the potential well becomes deeper and the binding becomes much tighter by placing two nitrogens close together. Therefore, the strongest binding energy occurs for the nearest pairs, NN_1, which gives the lowest emission energy. The photoluminescence (PL) spectrum of a highly doped GaP sample with a nitrogen concentration of ~10^{19} cm^{-3} measured at 77 K is shown in Fig. 8.22. The A line emission is barely seen in the magnified ($\times$170) plot, and the NN_1 line and its associated phonon lines dominate the PL spectrum.

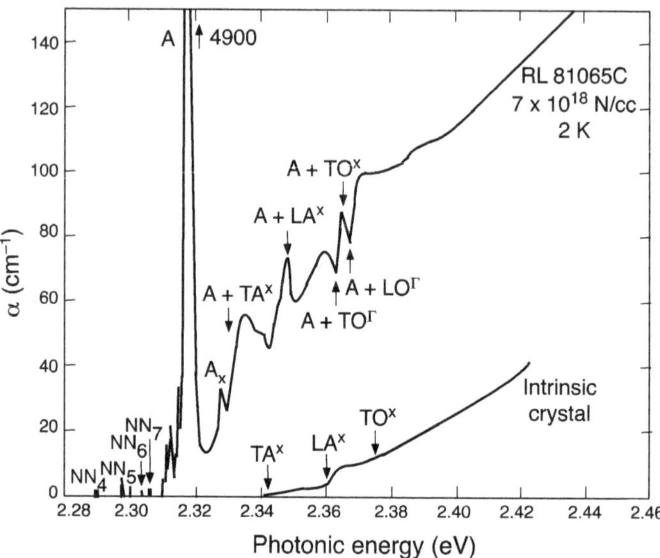

Fig. 8.21 Fundamental absorption edge at 2 K of undoped GaP and of GaP doped with 7×10^{18} cm^{-3} nitrogen. Reprinted with permission from [13] copyright American Physical Society

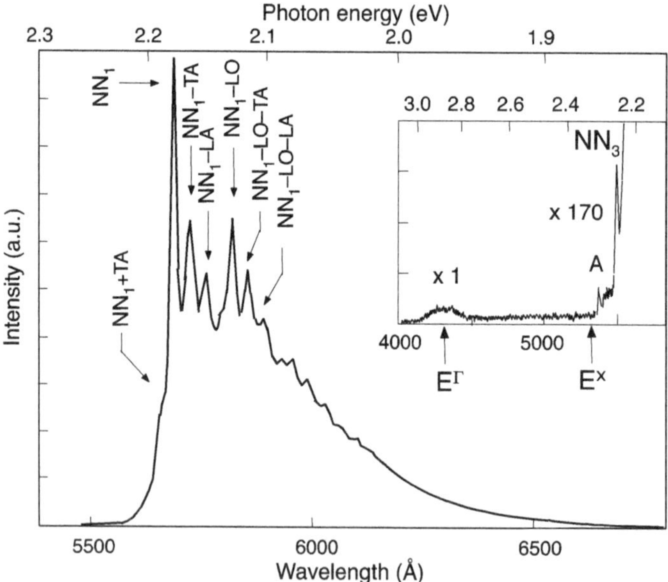

Fig. 8.22 Low-temperature (77 K) photoluminescence spectrum of a nitrogen-doped GaP at a concentration of 10^{19} cm^{-3}. The inset shows a wider scan of 77 K PL spectrum. The A line and direct gap (E_{Γ}) transition are shown. [14]

8.3 Radiative Transition Between Discrete States

8.3.1 Photon Density Distributions

The interaction of light and free-carrier electrons in a semiconductor can take three different pathways—absorption, spontaneous emission, and stimulated emission (Fig. 8.23). The semiconductor will absorb photons (upward transitions) with energy equal to or greater than the bandgap energy. The photo-generated electrons will not stay in the conduction band forever. They will recombine with holes, on average, after one carrier lifetime span. For a direct bandgap semiconductor, this process will generate a spontaneous emission (downward transition) with a wavelength equal to the bandgap energy. If an optical excitation is applied to the semiconductor before the excited electrons in the conduction band reach their lifetime, the excited electrons can recombine with holes prematurely to generate photons in phase with the incident photons. This process is called *stimulated emission.* The rates of absorption, stimulated emission, and spontaneous emission are interrelated and can be described by Einstein's relations. The external-excitation-generated photon emission and absorption are fundamentally connected through the *photon density distribution*, which describes the number of available photons participating in the interaction per unit energy/frequency ($E = hv$) and unit volume. Thus, it has a unit of (photon number/eV-cm^3). To calculate the photon density distribution $P(E)$, we need to know the number of allowed photon states or the photon density of states $D_p(E)$ and the probability that a photon will occupy that state determined by the photon distribution function $\langle n_p \rangle$.

(a) **Photon density of states, $D_p(E)$**

The photon density of states can be derived similarly to the carrier density of states in a crystal. In a cubic enclosure with a dimension L ($\gg\lambda$) on each side, the allowed modes in terms of wave vector can be determined from the electromagnetic wave equation. The allowed wave vectors are

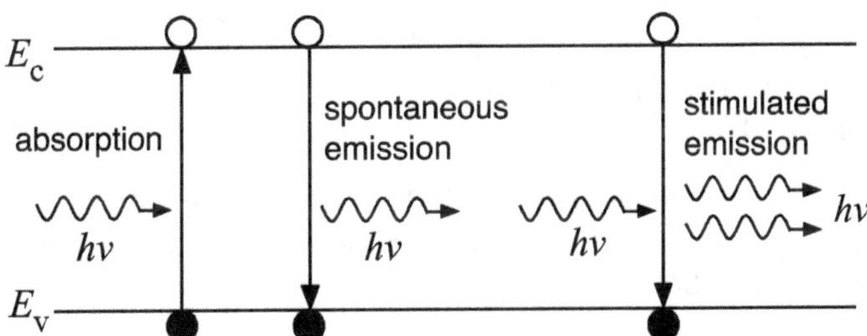

Fig. 8.23 Schematic of the absorption, spontaneous emission, and stimulated emission processes in a direct bandgap semiconductor

$$k_i = 2\pi m_i/L \quad i = x, y, z$$

where $m_i = 0, \pm 1, \pm 2, \pm 3, \ldots$ These discrete values (modes) of k_x, k_y, k_z determine discrete values of allowed electric fields and photon energies. The particle-like nature of electromagnetic radiation is represented by photons with energy

$$E = h\nu = \hbar\omega = pc$$

where $\omega = 2\pi\nu$ and the momentum of the photon is $p = \hbar k$. The photon energy and momentum have the same forms as those of carriers in a crystal. The number of states in k-space is found from the volume in a thin spherical shell of $2 \times (4\pi k^2 dk)$ divided by the unit volume, $(2\pi/L)^3$, which contains one allowed state. The factor of 2 counts for two different types of polarization, TE and TM modes, for a photon. The photon density of states within an interval of dk is the number of states per unit volume $V = L^3$.

$$\mathrm{d}D_\mathrm{p}(k) = \frac{2}{L^3}\frac{4\pi k^2}{(2\pi/L)^3}\mathrm{d}k = \frac{k^2}{\pi^2}\mathrm{d}k \tag{8.60}$$

In a dielectric medium with a refractive index $\overline{n}$, the wave vector is

$$k = \frac{2\pi}{\lambda} = \frac{2\pi}{c/(\overline{n}\nu)} = \frac{2\pi}{c}\frac{\overline{n}E}{h} \tag{8.61}$$

where ν is the oscillation frequency. Then, for $E = h\nu$,

$$\mathrm{d}k = \frac{2\pi\overline{n}}{ch}\left(1 + \frac{E}{\overline{n}}\frac{\mathrm{d}\overline{n}}{\mathrm{d}E}\right)\mathrm{d}E = \frac{2\pi\overline{n}}{c}\left(1 + \frac{\nu}{\overline{n}}\frac{\mathrm{d}\overline{n}}{\mathrm{d}\nu}\right)\mathrm{d}\nu \tag{8.62}$$

where the term in the parentheses is the refractive index dispersion and is unity for free space. The photon density of states within an energy interval dE becomes

$$\mathrm{d}D_\mathrm{p}(E) = \frac{8\pi\overline{n}^3E^2}{h^3c^3}\left(1 + \frac{E}{\overline{n}}\frac{\mathrm{d}\overline{n}}{\mathrm{d}E}\right)\mathrm{d}E \tag{8.63}$$

(b) **Photon distribution function, $\langle n_\mathrm{p}\rangle$**

Since photons are indistinguishable, and identical particles do not follow Pauli's exclusion principle, the average number of photons per state $\langle n_\mathrm{p}\rangle$ is given by the Bose–Einstein distribution law.

$$\langle n_\mathrm{p}\rangle = \left[\exp(E/kT) - 1\right]^{-1} \tag{8.64}$$

The total photon density of states within an energy interval dE is given as $\mathrm{d}D_\mathrm{p}(E) \times \langle n_\mathrm{p}\rangle$.

(c) **Photon density distributions, $P(E)$**

The photon density distribution is the number of photons per unit volume and per unit energy. Therefore, the photon density distribution $P(E)$ at a specific energy E is obtained by dividing $\mathrm{d}D_\mathrm{p}(E) \times \langle n_\mathrm{p} \rangle$ with the energy interval $\mathrm{d}E$.

$$P(E) = \frac{8\pi \overline{n}^3 E^2}{h^3 c^3} \frac{\left[1 + \dfrac{E}{\overline{n}} \dfrac{\mathrm{d}\overline{n}}{\mathrm{d}E}\right]}{\left[\exp(E/kT) - 1\right]} \quad \left(\text{photon number/eV-cm}^3\right) \tag{8.65}$$

8.3.2 *Einstein's* A *and* B *Coefficients*

In semiconductors, the available electron states have a continuous band of states within the conduction and valence bands. This is in contrast to two sharp levels for an atomic system. To start, we assume the absorption of a photon of energy $\hbar\omega = E_{21} \geq E_\mathrm{g}$ results in an electron transition from a state E_1 in the valence band to a state E_2 within the conduction band as seen in Fig. 8.24. The transition rate between the two discrete states depends on (1) the probability that the upward transition can occur, B_{12}; (2) the probability that the state E_1 contains an electron, f_1; (3) the probability that the state E_2 is empty, $(1 - f_2)$; and (4) the density of photons of energy E_{21}, $P(E_{21})$. Using similar parameters, we also describe the spontaneous emission and stimulated emission processes in the same semiconductor. The absorption, spontaneous emission, and stimulated emission processes are not isolated events but are interrelated through Einstein's A and B coefficients.

(a) **Upward transition rate, r_{12}**

When absorb a photon with energy E_{21}, the electron in the valence band can make an upward transition to an empty state within the conduction band. The transition

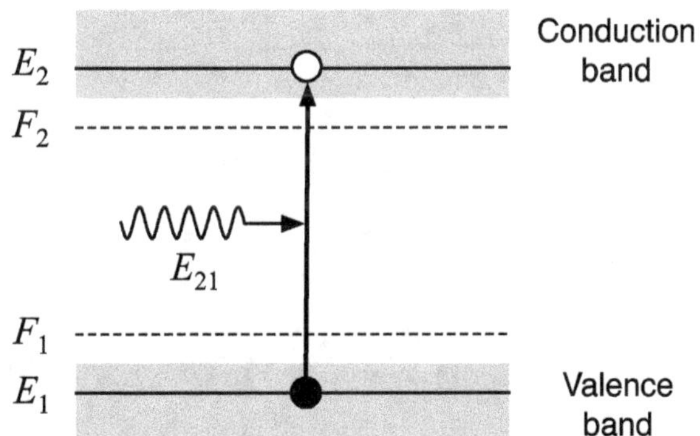

Fig. 8.24 Absorption process in a semiconductor. An electron in the valence band at E_1 is excited into the conduction band at E_2 when excited with a photon of energy $E_{21} = E_2 - E_1$. The quasi-Fermi levels of the conduction and valence bands is F_2 and F_1, respectively

rate is given as

$$r_{12} = B_{12} f_1 (1 - f_2) P(E_{21}) \tag{8.66}$$

where f_1 and f_2 are the Fermi–Dirac distribution functions.

$$f_1 = \frac{1}{1 + \exp[(E_1 - F_1)/kT]} \tag{8.67a}$$

$$f_2 = \frac{1}{1 + \exp[(E_2 - F_2)/kT]} \tag{8.67b}$$

where F_1 and F_2 are the quasi-Fermi levels for the valence band and the conduction band, respectively.

(b) **Downward transition rate, r_{21}**

The absorbed photons can stimulate the emission of a similar photon by the transition of an electron from E_2 to E_1. The downward transition rate for this process is

$$r_{21} = B_{21} f_2 (1 - f_1) P(E_{21}) \tag{8.68}$$

where B_{21} is the downward transition probability.

(c) **Spontaneous emission rate, s_{21}**

The conduction band electrons can spontaneously return to E_1 without interaction with the radiation field, $P(E_{21})$. The transition probability for this process is represented by A_{21}, and the spontaneous emission rate is given by

$$s_{21} = A_{21} f_2 (1 - f_1) \tag{8.69}$$

At thermal equilibrium, the absorption rate equals the total emission rate

$$r_{12} = r_{21} + s_{21} \tag{8.70a}$$

and

$$F_1 = F_2 \tag{8.70b}$$

This leads to

$$P(E) = \frac{A_{21} f_2 (1 - f_1)}{B_{12} f_1 (1 - f_2) - B_{21} f_2 (1 - f_1)} = \frac{8\pi \bar{n}^3 E_{21}^2}{h^3 c^3 \left[\exp(E_{21}/kT) - 1\right]} \tag{8.71}$$

It can be rewritten as

$$\frac{A_{21}}{B_{12}\exp(E_{21}/kT) - B_{21}} = \frac{8\pi\overline{n}^3 E_{21}^2}{h^3c^3\left[\exp(E_{21}/kT) - 1\right]}$$

or

$$A_{21}\exp(E_{21}/kT) - A_{21} = \frac{8\pi\overline{n}^3 E_{21}^2}{h^3c^3}\left[B_{12}\exp(E_{21}/kT) - B_{21}\right]$$

Equating the temperature independent terms gives

$$A_{21} = \left(\frac{8\pi\overline{n}^3 E_{21}^2}{h^3c^3}\right)B_{21} \tag{8.72}$$

Then, equating the temperature dependent terms with A_{21} given above gives

$$B_{21} = B_{12} \tag{8.73}$$

Equations (8.72) and (8.73) represent the Einstein's A and B coefficients. They indicate that the spontaneous emission rate is related to the absorption rate and the stimulated emission rate.

8.3.3 Absorption and Stimulated Emission

(a) **Necessary conditions for stimulated emission in semiconductors**

The necessary conditions for stimulated emission simply require that the photon-induced downward radiative transition rate exceed the upward photon absorption rate, $r_{21} > r_{12}$:

$$B_{21}f_2(1 - f_1)P(E_{21}) > B_{12}f_1(1 - f_2)P(E_{21}) \tag{8.74}$$

Since $B_{21} = B_{12}$, this condition becomes

$$f_2(1 - f_1) > f_1(1 - f_2) \quad \text{or } f_2 > f_1 \tag{8.75}$$

In other words, this is the condition required for population inversion. It further reduces to

$$\exp[(F_2 - F_1)/kT] > \exp[(E_2 - E_1)/kT] \tag{8.76a}$$

or

$$F_2 - F_1 > E_2 - E_1 \tag{8.76b}$$

The condition states that in order for the downward stimulated emission rate to exceed the upward absorption rate, the separation of the quasi-Fermi levels must exceed the photon emission energy.

(b) **Relation of net stimulated emission rate to the absorption coefficient**

The interaction of electrons and photons in the solid can be related to the absorption coefficient, $\alpha(E_{21})\,(\mathrm{cm}^{-1})$. The net absorption rate $R(abs)$ (photon number/eV-cm^3-s) is the difference between the upward transition rate r_{12} and the downward transition rate r_{21}. The net absorption rate is also equal to the absorption coefficient times the photon flux, $\varphi(E_{21}) = P(E_{21})\upsilon_g$.

$$R(abs) = r_{12} - r_{21} = B_{12}(f_1 - f_2)P(E_{21}) = \alpha(E_{21})P(E_{21})\upsilon_g \tag{8.77}$$

where υ_g is the group velocity. In the dielectric medium, the group velocity is expressed as

$$\upsilon_g = \frac{\mathrm{d}\omega}{\mathrm{d}k} = \frac{2\pi}{h}\frac{\mathrm{d}E}{\mathrm{d}k} = \frac{c/\overline{n}}{1 + (E/\overline{n})(\mathrm{d}\overline{n}/\mathrm{d}E)} \cong \frac{c}{\overline{n}} \tag{8.78}$$

Therefore, the absorption coefficient may be written as

$$\alpha(E_{21}) = \frac{R(abs)}{P(E_{21})\upsilon_g} = \frac{B_{12}(f_1 - f_2)}{c/\overline{n}} \tag{8.79}$$

The numerator of (8.79) is related to the net stimulated emission rate, R(st). The net stimulated emission rate is the difference between the downward transition (stimulated emission) rate r_{21} and the upward transition (absorption) rate r_{12}.

$$R(\mathrm{st}) = r_{21} - r_{12} = B_{12}(f_2 - f_1)P(E_{21}) \tag{8.80}$$

Using the Einstein's relations and the photon density distribution, this relation reduces to

$$R(\mathrm{st}) = \frac{A_{21}(f_2 - f_1)}{\exp(E_{21}/kT) - 1} = \frac{r_{\mathrm{stim}}(E_{21})}{\exp(E_{21}/kT) - 1} \tag{8.81}$$

Here we defined $r_{\mathrm{stim}}(E_{21}) \equiv A_{21}(f_2 - f_1)$ as the stimulated emission rate because of its similarity in form to the spontaneous emission rate in (8.69). The number of photons per state is represented by the term $\langle n_p \rangle = 1/[\exp(E_{21}/kT) - 1]$. The constant A_{21} may be replaced by B_{21} through Einstein's relation, and the stimulated emission rate is given as

$$r_{\mathrm{stim}}(E_{21}) = \left(\frac{8\pi\overline{n}^3 E_{21}^2}{h^3c^3}\right)B_{12}(f_2 - f_1) \tag{8.82}$$

The stimulated emission rate and the absorption coefficient (8.79) are different by a constant.

$$r_{\text{stim}}(E_{21}) = -\left(\frac{8\pi\overline{n}^2E_{21}^2}{h^3c^2}\right)\alpha(E_{21}) \tag{8.83}$$

The absorption coefficient is measurable and may be used to derive the stimulated emission rate.

(c) **Relation of spontaneous emission rate to the absorption coefficient**

The spontaneous emission rate may also relate to the absorption coefficient. It has been defined earlier as

$$s_{21} = A_{21}f_2(1 - f_1) \equiv r_{sp}(E_{21}) \tag{8.69}$$

Using (8.72) and (8.79), we may express s_{21} in terms of $\alpha(E_{21})$ given as

$$r_{\text{sp}}(E_{21}) = \left(\frac{8\pi\overline{n}^2E_{21}^2}{h^3c^2}\right)\alpha(E_{21})\frac{f_2(1 - f_1)}{(f_1 - f_2)} \tag{8.84}$$

Solving for the Fermi–Dirac functions, f_1 and f_2, we have

$$r_{\text{sp}}(E_{21}) = \left(\frac{8\pi\overline{n}^2E_{21}^2}{h^3c^2}\right)\frac{\alpha(E_{21})}{\exp\{[E_{21} - (F_2 - F_1)]/kT\} - 1} \tag{8.85}$$

Therefore, the spontaneous emission rate is also related to the absorption coefficient. We can also relate the spontaneous emission rate to the stimulated emission rate as

$$r_{\text{stim}}(E_{21}) = r_{\text{sp}}(E_{21})\{1 - \exp\{[E_{21} - (F_2 - F_1)]/kT\}\} \tag{8.86}$$

The spontaneous emission rate, the stimulated emission rate, and the absorption coefficient are all interrelated. The spontaneous and stimulated emission rates can be determined from experimentally measured absorption coefficients.

(d) **Transition probability, B_{12}**

Since both $r_{\text{stim}}(E_{21})$ and $r_{\text{sp}}(E_{21})$ are related to the absorption coefficient, it is only necessary to determine $\alpha(E_{21})$ to obtain the net spontaneous and stimulated emission rates. Also, calculated values of $\alpha(E_{21})$ may readily be verified by experimental measurements. Therefore, $\alpha(E_{21})$ becomes the most important parameter to be considered in the analysis of light emission processes. For photon absorption between two discrete levels, the absorption coefficient was derived as

$$\alpha(E_{21}) = \left(\frac{\overline{n}}{c}\right)B_{12}(f_1 - f_2) \tag{8.79}$$

In this equation, the transition probability B_{12} is unknown and relates the interaction of electrons in the solid with electromagnetic radiation. Using quantum mechanical techniques of time-dependent perturbation theory, the transition probability between a single initial state 1 and a single final state 2 is given as:

$$B_{12} = \left(\frac{\pi}{2\hbar}\right)\left|\int_V \Psi_1^*(\boldsymbol{r},t)H^I\Psi_2(\boldsymbol{r},t)d^3\boldsymbol{r}\right|^2 \tag{8.87}$$

In this equation, $\boldsymbol{r}$ is the three-dimensional spatial vector, $\Psi_1^*(\boldsymbol{r}, t)$ the complex conjugate of the wave function of the initial state, H^I the interaction Hamiltonian, and $\Psi_2(\boldsymbol{r}, t)$ the wave function of the final state.

In the case of the transition from a continuum of state 1 to a continuum of states 2, the transition probability takes a form of

$$B_{12} = \left(\frac{2\pi}{\hbar}\right)|H_{12}|^2\delta(E_2 - E_1 - \hbar\omega) \tag{8.88}$$

and

$$H_{12} = \int_V \Psi_1^*(\boldsymbol{r},t)H^I\Psi_2(\boldsymbol{r},t)d^3\boldsymbol{r} \tag{8.89}$$

This is called *Fermi's Golden Rule*. The delta function is used to specify that the energy separation between the two states equals $\hbar\omega$. In either case, the transition probability is proportional to a *transition matrix* $|M_T|^2$, which is a function of the bandgap energy and the conduction band effective mass of the semiconductor.

The transition probability B also can be evaluated by the method of van Roosbroeck and Shockley, using experimentally determined values of the absorption coefficient [15]. The van Roosbroeck and Shockley relation is that at equilibrium, the optical generation rate of electron–hole pairs is equal to the rate of radiative recombination. This relation has already been described by (8.85), which shows the fundamental relation between the emission spectrum and the observed absorption spectrum. One can also relate the rate of radiative recombination R_0, under equilibrium condition, by the product of the electron density in the upper state n_0, the hole density in the lower state p_0, and the transition probability B as

$$R_0 = n_0 p_0 B. \tag{8.90}$$

The total radiative recombination rate under non-equilibrium conditions becomes

$$R = \frac{np}{n_i^2}R_0 \tag{8.91}$$

This equation implies that the recombination rate increases with increasing carrier concentration, and that R reduces to R_0 when the product np approaches the equilibrium intrinsic value $n_0 p_0 = n_i^2$.

Under non-equilibrium conditions, where the carrier concentration deviates from its equilibrium concentration n_0 (or p_0) by a small amount $\Delta n = \Delta p$, then $n = n_0 + \Delta n$ and $p = p_0 + \Delta p$, the recombination rate becomes

$$R = R_0 + \Delta R = \frac{(n_0 + \Delta n)(p_0 + \Delta p)}{n_0 p_0} R_0 = \frac{n_0 p_0 + n_0 \Delta p + p_0 \Delta n + \Delta n \Delta p}{n_0 p_0} R_0 \quad (8.92)$$

Neglecting the very small term $\Delta n \Delta p$ from the above equation,

$$\frac{\Delta R}{R_0} \cong \frac{\Delta n}{n_0} + \frac{\Delta p}{p_0} \quad (8.93)$$

Assuming $\Delta n = \Delta p$, the radiative lifetime of excess carriers is obtained as

$$\tau = \frac{\Delta n}{\Delta R} = \frac{1}{R_0} \frac{n_0 p_0}{n_0 + p_0} \quad (8.94)$$

For an intrinsic material, $n_0 = p_0 = n_i$,

$$\tau = \frac{n_i}{2R_0} = \frac{1}{2n_i B} \quad (8.95)$$

Through this equation, one can relate the transition probability to the measured carrier lifetime. For direct bandgap semiconductors, e.g., GaAs and InP, in which recombination occurs by a direct radiative transition, $B \sim 10^{-10}$ cm^3/s. For indirect bandgap semiconductors such as Si, Ge, and GaP, the transition probability is smaller by a factor of 10^3–10^4, a quantity that may vary considerably from one material to another, being dependent upon the nature of the energy band structure. The large difference in transition probability between direct and indirect bandgap materials is also reflected in the shape of measured absorption coefficient curves shown in Fig. 8.10, where steep rise of the absorption coefficient at the bandgap energy is seen in direct bandgap semiconductors. Due to the low transition probability of the indirect material, the absorption coefficient increases slowly for energies near the bandgap energy.

8.4 Optical Transitions Between Energy Bands

8.4.1 Direct and Indirect Bandgaps

For efficient optical transitions between the conduction band minimum ($E_{\mathrm{c,min}}$) and the valence band maximum ($E_{\mathrm{v,max}}$), the energy conservation and momentum conservation conditions have to be observed.

$$\begin{cases} \text{Momentum conservation: } k_i \cong k_f \\ \text{Energy conservation: } \quad E_i - E_f = h\nu \end{cases} \tag{8.96}$$

where ν is the frequency of the optical radiation, and subscripts i and f represent the initial and final states of the transition, respectively.

As illustrated in Fig. 8.25, the momentum conservation condition requires the wave vectors of the initial (k_i) and the final (k_f) states to be the same as seen in a direct bandgap semiconductor, where both $E_{\mathrm{c,min}}$ and $E_{\mathrm{v,max}}$ are located at $k = 0$. To fulfill the second condition for energy conservation, the electron–hole recombination generated excess energy must be released by emitting a photon with the equivalent energy. Since the photon has a negligibly small momentum, the momentum conservation condition still holds. Therefore, light emission is possible from a direct bandgap semiconductor with an emission wavelength

$$\lambda(\mu\mathrm{m}) = \frac{1.2398}{E_i - E_f} = \frac{1.2398}{E_g(\mathrm{eV})} \tag{8.97}$$

For indirect semiconductors, the momentum associated with photon emission is too small to balance the large difference in momentum change between the initial

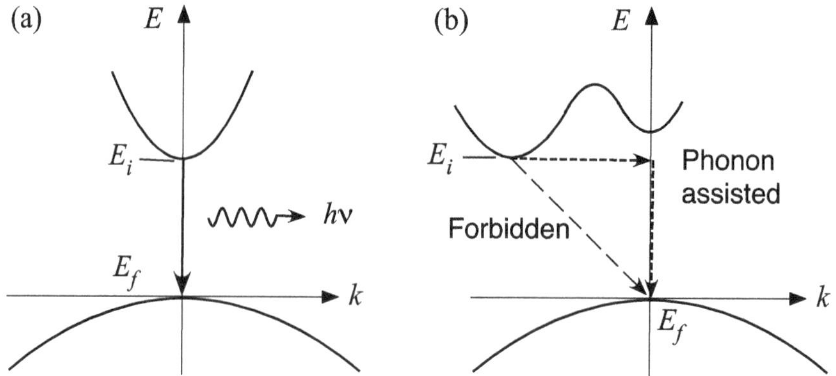

Fig. 8.25 Energy band diagrams and optical emission processes of **a** direct and **b** indirect semiconductors. The dashed lines indicate non-radiative transitions

and the final states; therefore, phonon emission is required to fulfill the momentum conservation requirement. The process is given by $h\nu_e = E_i - E_f - E_p$, where E_p is the phonon energy. Therefore, efficient light emission from indirect bandgap material is not allowed.

8.4.2 Band-to-Band Transitions in Bulk Semiconductors

For band-to-band transitions in a semiconductor, as shown in Fig. 8.26, we must consider all participating states in both the conduction and valence bands separated by the photon energy $\hbar\omega$. The number of states involved is determined by the density of states $D_i(E)$ and the occupation probability f_i. For example, the upward transition rate depends on the density of filled states in the valence band, $D_v(E)f_v$, and the density of empty states in the conduction band $D_c(E)(1 - f_c)$. Thus, the absorption coefficient in a semiconductor is the sum of the absorption coefficients at $\hbar\omega$ for all of the energy levels separated by $\hbar\omega$. It is given by

$$\alpha(\hbar\omega) = \int_{-\infty}^{\infty} \frac{B_{12}}{c/\overline{n}}(f_1 - f_2)D_v(E_v - E)D_c(E - E_c)\mathrm{d}E \tag{8.98}$$

and

$$\alpha(\hbar\omega) = \frac{\pi q^2 \hbar}{\epsilon_0 m^2 c\overline{n} E} \int_{-\infty}^{\infty} D_v(E'')D_c(E')|M_T|^2[f_v(E'') - f_c(E')]dE' \tag{8.99}$$

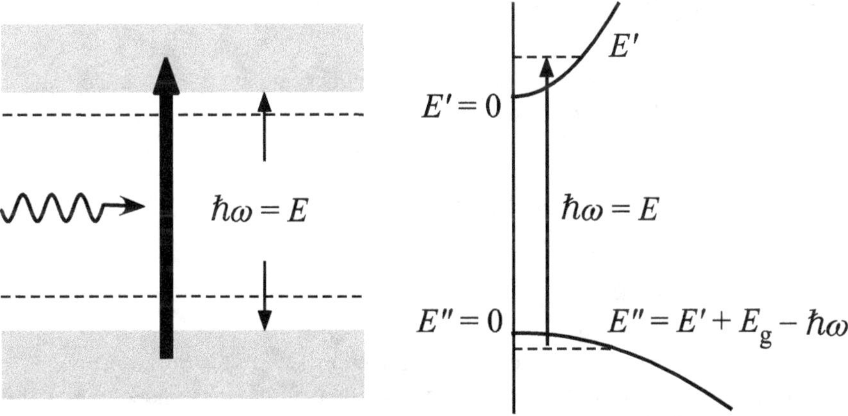

Fig. 8.26 Band-to-band optical absorption in a semiconductor. The right panel shows the *E-k* curves

Here, E' describes the conduction band edge with $E' = 0$ at E_c, and $E'' = E' + E_g - \hbar\omega$ for the valence band. The transition probability between two single states 1 and 2 is related to the transition matrix element as

$$B_{12} = \frac{\pi q^2}{m^2 \epsilon_0 \overline{n}^2 \omega} |M_T|^2 \tag{8.100}$$

We also obtain the stimulated emission rate from the absorption coefficient shown here:

$$r_{stim}(E) = \frac{4\pi \overline{n} q^2 E}{\epsilon_0 m^2 c^3 h^2} \int_{-\infty}^{\infty} D_v(E'') D_c(E') |M_T|^2 [f_c(E') - f_v(E'')] dE' \tag{8.101}$$

The problem of evaluating $\alpha(E)$ and $r_{stim}(E)$ becomes one of evaluating $D_c(E)$, $D_v(E)$, and $|M_T|^2$. Under high injection carrier densities or for high doping levels, such as in injection lasers, the density of states becomes concentration dependent and cannot be represented by the simple parabolic expression. The matrix element for transitions between band tail states in heavily doped semiconductors also differs from that involving free electron and free hole states. Fortunately, since the absorption coefficient is measurable for all energy ranges, we can fit (8.99) using the experimental results. Parameters used to fit experimental $\alpha(E)$ can be applied to evaluate $r_{stim}(E)$. Figure 8.27 is an example of using experimental absorption coefficient results to predict the photoluminescence spectrum of a heavily doped *p*-type GaAs (1.2×10^{18} cm^{-3}). Indeed, it shows that the absorption coefficient is closely related to the spontaneous and stimulated emission processes.

8.4.3 Fermi–Dirac Inversion Factor, Transparency, and Gain

The intrinsic absorption coefficient for allowed direct-transition in a semiconductor, whose valence band is completely full, where $f_v(E) = 1$, and whose conduction band is completely empty, where $f_c(E) = 0$, was given as

$$\alpha_0(h\nu) = A^* \sqrt{h\nu - E_g} \tag{8.44}$$

and $A^* = \dfrac{q^2 (2m_r)^{3/2} |M_T|^2}{2\pi m_0^2 c \overline{n} \epsilon_0 \hbar^2 (\hbar\omega)}$

The constant A^* was derived by John Bardeen in 1952. This equation applies to the situation where a low-level excitation is used such that the conduction band is almost empty and $f_c(E) = 0$ is fulfilled.

In contrast, semiconductors under strong excitations have partially filled parabolic conduction and valence bands, and there is stimulated emission allowed in addition

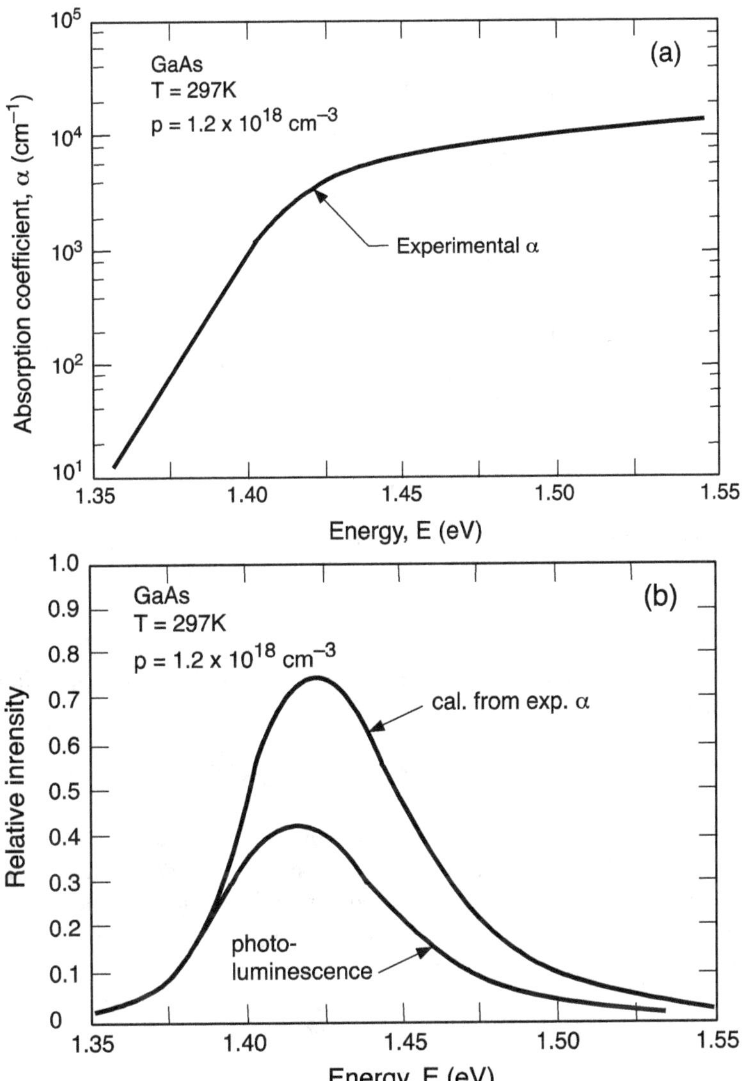

Fig. 8.27 **a** Experimental absorption coefficient of a *p*-type GaAs measured at room temperature. **b** Measured room temperature photoluminescence spectrum compared with the calculated stimulated emission rate using results in (**a**). Reprinted with permission from [16] copyright AIP Publishing

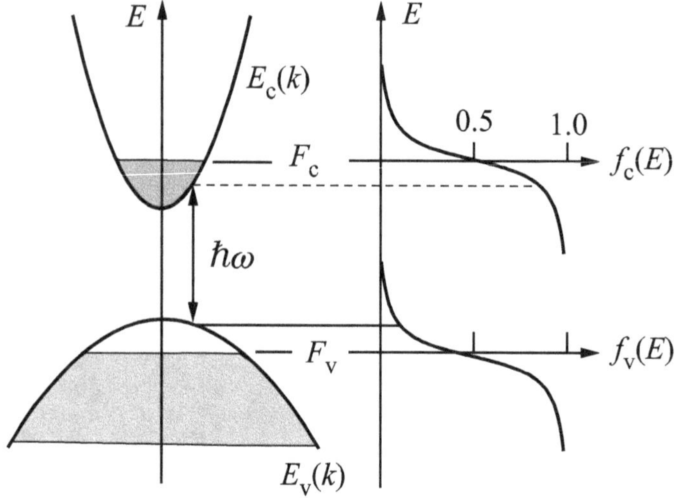

Fig. 8.28 Optical transition between the conduction and valence bands and related Fermi–Dirac distributions

to absorption. Under this circumstance, (8.99) has to be used to include the term $(f_v - f_c)$ in the calculation of the absorption coefficient and is related to the intrinsic absorption coefficient as

$$\alpha(\hbar\omega) = \alpha_0(\hbar\omega)[f_v(k) - f_c(k)] \tag{8.102}$$

The quantity $[f_v(k) - f_c(k)]$ is called the *Fermi–Dirac inversion factor*. Figure 8.28 shows the schematic of the optical transition of $\hbar\omega$ between the valence band and the conduction band. The corresponding Fermi–Dirac functions $f_v(k)$ and $f_c(k)$ are also shown. For $f_v(k) < f_c(k)$, one finds negative absorption, i.e., the gain coefficient.

$$\mathrm{g}(\hbar\omega) = -\alpha(\hbar\omega) \tag{8.103}$$

The condition of $f_v(k) - f_c(k) < 0$ satisfies the population inversion condition as shown here:

$$f_v(k) - f_c(k) = \frac{\exp[(E_c - F_c)/kT] - \exp[(E_v - F_v)/kT]}{\{1 + \exp[(E_c - F_c)/kT]\}\{1 + \exp[(E_v - F_v)/kT]\}} < 0 \tag{8.104}$$

and

$$\exp[(E_c - F_c)/kT] < \exp[(E_v - F_v)/kT] \tag{8.105}$$

or

$$F_c - F_v > E_c - E_v \tag{8.106}$$

where F_c and F_v are quasi-Fermi levels. The minimum of (8.106) determines a *transparency condition* where $(F_c - F_v)$ equals E_g. Above transparency, the semiconductor laser starts to produce gain, or negative absorption, which is essential for the lasing action.

In the following, we will evaluate the occupation probabilities in terms of the material parameters and excitation levels. Assuming a parabolic band-edge configuration in both conduction and valence bands, $E_c(k)$ and $E_v(k)$ are expressed as

$$E_c = E_g + \frac{\hbar^2 k^2}{2m_e^*} \tag{8.107a}$$

and

$$E_v = -\frac{\hbar^2 k^2}{2m_h^*} \tag{8.107b}$$

Here we assumed $E = 0$ at the valence band maximum. For direct transitions, $\Delta k = 0$, and the energy involved is expressed as

$$E = \hbar\omega - E_g = \frac{\hbar^2 k^2}{2m_e^*} + \frac{\hbar^2 k^2}{2m_h^*} = \frac{\hbar^2 k^2}{2m_r} \tag{8.108}$$

Thus,

$$E_c = E_g + \frac{\hbar^2 k^2}{2m_r}\left(\frac{m_r}{m_e^*}\right) = E_g + \left(\frac{m_r}{m_e^*}\right)E \tag{8.109a}$$

and

$$E_v = \left(\frac{m_r}{m_h^*}\right)E \tag{8.109b}$$

These band-edge energies are used to express $f_c(E)$ and $f_v(E)$ in terms of effective masses and quasi-Fermi levels as

$$f_c(E) = \frac{1}{1 + \exp[(E_c - F_c)/kT]} = \frac{1}{1 + \exp\{[E_g + (m_r/m_e^*)E - F_c]/kT\}} \tag{8.110a}$$

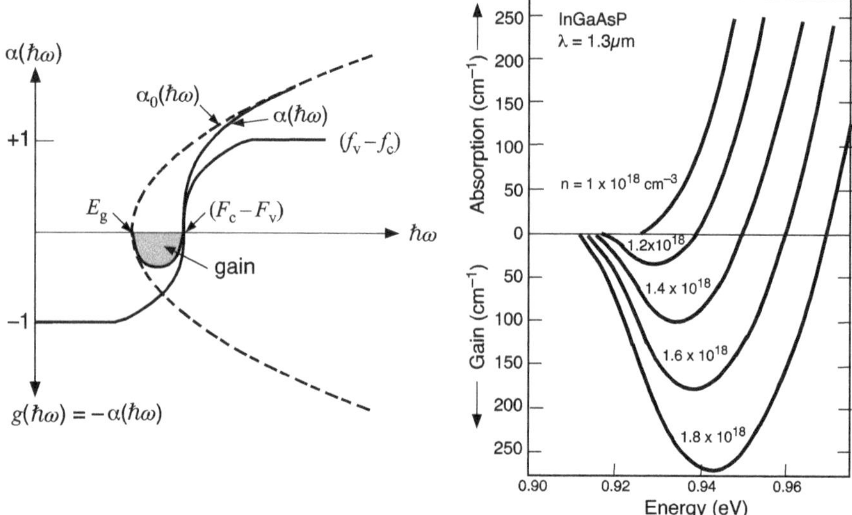

Fig. 8.29 **a** Gain-absorption curve as a function of energy. The dashed line is the intrinsic absorption curve. The Fermi–Dirac inversion factor is also plotted. **b** Gain curves of an InGaAsP laser under different injected carrier density conditions. The system reaches the transparency condition, $f_{\mathrm{v}} \leq f_{\mathrm{c}}$, at an injected density of $\geq 1 \times 10^{18}$ cm^{-1}. Reprinted with permission from [17] copyright AIP Publishing

$$f_{\mathrm{v}}(E) = \frac{1}{1 + \exp[(E_{\mathrm{v}} - F_{\mathrm{v}})/kT]} = \frac{1}{1 + \exp\{[-(m_{\mathrm{r}}/m_{\mathrm{h}}^{*})E - F_{\mathrm{v}}]/kT\}} \tag{8.110b}$$

Once the quasi-Fermi levels are determined from the injected carrier densities, both $f_{\mathrm{c}}(E)$ and $f_{\mathrm{v}}(E)$ can be solved as functions of E. Then, the gain $(-\alpha)$ of a specific energy $(\hbar\omega)$ is determined by multiplying the inversion factor by the intrinsic absorption coefficient, as shown in Fig. 8.29. As the quasi-Fermi level separation $(F_{\mathrm{c}} - F_{\mathrm{v}})$ increases with increasing carrier injection level, the peak of the gain curve grows quickly. In a properly designed cavity, once the peak gain value surpasses the total loss of the cavity, the lasing action starts.

8.4.4 Optical Absorption and Gain in Quantum Wells

The procedures of calculating gain curves in bulk semiconductors are applicable to quantum-well materials with two important modifications. First, the 2D system of quantum well has a step-like DOS profile, which should be included in the absorption coefficient calculation. The other major difference is the restrictions on allowed optical transitions. In bulk semiconductors, all vertical transitions between valence and conduction bands are allowed as long as $\Delta k = 0$. The optical transition in quantum wells has to follow the quantum mechanical 'selection rule' such that the quantum numbers of the initial and final states are the same, $n_i - n_f = \Delta n = 0$. For example (Fig. 8.30), from $n = 1$ heavy-hole (HH) state in the valence band, only the transition to $n = 1$ electron state in the conduction band is allowed. All other transitions are prohibited, since $\Delta n \neq 0$.

The intrinsic absorption coefficient of a multiple quantum-well system has the form of

$$\alpha_0^{2D}(\hbar\omega) \propto \frac{m_r}{\pi\hbar^2 L} N_w \sum_{m,n} \Theta(\hbar\omega - E_{mn}) \tag{8.111}$$

where L is the width of each quantum well, N_w is the total number of quantum wells, and Θ is the Heaviside step function. Figure 8.31 shows the experimental absorption spectrum of an undoped InGaAs/InAlAs, quantum well measured at 12, 100, and 300 K. The step-like features are clearly illustrated with pronounced exciton peaks near each step edge. Due to the strong binding energy, the exciton-related peaks are visible in the absorption spectrum even at room temperature. Another interesting feature is the observation of both HH and LH related excitons near the absorption step edges. In bulk semiconductors, due to the degeneracy of the HH and LH bands near the zone center, only a single exciton peak will be observed in the low-temperature absorption spectrum (Fig. 8.16). However, in a quantum well, the subbands of the heavy and light holes at the zone center are no longer degenerate and show two separate exciton peaks associated with the HH and LH bands, respectively.

Once we have determined the intrinsic absorption coefficient, the absorption in a quantum well can be calculated as

$$\alpha(\hbar\omega) = \alpha_0^{2D}(\hbar\omega)[f_v(E) - f_c(E)] \tag{8.112}$$

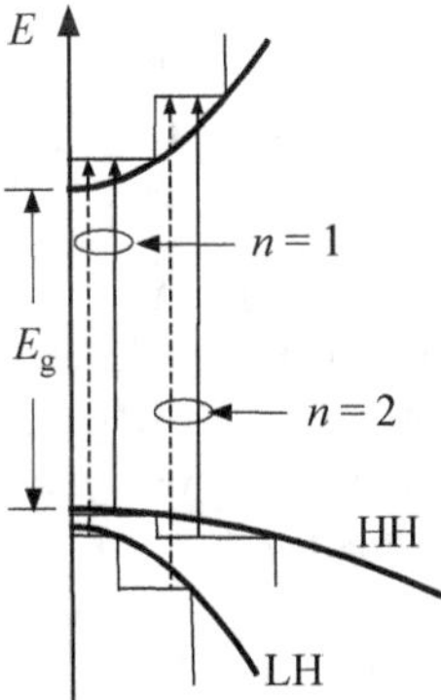

Fig. 8.30 Density of states and direct electronic transitions in a quantum well

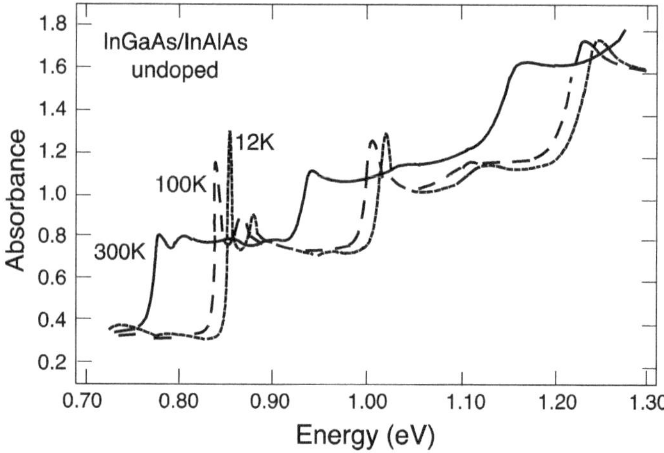

Fig. 8.31 Absorption spectra of an InGaAs/InAlAs quantum well measured at various temperatures. Reprinted with permission from [18] copyright IEEE

The construction of the absorption curve in a quantum well is illustrated in Fig. 8.32a. Because of the step-like feature of the intrinsic absorption coefficient, the ideal absorption/gain curve also contains sharp step edges. However, the intra-band carrier relaxation due to scattering within the quantum well can cause the broadening of the energy states and, thus, spectral broadening (Fig. 8.32b). The relaxation process is characterized by τ_{in}, the intra-band relaxation time of *e-e* or *h-h* interaction, or the reciprocal scattering probability. The energy peak broadening due to intra-band scattering, δE_{in}, is related to τ_{in} using the uncertainty principle, $\delta E_{\text{in}} = \hbar/\tau_{\text{in}}$. In the ideal relaxation-free case, $\tau_{\text{in}} \rightarrow \infty$. For GaAs, a short intra-band relaxation time of 0.1 ps is typically used. Therefore, deformation and reduction of the peak values take place in the gain spectrum. The spectral shape becomes smooth and broad in spite of the sharp step-like DOS.

8.5 Non-radiative Auger Recombination Processes

Electrons and holes in a direct bandgap semiconductor can also recombine non-radiatively. Non-radiative mechanisms include recombination at defects, surface recombination, and Auger recombination, among others. For long-wavelength semiconductor lasers with a narrow gap, however, the Auger process is generally the predominant non-radiative mechanism.

The Auger recombination process involves three particles (two electrons and one hole, or one electron and two holes) and four states as described below. In this process, the energy released during the band-to-band electron–hole recombination is transferred to another electron (or hole) which gets excited to a high-energy state

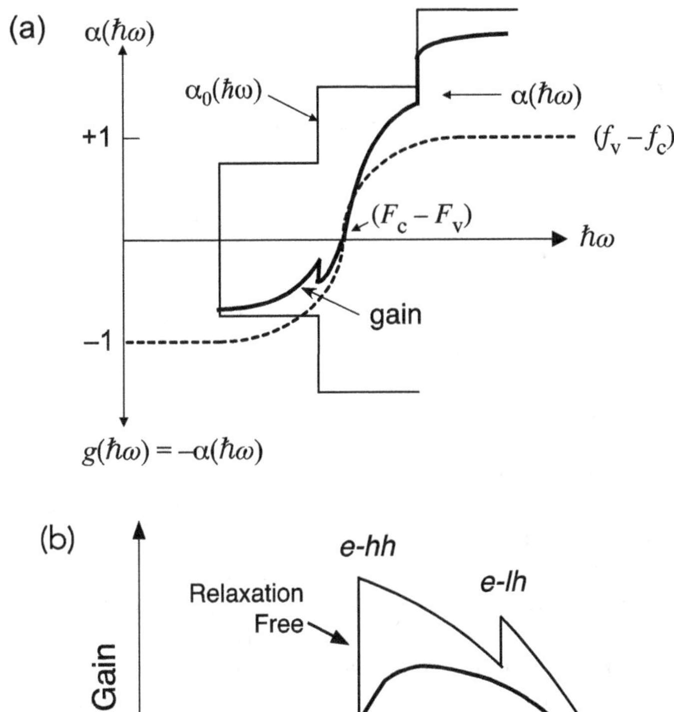

Fig. 8.32 **a** Gain-absorption curve as a function of energy in a quantum well. The staircase line is the intrinsic absorption curve. The Fermi–Dirac inversion factor is also plotted. **b** Gain curve of an ideal relaxation-free quantum well has sharp steps. When a short relaxation time of 0.1 ps is incorporated, the gain curve becomes smooth without sharp edges

in the band. This electron or hole then relaxes back to achieve thermal equilibrium by losing its energy to lattice vibrations or phonons. Thus, the Auger process is a non-radiative recombination process. Since there are three particles involved, the Auger recombination rate R_A may be approximately written as

$$R_A = Cn^3 \tag{8.113}$$

where n is the injected carrier density in an undoped semiconductor. It is useful to define the carrier lifetime τ_A for the Auger process as

$$\tau_A = n/R_A = 1/(Cn^2) \tag{8.114}$$

The quantity C is called the Auger coefficient. The inverse-square dependence of carrier lifetime on carrier density is used to identify the Auger effect experimentally.

In general, band-to-band Auger processes are characterized by a strong temperature dependence and bandgap energy dependence following $(kT/E_g)^{3/2}\exp(-E_g/kT)$. The Auger rate increases rapidly either for high temperatures or for narrow bandgap materials. In narrow bandgap semiconductors such as InAs and GaSb, where $E_g \approx \Delta_{so}$, Auger process should be strongly temperature dependent. For example, in an InGaAsP laser (~1.3 μm), the band-to-band Auger process dominates for temperatures exceeding $T = 160$ K. In large bandgap materials, the Auger process depends on the doping level and become significant in degenerated semiconductors. These dependencies arise from the laws of energy and momentum conservation that the four free particle states involved must satisfy. The energy taken by the third particle in Auger recombination is on the order of the bandgap, which corresponds to a large momentum close to $\sqrt{2mE_g}$. Conservation of momentum requires that the three-particle system must have large momentum before the interaction can take place, or a threshold energy E_T for the processes.

The band-to-band Auger processes in a direct-gap semiconductor are shown in Fig. 8.33. The three processes are labeled CHCC, CHLH, and CHHS where C stands for the conduction band and H, L, and S stand for heavy-hole, light-hole, and split-off valence bands, respectively. For example, the CHCC process involves two electrons and one hole and is dominant in n-type material. In this process, electron 1 (of energy E_1) makes a transition to empty state 1′ (of energy E'_1) and the excess energy

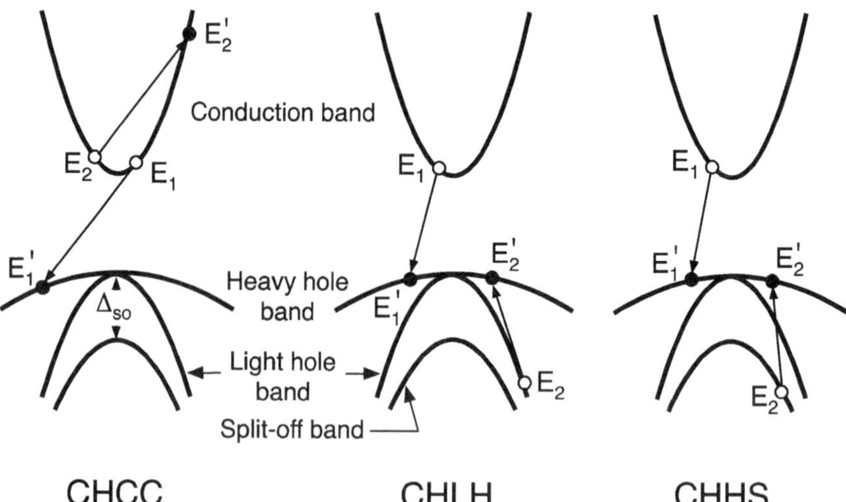

Fig. 8.33 Three types of band-to-band Auger processes. Electrons and holes are represented by closed circles and open circles, respectively

is transferred to electron 2 (of energy E_2), which is excited to state 2′ (energy E'_2). Electron 2′ loses its energy to optical phonons when it relaxes back to thermal equilibrium. On the other hand, CHHS and CHLH Auger processes are dominant in *p*-type semiconductors. The CHHS process involves one electron, one heavy hole, and a split-off-band hole. CHLH is similar to CHHS except that it involves a light hole. Under high-injection conditions commonly present in semiconductor lasers, all three mechanisms must be considered.

Problems

1. (a) The direct allowed transition in a direct bandgap semiconductor is described by a general equation of $\alpha(h\nu) = A^*\sqrt{h\nu - E_g}$. The proportional constant A^* can be expressed as a function of effective masses.

$$A^* \equiv \frac{q^2\left[2m_e^* m_{hh}^*/\left(m_e^* + m_{hh}^*\right)\right]^{3/2}}{\overline{n}ch^2 m_e^*}$$

   where $\overline{n}$ is the index of refraction and c is the speed of light. Calculate A^* for GaAs.

   (b) Using the absorption spectra shown in Fig. 8.12b, calculate the phonon energy near room temperature (291 K) in a high-purity Si crystal.

   (c) Using the absorption spectrum shown in Fig. 8.16, calculate the bandgap energy of GaAs at 186 K.

2. The active region of an n-$Al_{0.3}Ga_{0.7}As$/GaAs/p-$Al_{0.3}Ga_{0.7}As$ double-heterostructure (DH) diode laser has a 0.1 μm thick undoped GaAs active region. Under a forward bias, electrons and holes are injected into the GaAs active region from *n*- and *p*-sides of the DH structure, respectively. The increasing carrier concentration causes the quasi-Fermi level of electrons and holes to move toward the conduction band and valence band, respectively. Under the strong carrier injection, the *transparency condition* is reached when the separation in quasi-Fermi levels $(F_c - F_v) = E_g$.

   (a) Find the transparency carrier density n_{tr} for GaAs under the condition of equal injection of both types of carriers ($n = p$). You have to use Fermi–Dirac integral for both electrons and holes.

   (b) Plot the normalized gain, $g_n = \alpha(\hbar\omega)/\alpha_0(\hbar\omega)$, or Fermi–Dirac inversion factor $(f_v - f_c)$ as a function of energy between 1.3 and 1.6 eV at injection concentrations ($n = p$) of $n_0 = 1 \times 10^{18}$, 2×10^{18}, 4×10^{18}, and 6×10^{18} cm^{-3}. Note that the Fermi integral is required to determine quasi-Fermi levels of the active region.

   (c) Plot the gain as a function of energy between 1.3 and 1.6 eV at injection concentrations ($n = p$) similar to those listed in Part (b). Identify the injection concentration(s) for which the transparency condition has not been reached.

   (d) If the GaAs active region of the same DH laser is heavily doped, calculate the transparency condition by considering only electron injection with hole density fixed at the doping level of 6×10^{18} cm^{-3}. Plot the gain curve for injected electron density at 1.5× the transparency condition. Comment on whether this approach is advantageous over the structure with an undoped active region.

(e) Plot the equilibrium energy band diagram of the DH structure using *SimWindows*. Both the doping concentration and thickness of the n- and p-doped AlGaAs are at the same level of 10^{18} cm^{-3} and 1000 nm, respectively.

3. For erbium-doped fiber communication application, a 0.98 μm diode laser is used as the pumping source. The active region of the laser has a GaAs-$Ga_{0.8}In_{0.2}$As-GaAs strained quantum well. The bandgap energy of $Ga_xIn_{1-x}As$ is expressed as

$$E_g(\Gamma) = 0.358 + 0.589x + 0.477x^2$$

(a) Find the transparency carrier density n_{tr} for $Ga_{0.8}In_{0.2}As$ under the condition of neutral carrier injection ($n = p$). You have to use the Fermi–Dirac integral for both electrons and holes. Both the electron and hole effective masses of GaInAs vary linearly between GaAs and InAs.

(b) Plot the intrinsic absorption coefficient as a function of energy for the undoped $Ga_{0.8}In_{0.2}As$ up to 1.4 eV using $\alpha_0(\hbar\omega) = 10^4\sqrt{\hbar\omega - E_g}$.

(c) Plot the normalized gain, $g_n = \alpha(\hbar\omega)/\alpha_0(\hbar\omega)$, or Fermi–Dirac inversion factor $(f_v - f_c)$, as a function of energy between 1 and 1.4 eV at injection concentrations ($n = p$) of $n_0 = 5 \times 10^{17}$, 1.5×10^{18}, and 5×10^{18} cm^{-3}.

(d) Under steady-state conditions, the current density of a laser diode for carrier neutral injection ($n = p$) is $J \approx qL_zR(n)$, where L_z is the thickness of the undoped active region and $R(n) = B_rn^2$. Plot the normalized gain versus $(J/J_0) = (n/n_{tr})^2$ for $L_z = 100$ Å. J_0 is the transparency current density where $n = n_{tr}$.

(e) Evaluate J_0 for $B_r = 1.5 \times 10^{-10}$ cm^3/s.

References

1. M. Fox, *Optical Properties of Solids* (Oxford, 2001), p. 32
2. S. Adachi, J. Appl. Phys. **53**, 5863 (1982)
3. S. Adachi, J. Appl. Phys. **58**, R1 (1985); or H. C. Casey Jr., D.D. Sell, M.B. Panish, Appl. Phys. Lett. **24**, 63 (1974)
4. G.M. Laws, E.C. Larkins, I. Harrison, C. Molloy, D. Somerford, J. Appl. Phys. **89**, 1108 (2001)
5. D. Brunner et al., J. Appl. Phys. **82**, 5090 (1997)
6. U. Tisch, B. Meyler, O. Katz, E. Finkman, J. Salzman, J. Appl. Phys. **89**, 2676 (2001)
7. G.E. Stillman, V.M. Robbins, N. Tabatabaie, IEEE Trans. Electron Devices **31**, 1643 (1984)
8. G.G. MacFarlane, T.P. McLean, J.E. Quarrington, V. Roberts, Phys. Rev. **111**, 1245 (1958)
9. H.C. Casey Jr., D.D. Sell, K.W. Wecht, J. Appl. Phys. **46**, 250 (1975)
10. M.D. Sturge, Phys. Rev. **127**, 768 (1962)
11. P.J. Dean, D.G. Thomas, Phys. Rev. **150**, 690 (1966)

12. R. Dingle, W. Wiegmann, C.H. Henry, Phys. Rev. Lett. **33**, 827 (1974)
13. J.J. Hopfield, P.J. Dean, D.G. Thomas, Phys. Rev. **158**, 748 (1967)
14. J.N. Baillargeon, *Gas Source Molecular Beam Epitaxy of Aluminum Gallium Indium Phosphide for Visible Spectrum Light Emitting Diode*. Ph.D. thesis, University of Illinois at Urbana-Champaign, 1990
15. W. van Roosbroeck, W. Shockley, Phys. Rev. **94**, 1558 (1954)
16. H.C. Casey Jr., F. Stern, J. Appl. Phys. **47**, 631 (1976)
17. N.K. Dutta, J. Appl. Phys. **51**, 6095 (1981)
18. G. Livescu, D.B.A. Miller, D.S. Chemla, M. Ramaswamy, T.Y. Chang, N. Sauer, A.C. Gossard, J.H. English, IEEE Quantum Electron. **24**, 1677 (1988)

Further Reading

1. H.C. Casey, M.B. Panish, *Heterostructure Lasers: Part A, Fundamental Principles* (Academic Press, 1978)
2. S. Wang, *Fundamentals of Semiconductor Theory and Device Physics* (Prentice-Hall, 1989)
3. J.I. Pankove, *Optical Processes in Semiconductors* (Prentice-Hall, 1971)
4. M. Fox, *Optical Properties of Solids* (Oxford, 2001)
5. N.K. Dutta, R.J. Nelson, J. Appl. Phys. **53**, 74 (1982)

Chapter 9
Heterostructure Electronic Devices

Abstract Compound semiconductors are excellent candidates for high-speed device applications due to their high electron mobility, the ability to form heterostructures, and the availability of semi-insulating substrates. However, the multicomponent nature and the lack of robust native oxides make the fabrication of compound semiconductor-based metal-oxide-semiconductor field-effect transistors (MOSFETs) an extremely difficult task. Instead, GaAs-based metal–semiconductor field-effect transistors (MESFET) were developed initially. A MESFET utilizes a metal–semiconductor Schottky barrier to replace the MOS gate structure. The charge carriers in the active region (channel) are separated spatially from the control (gate) electrode by a depletion layer formed on the semiconductor surface. Nevertheless, due to the nature of the Schottky barrier and high electron mobility, only *n*-channel MESFETs were studied and the devices are leaky at high positive biases. With the advancements of epitaxial growth technology and new device concepts, efficient advanced devices were developed. Among many three-terminal device structures, the modulation-doped field-effect transistor (MODFET), or high-electron-mobility transistor (HEMT), and the heterojunction bipolar transistor (HBT) are the most widely used high-speed devices. In this chapter, after a brief review of the MESFET, the development and working principles of HEMTs and HBTs are discussed in great detail. Then the key concepts of achieving an oxide-semiconductor interface with low interface state density necessary for the fabrication of high performance III–V MOSFETs are discussed.

The HEMT is a voltage-controlled majority carrier device. It shares the same device geometry as the MESFET except for the details of the active region. The control electrode (gate) is coupled to the active region of the device through a capacitor. The conduction carriers, in the form of the two-dimensional electron gas (2DEG), in the active region are generated by modulation doping at the heterostructure interface.

The structure of the HBT is similar to that of the homostructure bipolar junction transistor (BJT), with the emitter–base junction replaced by a heterojunction. With properly designed heterojunctions, the device parameters such as doping level and layer thickness can be individually optimized for high-speed and high-gain operations. The HBT is a current-controlled minority carrier device with the charged

K. Y. Cheng, *III–V Compound Semiconductors and Devices*, Graduate Texts in Physics,
https://doi.org/10.1007/978-3-030-51903-2_9

Table 9.1 Comparison of BJT and FET

	Carriers	Transport	Output control	Input impedance
BJT	Minority	Diffusion	Base current	Low
FET	Majority	Drift	Gate voltage	High

carriers being separated energetically by the energy barriers. The control electrode (base) is resistively coupled to the active device region. The major device attributes of FETs and BJTs are compared in Table 9.1.

In late 1990, the inversion-mode GaAs MOSFET with a low interface trap density was finally achieved using in situ electron beam evaporated $Ga_2O_3(Gd_2O_3)$ dielectric film on MBE GaAs. Next, using ex situ atomic layer deposited high-κ Al_2O_3 and HfO_2 films as the gate dielectric, similar devices were demonstrated on GaAs and other III–V materials. Further refinement of oxide deposition methods and development of additional high-κ gate dielectrics were exploited in the following two decades. At the moment, III–V MOSFET manufacturing technologies are still under development.

9.1 Metal–Semiconductor Field-Effect Transistors (MESFETs)

9.1.1 Basic Operation Principles

As a prelude to the HEMT, the device characteristics of the MESFET are discussed first. The operating principle of the MESFET is identical to that of a junction FET, and its structure is also similar except that the *p*–*n* junction is replaced by the Schottky barrier. Most MESFETs use *n*-type III–V compound semiconductors for the conduction channel because of their high electron mobility and large Schottky barrier height. To minimize the parasitic capacitance, semi-insulating substrates are usually used. A schematic diagram of the MESFET is shown in Fig. 9.1. The active region of the device has a conduction channel thickness a, a gate length L, and a gate width Z. The depletion region has a non-uniform thickness of $W(x)$ along the channel under a large drain bias voltage.

Under normal operation, the source is grounded, the gate is reverse-biased to modulate the conduction channel, and the drain contact is biased at a proper voltage to collect majority carriers. The effect of the applied drain voltage, V_D, with the gate open, on the current–voltage relationship is investigated first. When a small positive V_D is applied with $V_{GS} = 0$, the channel of the MESFET is essentially a uniform resistor with a cross-sectional area determined by the width of the device (Z) and the non-depleted channel thickness ($a - W$). The channel current I_D is a linear function

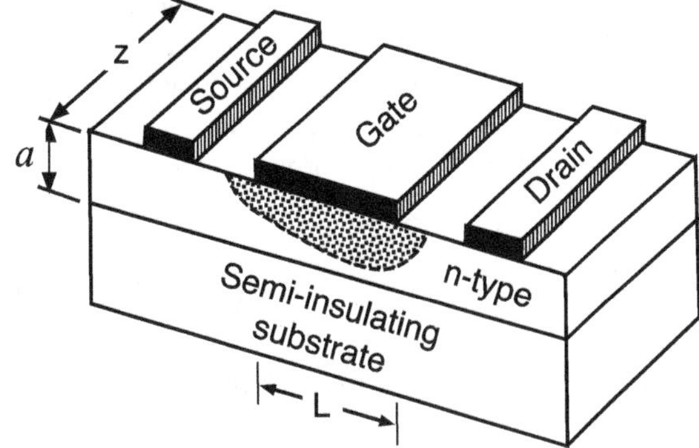

Fig. 9.1 Schematic structure and dimension of a metal–semiconductor FET. The shaded region is the depletion region under the gate electrode

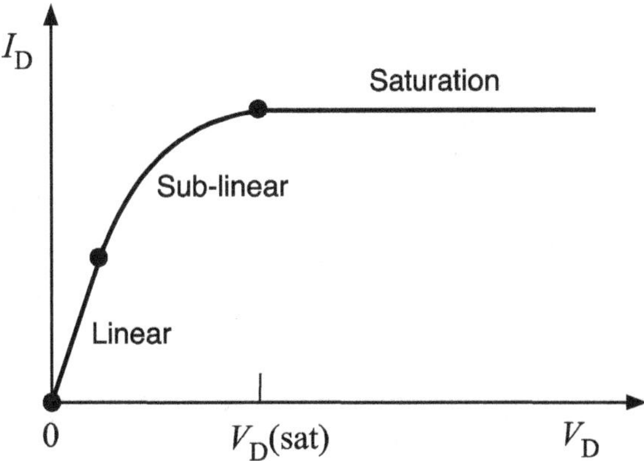

Fig. 9.2 General form of the current–voltage characteristics of a MESFET. The pinch-off point defines the saturation drain voltage, V_D(sat)

of the drain voltage as shown in Fig. 9.2. The depletion width along the conduction channel has a constant value of

$$W = \sqrt{\frac{2\epsilon}{qN_d} V_0} \tag{9.1}$$

where V_0 is the contact potential of the Schottky barrier gate, N_d is the carrier concentration of the channel, and ϵ is the permittivity of the semiconductor.

When a moderate V_D of a few tenths of a volt is applied with $V_{GS} = 0$, the V_D simultaneously reverse biases the gate–drain metal–semiconductor junction and widens the depletion region near the drain contact. The depletion region width increases along the conduction channel from source to drain. The depletion width closest to the drain contact becomes

$$W = \sqrt{\frac{2\epsilon}{qN_d} (V_0 + V_D)} \tag{9.2}$$

The narrowing of the conduction channel leads to an increased channel resistance. The increasing of channel current I_D deviates from the linear relationship with increasing V_D. Continuing to increase V_D causes an ever-increasing depletion of the channel. Eventually, the conduction channel vanishes near the drain contact, where the depletion width approaches the channel width for $V_{GS} = 0$.

$$W = \sqrt{\frac{2\epsilon}{qN_d}[V_0 + V_D(\text{sat})]} = a \tag{9.3}$$

The drain voltage V_D becomes the saturation drain voltage $V_D(\text{sat})$ and can be represented as

$$V_D(\text{sat}) = \frac{qa^2N_d}{2\epsilon} - V_0 \tag{9.4}$$

At the saturation voltage and beyond, the drain current becomes saturated, $I_D(\text{sat})$ and does not increase with V_D.

Next, we will investigate the gate bias effect. When the gate-to-channel junction of a MESFET is reverse-biased, the depletion width gets wider, and $V_D(\text{sat})$ will decrease.

$$V_D(\text{sat}) = \frac{qa^2N_d}{2\epsilon} - V_0 + V_{GS} \tag{9.5}$$

Since V_{GS} is negative for reverse bias of the gate–channel junction, $V_D(\text{sat})$ is decreased as shown in Fig. 9.3.

The depletion width W eventually will reach across the channel thickness a when the gate reverse bias is further increased. The gate voltage at which the conduction channel is pinched off when $V_D = 0$ is called the *threshold voltage* V_T. For a gate bias of V_{GS} and $V_D = 0$, the depletion width is expressed as

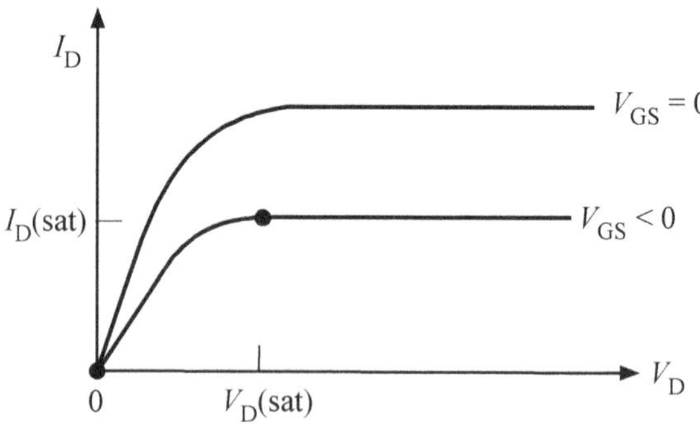

Fig. 9.3 Modification of current–voltage characteristics for $V_G < 0$

$$W = \sqrt{\frac{2\epsilon}{qN_d}(V_0 - V_{GS})} \tag{9.6}$$

At the threshold voltage $V_{GS} = V_T$, $W = a$. This leads to

$$a^2 = \frac{2\epsilon}{qN_d}(V_0 - V_T) \tag{9.7}$$

$$V_T = V_0 - \frac{qa^2N_d}{2\epsilon} = V_0 - V_P \tag{9.8}$$

We define V_P as the *pinch-off voltage*, which is the potential across the depletion region at the threshold condition ($W = a$).

$$V_P = \frac{qa^2N_d}{2\epsilon} \tag{9.9}$$

The depletion width in a MESFET with $V_{GS} = 0$ can be either greater than or less than the channel thickness because of the contact potential. If $W < a$, a channel exists with zero gate bias. The device is operated in the *depletion mode*. On the other hand, if $W > a$, the depletion region extends across the channel in equilibrium, and no significant channel current can flow. A positive gate voltage must be applied to reduce the depletion width to a value smaller than the channel thickness a. The device is operated in the *enhancement mode*.

Figure 9.4 shows the energy band diagrams of a depletion-mode MESFET under $V_D = 0$ and various V_{GS}. A reverse bias on the gate ($V_{GS} < 0$) is required to modulate the channel width. The threshold voltage V_T can also be termed the 'turn-off' voltage of the device. For enhancement-mode MESFETs, a positive gate bias is required to open the channel. The energy band diagrams for various bias conditions ($V_D = 0$) are shown in Fig. 9.5.

9.1.2 Current–Voltage Characteristics

A one-dimensional model is used to develop the DC current–voltage (I–V) characteristics of a MESFET. It is assumed that no breakdown will be induced before the channel is pinched off and the device has a doping density of N_d in the channel. Furthermore, the active region of the MESFET has a channel length L, junction width Z, channel depth a, depletion width $W(x)$, and conducting channel width $h(x)$. The origin of the x-axis is located at the edge of the gate adjacent to the source. A schematic of the MESFET structure and the corresponding voltage drop due to the drain bias is shown in Fig. 9.6.

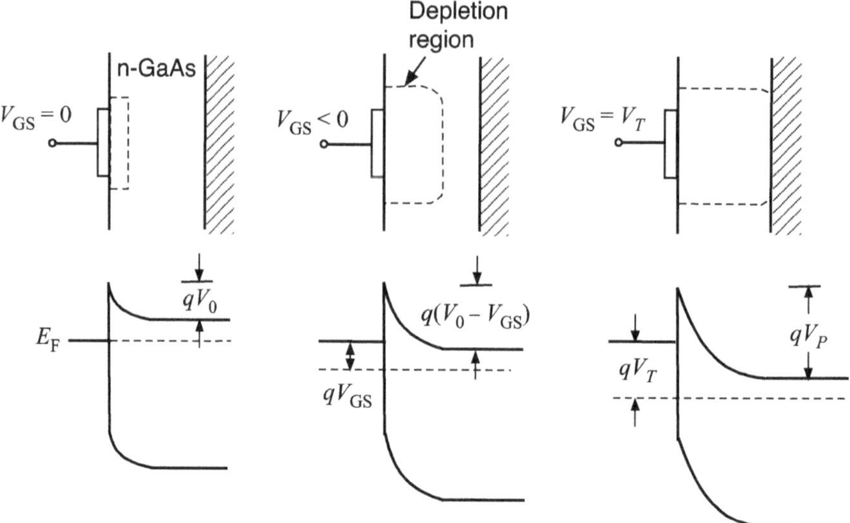

Fig. 9.4 Depletion region profile under the gate and energy band diagrams of depletion-mode MESFET under $V_D = 0$ and various V_{GS} (negative)

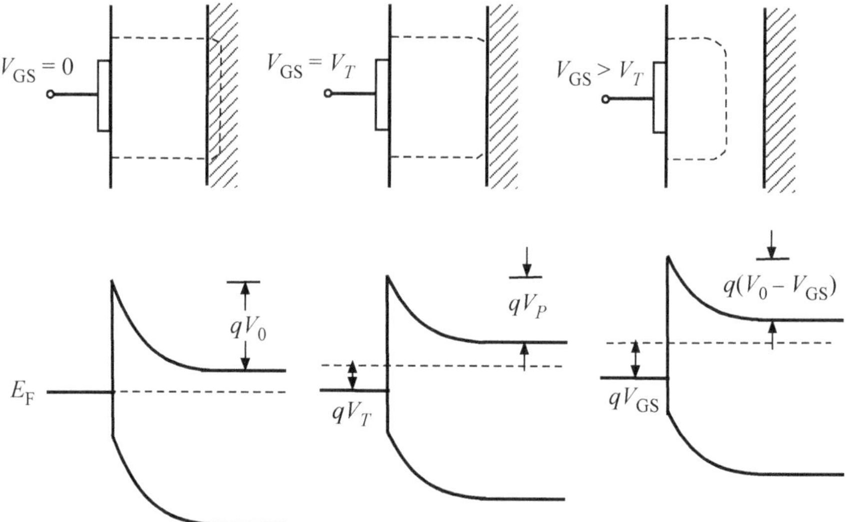

Fig. 9.5 Depletion region profile under the gate and energy band diagrams of enhancement-mode MESFET under $V_D = 0$ and various V_{GS} (positive)

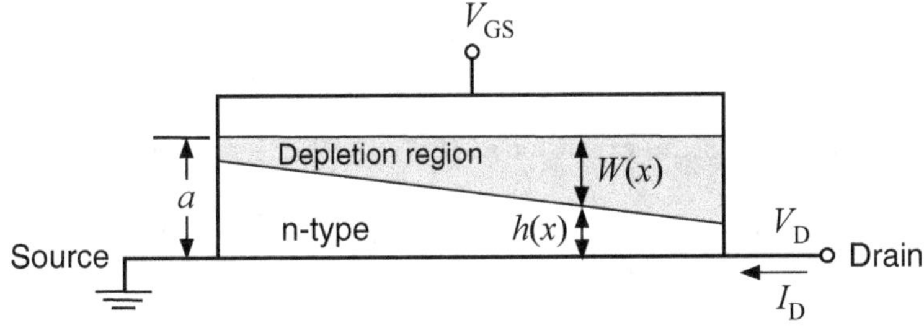

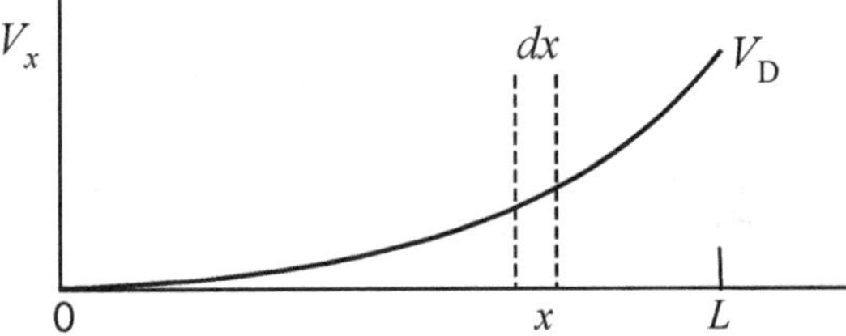

Fig. 9.6 Schematic diagram of the conducting channel, dimensions, and potential drop along the channel

The current at position x in the channel is described as

$$I_D(x) = A(x)\sigma F(x) = \frac{Zh(x)}{\rho}\frac{dV_x}{dx} \tag{9.10}$$

where σ is the channel conductivity, ρ is the resistivity of the channel material, $h(x) = a - W(x)$, $W = \sqrt{(2\epsilon/qN_d)(V_0 - V_{GS} + V_x)}$ and $V_P = qa^2N_d/2\epsilon$. Since the drain current is a constant and independent of x, we can solve I_D by integration.

$$I_D\int_0^L dx = I_D L = \frac{Za}{\rho}\int_0^{V_0}\left[1 - \sqrt{\frac{V_0 + V_x - V_{GS}}{V_P}}\right]dV_x$$

$$= \frac{Za}{\rho}\left[V_x - \frac{2}{3}\frac{(V_0 + V_x - V_{GS})^{3/2}}{\sqrt{V_P}}\right]_0^{V_D}$$

$$\therefore \quad I_D = G_0V_P\left[\frac{V_D}{V_P} - \frac{2}{3}\left(\frac{V_0 + V_D - V_{GS}}{V_P}\right)^{3/2} + \frac{2}{3}\left(\frac{V_0 - V_{GS}}{V_P}\right)^{3/2}\right] \tag{9.11}$$

where G_0 is the channel conductance at $W(x) = 0$ and $G_0 \equiv Za/\rho L$. This is the complete I–V characteristic equation for all bias conditions.

In the low bias region, where $V_D \ll V_0 - V_{GS}$, the I–V characteristics have a linear relationship. We can first rewrite I_D in the following form:

$$I_{\mathrm{D}} = G_0\left\{V_{\mathrm{D}} - \frac{2}{3\sqrt{V_P}}(V_0 - V_{\mathrm{GS}})^{3/2}\left[\left(1 + \frac{V_{\mathrm{D}}}{V_0 - V_{\mathrm{GS}}}\right)^{3/2} - 1\right]\right\} \tag{9.12}$$

Then, using Taylor series expansion of $(1+x)^n \cong (1+nx)$ for $x \ll 1$, we have

$$\begin{aligned} I_{\mathrm{D}} &\approx G_0\left\{V_{\mathrm{D}} - \frac{2}{3\sqrt{V_P}}(V_0 - V_{\mathrm{GS}})^{3/2}\left[\left(1 + \frac{3V_{\mathrm{D}}/2}{V_0 - V_{\mathrm{GS}}}\right) - 1\right]\right\} \\ &= G_0 V_{\mathrm{D}}\left[1 - \frac{\sqrt{V_0 - V_{\mathrm{GS}}}}{\sqrt{V_P}}\right] \propto V_{\mathrm{D}} \end{aligned} \tag{9.13}$$

The drain current simply increases linearly with drain bias voltage.

At the other end of the spectrum, the conduction current saturates at high drain bias when the channel is pinched off. The I–V relationship described above is only valid up to pinch-off, where $V_P = V_{\mathrm{D}}(\mathrm{sat}) + V_0 - V_{\mathrm{GS}}$. Using this pinch-off voltage relation in (9.11), the saturation current is obtained as

$$I_{\mathrm{D}}(\mathrm{sat}) = G_0 V_P\left[\frac{V_{\mathrm{D}}(\mathrm{sat})}{V_P} - \frac{2}{3} + \frac{2}{3}\left(\frac{V_0 - V_{\mathrm{GS}}}{V_P}\right)^{3/2}\right] \tag{9.14}$$

This equation can be further reduced by replacing $V_{\mathrm{D}}(\mathrm{sat})/V_P$ with $1 - (V_0 - V_{\mathrm{GS}})/V_P$.

$$I_{\mathrm{D}}(\mathrm{sat}) = G_0 V_P\left[\frac{1}{3} - \left(\frac{V_0 - V_{\mathrm{GS}}}{V_P}\right) + \frac{2}{3}\left(\frac{V_0 - V_{\mathrm{GS}}}{V_P}\right)^{3/2}\right] \tag{9.15}$$

The typical I–V characteristics of an n-channel depletion-mode MESFET are shown in Fig. 9.7. The drain current flows for V_{GS} are less negative than the threshold voltage V_T. At sufficiently high drain bias, V_B, avalanche breakdown occurs between the gate and drain and the drain current increases abruptly.

9.1.3 Transconductance and Equivalent Circuit of MESFET

One of the figures of merit of the MESFET is the transconductance, g_m, which describes the effect of the input voltage change on the output current. In the equivalent circuit model, the intrinsic transconductance g_{mi} times the input voltage represents a current source. Under a fixed drain bias, the change in gate voltage leads to a depletion layer width change. Because the space charge in the depletion region changes by ΔQ_{G} in a time interval of τ, a drain current change is induced.

$$\Delta I_{\mathrm{DS}} = \frac{\Delta Q_{\mathrm{G}}}{\tau} \quad \text{or} \quad \left.\frac{\partial I_{\mathrm{DS}}}{\partial Q_{\mathrm{G}}}\right|_{V_{\mathrm{DS}}} = \frac{1}{\tau} \tag{9.16}$$

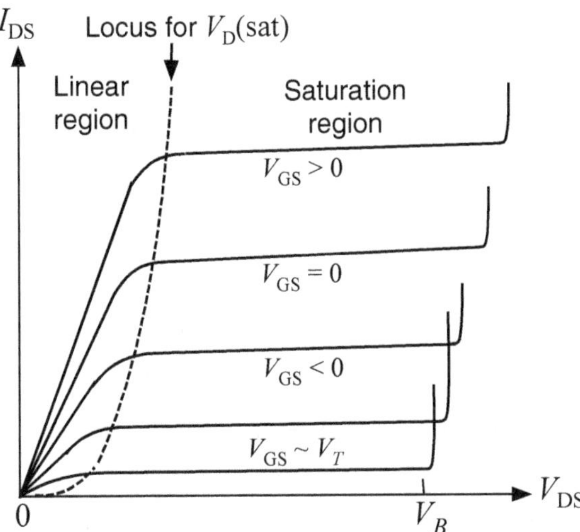

Fig. 9.7 Typical current–voltage characteristics of a depletion-mode MESFET. Breakdowns occur at the breakdown voltages V_B

The rate of conduction current change with respect to the charge change in the depletion region defines a carrier transient time (τ) crossing the conduction channel. The intrinsic transconductance, g_{mi}, is then defined as

$$g_{mi} \equiv \left.\frac{\partial I_{DS}}{\partial V_{GS}}\right|_{V_{DS}} = \left.\frac{\partial I_{DS}}{\partial Q_G}\right|_{V_{DS}} \cdot \left.\frac{\partial Q_G}{\partial V_{GS}}\right|_{V_{DS}} = \frac{C_G}{\tau} \tag{9.17}$$

C_G is the total gate capacitance and $C_G = C_{GS} + C_{DG}$, where C_{GS} and C_{DG} are the gate-to-source capacitance and drain-to-gate capacitance, respectively. To increase g_{mi}, we have to reduce τ. The minimization of τ can be achieved either by using materials with high drift velocity or by minimizing the channel length through the reduction of gate length. The approach of increasing the total gate capacitance to achieve high g_{mi} is not practical, since it will limit the frequency response of the device.

Next, we examine the property of the drain terminal. For a fixed gate bias, a change in V_{DS} will also change the space charge in the depletion region, ΔQ_G, which defines a current change, ΔI_{DS}, and leads to a drain-to-gate capacitance C_{DG}.

$$C_{DG} \equiv \left.\frac{\partial Q_G}{\partial V_{DS}}\right|_{V_{GS}} \tag{9.18}$$

The current–voltage change further defines a drain-to-source output conductance (G_o):

$$G_o \equiv \left.\frac{\partial I_{DS}}{\partial V_{DS}}\right|_{V_{GS}} = \left.\frac{\partial I_{DS}}{\partial Q}\right|_{V_{GS}} \cdot \left.\frac{\partial Q}{\partial V_{DS}}\right|_{V_{GS}} \tag{9.19}$$

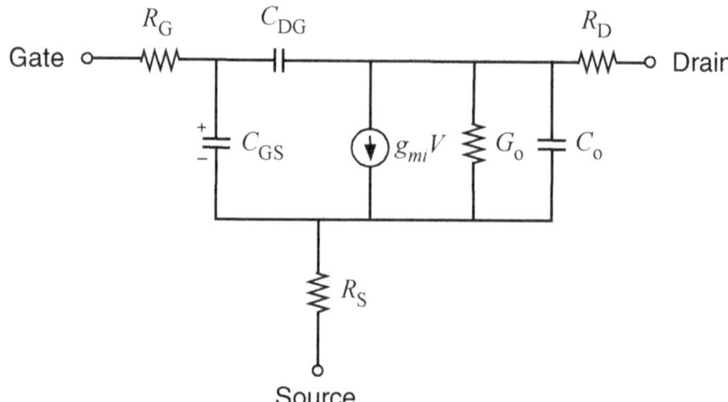

Fig. 9.8 Intrinsic equivalent circuit model of the MESFET including terminal series resistances

where Q is the total charge in conduction channel.

Assuming the carrier velocity υ is constant in the conduction channel of an effective length L_{eff}, $\Delta I_{\mathrm{DS}} \sim \Delta Q/\Delta t$, and $\Delta t \sim L_{\mathrm{eff}}/\upsilon$, the right-hand side of the output conductance (9.19) can be expressed as

$$\left.\frac{\partial I_{\mathrm{DS}}}{\partial Q}\right|_{V_{\mathrm{GS}}} \cong \frac{\upsilon}{L_{\mathrm{eff}}} \quad \text{and} \quad \left.\frac{\partial Q}{\partial V_{\mathrm{DS}}}\right|_{V_{\mathrm{GS}}} \equiv C_{\mathrm{o}} \tag{9.20}$$

C_{o} defines an output capacitance.

$$\therefore\ G_{\mathrm{o}} = \frac{C_{\mathrm{o}}\upsilon}{L_{\mathrm{eff}}} \tag{9.21}$$

To achieve large G_{o}, high carrier mobility and small gate length are necessary.

Figure 9.8 shows a simple intrinsic equivalent circuit model of the MESFET that ignores all parasitic elements except the terminal series resistances R_{G}, R_{S}, and R_{D}.

9.1.4 High-Speed Figure of Merit

The high-frequency performance of a MESFET is usually characterized by parameters related to its ultimate operation frequency. However, these parameters are difficult to measure due to the fact that the performance of the device under testing can depend as much on the measurement circuit as on the device itself. As a result, high-frequency transistors are commonly characterized by frequency-dependent gains, deduced from network-analyzer measurements. The high-frequency performance of the device measured in this way is reasonably independent of the circuit in which it is placed. There are two gains commonly used to deduce the small signal, linear-operation figure of merit of the device. One of these gains is h_{21}, which is the forward

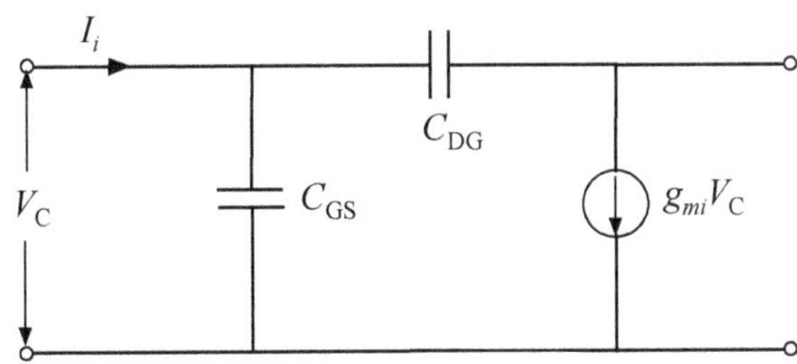

Fig. 9.9 Intrinsic equivalent circuit of a MESFET for frequency response analysis

current gain with the output short-circuited. The frequency at which this gain extrapolates to one is the *unity short-circuit current gain frequency* f_T. Consider the simple intrinsic equivalent circuit of Fig. 9.9. This frequency can be derived as the frequency at which the magnitude of the input current I_i equals the magnitude of the ideal output current $g_{mi}V_C$ of the transistor. When the output is short-circuited,

$$|I_i| = |i\omega(C_{GS} + C_{DG})V_C| = 2\pi f_T(C_{GS} + C_{DG})V_C = g_{mi}V_C \tag{9.22}$$

$$f_T = \frac{1}{2\pi}\left(\frac{g_{mi}}{C_{GS} + C_{DG}}\right) = \frac{g_{mi}}{2\pi C_G} = \frac{1}{2\pi\tau} \tag{9.23}$$

This equation shows that f_T is a measure of a quantity that is fundamental to the device—the delay time of the current through the conduction channel.

The other important figure of merit of the transistor is the *maximum frequency of oscillation*, f_{max}. At this frequency, the unilateral power gain (U) of the device becomes unity. It is the highest frequency at which the transistor can amplify the power of a signal and can be expressed as

$$f_{max} \approx \frac{f_T}{2\sqrt{G_o(R_G + R_S) + 2\pi f_T C_{DG} R_G}} \tag{9.24}$$

The f_{max} of a MESFET may be either above or below its f_T, depending on the specific circuit values in the above equation.

9.1.5 MESFET Fabrication and Performance

A typical fabrication sequence for a high-speed MESFET is shown in Fig. 9.10. The process was designed to maximize the cutoff frequency of the transistor by minimizing the contact resistances, reducing the parasitic in the active conduction channel, and shortening the gate length. An n^+ contact layer is added on top of the conduction channel layer outside the gate region to reduce the source and drain contact resistances R_S and R_D. In addition, a thick channel layer outside the gate area is used to minimize the surface depletion-induced additional resistance. To maintain a

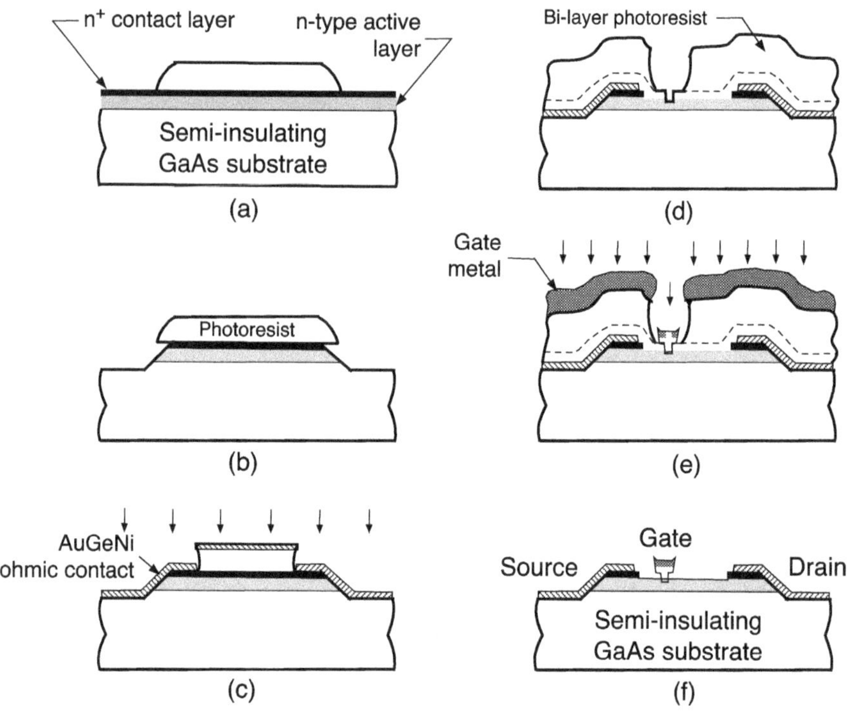

Fig. 9.10 Typical fabrication sequence of submicron GaAs MESFET using the double-recess process. **a** Epitaxy and using photoresist (PR) to define device active area, **b** mesa etch for isolation, **c** source and drain ohmic contacts deposition, **d** channel recess etch and gate recess etch using bilayer PR process, **e** T-gate metal deposition, and **f** PR liftoff to complete MESFET process

short gate length while minimizing the gate contact resistance R_{G}, the usual practice is to use a 'T' gate design where a thick top metal contact with a narrow base is adopted. The gate is situated in a recess of the channel layer in order to control the threshold voltage, V_T. The gate recess is usually off-centered and located near the source contact to minimize R_{S}. The gate–drain separation, L_{GD}, is designed to be greater than the depletion width at the breakdown voltage.

The fabricated MESFETs are characterized by measuring their I–V characteristics, g_m, f_{T}, and $f_{\max}$. Figure 9.11 shows the I–V characteristics of a $0.12 \times 50\ \mu\mathrm{m}^2$ dual-gate GaAs depletion-mode MESFET. The drain currents show weak saturation for the negative gate bias voltages. This is a common behavior called *short-channel effect* in MESFETs with gate length less than $0.25\ \mu$m.

The short-channel effect is due to the less efficient inter-valley transfer of electrons from the Γ-valley to the L-valley at high fields in short-channel devices. Thus, most electrons stay in the Γ-valley when reaching the drain and show as a weaker

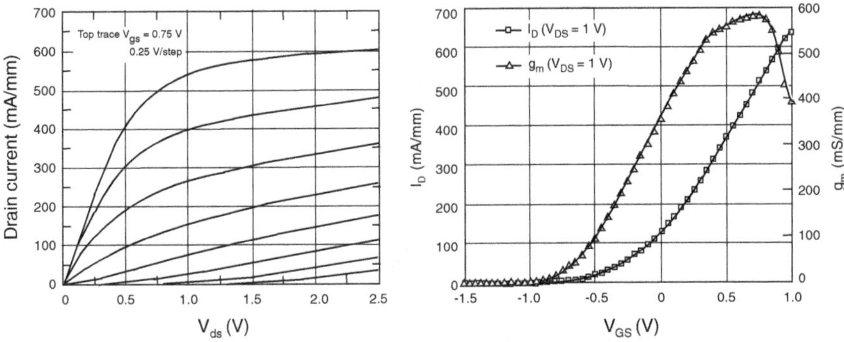

Fig. 9.11 Measured I_D-V_{DS} characteristics and channel current and transconductance versus gate bias of a dual-gate (0.12 μm × 50 μm) GaAs MESFET. *Courtesy* M. Feng, University of Illinois at Urbana-Champaign

velocity saturation. The parasitic leakage through the semi-insulating substrate also contributes to the lack of current saturation for the more negative gate-voltage region. The transconductance g_m is deduced from the drain current variation under various gate bias voltages. Note that the measured g_m is the extrinsic transconductance including the effect of R_S and expressed as

$$g_m = \frac{g_{mi}}{1 + g_{mi} R_S} \tag{9.25}$$

The maxima f_T and f_{max} of the MESFET are estimated to be 110 and 235 GHz, respectively, and shown in Fig. 9.12. These values are deduced from the measured low-frequency values of h_{21} and U_{max} up to 40 GHz using a roll-off of −6 dB/octave.

9.2 Modulation Doping and Two-Dimensional Electron Gas (2DEG)

As shown in Fig. 4.7, the carrier mobility of a semiconductor is determined by the sum of inverse mobility components (μ_i) due to different scattering mechanisms, i.e., $\mu^{-1} = \Sigma(\mu_i)^{-1}$. At a fixed temperature, the lowest mobility component dominates the total carrier mobility. In binary compound semiconductor alloys, the major scattering processes are the optical phonon (polar) scattering at high temperature, the ionized impurity scattering at low temperature, the acoustic phonon scattering due to piezoelectric field, and the acoustic phonon scattering due to deformation potential. With the exception of the impurity scattering, the effect of all scattering mechanisms on electron mobility can be reduced at low temperatures. The total temperature-dependent electron mobility increases with decreasing temperature and reaches a maximum value before decreasing with further reducing temperature. One

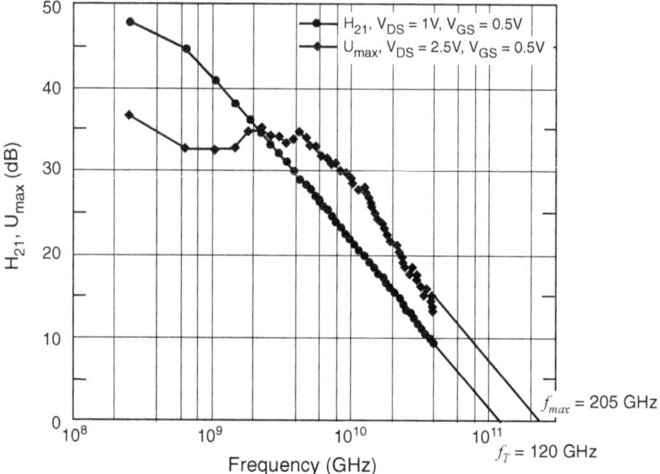

Fig. 9.12 Measured unity current gain frequency f_T (H_{21}) and maximum cutoff frequency f_{max} of the same dual-gate (0.12 μm × 50 μm) GaAs MESFET characterized in Fig. 9.11. *Courtesy* M. Feng, University of Illinois at Urbana-Champaign

way to reduce the ionized impurity scattering at low temperature is to use high-purity materials, as evidenced in Fig. 4.7, where there are fewer ionized impurities. However, for most high-speed device applications, certain doping concentrations have to be maintained. Therefore, this option is not practical. In 1978, R. Dingle and colleagues at Bell Laboratories invented a modulation-doping technique to eliminate the influence of ionized impurity scattering in heterojunction superlattices [1]. By separating the physical location of ionized dopants from generated electrons across the heterojunction, with ionized impurities on the large bandgap semiconductor side and generated conduction electrons on the undoped small bandgap semiconductor side, electron mobility much higher than the equivalent bulk material was achieved at low temperature without sacrificing the carrier concentration. Since these conduction electrons occupy a thin layer ($\leq$100 Å) in the triangular quantum well formed at the heterojunction interface, they can only move easily close to and parallel to the heterojunction interface. These electrons form an electron cloud, or a pseudo-two-dimensional electron gas (2DEG), in the modulation-doped heterostructure. Based on this concept, a new type of FET, the high-electron-mobility transistor (HEMT), was derived.

9.2.1 Modulation-Doped (MD) Heterostructures

The modulation-doping concept was first developed utilizing heterojunction superlattices as shown in Fig. 9.13. In a superlattice structure with type-I band alignment, quantum wells are formed in both conduction and valence bands. Under equilibrium

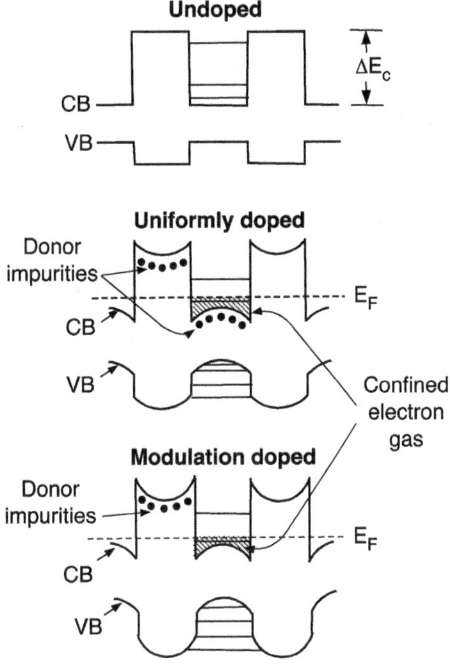

Fig. 9.13 Energy band diagrams for *n*-doped and undoped GaAs-$Al_xGa_{1-x}As$ superlattices. Reprinted with permission from [1], copyright AIP Publishing

conditions, a continuous Fermi level throughout the superlattice causes appreciable band bending, leading to the formation of pseudotriangular quantum wells at heterojunctions. When the superlattice is either uniformly doped (UD) or modulation-doped (MD) with donor impurities, due to the high potential energy associated with the barrier layer relative to the quantum well, electrons associated with donor impurities in barriers will move into quantum wells. In both UD and MD superlattice structures, the carriers confined to the pseudotriangular quantum well form a 2DEG. The ionized impurity scattering in the conduction channel is removed in MD structures, where 2DEG conduction electrons and their parent donor impurities are spatially separated from each other in an irreversible manner. The electron mobility of the 2DEG suffers no ionized impurity scattering, leading to a much higher mobility, especially at low temperatures. Figure 9.14 shows the temperature-dependent electron mobility of bulk, UD and MD GaAs structures. The UD and MD structures used the AlGaAs/GaAs material system. The electron mobility enhancement at low temperatures clearly demonstrates the reduction (or even a lack) of impurity scattering as expected. Since the doped barrier layers are totally depleted in MD superlattice structures, the electron density in the quantum well may greatly exceed the unintentionally doped impurity density. Thus, the concentration of 2DEG is independent of the doping level in the conduction channel. By the same token, an enhanced hole mobility of the two-dimensional hole gas (2DHG) is achievable in *p*-type MD heterostructures.

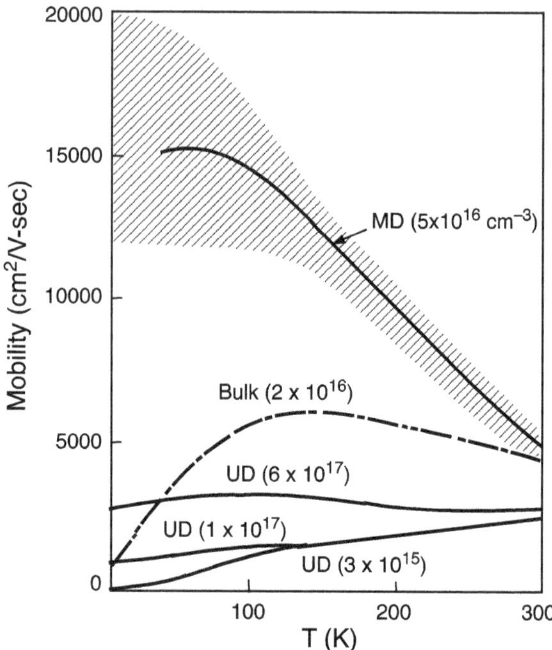

Fig. 9.14 Temperature-dependent electron mobility for bulk GaAs and several undoped UD and MD superlattices. The cross-hatched region includes most of the MD data. Reprinted with permission from [1], copyright AIP Publishing

Of course, a superlattice heterostructure with multiple MD heterojunctions is not necessary for device implementation. With a proper energy band discontinuity, a single MD heterojunction is sufficient to form a 2DEG at the heterostructure interface and to achieve the electron mobility enhancement. In fact, for FET applications, the single MD heterojunction design makes it easy to control the channel charge carriers. The first FET utilizing the high-electron-mobility feature of the MD heterostructure has just one hetero-interface as shown in Fig. 9.15. This design first demonstrated by engineers at Fujitsu Laboratories in Japan—a high-electron-mobility transistor (HEMT) design, as it was called—became the standard structure of future FETs utilizing 2DEG in the conduction channel [2].

9.2.2 *Scattering Mechanisms in MD Heterostructures*

It is important to analyze various scattering mechanisms limiting the mobility in the 2DEG in order to be able to provide interface design rules for yielding high-performance devices. In addition to the bulk scattering mechanisms, some additional scattering mechanisms unique to heterojunctions are introduced here. A GaAs/AlGaAs MD heterostructure similar to the HEMT structure shown in Fig. 9.15 is used to illustrate the properties of added scattering mechanisms. On the semi-insulating GaAs substrate, an undoped GaAs channel layer was first grown, followed

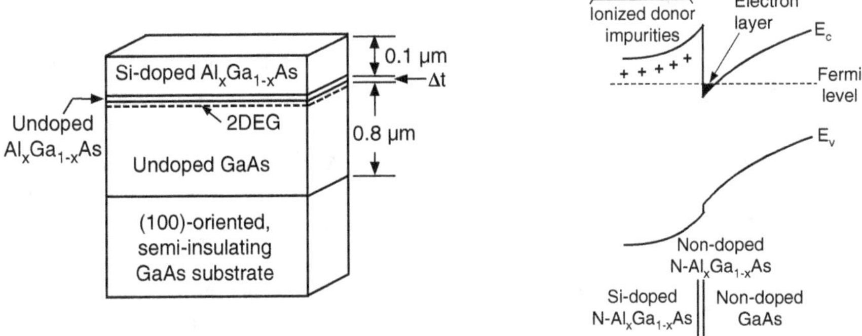

Fig. 9.15 Layer structure and energy band diagram of an MD GaAs-$Al_xGa_{1-x}As$ heterostructure grown by MBE with a spacer layer inserted between the barrier layer and the 2DEG channel. Reprinted with permission from [2, 3], copyright Japanese Society of Physics

by a silicon-doped *n*-type AlGaAs barrier layer to complete the MD heterostructure. The 2DEG channel will be formed at the GaAs/AlGaAs interface just inside the undoped GaAs layer.

(a) **Coulomb interaction from ionized impurities located in the barrier**

Since the 2DEG is located right next to the hetero-interface, the physical quality of the interface as well as any force exerted that diverts the transport direction toward the interface will affect the nature of carrier scattering. The attraction force due to Coulomb interaction between 2DEG in the conduction channel and its ionized parent donor impurities in the barrier layer will cause electrons to scatter with the hetero-interface. The electron mobility is reduced due to this added scattering mechanism. The carrier mobility is thus a function of the 2DEG sheet charge density. At a low barrier doping density (small 2DEG sheet charge density), the Coulomb interaction is weak and the electron mobility is limited by the background scattering in the undoped GaAs channel. For a heavily doped barrier layer structure, the Coulomb interaction dominates the carrier scattering process.

The Coulomb scattering problem cannot be totally eliminated but can be reduced by setting back the doping region of the barrier layer away from the hetero-interface. This undoped barrier layer inserted between the 2DEG and doped barrier layer is called a *spacer layer* [4]. Remember, the Coulomb attraction force is inversely proportional to the square of the charge separation distance. For a fixed doping density in the barrier, the Coulomb interaction between the 2DEG and ionized donors weakens with increasing spacer layer thickness, thus increasing electron mobility. However, due to the decreased electron transfer rate from the thicker barrier layer to the conduction channel, for a fixed barrier doping density, the sheet density of the 2DEG will decrease with increasing spacer layer thickness. In practical devices, a compromised spacer layer thickness of 25–30 Å is used to optimize the electron mobility while maintaining a high 2DEG sheet charge density.

(b) Interface-roughness-induced scattering

Since the 2DEG in the conduction channel of an MD heterostructure is located very close to the hetero-interface, any minute interface roughness could induce additional scattering and decrease the carrier mobility. Normally, in MD heterostructures, the AlGaAs barrier layer is grown on top of the GaAs 2DEG channel layer using MBE. The surface morphology of the GaAs channel layer at the hetero-interface is critically important. An extremely flat and smooth GaAs/AlGaAs interface is usually achieved under proper growth conditions including a wide range of V/III flux ratio, substrate temperature, and growth rate. This layer structure is convenient for FET-like device fabrication, including HEMT, and is called the *normal* MD heterostructure. If the layer sequence is reversed such that the GaAs 2DEG channel layer is placed on top of the AlGaAs barrier layer, the structure is called an *inverted* MD heterostructure. To achieve high electron mobility in such structures, an extremely smooth AlGaAs surface morphology is required. However, only under a very narrow MBE growth window does the AlGaAs show a less rough surface. Figure 9.16 shows the difference in the 77 K electron mobility for normal and inverted MD heterostructures grown at various substrate temperatures with a 75 Å spacer layer. The 2DEG mobilities for normal and inverted MD structures are shown at different scales in the figure. Note that, even using the optimal AlGaAs growth temperature, the electron mobility of the inverted structure is still an order of magnitude smaller than that of the normal structure.

(c) Inter-subband scattering at high carrier density

Since the 2DEG is confined in the quasi-triangular QW at the hetero-interface, electrons are occupying the quantized states of the QW. The ground state gets populated

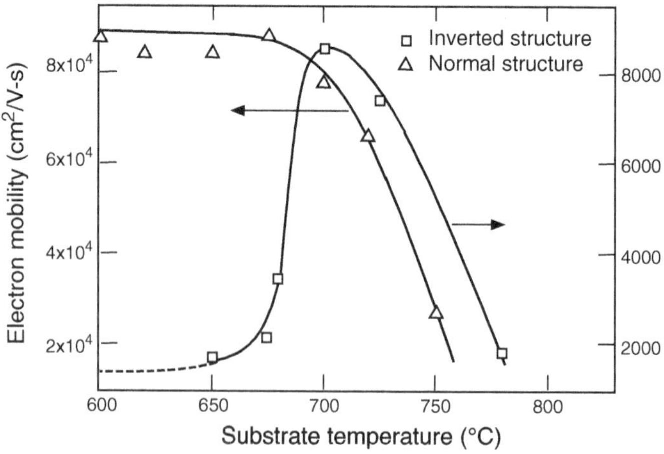

Fig. 9.16 Measured 2DEG mobility of normal and inverted MD GaAs/AlGaAs heterostructures grown at different temperatures. Reprinted with permission from [4], copyright AIP Publishing

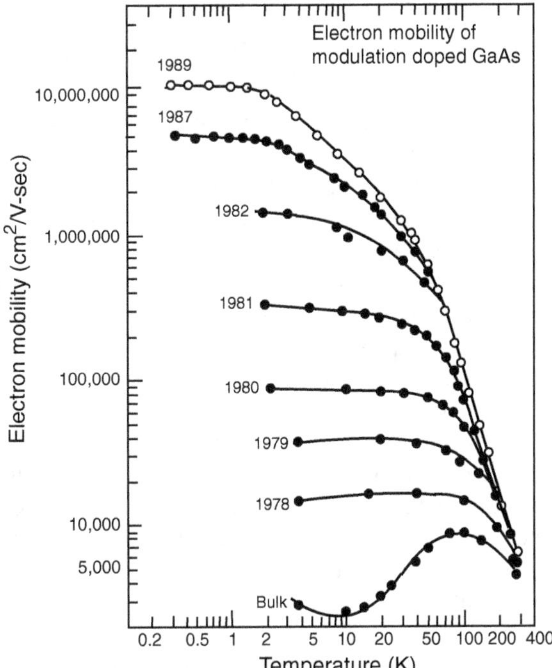

Fig. 9.17 Progress of peak mobility achieved in 2DEG as a function of temperature over a span of 12 years Reprinted with permission from [5], copyright AIP Publishing

first with electrons transferred from the barrier layer. When the ground state is fully populated, any further increase of the transferred electron density will have to occupy the higher energy states. The scattering probability is increased for the multiple-state system, and it reduces the total electron mobility.

The electron mobility values of the modulation-doped heterostructures have progressed steadily since 1978 (Fig. 9.17). In just three years, low-temperature electron mobility of over one million (cm^2/V s) was achieved in many laboratories. Further improvements of MD heterostructure designs and growth techniques pushed the electron mobility of the 2DEG over the 10 million (cm^2/V s) mark in 1989 at Bell Laboratories. A very thick spacer layer (70 Å) and planar (delta) doping scheme were used to minimize the Coulomb-interaction-induced interface scattering. The superlattice buffer layer beneath the GaAs channel was extensively used to guarantee minimized interface roughness scattering. Furthermore, to minimize the background doping concentration in the GaAs channel layer, the MBE system was thoroughly baked at high temperature (>200 °C) to get rid of volatile impurities inside and kept under liquid nitrogen cooling all the time. These grown samples provided a wonderful playground for some solid-state physicists, but have no immediate device application due to the low 2DEG sheet charge density achieved.

9.3 High-Electron-Mobility Transistor Basics—A Triangular Quantum Well Approach

The FET structure based on a 2DEG in the conduction channel was first demonstrated in 1980 by Fujitsu Laboratory of Japan, less than two years after the inception of the MD concept at Bell Laboratories. Therefore, there were different names used for this newly developed FET. The Fujitsu group called it the HEMT (*high-electron-mobility transistor*), Bell Laboratories named it SDHT (*selectively-doped hetero-field-effect transistor*), and the group at Thomson CSF in France called it TEGFET (*two-dimensional electron gas field-effect transistor*). The name MODFET (*modulation-doped field-effect transistor*) has also been used. Today, HEMT is the most commonly used name for this kind of FET.

The basic device structure of the HEMT is very similar to a MESFET, as shown in Fig. 9.18. Similar to a MESFET, the conduction current flows between the source and drain, and the amount of current flow is controlled by the gate bias. However, a HEMT does not use the gate bias to adjust the width of the conduction channel. Rather, the gate bias modulates the Fermi level position and, thus, controls the 2DEG sheet carrier density in the conduction channel. In this section, we will use a simple pseudotriangular QW approach to examine the equilibrium 2DEG sheet carrier density of a heterostructure.

An ideal triangular QW structure has an infinitely high vertical barrier on one side and a constant slope on the other side. The eigenenergy solutions of the ideal triangular QW are in the form of the Airy functions. If an ideal triangular QW was assumed for the HEMT structure, electrons transferred from the barrier layer fill the first allowed state (E_0) in the QW up to the Fermi level (E_F). Under ideal conditions, carriers of the 2DEG are solely transferred from the doped barrier layer with a doping concentration N_d and a depletion layer thickness W_d. Thus, the sheet carrier concentration of the 2DEG, N_S, simply equals to $N_d W_d$ (cm^{-2}). However, one should notice that the conduction band structure of the HEMT is far from ideal—the pseudotriangular QW has a finite band discontinuity, ΔE_c, on one side and a sublinear energy barrier on the channel side as shown in Fig. 9.19 for an AlGaAs/GaAs HEMT.

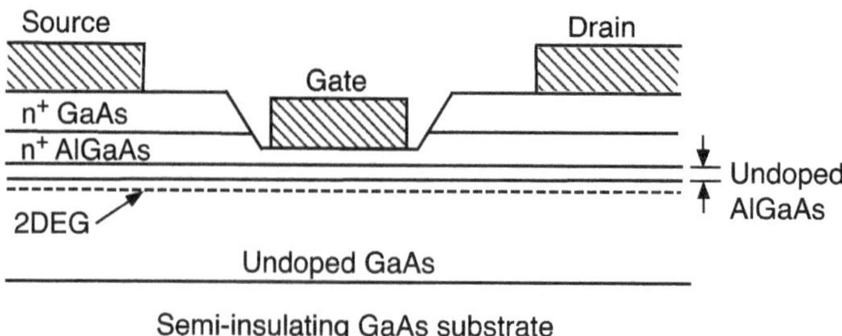

Fig. 9.18 Schematic structure of a GaAs/AlGaAs high-electron-mobility transistor

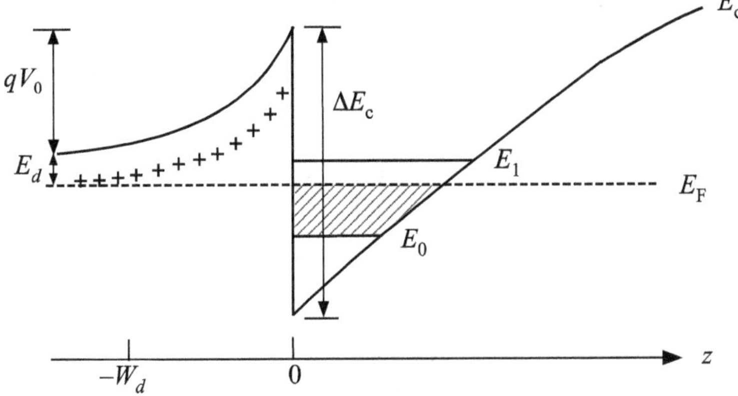

Fig. 9.19 Heterojunction energy band diagram of the HEMT structure shown in Fig. 9.18. The spacer layer in AlGaAs is removed from the heterojunction interface

Using an ideal triangular QW model, the possible sheet charge density (N_S) is estimated first. From Gauss' law for an ideal triangular QW, the displacement flux density (D) relates the sheet charge density and the electric field intensity (F) by

$$D = \epsilon F = \rho_S = qN_S \quad \text{or} \quad F = qN_S/\epsilon \tag{9.26}$$

Therefore, the ground state inside the QW, E_0, can be expressed in terms of N_S by

$$E_0 = \left(\frac{\hbar^2}{2m^*}\right)^{1/3}\left(\frac{9\pi q^2 N_S}{8\epsilon}\right)^{2/3} \approx \gamma N_S^{2/3} \tag{9.27}$$

At $T = 0$ K, where $f(E) = 1$, the equilibrium sheet carrier density is given by

$$N_S \approx \int_{E_0}^{E_F} D_{2D}(E) f(E) \mathrm{d}E = \frac{m^*}{\pi\hbar^2}(E_F - E_0) \tag{9.28}$$

At a finite temperature,

$$\begin{aligned} N_S &\approx \int_{E_0}^{E_F} D_{2D}(E) \frac{1}{1 + \exp[(E_0 - E_F)/kT]} \mathrm{d}E \\ &= \frac{m^*}{\pi\hbar^2} kT \ln\{1 + \exp[(E_F - E_0)/kT]\} \end{aligned} \tag{9.29}$$

If we include a thin undoped spacer layer with a thickness of W_{sp} at the interface, a voltage drop V_{sp} will appear across this layer. According to the energy band diagram shown in Fig. 9.20, the conduction band discontinuity ΔE_c at 0 K can be expressed as the following:

$$\Delta E_c = qV_0 + qV_{sp} + E_d + E_0 + \delta E \tag{9.30}$$

where

$$\begin{cases} \delta E = E_F - E_0 = \dfrac{\pi\hbar^2}{m^*} N_S \\ V_0 = \dfrac{qN_d W_d^2}{2\epsilon} = \dfrac{qN_S W_d}{2\epsilon} \\ V_{sp} = \dfrac{qN_S W_{sp}}{\epsilon} \end{cases} \tag{9.31}$$

Here the bottom tip of the triangular QW is set as $E = 0$. We notice that each term, except E_d, in this equation is a function of N_S and has a unique solution for each E_F. By moving the E_d term to the left of the equation, the right-hand side of the equation can be expressed in terms of N_S.

$$\Delta E_c - E_d = qV_0 + qV_{sp} + E_0 + \frac{\pi\hbar^2}{m^*} N_S \tag{9.32}$$

$$\begin{aligned} \Delta E_c - E_d = \frac{q^2 N_S W_d}{2\epsilon} + \frac{q^2 N_S W_{sp}}{\epsilon} + E_0 + \frac{\pi\hbar^2}{m^*} N_S = \frac{q^2}{2\epsilon N_d} N_S^2 \\ + \left(\frac{q^2 W_{sp}}{\epsilon} + \frac{\pi\hbar^2}{m^*}\right) N_S + E_0 \end{aligned} \tag{9.33}$$

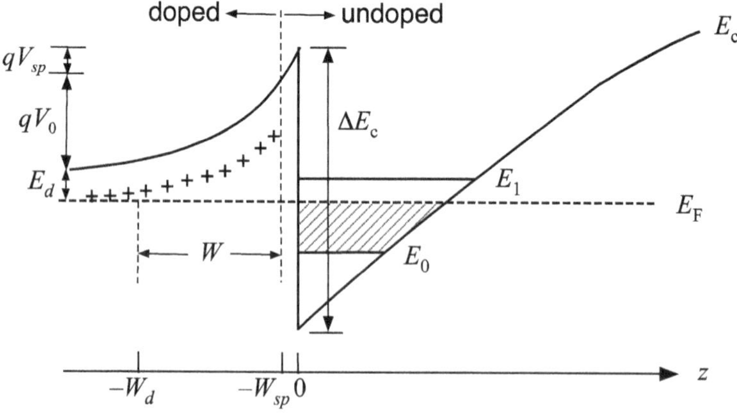

Fig. 9.20 Heterojunction energy band diagram of the HEMT structure shown in Fig. 9.18. The spacer layer in AlGaAs at the heterojunction interface is included

Assuming an Airy function for the ideal triangular QW and using the appropriate material parameters, one can solve N_S self-consistently.

However, the difficulty in solving N_S from the above equation lies in the determination of the eigenenergy E_0, since E_0 depends on the shape of the potential well. In a realistic quasi-triangular QW, E_0 cannot be determined exactly. Therefore, instead of solving N_S rigorously, one can qualitatively understand the factors that control N_S using the following extreme conditions. Assuming the density of states of the ground state subband is so large that the Fermi level is close to the bottom of the well, $E_F \approx 0$. For a very small spacer layer, $W_{sp} \approx 0$, the charges in the AlGaAs barrier are totally depleted by the conduction band discontinuity and transferred to the QW, forming a 2DEG.

$$\Delta E_c - E_d \cong \frac{q^2}{2\epsilon N_d} N_S^2 \quad \text{and} \quad N_S \cong \sqrt{\frac{2\epsilon N_d}{q^2}(\Delta E_c - E_d)} \tag{9.34}$$

For a thick spacer layer, $W_{sp} \gg W_d$, the transferred N_S is small and we can neglect the term contains N_S^2.

$$N_S = \frac{\epsilon}{q^2 W_{sp}}(\Delta E_c - E_d) \tag{9.35}$$

For a GaAs/AlGaAs HEMT with $N_d = 10^{18}$ cm^{-3}, $\Delta E_c = 0.2$ eV, and $\epsilon = 12.85\epsilon_0$, the sheet electron concentrations under these two extreme conditions are calculated as

$$N_S = \begin{cases} 1.6 \times 10^{12}\ \text{cm}^{-2} & \text{for} \quad W_{sp} = 0 \\ 7.2 \times 10^{11}\ \text{cm}^{-2} & \text{for} \quad W_{sp} = 200\ \text{Å} \end{cases} \tag{9.36}$$

These numbers are ~30% overestimated: In the first case, for $W_{sp} = 0$, $E_F \neq 0$ and requires a lot more electrons to fill the subband. In the second case, due to the large extent of the spacer layer, only a portion of the doped barrier is depleted. Even at these overestimated values, they are still much smaller than that of a comparable GaAs MESFET with a 0.2 μm conduction channel doped to 10^{17} cm^{-3} (2–3$\times 10^{12}$ cm^{-2}). Since a large N_S can reduce the channel resistance and increase current drive capability of the FET, a large N_S is desirable for device applications. Therefore, it is critical to maximize the carrier density of the HEMT. Empirically, N_S is proportional to $N_d \Delta E_c / W_{sp}$ in a HEMT. Based on this relation, we should select material systems with a large ΔE_c, a small W_{sp}, and a maximum N_d.

9.4 Operation Properties of the HEMT

The operation of the HEMT can be best understood through the analysis of its energy band diagram. Unlike MESFETs, the sheet charge density in a HEMT is controlled by the Fermi level position relative to the QW subbands shown in Fig. 9.21. The Fermi level position in the quasi-triangular QW is determined by the work function of the gate metal, barrier layer thickness, doping concentration, conduction band discontinuity, and gate bias. In a properly designed HEMT, through the modulation of the Fermi level, the gate bias voltage V_{GS} directly controls the sheet charge density similar to that of MESFETs. Depending on the device parameters, depletion-mode and enhancement-mode operations of HEMTs are also expected.

9.4.1 Isolated Heterojunction Under Equilibrium

To understand the relationship between the applied gate bias and the Fermi level position or sheet charge density, we first examine the 2DEG heterojunction under equilibrium conditions. In the band diagram shown in Fig. 9.22a, the undoped conduction channel layer and the n-type-doped barrier layer are assigned as regions 1 and 2, respectively. The origin of this 1D diagram is set at the hetero-interface with a depletion width of W_d into the infinite long barrier layer. We also assumed an undoped spacer layer of W_{sp} on the barrier layer side of the heterojunction, a dopant ionization energy of E_d and a potential barrier height of qV_{b0} formed by the depletion layer at the hetero-interface.

Under equilibrium, according to Gauss' law, at the hetero-interface, it follows that

$$D_1 = D_2 \tag{9.37}$$

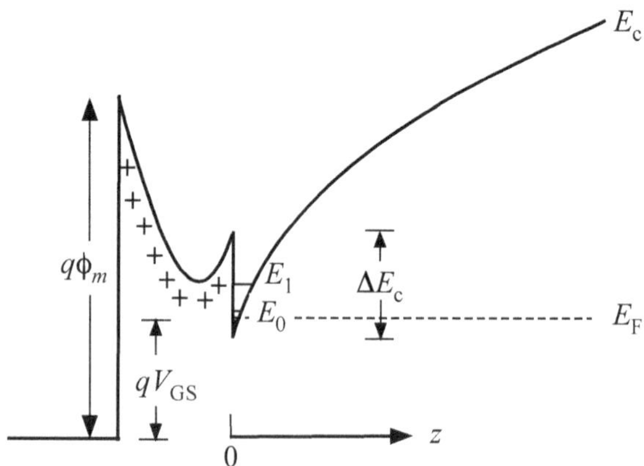

Fig. 9.21 Conduction band diagram of a depletion-mode HEMT under a gate bias of $V_{GS} < 0$. The work function of the metal contact, $q\phi_m$, is also shown

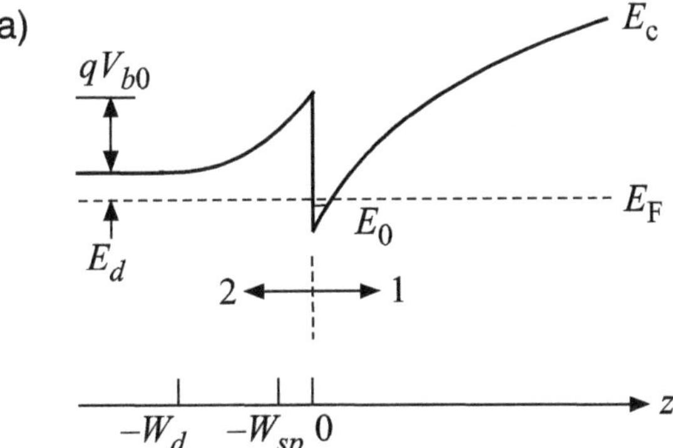

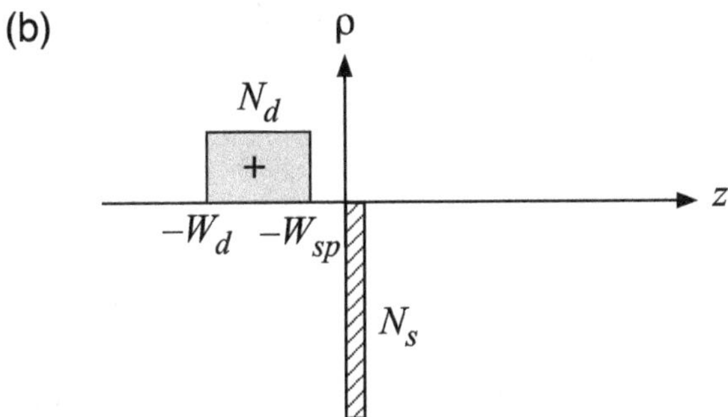

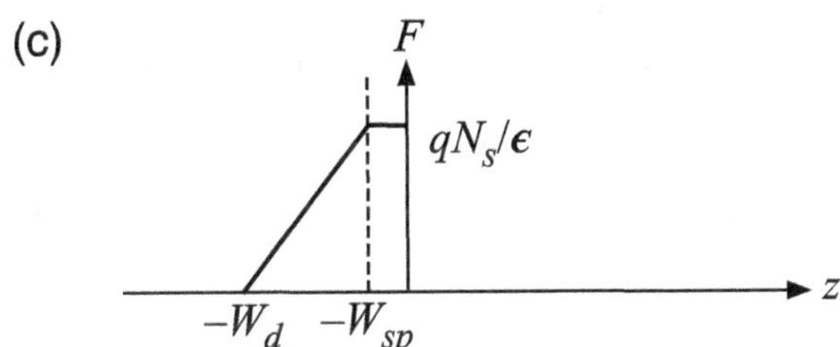

Fig. 9.22 **a** Energy band diagram in conduction band, **b** charge distribution, and **c** electric field distribution at the heterojunction of a HEMT

$$\epsilon_1 F_1 = \epsilon_2 F_2 = q N_S \tag{9.38}$$

$$N_S = \epsilon_2 F_2 / q \tag{9.39}$$

where D, F, and ϵ are the displacement flux density, electric field intensity, and permittivity, respectively. In this section, all discussions concern material 2 only; we will use ϵ instead of ϵ_2 for simplicity.

Assuming donors in the barrier layer are fully ionized between $-W_d \leq z \leq -W_{sp}$ and remain neutral beyond $-W_d$, Poisson's equation can be written as

$$\frac{d^2V}{dz^2} = -\frac{dF}{dz} = -\frac{qN(z)}{\epsilon} \tag{9.40}$$

And, as shown in Fig. 9.22b, the charge distribution is

$$N(z) = \begin{cases} 0, & -W_{sp} < z < 0 \\ N_d, & -W_d < z < -W_{sp} \end{cases} \tag{9.41}$$

Thus, in the depletion region of the doped barrier

$$F = \frac{q}{\epsilon}\int N(z)dz = \frac{qN_d}{\epsilon}z + C \tag{9.42}$$

Using the boundary condition of $F = 0$ at $z = -W_d$, we found (Fig. 9.22c)

$$F = \frac{qN_d}{\epsilon}(z + W_d) \quad \text{for} \quad -W_d < z < -W_{sp} \tag{9.43}$$

In the spacer layer, F is a constant of

$$F = \frac{qN_S}{\epsilon} \quad \text{for} - W_{sp} < z < 0 \tag{9.44}$$

The electric field intensity distribution across the depletion region of the heterojunction is shown in Fig. 9.22c.

The electric field intensity in the barrier has a slope of qN_d/ϵ. The potential drop across the doped barrier layer (layer thickness $\gg W_d$) can then be calculated as

$$V_{b0} = -\int F dz = -\frac{qN_d}{\epsilon}\int_{-W_{sp}}^{-W_d}(z + W_d)dz =$$

$$-\frac{qN_d}{\epsilon}\left(\frac{z^2}{2} + W_d z\right)\Bigg|_{-W_{sp}}^{-W_d} = \frac{qN_d}{2\epsilon}\left(W_d - W_{sp}\right)^2 \tag{9.45}$$

By the same token, the potential drop across the spacer layer becomes

$$V_{sp} = -\int_0^{-W_{sp}}\left(\frac{qN_S}{\epsilon}\right)dz = \frac{qN_S}{\epsilon}W_{sp} \tag{9.46}$$

The total potential drop across the barrier is the sum of V_{b0} and V_{sp}.

$$V = V_{b0} + V_{sp} = \frac{qN_d}{2\epsilon}\left(W_d^2 + W_{sp}^2\right) \tag{9.47}$$

9.4.2 Charge Control by Gate Bias in HEMT

In HEMT structures, the doped barrier layer has a finite thickness and a metal Schottky gate is applied on the surface. The Schottky barrier also creates a surface depletion layer. It substantially alters the conduction band shape of the barrier layer and plays an important role in determining the Fermi level position. To simplify the discussion, assume the doped barrier layer is completely depleted. Relative to the surface, as shown in Fig. 9.23, the potential drop across the barrier is defined as V_i.

$$V_i = \phi_m - V_{\mathrm{GS}} - \frac{1}{q}(\Delta E_{\mathrm{c}} - E_{\mathrm{F}}) \tag{9.48}$$

This potential difference V_i can also be derived from Poisson's equation.

$$\frac{\mathrm{d}^2 V}{\mathrm{d}z^2} = -\frac{\mathrm{d}F}{\mathrm{d}z} = -\frac{qN(z)}{\epsilon} \tag{9.49}$$

and

$$N(z) = \begin{cases} N_d, & -W_d \le z \le -W_{sp} \\ 0, & -W_{sp} \le z \le 0 \end{cases} \tag{9.50}$$

Thus, the electric field intensities in the depleted doped barrier and spacer layers are

$$F = \frac{q}{\epsilon}\int N(z)\mathrm{d}z = \begin{cases} \dfrac{qN_d}{\epsilon}z + C_1, & -W_d \le z \le -W_{sp} \\ \dfrac{qN_S}{\epsilon}, & -W_{sp} \le z \le 0 \end{cases} \tag{9.51}$$

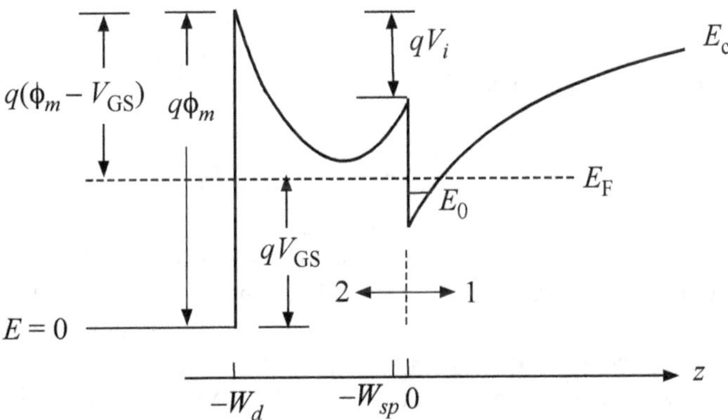

Fig. 9.23 Energy band model of a HEMT under a gate bias $V_{\mathrm{GS}} < 0$. In this model, it is assumed that, under all bias conditions, the barrier layer beneath the gate is completely depleted

Assuming $F' = qN_S/\epsilon = F$ at $z = -W_{sp}$, then $C_1 = F' + qN_d W_{sp}/\epsilon$. The potential drop across the depleted barrier layer can then be calculated as

$$\begin{aligned} V &= -\int F \mathrm{d}z = -\frac{qN_d}{\epsilon}\int_{-W_{sp}}^{-W_d} z\mathrm{d}z - \frac{qN_d W_{sp}}{\epsilon}\int_{-W_{sp}}^{-W_d} \mathrm{d}z - F'\int_{-W_{sp}}^{-W_d} \mathrm{d}z \\ &= -\frac{qN_d}{\epsilon}\left(\frac{z^2}{2}\right)_{-W_{sp}}^{-W_d} - \frac{qN_d W_{sp}}{\epsilon} z|_{-W_{sp}}^{-W_d} - F' z|_{-W_{sp}}^{-W_d} \\ &= -\frac{qN_d}{2\epsilon}\left(W_d^2 - W_{sp}^2 - 2W_d W_{sp} + 2W_{sp}^2\right) + F'\left(W_d - W_{sp}\right) \end{aligned} \tag{9.52}$$

$$V = -\frac{qN_d}{2\epsilon}\left(W_d - W_{sp}\right)^2 + F'\left(W_d - W_{sp}\right) \tag{9.53}$$

The potential drop across the spacer layer is simply $F'W_{sp}$. Letting $V(0) = 0$, the potential drop V_i can be expressed as

$$\begin{aligned} V_i = -V - F'W_{sp} &= \frac{qN_d}{2\epsilon}\left(W_d - W_{sp}\right)^2 - F'\left(W_d - W_{sp}\right) \\ &- F'W_{sp} = V_P - \frac{qN_S}{\epsilon} W_d \end{aligned} \tag{9.54}$$

Here we defined a pinch-off voltage of $V_P = (qN_d/2\epsilon)\left(W_d - W_{sp}\right)^2$ just like a MESFET. Since $V_i = \phi_m - V_{\mathrm{GS}} - (\Delta E_c - E_F)/q$, we can write the sheet charge density in (9.54) as a function of voltage across the heterostructure.

$$N_S = \frac{\epsilon}{qW_d}\left[V_{\mathrm{GS}} - \left(\phi_m - V_P - \frac{\Delta E_{\mathrm{c}} - E_{\mathrm{F}}}{q}\right)\right] \tag{9.55}$$

The terms in the parenthesis are all fixed for a particular device structure and are defined as a *threshold voltage* or off-voltage V_T. At an applied gate bias V_{GS} equal to V_T, where $E_{\mathrm{F}} \leq E_0$, the 2DEG vanishes.

$$V_T \equiv \phi_m - V_P - \frac{1}{q}(\Delta E_{\mathrm{c}} - E_{\mathrm{F}}) \tag{9.56}$$

The Fermi level energy E_{F} is usually small compared with other terms and can be neglected. Finally, the distribution of the 2DEG in the triangular QW has a peak at a distance of ΔW from the heterojunction interface. The effective depletion layer thickness should be corrected as $(W_d + \Delta W)$. For the values of N_S between 5×10^{11} cm^{-2} and 1.5×10^{12} cm^{-2} in AlGaAs/GaAs HEMTs, ΔW is about 80 Å. The final expression of the sheet carrier density becomes

$$N_S \cong \frac{\epsilon}{q(W_d + \Delta W)}(V_{GS} - V_T) \tag{9.57}$$

This equation indicates that the 2DEG density is linearly proportional to the gate bias, V_{GS}. This linear relationship is limited by two extremes, $N_S = 0$ and $N_S = N_d(W_d - W_{sp})$, as illustrated in Fig. 9.24.

In a HEMT structure with a 2DEG created at zero gate bias, the Fermi level is positioned above the ground state, E_0. By applying a reverse gate bias, as shown in Fig. 9.24a, the rise of the total Schottky barrier potential energy at the surface leads to a lowering of E_F, which reduces the sheet charge density. When the gate bias is equivalent to the threshold voltage, V_T, the conduction channel is turned off. On the other hand, as shown in Fig. 9.24b, under forward gate bias condition, the rising Fermi level allows more electrons to be transferred into the triangular QW, contributing to the current conduction. This process will continue until all donors in the barrier are ionized. The 2DEG sheet carrier density becomes saturated with further increasing of V_{GS}.

Figure 9.25 shows the experimental results of N_S as a function of gate bias in several $Al_{0.3}Ga_{0.7}As$/GaAs HEMT samples with different spacer layer thicknesses. All samples have a doping concentration of $N_d = 4.6 \times 10^{17}$ cm^{-3} in the barrier layer, except sample No. 76 ($N_d = 9 \times 10^{17}$ cm^{-3}). It is clear, in all samples, that N_S increases linearly with increasing V_{GS} and saturates at a higher gate bias voltage. The flat saturated region indicates the maximum achievable sheet carrier density in that sample. By extrapolating the linear part of the $V_{GS} - N_S$ curve to zero sheet concentration, the threshold voltage is determined. Obviously, the threshold voltage is dependent on the spacer layer thickness due to the V_P term involved, which depends on the spacer layer thickness. As expected, for the same device structure, the thinner the spacer layer, the higher the saturated 2DEG density.

The channel conduction current density of a HEMT is primarily determined by the sheet charge density, which is controlled by the difference between gate bias voltage and threshold voltage. A closer examination of the threshold voltage reveals that it depends on the layer thickness. For a HEMT structure, ϕ_m and ΔE_c are fixed and E_F

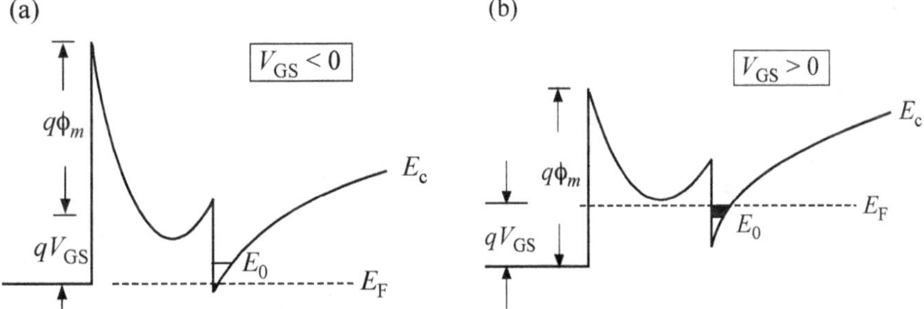

Fig. 9.24 Energy band diagram of a depletion-mode HEMT under different gate biases. **a** $V_{GS} < V_T$ such that the channel is turned off. **b** Under a high forward gate bias, the 2DEG is saturated

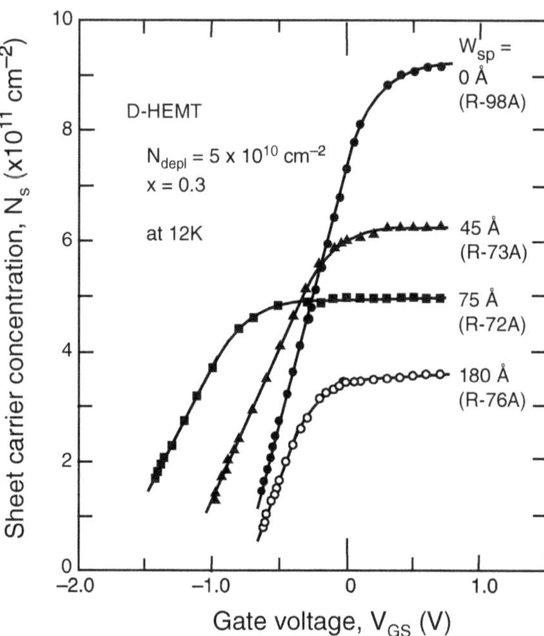

Fig. 9.25 Measured gate-voltage dependence of the sheet carrier density of 2DEG in the GaAs/AlGaAs system. W_{sp} represents the spacer layer thickness in each sample. All samples have $N_d = 4.6 \times 10^{17}$ cm^{-3} except sample No. 76, which has $N_d = 9.2 \times 10^{17}$ cm^{-3}. Reprinted with permission from [6], copyright AIP Publishing

is relatively small. The major term controlling V_T is $V_P = (qN_d/2\epsilon)\left(W_d - W_{sp}\right)^2$, which mainly depends on W_d. Therefore, one can design the operation mode of the device by selecting the proper barrier thickness as shown in Fig. 9.26. For example, a thick barrier (large W_d) leads to a larger V_P and a more negative V_T. A conduction channel is formed in such a HEMT structure even at zero gate bias (Fig. 9.26a). One has to apply a gate bias equal to V_T to completely deplete the 2DEG, thus forming a depletion-mode (normally-on) HEMT. When the barrier layer is reduced to a point such that the Fermi level is well below the ground state (E_0) of the triangular QW at

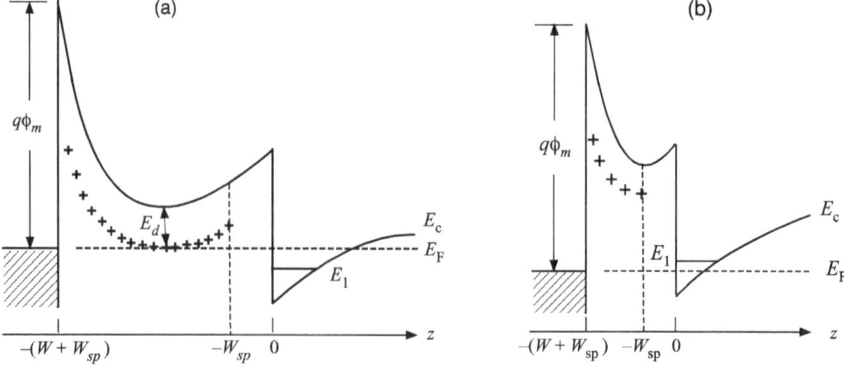

Fig. 9.26 Schematic band diagram at $V_{GS} = 0$ for **a** normally-on and b normally-off HEMTs

zero bias, V_T becomes positive. A positive gate bias is required to induce the 2DEG for current conduction, and the HEMT is an enhancement-mode (normally-off) FET (Fig. 9.26b).

9.4.3 Current and Voltage Characteristics of HEMT

The drain–source conduction current can be expressed similarly as in a MESFET.

$$I_{\mathrm{DS}} = qZN_S\upsilon_d(F) \tag{9.58}$$

where Z is the gate width, N_S is the 2DEG density, and $\upsilon_d(F)$ is the electron drift velocity. Since there is a voltage drop, V_x, between source and drain, it should be included in the N_S expression of (9.57) as

$$N_S = \frac{\epsilon}{q(W_d + \Delta W)}(V_{\mathrm{GS}} - V_T - V_x) \tag{9.59}$$

Depending on the field strength, the drift velocity along the conduction channel can be modeled with a linear region and a saturation region with the critical field intensity, F_C, as the deflection point in the υ_d versus F plot.

$$\upsilon_d(F) = \begin{cases} \mu F & \text{for} \quad F < F_C \\ \upsilon_{\mathrm{sat}} & \text{for} \quad F > F_C \end{cases} \tag{9.60}$$

In the linear region ($F < F_C$), let $F = \dfrac{\mathrm{d}V_x}{\mathrm{d}x}$:

$$I_{\mathrm{DS}} = qZN_S\mu F = \frac{\mu\epsilon Z}{(W_d + \Delta W)}(V_{\mathrm{GS}} - V_T - V_x)\frac{\mathrm{d}V_x}{\mathrm{d}x} \tag{9.61}$$

Since the conduction current in the channel remains constant under a fixed drain bias, the total current can be calculated through integration along the channel length, L.

$$I_{\mathrm{DS}}\int_0^L \mathrm{d}x = \int_0^{V_{\mathrm{DS}}} \frac{\mu\epsilon Z}{(W_{\mathrm{d}} + \Delta W)}(V_{\mathrm{GS}} - V_T - V_x)\mathrm{d}V_x \tag{9.62}$$

Thus, the drain current as a function of drain voltage under fixed V_{GS} is expressed as

$$I_{\mathrm{DS}} = \frac{\mu\epsilon Z}{L(W_d + \Delta W)}\left[(V_{\mathrm{GS}} - V_T)V_{\mathrm{DS}} - \frac{V_{\mathrm{DS}}^2}{2}\right] \tag{9.63}$$

At very low drain bias, the $V_{DS}^2/2$ term is very small and can be neglected. The drain current is linearly proportional to the drain bias. At a sufficiently high drain bias, the $V_{DS}^2/2$ term becomes significant and leads to a sublinear I_{DS}-V_{GS} characteristic. Further increasing the drain bias, the drain current becomes saturated.

The effectiveness of current control by the gate bias voltage is determined by the transconductance, g_m. At a fixed V_{DS}, g_m is calculated as the variation of drain current per gate bias change.

$$g_m = \left.\frac{dI_{DS}}{dV_{GS}}\right|_{V_{DS}} = \frac{\epsilon Z V_{DS}}{(W_d + \Delta W)}\frac{\mu}{L} \propto \frac{\mu}{L} \tag{9.64}$$

It is clear that large low-field electron mobility and a short channel length are required for a HEMT to achieve a high g_m at a fixed V_{DS}. Within the linear region, g_m increases with increasing V_{DS}.

In the saturation region ($F \geq F_C$), the drift velocity does not increase with the electric field but has a constant value called the saturation velocity, υ_{sat}. Therefore, the current becomes a constant, $I_{DS}(\text{sat}) = qZN_s\upsilon_{sat}$ and

$$\frac{dI_{DS}(\text{sat})}{dV_{DS}} = \frac{\mu\epsilon Z}{L(W_d + W)}[(V_{GS} - V_T) - V_{DS}] = 0 \tag{9.65}$$

From this equation, the threshold saturation voltage is calculated as $V_{DS}(\text{sat}) = V_{GS} - V_T$. Using this relation in I_{DS}, the I–V characteristic in the saturation region is expressed as

$$I_{DS}(\text{sat}) = \frac{\mu\epsilon Z}{L(W_d + \Delta W)}\frac{(V_{GS} - V_T)^2}{2} = \frac{\mu C_b Z}{L}\frac{(V_{GS} - V_T)^2}{2} \tag{9.66}$$

where $C_b = \epsilon/(W_d + \Delta W)$ is the capacitance per unit area in the barrier. The transconductance in the saturation region can also be derived as

$$g_m = \left.\frac{dI_{DS}(\text{sat})}{dV_{GS}}\right|_{V_{DS}} = \frac{\mu C_b Z}{L}(V_{GS} - V_T) \propto \frac{\mu}{L} \tag{9.67}$$

Again, large low-field electron mobility and a short channel length are required to achieve a high g_m in the saturation region.

9.4.4 Microwave Noise Performance

The essential elements of the high-speed performance of HEMTs are very similar to those of conventional MESFETs. Therefore, a common microwave equivalent circuit model can be used for both devices. Consider the case where the FET is operating at a frequency below its cutoff frequency, f_T, at room temperature. The microwave noise performance of the HEMT can be derived from the equivalent circuit by the same classical noise model of the MESFET. The minimum noise figure $F_{\min}$ for the HEMT is described as

$$F_{\min} = 1 + K_f(f/f_{\mathrm{T}})\sqrt{g_m(R_{\mathrm{G}} + R_{\mathrm{S}})} \tag{9.68}$$

where g_m, R_{G}, and R_{S} are the transconductance, gate resistance, and source resistance, respectively. K_f is a fitting factor and represents the quality of the channel material. Since all external device parameters for both HEMT and MESFET can be made roughly equal, the noise performance difference between the two devices originates in the K_f difference. It was suggested that the size of K_f is related to excess noise in the drain current, which originates as diffusion noise in the channel. For GaAs MESFETs, $K_f \approx 2$. In HEMTs, the electrons are narrowly confined to a quantized state along the interface rather than in a 3D configuration. Monte Carlo simulations indicated much lower diffusion coefficients as a function of electric field than those of MESFETs. Experimental K_f for HEMTs have values between 1 and 2. Therefore, HEMTs have a better microwave noise performance than MESFETs. In addition, since f_{T} can be much higher for HEMTs, we also expect a lower noise figure in HEMTs.

9.5 Optimal Design of the HEMT

Based on discussions above, to achieve high performance, the optimized HEMT structure should incorporate a large ΔE_{c} ($\geq$0.5 eV) for high sheet carrier density N_S and high current carrying capability; furthermore, materials with high electron mobility should be selected for efficient high-speed operation, and small gate length ($L_g < 0.1\ \mu$m) should be used to increase the cutoff frequency. In addition, to enhance electron transfer from the barrier into the triangular QW, a small spacer layer thickness (W_{sp} ~ 25 Å) is preferred. However, in practice, it was not straightforward to realize all these requirements in the early development of HEMTs. Specifically, the DX center problem severely limited the use of $\mathrm{Al}_x\mathrm{Ga}_{1-x}\mathrm{As}$ with high aluminum content ($x \leq 25\%$) in the HEMT design. In addition, when increasing the barrier doping level to achieve higher 2DEG density, parallel conduction from free electrons trapped in the barrier and 2DEG could form. To overcome these problems, a pseudomorphic layer and delta-doping designs are incorporated to increase ΔE_{c} and 2DEG density.

9.5.1 DX Center in Si-Doped AlGaAs/GaAs HEMT and Delta Doping

As discussed in Chap. 4, Si-dopants in AlGaAs can take different crystal configurations: interstitial (relaxed state, DX center) or substitutional (excited state). The DX center is a deep trap that requires different energies for capture and emission of electrons. Further, the finite capture barrier makes it difficult to achieve equilibrium at low temperature (<150 K). Depending on the external excitation, the free electron concentration can have different values. For example, in an $Al_{0.32}Ga_{0.68}As$/GaAs MD structure, the carrier concentration determined by Hall measurement under dark as a function of temperature is shown as solid circles in Fig. 9.27. Upon illumination with a light of energy below the fundamental bandgap of $Al_{0.32}Ga_{0.68}As$ at low temperature, the carrier concentration increases to a high value shown as the open circles. The increase in carrier concentration persists and does not return to its dark value even after the illumination has been turned off. This phenomenon is known as the persistent photoconductivity (PPC) effect. The carrier concentration remains high even hours or days after termination of the photo-excitation. The high

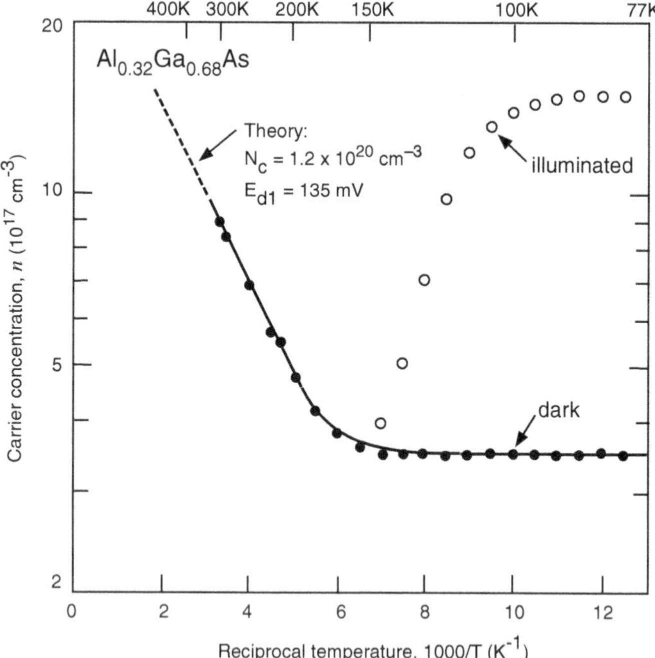

Fig. 9.27 Dependence of the Hall electron concentration in n-type $Al_{0.32}Ga_{0.68}As$:Si on reciprocal temperature. Solid and open circles indicate experimental data measured in the dark and after illumination at low temperatures, respectively. Reprinted with permission from [7], copyright AIP Publishing

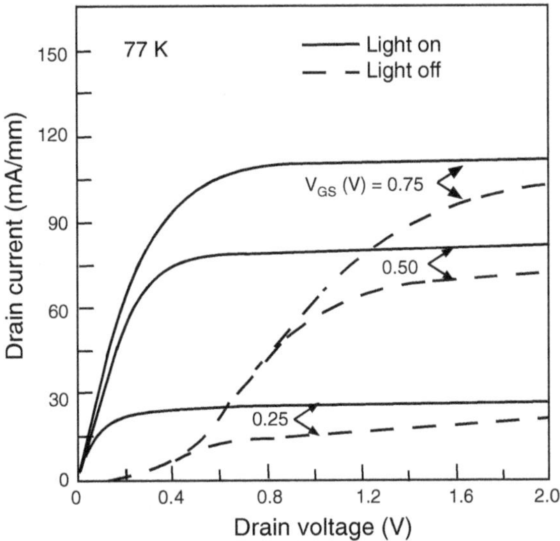

Fig. 9.28 *I–V* characteristics of a HEMT at 77 K under illumination and in darkness. The degradation of the characteristics upon the elimination of light occurred gradually over a period of 20–30 s. Reprinted with permission from [8], copyright IEEE

PPC carrier concentration can be quenched by heating the sample to a temperature higher than 150 K. In Si-doped AlGaAs/GaAs HEMT structures, DX centers can cause anomalous device behaviors including threshold voltage shifts and 'collapse' of *I–V* characteristics (Fig. 9.28). These properties degrade the HEMT performance severely and make the device unreliable.

The origin of the DX center problem is closely related to the ionization energy of Si donors in $Al_xGa_{1-x}As$, which is shown in Fig. 9.29. For Al-composition of $x \leq 20\%$, the Si donor behaves like a shallow donor ($E_d \leq 5$ meV) and no DX center related problem is observed. For $x \geq 30\%$, the ionization energy of the Si donor rises rapidly to $\geq$120 meV such that it acts like a deep donor and causes DX center problems.

For example, in an $Al_xGa_{1-x}As$/GaAs HEMT with a composition of $x = 30\%$, the donor level, E_d, is 0.12 eV below the conduction band. Since E_d is very deep, under dark, a portion of the donor level inside the barrier is below the Fermi level and stays neutral (unionized). Only a thin-doped barrier layer adjacent to the spacer layer is contributing electrons to the 2DEG channel and the sheet charge density is very low. The neutral layer in the barrier also decouples the 2DEG and the surface. Therefore, N_S in 2DEG is independent of bias! When illuminated with a light source having energy less than the bandgap, electrons are excited out of DX centers. All the donors are ionized and the 2DEG has risen. Due to the finite electron trapping barrier height, there are free electrons remaining in the barrier at low temperatures and these form 'parallel conduction' channel in addition to the 2DEG conduction.

Since the ionization energy of Si dopants increases rapidly with increasing Al-composition in AlGaAs-based HEMT structures, ΔE_c cannot be increased much beyond 20%. To achieve a high sheet charge density, increasing the doping level of the AlGaAs barrier layer is the other alternative. But a large and excessive N_d can lead

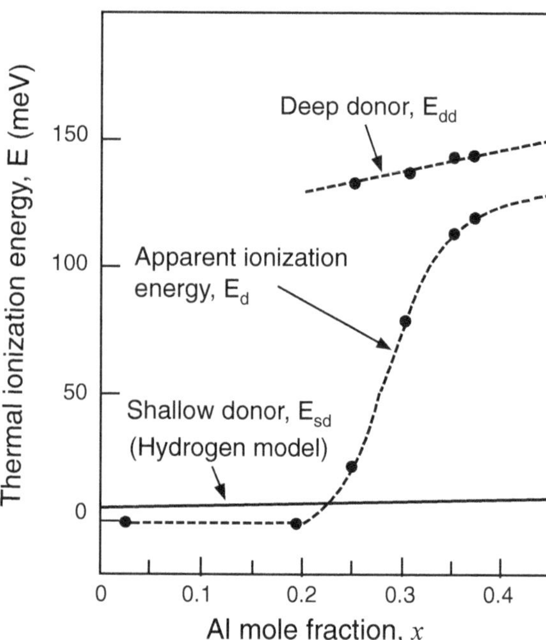

Fig. 9.29 Thermal ionization energy of the Si donor as a function of the Al mole fraction in $Al_xGa_{1-x}As$. Reprinted with permission from [7], copyright AIP Publishing

to gate leakage and parallel conduction, which degrade the device performance. One simple solution to this problem is to use a planar (pulse or delta) doping scheme. The doping impurities are all deposited on the same growth plane such that the impurity distribution is confined in a sheet layer. Since the barrier is undoped except for the doping sheet, the band profile shows a straight line-like variation (Fig. 9.30). The planar-doped sheet locates at the minimum of the conduction band profile in the barrier. Since the AlGaAs barrier layer is undoped, it not only solves the gate leakage and parallel conduction problems but also minimizes the DX center-induced complications of the HEMT structure.

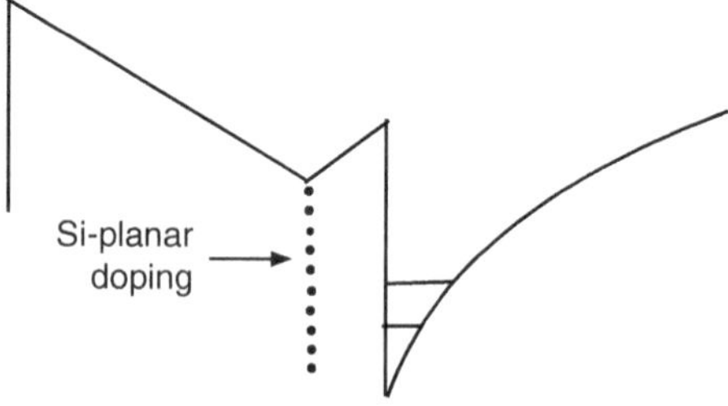

Fig. 9.30 Schematic energy band diagram of Si-planar doping in a MD structure

9.5.2 *Pseudomorphic High-Electron-Mobility Transistors (pHEMT)*

Since N_S in 2DEG is in proportion with the conduction band discontinuity (ΔE_c) at the heterojunction, a large ΔE_c is essential for a high sheet electron density. In AlGaAs/GaAs HEMTs, we cannot increase ΔE_c by increasing the Al-composition much beyond 20% due to the DX center problem. This limitation can be relaxed by using a thin layer of strained InGaAs to replace GaAs at the heterojunction as the 2DEG conduction channel layer (Fig. 9.31). The ΔE_c increases almost linearly with increasing In composition. The amount of indium composition in the InGaAs layer grown on GaAs is limited only by the critical layer thickness such that no dislocations are generated in the channel layer. This type of strained layer HEMT structure is called the *pseudomorphic* HEMT or pHEMT. The other advantage of using InGaAs as the conduction channel is the associated electron mobility which is higher than that of GaAs. These advantages make AlGaAs/InGaAs pHEMT the most commonly used device in high-speed applications. To further enhance the high-speed operation properties beyond AlGaAs/InGaAs pHEMT, the AlInAs/InGaAs/InP material system is preferred due to the high electron mobility of the In-rich InGaAs channel material.

9.6 GaN-Based HEMT Structures

The rapid expansion of wireless communications worldwide has increased the demand for high-power electronic devices in both terrestrial base stations and satellite communication systems. Current GaAs-based FETs have been dominating the market for many years, but the increase in demand calls for devices with higher power

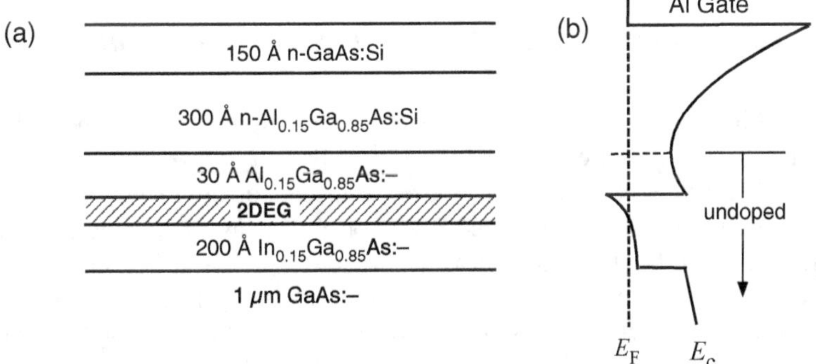

Fig. 9.31 **a** Schematic of InGaAs/AlGaAs pseudomorphic HEMT. **b** Conduction band diagram of the pHEMT

Table 9.2 Material parameters of GaN and GaAs

Material parameters	GaN	GaAs
Bandgap energy (eV)	3.438	1.424
Maximum sheet electron density (cm^{-2})	1–5 × 10^{13}	2–3 × 10^{12}
Breakdown field strength (10^5 V/cm)	50	4
2DEG mobility (cm^2/V s)	2000	8500
Saturation electron velocity (10^7 cm/s)	2.5	1.0
Thermal conductivity (W/cm K)	≥2.1	0.44

operation capacity at microwave frequencies. Among all semiconductors, the GaN-based material system has shown great potential in high-power and high-frequency device applications. Some of the key device properties of GaN- and GaAs-based HEMTs are compared in Table 9.2. Because of their high breakdown voltages, GaN-based HEMTs are ideal for high voltage operation, reducing the need for voltage conversion. This makes device integration much simpler. Another attractive fact is that GaN has a high saturation electron velocity suitable for high-frequency operation. Its great thermal conduction properties and large maximum 2DEG density ($\geq 10^{13}/cm^2$) translate into much higher power densities, making it ideal for high-power operation. In addition, GaN devices can also operate at higher temperatures than their GaAs counterparts because of the higher energy bandgap (~3.4 eV). All these properties make the GaN-based HEMT an ideal candidate for high-power operation at microwave frequencies. However, there are major differences between GaN-HEMT device structures and those of other III–V HEMTs. First, no doping in the AlGaN barrier layer is needed to generate the 2DEG in the GaN channel layer and, second, the generation of the 2DEG is heterostructure dependent.

9.6.1 Polarization-Induced Sheet Charges at Heterojunctions

GaN-based HEMT layer structures are currently grown using either MBE or MOCVD. MBE layers are usually grown directly on (0001) sapphire or c-Al_2O_3 under Ga-rich surface conditions to achieve a high layer quality. This particular MBE growth condition yields N-terminated B-face or N-face layers, i.e., the GaN growth direction is along $[000\bar{1}]$ capped with a top nitrogen plane (Fig. 9.32). In the MOCVD approach, a low-temperature-grown AlGaN nucleation layer on (0001) sapphire is needed before the growth of GaN. The nitrogen-rich growth condition for the nucleation layer yields a Ga-terminated A-face or Ga-face layer and a growth direction along [0001] capped with a top gallium plane. This growth direction is opposite to the MBE-grown nitrides and causes a sign change in spontaneous polarization.

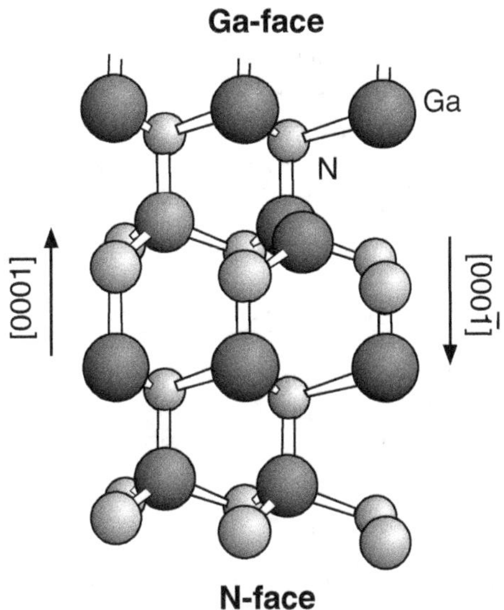

Fig. 9.32 Schematic diagram of the crystal structure of wurtzite Ga-face GaN. Reprinted with permission from [9], copyright AIP Publishing

Using an III–N heterostructure grown by MOCVD as an example, we can examine contributions of spontaneous polarization and piezoelectric polarization in the structure to the generation of 2DEG at hetero-interfaces. An AlGaN/GaN heterostructure is grown on c-Al_2O_3 with a thin AlGaN nucleation layer deposited first. The GaN channel layer is grown next and finished by depositing a thin $Al_xGa_{1-x}N$ barrier layer. The whole heterostructure is intentionally undoped. Since MOCVD-grown AlGaN has a Ga-face surface, P_{SP} is pointing in the [000$\bar{1}$] direction toward the substrate and has a negative value for all compositions. Here we assumed the polarization along the c-axis or [0001] toward surface is positive. The AlGaN/GaN heterostructures grown by MOCVD and MBE and their polarization components are shown in Fig. 9.33. The magnitude and sign of P_{PE} depend on the strain condition. The thick GaN channel layer is relaxed and has a zero piezoelectric polarization component. Since AlGaN has a smaller lattice constant than GaN, the thin AlGaN barrier layer is under tensile strain and P_{PE} has a negative value. Therefore, the net polarization P is negative at the AlGaN/GaN interface. If the polarity flips over from Ga-face to N-face in structures grown by MBE, both P_{SP} and P_{PE} change their signs.

Associated with a nonzero polarization $\boldsymbol{P} = P\hat{z}$ in a dielectric slab is an induced internal electric field $\boldsymbol{F}$, which is produced by the fictitious surface charge density $\sigma = \hat{\boldsymbol{n}} \cdot \boldsymbol{P}$, where $\hat{\boldsymbol{n}}$ is the surface unit vector. By Gauss' law, the electric field between the plates produced by these charges is $\boldsymbol{F} = -\boldsymbol{P}/\epsilon$. In analog, the net polarization change within the AlGaN/GaN bilayer heterostructure under discussion induces sheet charge densities at the hetero-interface and on the surface of AlGaN defined as

$$\sigma = P(\text{bottom}) - P(\text{top})$$

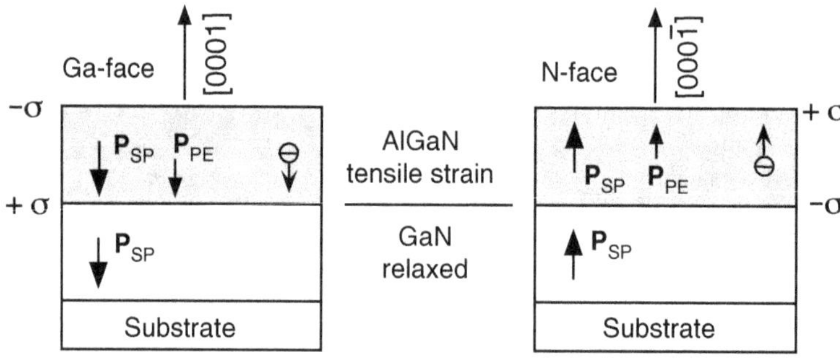

Fig. 9.33 Polarization-induced sheet charge density (σ) and directions of the spontaneous (P_{SP}) and piezoelectric (P_{PE}) polarization in Ga- and N-face tensile strained AlGaN/GaN heterostructures grown by MOCVD (left) and MBE (right). Reprinted with permission from [9], copyright AIP Publishing

$$= [P_{\mathrm{SP}}(\text{bottom}) + P_{\mathrm{PE}}(\text{bottom})] - \left[P_{\mathrm{SP}}(\text{top}) + P_{\mathrm{PE}}(\text{top})\right] \tag{9.69}$$

In this heterostructure, the top surface of AlGaN and the underneath AlGaN/GaN hetero-interface carry $-\sigma$ and $+\sigma$, respectively. During the cooling process, after the epitaxial growth, the polarization-induced positive sheet charge, $+\sigma$, will attract free electrons to compensate it and form a 2DEG at the hetero-interface. If the heterostructure is grown by MBE, resulting in an N-face material, the piezoelectric polarization, along with the spontaneous polarization, changes its sign. A negative sheet charge, $-\sigma$, will form at the AlGaN/GaN hetero-interface and attract holes. To utilize the fictitious positive sheet charge to induce 2DEG on the AlGaN surface, a GaN layer is needed on top of the AlGaN/GaN heterostructure.

9.6.2 *Sheet Carrier Concentration of 2DEG*

When Schottky gate contact is made to the surface of the AlGaN barrier layer, the heterostructure with a Ga-face surface forms a HEMT structure. The equilibrium energy band diagram and sheet charge density (σ_i) of the HEMT structure are shown in Fig. 9.34. At equilibrium, due to the band bending across the barrier layer, the Schottky barrier induces an electric field in AlGaN and a net charge density is required to terminate the electric field. The size of the electric field is fixed by the potential drop in the AlGaN determined by the Fermi level position at the surface and at the hetero-interface. Thus, the 2DEG density at the hetero-interface departs somewhat from the polarization-induced electron density. The modified 2DEG carrier density is expressed as

$$n_S(x) = \frac{\sigma_{\mathrm{2DEG}}}{q} = \frac{\sigma_p(x)}{q} - \frac{\epsilon(x)}{q^2 d}[q\phi_m(x) - \Delta E_{\mathrm{c}}(x) + E_{\mathrm{F}}(x)] \tag{9.70}$$

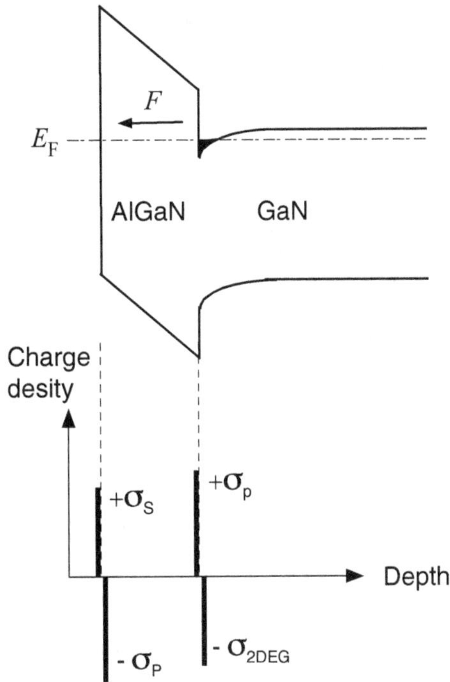

Fig. 9.34 Band diagram of an AlGaN/GaN HEMT structure grown on Ga-face GaN. The sheet charge density distributions in the structure are also shown. σ_p, σ_{2DEG}, and σ_s are induced piezoelectrical, 2DEG, and surface charge densities, respectively

where σ_p is a piezoelectrically induced charge density at the AlGaN/GaN interface, ϵ is the dielectric constant of AlGaN barrier, d is the barrier thickness, $q\phi_m$ is the Schottky barrier height of a gate contact, ΔE_c is the conduction band offset at the AlGaN/GaN interface, and E_F is the Fermi level with respect to the GaN conduction band-edge energy. To determine the net sheet charge density from the polarization-induced 2DEG, we use the following approximations.

For $Al_xGa_{1-x}N$, $\Delta E_c(x) \approx 2.12x$ and $E_g(x) = xE_g(\text{AlN}) + (1\text{–}x)E_g(\text{GaN}) - 0.7x(1\text{–}x)$. The dielectric constant is expressed as $\epsilon(x) = 9.5 - x$, and $q\phi_m(x) = 1.3x + 0.84$ (eV). The Fermi energy has the form of

$$E_F(x) = E_0(x) + \frac{\pi\hbar^2}{m^*(x)} n_S(x) \tag{9.71}$$

where $E_0(x)$ is determined from the Airy function and $m^*(x) = 0.22m_e$. Using the piezoelectric-induced sheet charge density expression of

$$\sigma_p(x) = [P_{SP}(\text{GaN})] - [P_{SP}(\text{AlGaN}) + P_{PE}(\text{AlGaN})],$$

and polarizations from (7.128) to (7.131)

$$P_{SP}(Al_xGa_{1-x}N/GaN) = -0.090x - 0.034(1-x) + 0.021x(1-x) \text{ c/m}^2$$

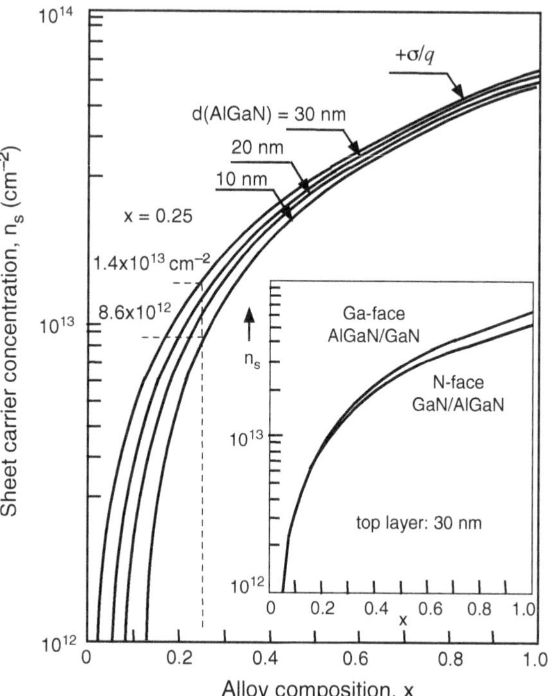

Fig. 9.35 Sheet carrier concentration of the 2DEG confined at a Ga-face GaN/AlGaN/GaN or N-face GaN/AlGaN/GaN interface for different thicknesses of the $Al_xGa_{1-x}N$ barrier. The inset shows the maximum sheet carrier concentration of a pseudomorphically grown Ga-face AlGaN/GaN and an N-face GaN/AlGaN heterostructure. Reprinted with permission from [9], copyright AIP Publishing

$$P_{\mathrm{PE}}(\mathrm{Al}_x\mathrm{Ga}_{1-x}\mathrm{N/GaN}) = -0.0525x + 0.0282x(1-x)\ \mathrm{c/m^2}$$

the 2DEG density can be obtained.

Figure 9.35 displays the calculated 2DEG sheet charge density as a function of Al-composition x for 10, 20, and 30 nm AlGaN barriers using parameters available at the time of report [9]. Nevertheless, the results give a meaningful insight into and design guidelines for the AlGaN/GaN HEMT. They show that the piezoelectric effect leads to the formation of a high-density 2DEG (~10^{13} cm^{-2}) in AlGaN HEMT structures without intentional doping. In fact, the sheet charge densities in nominally undoped nitride HEMTs can be comparable to those achievable in doped-channel FET structures, but without the degradation in mobility resulting from the presence of ionized impurities in the channel. The 2DEG sheet charge density can go even higher by increasing the Al-composition x in the barrier layer. However, for high crystal quality, the large lattice-mismatch between AlN and GaN prevents it from using $x \geq 40\%$ in the $Al_xGa_{1-x}N$ barrier. On the other hand, for $x < 15\%$, the conduction band offset becomes too low to confine the 2DEG. Therefore, an optimal composition of $Al_xGa_{1-x}N$ barrier for maximum 2DEG sheet charge density is limited to $0.15 < x < 0.4$. The inset compares the maximum sheet charge density of a pseudomorphic grown Ga-face AlGaN/GaN and an N-face GaN/AlGaN heterostructure. The figures show similar trends in the useful composition range.

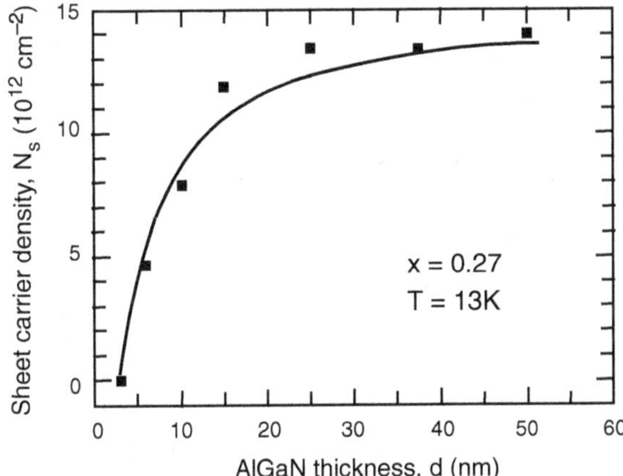

Fig. 9.36 2DEG sheet carrier density in the $Al_{0.27}Ga_{0.73}N$/GaN structures as a function of AlGaN barrier width. Reprinted with permission from [10], copyright AIP Publishing

Since the 2DEG in AlGaN HEMT structure is generated by polarization, it requires a minimum strain in the AlGaN barrier, which can be controlled by adjusting the composition and/or the layer thickness. The higher Al composition x in AlGaN causes a larger strain leading to a greater polarization as well as a greater conduction band-edge discontinuity. Therefore, the resultant 2DEG density increases with increasing x for a fixed barrier thickness. On the other hand, for a fixed x, the lattice-mismatch-induced strain increases with the barrier layer thickness, which, in turn, increases the polarization. As shown in Fig. 9.36, for $x = 0.27$, the induced 2DEG sheet charge density becomes significant only after the barrier thickness is larger than ~5 nm and increases rapidly thereafter. However, the 2DEG density also depends on the efficiency of carrier transfer across the undoped AlGaN barrier, and it prefers a thin AlGaN for easy transfer of sheet charges. Thus, the barrier layer thickness increases beyond ~15 nm, the 2DEG sheet charge density does not increase appreciably and saturates gradually.

Now, we have to answer the question: Where do those high-density 2DEG carriers come from in such an undoped structure? Experiments have been carried out to rule out the possibility of the residual impurities in the undoped heterostructure as the source. It turns out that the donor-like surface states E_{ss} located ~1.5 eV below the conduction band edge of the AlGaN are the source of the 2DEG, as seen from the band diagrams shown in Fig. 9.37. In an AlGaN HEMT structure with a very thin barrier, the energy band edge maintains a near flat band configuration and $E_F \gg E_{ss}$. No 2DEG is formed. When increasing the AlGaN barrier thickness, a greater potential drop is developed and the surface E_{ss} increases accordingly. For thin AlGaN barriers, E_{ss} is still below E_F and no 2DEG forms (Fig. 9.37b). Only after the AlGaN barrier thickness reaches a threshold such that $E_{ss} \geq E_F$ and electrons associated with the donor-like surface states are ready to transfer to the hetero-interface does the 2DEG form (Fig. 9.37c).

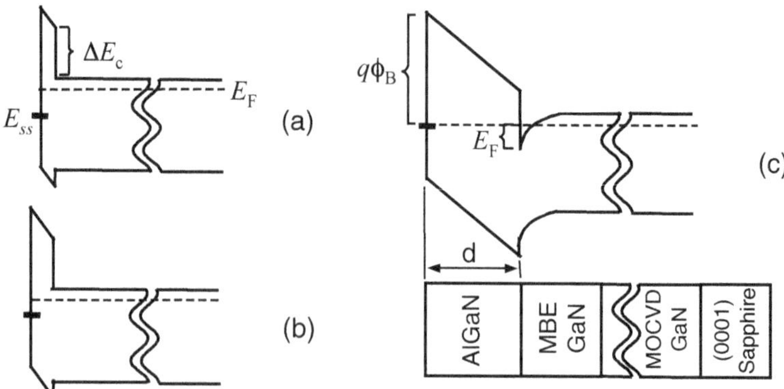

Fig. 9.37 Schematic diagram showing the development of the band structure in AlGaN/GaN samples with increasing AlGaN barrier width. Reprinted with permission from [10], copyright AIP Publishing

9.7 Heterojunction Bipolar Transistors (HBTs)

9.7.1 Introduction

The idea of using heterostructure at the emitter–base junction in a bipolar junction transistor (BJT) is not new. It was pointed out and patented by William Shockley at the time of the invention of the bipolar transistor at Bell Labs that a large bandgap emitter can be used to increase the efficiency of the device. US Patent 2,569,347 issued to W. Shockley (filed on June 26, 1948; expired on Sept. 24, 1968) claimed.

Claim 2 *"A device as set forth in claim 1 in which one of the separated zones is of a semiconductor material having a wider energy gap than that of the material in the other zones"*

Herb Kroemer proposed the first heterojunction bipolar transistor (HBT) structure in 1957 [11]. Like other heterostructure devices, the HBT technology has been progressing rapidly only after the material growth technologies were matured. Currently, the highest operation frequency of $f_T > 800$ GHz has been demonstrated in InP/InGaAs HBTs [12]. New functions, such as high-speed spontaneous and stimulated light emission, have also been demonstrated in HBTs (light-emitting transistors and transistor lasers) [13].

The HBT inherits all the advantages of the BJT including high cutoff frequency due to the short transient time across the thin base region, uniform turn-on voltage which is determined by the built-in potential of the junction, high power capability resulting from the entire emitter area conducting current, and high transconductance which enables low-power applications. By adding a heterojunction to the BJT, additional advantages emerge. The energy band discontinuity at the emitter–base (EB) heterojunction of a HBT suppresses the diffusion of majority carriers from the base into emitter for higher gain. The EB heterojunction also allows the use of a highly doped base region for low base sheet resistance without lowering gain. This means a very thin base layer can be used for a shorter transient time to achieve a higher cutoff frequency. It is expected that HBTs have higher current gain and cutoff frequency than BJT.

At the onset of HBT development, only one heterojunction was used at the EB junction of a single heterojunction bipolar transistor (SHBT). The emitter region of the GaAs-based SHBT is replaced by a larger bandgap AlGaAs, and the base–collector (BC) junction remains a GaAs homojunction. To enhance the breakdown voltage performance of a HBT, a larger bandgap material, such as AlGaAs, is used in the collector region. In this case, both EB and BC junctions are replaced by AlGaAs/GaAs heterojunctions to form a double-heterojunction bipolar transistor (DHBT). Currently, HBTs using different material systems have been developed including $In_xGa_{1-x}P$/GaAs, InP/$In_xGa_{1-x}As$, Si/Si_xGe_{1-x}, and InP/$GaAs_xSb_{1-x}$ for various applications. Obviously, heterojunctions with either type-I or type-II energy band discontinuity can be used in HBT designs.

9.7.2 Basic Theory of Heterojunction Bipolar Transistors

The energy band diagram of a type I N-p^+-n SHBT under normal operation is shown in Fig. 9.38. To generalize the discussion, the hetero-interface of the HBT is graded such that the effect of the EB junction energy spike in the conduction band is neglected. The energy bandgap of the emitter and base are E_{gN} and E_{gp}, respectively. A forward

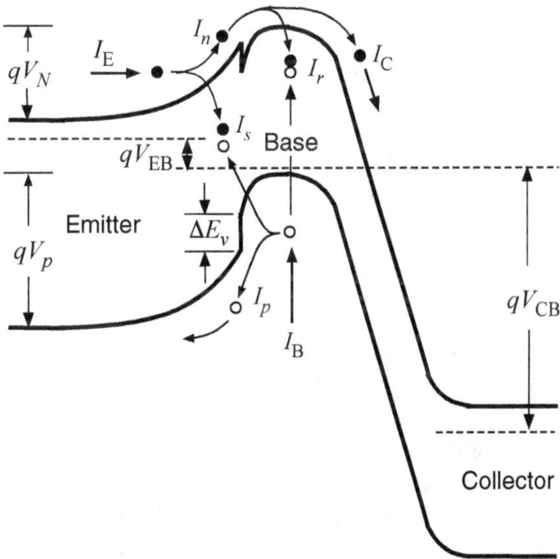

Fig. 9.38 Energy band diagram of an N-p-n HBT, under normal mode operation, showing the various current components and the hole-blocking effect of the valence band discontinuity ΔE_v at the EB junction

bias voltage of V_{EB} and a reverse bias voltage of V_{CB} are applied to EB and BC junctions, respectively. The terminal currents of the SHBT are the emitter current I_E, base current I_B, and collector current I_C. These currents are related to internal current components of the injected electron current into base (I_n), the injected hole current into emitter (I_p), the saturation current at the EB junction (I_s), and the recombination current in the base region (I_r) through the following equations:

$$\begin{cases} I_E = I_n + I_s + I_p \\ I_B = I_p + I_r + I_s \\ I_C = I_n - I_r \end{cases} \tag{9.72}$$

The above discussions can also apply to BJTs by simply setting $E_{gN} = E_{gp}$. The ratio of the collector current and base current defines the current gain of the junction transistor.

$$\beta = \frac{I_C}{I_B} = \frac{I_n - I_r}{I_p + I_r + I_s} \tag{9.73}$$

To maximize β, the components in the numerator and denominator need to be maximized and minimized, respectively. The maximum current gain (β_{max}) is simply the ratio I_n/I_p in the limit of negligible recombination losses. Assuming a non-degenerate case, the approximate injection electron and hole current densities at the EB junction are

$$J_n = qn_B\upsilon_{nB} = qN_E\upsilon_{nB}\exp(-qV_N/kT) \tag{9.74a}$$

$$J_p = qp_E\upsilon_{pE} = qp_B\upsilon_{pE}\exp\left(-qV_p/kT\right) \tag{9.74b}$$

where N_E and p_B are the doping level in the emitter and base, respectively; υ_{nB} is the electron mean velocity in the base near the EB junction; υ_{pE} is the hole mean velocity in the emitter near the EB junction; and qV_N and qV_p are the potential barrier height for electrons and holes, respectively. At the heterojunction, $q(V_p - V_N) = E_{gN} - E_{gp} = \Delta E_g$, we obtain the maximum current gain (β_{max})

$$\beta_{max} = \frac{I_n}{I_p} = \frac{N_E}{p_B}\frac{\upsilon_{nB}}{\upsilon_{pE}}\exp\left(\Delta E_g/kT\right) \tag{9.75}$$

The mean velocity ratio of ($\upsilon_{NB}/\upsilon_{pE}$) is the least subject to manipulation. To achieve large β_{max} (>100), it is necessary that either $N_E \gg p_B$ or $\Delta E_g > kT$. Energy gap differences that are several kT are readily available. Due to the exponential nature of the ($\Delta E_g/kT$) term, it dominates β_{max} even if $N_E < p_B$. Therefore, using heterojunction structures with large ΔE_g, very high maximum current gain can be achieved regardless of the emitter-to-base doping ratio. It simply means that the hole injection current I_p is sufficiently suppressed by a large ΔE_v and becomes a negligible part of the base current compared to the two recombination currents: $I_B \approx I_s + I_r$. If we approximate I_E by I_n, we obtain

$$\beta \cong \frac{I_n}{I_r + I_s} \tag{9.76}$$

In a HBT with a highly doped base, the recombination current I_r is the dominant current component since the saturation current $I_s \ll I_n$ for a high-quality interface. At the same time, the very high doping density translates into a rather short minority carrier lifetime in the base region. Therefore, the bulk recombination current I_r, rather than the interface recombination current I_s, will dominate the base current. Thus, $I_B \approx I_r$ and equals the total minority carriers (Q_n) stored in the base divided by the minority carrier lifetime (τ_n):

$$I_r \approx \frac{Q_n}{\tau_n} = \frac{qAn_B W_B}{2\tau_n} = \left(\frac{qAW_B}{2\tau_n}\right) N_E \exp(-qV_N/kT) \tag{9.77}$$

where A is the cross-sectional area of the EB junction and W_B is the base width of the device. The current gain of a HBT structure with a large ΔE_g and high quality interfaces becomes

$$\beta \equiv \frac{I_n}{I_r} = \frac{N_E \upsilon_{nB} \exp(-qV_N/kT)}{(W_B/2\tau_n) N_E \exp(-qV_N/kT)} = \frac{2\tau_n \upsilon_{nB}}{W_B} \tag{9.78}$$

The effect of the base doping on the minority carrier lifetime and the base width in turn directly affects the current gain. Even if for heavy base doping levels the lifetimes may be short, high β's should be achievable in HBTs with sufficiently thin base region. Furthermore, in a thin base HBT, the electron velocity is likely to approach its saturation velocity of ~10^7 cm/s, enhancing the current gain. As a result, no serious concerns arise from reduced minority carrier lifetime in HBTs with heavily doped thin ($\leq$ 100 nm) base region.

Now we can examine the current gain property of a BJT. In homojunction bipolar junction transistors, $\Delta E_g = 0$. The only way to achieve high current gain is to make $N_E \gg p_B$. Since the base doping level is kept low, it requires a thick base layer to minimize the base sheet resistance. A thick base layer will eventually limit the cutoff frequency of the BJT. On the other hand, the doping level in the emitter cannot exceed certain limits. A heavily doped emitter will cause other problems. When the emitter degenerates, the Fermi level can move into the conduction band and cause bandgap shrinkage. This is equivalent to a reduction of conduction band discontinuity at the EB junction or $\Delta E_g < 0$. A lowering of β_{max} is expected. Therefore, the HBT takes advantage of the exponential dependency of ΔE_g to achieve high gain with high base doping concentration. Furthermore, the doping profile of the BJT is graded due to diffusion and/or ion implantation processes used in device fabrication. For HBTs, a much sharper doping profile is achieved by controlling doping properties during epitaxy. The abrupt and heavily doped base region is desirable for achieving high current gain and high cutoff frequency in a HBT. Thus, it is critical to use a suitable *p*-type dopant that possesses a high doping efficiency and a low diffusivity. For GaAs-based HBTs, carbon has proved to be an ideal *p*-type dopant among other common candidates such as Zn and Be. It can be easily doped to over 10^{20} cm^{-3} in many III–V compound semiconductors with negligible out-diffusion. However, carbon atoms are hard to generate from a solid source, such as graphite, due to their extremely low vapor pressure. Currently, the heavily *p*-type carbon doping is conveniently achieved using gaseous carbon sources such as CCl_4 (in MOCVD) and CBr_4 (in MBE).

The current gain of a BJT and HBT is commonly measured using the Gummel plot scheme [14]. A Gummel plot shows the base and collector currents as functions of V_{BE} when $V_{CB} = 0$. The ratio of the measured I_C/I_B at various V_{BE} gives the current gain. For the ideal case, both I_B and I_C are dominated by diffusion current component and proportional to $\exp(qV_{BE}/kT)$. We expect a constant gain independent of bias. In real devices, defects and other recombination centers at the EB hetero-interface and in the bulk of the base lead to a non-ideal base current with an $\exp(qV_{BE}/mkT)$ dependence. The parameter m is referred to as the *ideality factor* and has a value between 1 (dominated by diffusion current) and 2 (dominated by generation-and-recombination current I_s). Therefore, the EB junction property can be determined by the ideality factor (m) of I_B. For example, Fig. 9.39 shows the Gummel plot of an InGaP/GaAs DHBT with an emitter size of 60 × 60 μm^2. At low current, $m =$ 1.84 (~2), I_B is mostly due to the generation-recombination current in the depletion region of the EB junction and I_s is larger than I_r. At mid-to-high current, $m = 1.15$ (~1), showing that the bulk recombination current (I_r) in the base region becomes dominant due to its high doping level. The base current eventually departs from $m \sim 1$ due to the large voltage drops across the parasitic base and emitter resistances, causing the available junction voltage to decrease. For the collector current, it increases with V_{BE} with an ideality factor of $m = 1.07$, a value which is close to unity.

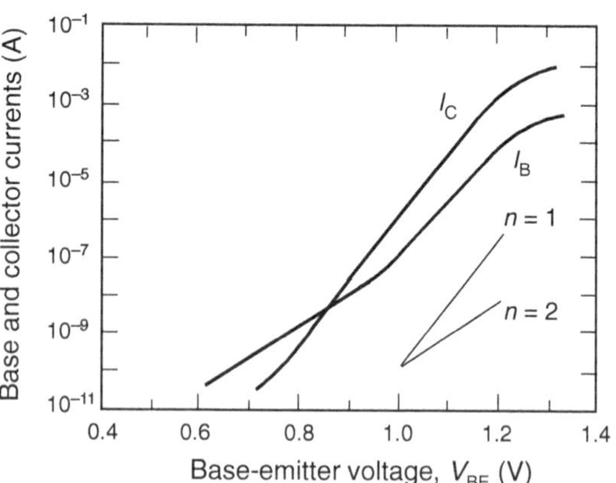

Fig. 9.39 Gummel plot for InGaP/GaAs HBT. The emitter size of the HBT is 60 × 60 μm^2. Reprinted with permission from [15], copyright IEEE

9.7.3 Band Discontinuity of Heterostructures

In a type-I heterostructure, a 'spike-and-notch' energy band diagram generally appears at the abrupt heterojunction due to the conduction band discontinuity ΔE_c as shown in Fig. 9.40 for a SHBT. This energy spike can be removed by using a graded composition in the large bandgap emitter near the hetero-interface as used for discussions in the previous section. The existence of the energy spike at the EB junction of an *abrupt* HBT can modify the current conduction across the heterojunction. Using the simple thermionic emission model, the magnitude of forward current depends on the amount of electrons having enough energy to surmount the potential barrier. In an *abrupt* HBT, the potential barrier for electrons is simply V_N. However, in a *graded* HBT, the effective potential barrier becomes $(V_N - \Delta E_c/q)$ since the energy spike, ΔE_c, has been eliminated by the graded composition design. Therefore, under the same bias condition, the collector current of the abrupt HBT, I_n(abrupt), is reduced from the graded HBT, I_n(graded), by a factor of

$$\frac{I_n(\text{abrupt})}{I_n(\text{graded})} = \frac{\exp(-qV_N/kT)}{\exp[-(qV_N - \Delta E_c)/kT]} = \exp(-\Delta E_c/kT) \tag{9.79}$$

Since the current gain is proportional to $\exp(\Delta E_g/kT)$ in the graded junction case, the current gain of abrupt HBTs is modified by reducing ΔE_c from ΔE_g in the injection current relation (9.75). Thus

$$\beta_{\max} = \frac{N_E \upsilon_{nB}}{p_B \upsilon_{pE}} \exp\left(\frac{\Delta E_g - \Delta E_c}{kT}\right) = \frac{N_E \upsilon_{nB}}{p_B \upsilon_{pE}} \exp\left(\frac{\Delta E_v}{kT}\right) \tag{9.80}$$

It is clear that a large ΔE_v is preferred in abrupt N-p^+-n HBT structures for a high current gain.

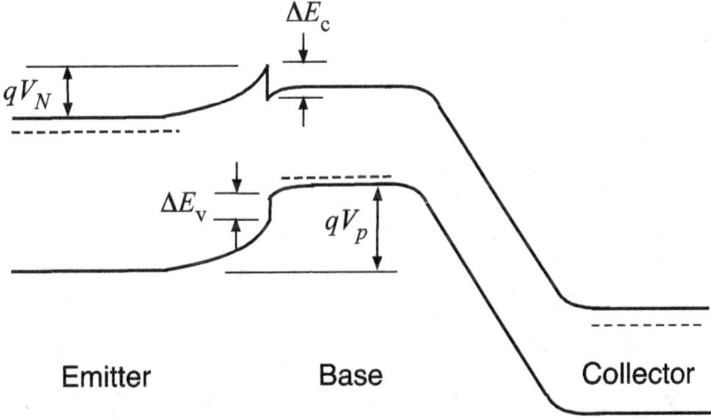

Fig. 9.40 Energy band diagram for HBT with abrupt emitter–base junction

The existence of the energy spike in the conduction band at the EB heterojunction has other effects on device performance. A beneficial effect is that electrons injected from emitter into the base have to overcome the EB junction barrier with higher potential energy and become hot electrons with high saturation velocity. The improved travel speed across the base region leads to a reduced base transient time for high frequency operation. A detrimental effect is that the energy spike can block current flow at bias voltages $V_{CE} < \Delta E_c/q = V_{off}$, which is the offset voltage. Also, the triangular QW formed on the base side of the EB junction can trap electrons that lead to an enhanced recombination at the junction. Therefore, the current gain will be reduced due to an increasing I_r.

Next, we turn our attention to the BC junction. In a SHBT, the same semiconductor is used for both base and collector and the BC junction is a homojunction. Therefore, the same arguments about the BC junction of a BJT can be applied to a SHBT. In DHBTs, the BC homojunction is replaced with a heterojunction. The use of a wide bandgap collector material can improve the power handling capability by the enhanced breakdown voltage. Again, due to the existence of a finite ΔE_c, an energy spike forms at the BC junction as shown in Fig. 9.41. This energy barrier (ΔE_c) impedes free collection of electrons by the reverse-biased BC junction. It will suppress the collector current until a large V_{CE} is reached to allow collection of electrons by the collector. Thus, it will increase the minority carrier resident time in the base region, which reduces the bandwidth of the device. Therefore, a graded collector junction in the form of composition grading or chirped superlattice structure is usually used to eliminate this problem. For a lattice-mismatched material system such as InGaAs/InP, this will complicate the growth procedures and may cause interface problems leading to a lower current gain.

The collector current blocking mentioned above can be totally avoided by using a type-II DHBT design. This design simplifies the growth of the BC junction without using composition grading. In particular, for the InP/GaAs$_{0.5}$Sb$_{0.5}$/InP DHBT design

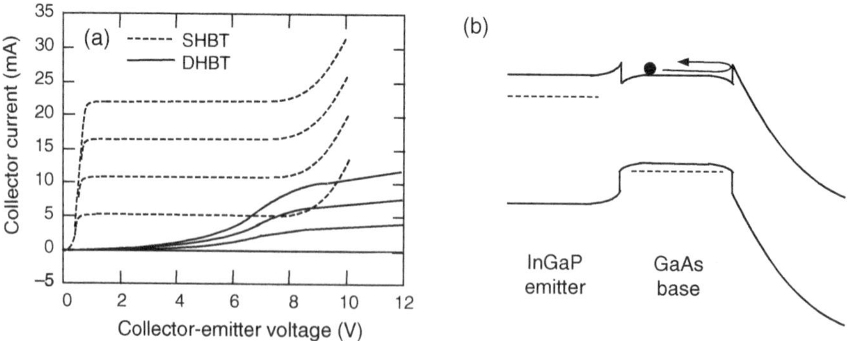

Fig. 9.41 **a** Comparison of the *I–V* characteristics of InGaP/GaAs SHBT and DHBT. The base current was stepped from 0 to 400 μm in 100 μA steps. Reprinted with permission from [16], copyright AIP Publishing. **b** Current blocking at the base–collector junction of an InGaP/GaAs DHBT

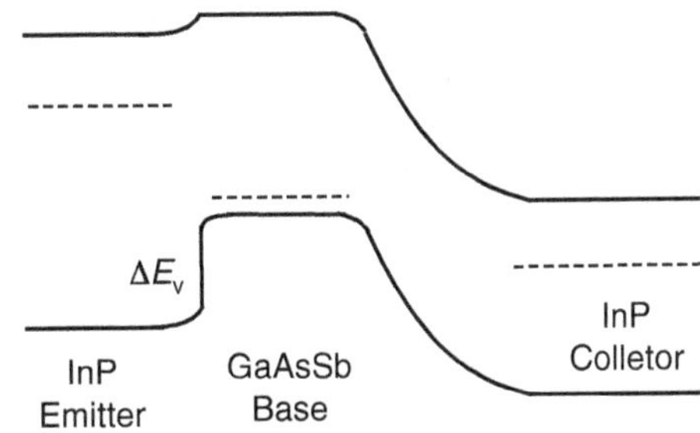

Fig. 9.42 Energy band diagram of an InP/GaAsSb type-II double-heterojunction bipolar transistor

shown in Fig. 9.42, it provides a nearly zero conduction band offset ($\Delta E_c = -0.06$ eV, E_c(InP) < E_c(GaAsSb)) for a negligible turn-on voltage and no current blocking effect at the BC junction. The large ΔE_v of ~0.62 eV and the use of InP collector material provide a large current gain and large breakdown voltage, respectively.

9.7.4 High-Frequency Operation of HBT

The high-frequency performance of HBTs is fundamentally determined by the stored minority carrier charge in the transistor that has to be removed from (added to) the transistor before it can turn off (on). Hence, the maximum frequency at which the transistor is capable of operating depends on the junction capacitance charging time and charge carrier transit time. In forward active operation, the forward transit time τ_F or the emitter-to-collector transit time τ_{EC} represents a fundamental limit for the switching speed and maximum frequency of operation of a HBT. The parameter most commonly used to define this maximum frequency of operation is the *cutoff frequency* f_T. This is the frequency at which the common emitter, small-signal current gain drops to unity under conditions of a short-circuit load.

In practice, parasitic capacitance and resistance of the HBT will slow down the switching of digital circuits and limit the frequency of operation of analog circuits. To include the effect due to parasitic resistances and junction capacitances, the *maximum oscillation frequency* f_{max} is a good predictor of transistor performance.

(a) **Emitter-to-collector transit time**

The emitter-to-collector transit time models the excess charge stored in the HBT under normal operation of a common-emitter configuration. This is an extremely important device parameter, since it sets a fundamental physical limit to the switching speed and maximum oscillation frequency of operation of a HBT. The emitter-to-collector transit time τ_{EC} can be written as the sum of the individual delay times as well as charging times of junction capacitances in the various regions of the HBT.

$$\tau_{EC} = \tau_E + \tau_B + \tau_C + \tau_{BC} \tag{9.81}$$

where τ_E, τ_B, τ_C, and τ_{BC} are the emitter delay time, base transit time, collector transit time, and base–collector space-charge layer delay time, respectively. The emitter delay τ_E associates the emitter junction charging time involving the emitter–base junction capacitance C_E and the dynamic emitter resistance kT/qI_C.

$$\tau_E = \frac{kT}{qI_C} C_E \tag{9.82}$$

The base delay τ_B represents the transport of electrons through the base layer via drift, diffusion, or ballistic transport. In the limit of diffusion current only,

$$\tau_B \equiv \frac{Q_B}{I_C} \tag{9.83}$$

with

$$\begin{cases} Q_B = \dfrac{qAW_B n_{B0}}{2} \exp\left(\dfrac{qV_{BE}}{kT}\right) \\ I_C = qAD_{nB}\dfrac{dn}{dx} = qAD_{nB}\dfrac{n_{B0}}{W_B} \exp\left(\dfrac{qV_{BE}}{kT}\right) \end{cases}$$

Thus,

$$\tau_B = \frac{W_B^2}{2D_{nB}} \tag{9.84}$$

W_B is the base width, D_{nB} is the electron diffusion constant, and n_{B0} is the equilibrium electron concentration. This result of τ_B is valid for HBTs with a uniform base. If the base is non-uniformly doped or has a composition gradient, then the variation in doping and/or bandgap energy gives rise to a built-in electric field across the neutral base region. This built-in field would aid electron transport across the base and hence reduce the base transient time. This situation can be taken into account by modifying the equation as

$$\tau_B = \frac{W_B^2}{\gamma D_{nB}} \tag{9.85}$$

where γ is a constant that has a value of $2 \le \gamma \le 4$.

The collector transit time τ_C corresponds to the transport time through the collector layer, is classically found to be proportional to the collector thickness W_C, and is expressed as

$$\tau_C = \frac{W_C}{2v_{sat}} \tag{9.86}$$

where υ_{sat} is the electron saturation velocity in the collector, typically measured as 3–4 × 10^7 cm/s.

The last term τ_{BC} represents the charging time delay associated with the parasitic BC junction capacitance C_{BC}. For an incremental input voltage change dV_{BE}, the BC junction voltage change is

$$dV_{BC} = dI_C(R_C + R_E + kT/qI_C) \tag{9.87}$$

where R_C and R_E are the parasitic collector and emitter resistances, respectively. Therefore,

$$\tau_{BC} = (R_C + R_E + kT/qI_C)C_{BC} \tag{9.88}$$

The total delay associated with a HBT is directly related to the cutoff frequency and can be described as

$$f_T = \frac{1}{2\pi\tau_{EC}} = \left\{2\pi\left[\tau_B + \tau_C + (R_E + R_C)C_{BC} + \frac{kT}{qI_C}(C_E + C_{BC})\right]\right\}^{-1} \tag{9.89}$$

Since f_T is defined for small-signal conditions, a small-signal circuit model, such as the hybrid-π model, as shown in Fig. 9.43, can be used to derive an expression for f_T. To a first order, the series resistances of three terminals and recombination in the depletion region were neglected. Under the conditions of a short-circuit load, the small-signal collector and base currents can be written as

$$i_c = g_m \upsilon_{be} - i\omega C_\mu \upsilon_{be} \tag{9.90}$$

$$i_b = g_\pi \upsilon_{be} + i\omega(C_\mu + C_\pi)\upsilon_{be} \tag{9.91}$$

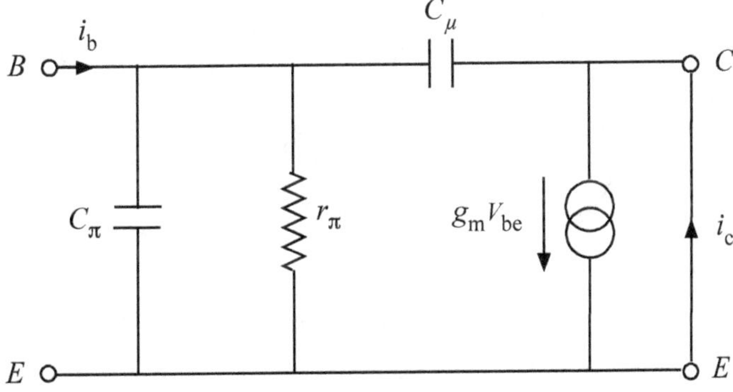

Fig. 9.43 Small-signal hybrid-π circuit model of HBT

where $g_\pi = 1/r_\pi$ (r_π is the equivalent input resistance), $\omega = 2\pi f$, g_m is the transconductance, C_π is the BE junction diffusion capacitance, and C_μ is the BC junction diffusion capacitance.

The common-emitter current gain can be written as

$$\beta = \frac{i_c}{i_b} = \frac{g_m - i\omega C_\mu}{g_\pi + i\omega(C_\mu + C_\pi)} \cong \frac{g_m}{g_\pi + i\omega(C_\mu + C_\pi)} \tag{9.92}$$

If we define $\beta_0 = g_m r_\pi$, the equation can be simplified as

$$\beta = \frac{\beta_0}{1 + i\omega r_\pi (C_\mu + C_\pi)} \tag{9.93}$$

At low frequencies, the current gain maintains a constant value of $\beta = \beta_0$. As the frequency increases, the second term in the denominator of the equation becomes large with respect to unity, and β can be approximated by

$$|\beta| = \frac{\beta_0}{2\pi f r_\pi (C_\mu + C_\pi)} \tag{9.94}$$

At high frequencies, β begins to roll off with a limiting roll-off rate of −20 dB/decade ($\propto 1/f$). The frequency-dependent gain values are determined from S-parameters measured with microwave network analyzers. The gain values (in dB) as functions of frequency are usually plotted in a semi-log fashion under constant $V_{CE} = V_{BE}$ and I_C as shown in Fig. 9.44a. For very high-speed devices, the gain at frequencies exceeding the limit of instrument capacity was obtained by extrapolating −20 dB/decade lines from a best-fit average of the measured current gain.

Since τ_{EC} is a function of the collector current, the cutoff frequency is also a function of the collector current as illustrated in Fig. 9.44b. At low collector currents, the

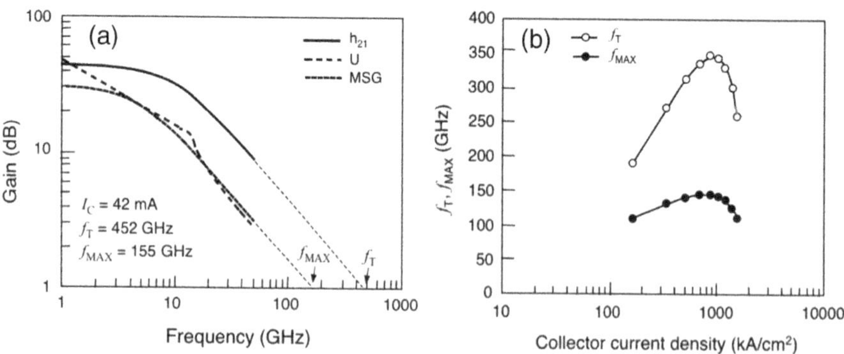

Fig. 9.44 **a** RF response and performance figure extrapolations for 0.25 × 16 μm² InP/InGaAs SHBT. Reprinted with permission from [17], copyright IEEE. **b** f_T and f_{max} as a function of collector current density of a 0.35 × 8 μm² type-II InP/GaAsSb DHBT. Reprinted with permission from [18], copyright American Vacuum Society

depletion capacitance term (τ_{BC}) in (9.89) dominates and hence f_T increases with I_C. At medium currents, the transit time terms become larger than the depletion capacitance term, and f_T ceases to rise with collector current. At high collector currents, the cutoff frequency decreases markedly due to high current effects, especially the Kirk effect which will be described later.

The other important high-frequency parameter for a HBT is the maximum oscillation frequency f_{max}, where the unilateral power gain (U) drops to unity. The unilateral or Mason gain is the maximum gain obtained by conjugate matching of inputs and outputs of the device, together with the use of a lossless feedback network to tune out any internal device feedback. The derived expression of the maximum oscillation frequency is

$$f_{max} = \sqrt{\frac{f_T}{8\pi C_{BC} R_B}} \tag{9.95}$$

which shows that the f_{max} is determined not only by the f_T but also by BC junction capacitance C_{BC} and the base resistance R_B. It is clear that a high *p*-type doping concentration in the base region leads to a low base sheet resistance and an improved f_T, both of which enhance the f_{max}. On the other hand, the increase in f_T due to reduced τ_{BC} by higher collector doping could increase C_{BC} and may cause a decrease in f_{max}, depending on the relative importance of τ_{BC} in f_T. Therefore, for high-frequency circuit applications, a careful optimization of these two parasitics is required.

(b) **Kirk effect**

The reduction of f_T at high collector current mentioned in the last section is due to a base widening or Kirk effect. Consider Poisson's equation in the depletion region of the BC junction:

$$\frac{dF}{dx} = \frac{\rho}{\epsilon} = \frac{q}{\epsilon}(p - n + N_d) \tag{9.96}$$

where p and n are mobile charges and N_d is the fixed (ionized) charges. In the reverse-biased collector of *N-p-n* HBT structures, $p \approx 0$ and electrons injected from the emitter are transported across the BC junction by drift under high field situations. The electron concentration relates the drift current density J_n by $n = J_n/qv_{sat}$, where v_{sat} is the electron saturation velocity. Thus,

$$\frac{dF}{dx} = \frac{1}{\epsilon}\left(qN_d - \frac{J_n}{v_{sat}}\right) \tag{9.97}$$

This equation indicates that the slope of the electric field intensity depends on the difference of qN_d and J_n/v_{sat}. Under low current density, the contribution from the mobile electrons can be neglected and dF/dx is unchanged (Fig. 9.45). At sufficiently

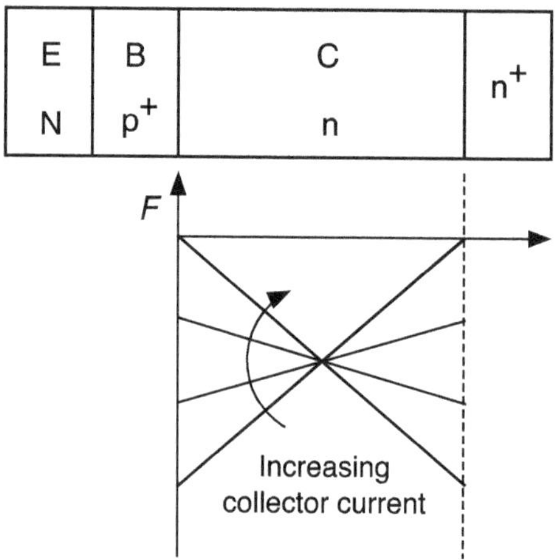

Fig. 9.45 Electric field intensity distribution in the fully depleted collector region under large collector current density. With further increase of collector current density, the slope of the electric field intensity curve changes from positive to negative as indicated by the arrow

large collector current density, the BC depletion region can extend all the way to the n/n^+ subcollector boundary. For a fixed V_{CB}, the depleted collector region where the electric field F exists is fixed and dF/dx still maintains a positive slope. With further increasing the collector current density (J_C), the mobile charge in the depletion region becomes significant and cannot be neglected. This leads to a decreasing F with increasing J_C or a decreasing dF/dx slope. At certain J_C, $F = 0$ and the depleted collector layer turns into a neutral layer. Holes can be injected freely from the base into the adjacent neutral collector region. With the injected holes, the conductivity type of the collector adjacent to base changes from n to p. Thus, the effective base thickness is increased at high collector current density, which degrades the cutoff frequency of the HBT as shown in Fig. 9.44b.

(c) **Device scaling and surface passivation**

Earlier studies of HBTs have shown that a significant component (>75%) of the total device delay was associated with the intrinsic forward delay of the device, primarily the base and collector transit times. Therefore, an effective method to increase the cutoff frequency f_T of the HBT is through *vertical scaling* of the base and collector epitaxial layers. Vertical scaling strictly involves the design of layer thicknesses, doping levels, and bandgap engineering to reduce device transit time, thereby increasing device bandwidth. Figure 9.46 shows the experimental relationship between vertical scaling and base/collector transit times. The base transit time is clearly illustrated by two slopes. For thicker bases, the classic diffusion transport dominates and is proportional to $(W_B)^2$. As the base thickness is scaled below the mean free path for electrons (~30 nm), the ballistic transport dominates with a linear dependence on W_B. However, as the base and collector layers are thinned, parasitic resistances and capacitances also increase—specifically the base resistance, R_B, and the base/collector capacitance, C_{BC}. The increase of these parasitic components acts to severely degrade the maximum oscillation frequency f_{max} of the device.

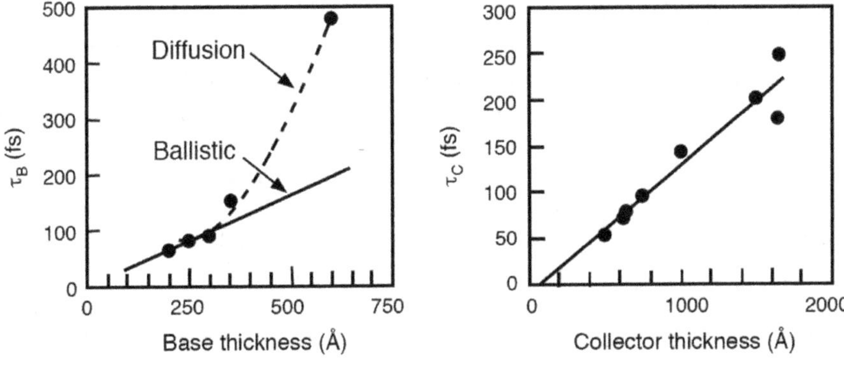

Fig. 9.46 Experimental results on the relationship between vertical scaling of InGaAs/InP SHBT and base/collector transit times

For effective reduction of transit times in HBTs without incurring significant increases in charging delays, vertical and lateral scaling must be simultaneously employed in order for submicron devices to achieve high performance. *Lateral scaling* refers to device topology design and processing techniques used to reduce parasitic capacitances and resistances to further enhance device speed. Lateral scaling also has the benefit of providing lower power operation and more efficient heat removal; reducing the emitter length lowers the thermal resistance of the HBT, allowing for higher current densities and lower junction temperatures. Figure 9.47 shows the peak f_{max} and f_T values versus emitter dimensions of InP/InGaAs SHBTs. For the 0.25 μm emitter width (W_E), a 76% improvement in f_{max} is observed from reducing emitter lengths (L_E) from 12 μm to 1 μm; shrinking the emitter width from 0.4 μm to 0.25 μm results in an 18% f_{max} increase.

Beyond a certain point, however, scaling has detrimental effects on device performance due to physical limitations imposed by shrinking device dimensions. The

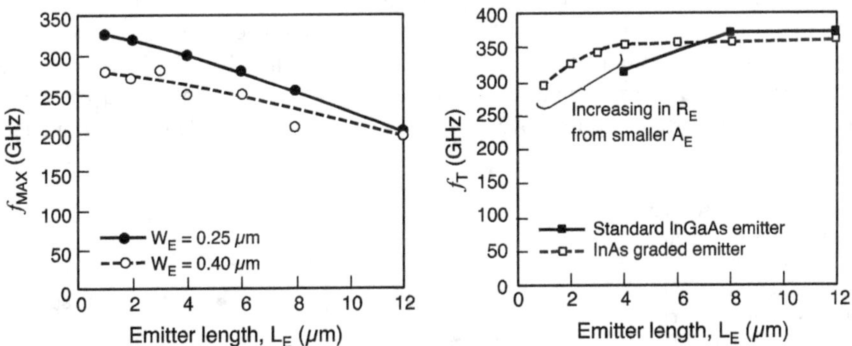

Fig. 9.47 Experimental results on the relationship between emitter lateral scaling of InGaAs/InP SHBT and cutoff frequency

significant drop in f_T observed in Fig. 9.47 for emitter lengths 8 μm and smaller on a standard InGaAs emitter cap is attributed to an increase in emitter resistance due to smaller contact areas. Ideally, scaling theory predicts that f_T should remain constant regardless of emitter length. However, the finite cap doping causes large increases in contact resistance for these emitter areas. The emitter cap cannot be doped higher due to solid solubility limits, but emitter contact resistance can be reduced through emitter cap engineering by using a narrow-bandgap cap material such as InAs. InAs provides a better contact material because of the smaller bandgap, higher doping capability, and higher thermal conductivity when compared to a standard InGaAs cap. The utilization of an InAs emitter cap is shown to reduce the emitter contact resistance by a factor of 2 in submicron HBT devices. The contact resistance reduction enhances the scalability of the emitter by allowing the emitter length to scale to dimensions smaller than 4 μm before the f_T degradation occurs. Overall, lateral scaling is often dependent on process maturity, relying on advanced fabrication techniques to counteract the effects of vertical scaling on f_{max}. Therefore, for HBT development using a new material system, the best indication of the system's potential is determined by the f_T of the transistor. It is assumed that as the process matures, the power gain of the device will be increased through technological advances.

In general, the HBT has a vertical geometry with an intrinsic device size defined by the emitter contact area. The base contact is designed to surround the intrinsic base region underneath the emitter. Unavoidably, there is an exposed extrinsic base region between the base contact and the intrinsic base area. Consequently, in HBTs made of materials with a high surface recombination velocity (SRV), such as GaAs (SRV > 10^6 cm/s), some minority carriers injected from the emitter recombine with the base majority carriers on the exposed surfaces resulting in the extrinsic base surface recombination current. This base current component is proportional to the magnitude of the emitter periphery rather than the emitter area. As the size of the emitter is reduced, the emitter perimeter-to-area ratio increases and surface recombination current becomes a major part of the overall base current. The current gain of such a scaled down GaAs–base HBT is substantially reduced from that of a large device whose perimeter-to-area ratio is small. The surface recombination current problem can be reduced by passivating the exposed extrinsic base region that is not directly under the emitter with an emitter ledge as shown in Fig. 9.48. A thin ledge of emitter material left surrounding the emitter mesa can dramatically reduce surface recombination current and improve device scaling. The exposed ledge layer outside

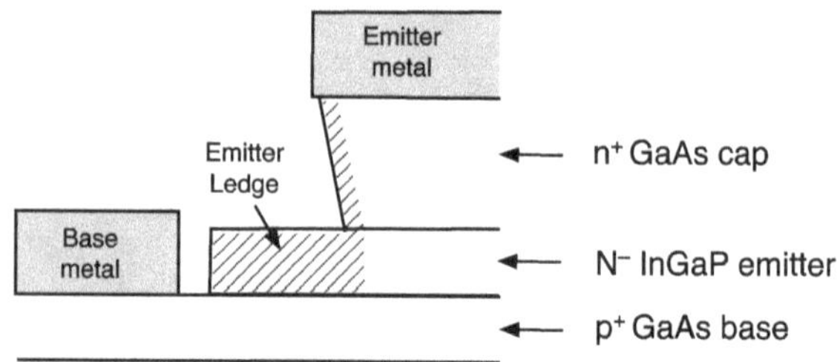

Fig. 9.48 Cross-sectional drawing of a HBT with emitter ledge passivation. The emitter ledge (hatched area) is totally depleted such that the base current underneath is not affected

the emitter mesa is totally depleted, and no carriers are available for recombination at the surface. Therefore, the base current is not affected and current gain is preserved. On the other hand, the free $In_xGa_{1-x}As$ surface shows more ideal surface characteristics with a small SRV of 10^3 cm/s for an In composition $x \sim 0.5$. Therefore, for HBT materials with low SRV such as in InP/InGaAs HBTs, the surface recombination current is not significant even in small devices and the surface passivation is not required.

9.7.5 Basics of HBT Processing

The design of HBTs is intimately interwoven with the technologies used for their fabrication. Therefore, any study of the HBT would be incomplete without understanding the fabrication technology. A typical HBT process developed at the University of Illinois at Urbana-Champaign (UIUC) is discussed below for either InGaP/GaAs or InP/InGaAs high-frequency HBTs. Other research groups may have developed their own HBT process steps, but the principal ideas discussed below remain the same. The standard InGaP/GaAs HBT structure is a single heterojunction device grown on a semi-insulating GaAs substrate that employs a carbon-doped GaAs base, a lattice-matched InGaP emitter, and a thin InGaAs emitter contact layer. Figure 9.49 shows a series of device profiles detailing the high-speed HBT process

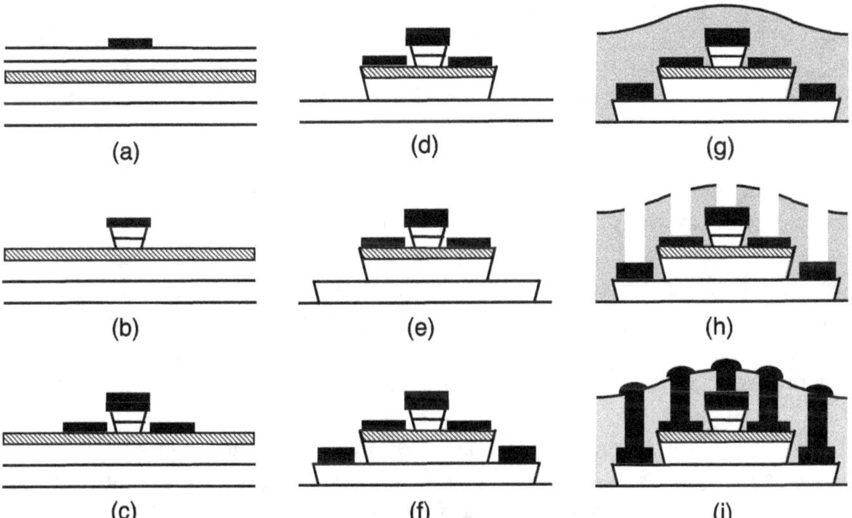

Fig. 9.49 Cross-sectional diagrams of UIUC HBT process: **a** post-emitter contact liftoff, **b** post-base etch, **c** post-base contact liftoff, **d** post-collector etch, **e** post-isolation etch, **f** post-collector contact liftoff, **g** post-polyimide/SiN_x passivation, **h** post-SiN_x via etch, and **i** post-overlay metal liftoff

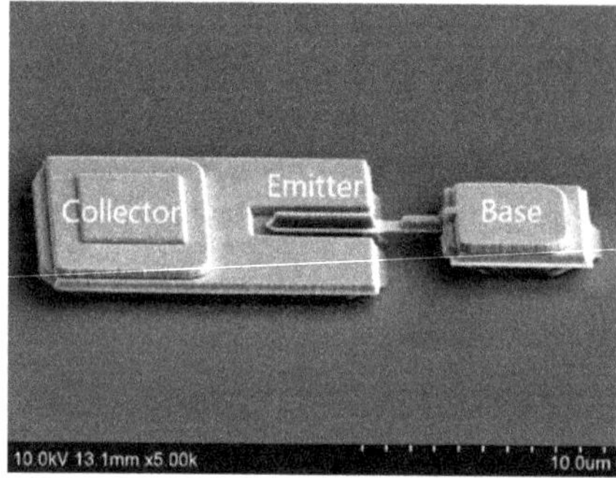

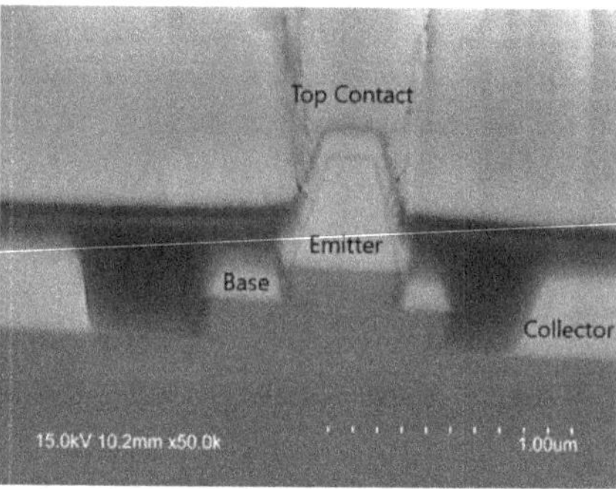

Fig. 9.50 Left: Top view SEM micrograph of a fully fabricated GaAsSb/InP DHBT before planarization with an emitter size of 0.25 × 5 μm^2. Right: Cross-sectional SEM micrograph of a completed GaAsSb/InP DHBT with an emitter size of 0.35 × 5 μm^2. *Courtesy* M. Feng, University of Illinois at Urbana-Champaign

sequence. Initially, in Fig. 9.49a, non-alloyed metal contacts are deposited on the InGaAs emitter contacting layer. The cap and emitter material are then selectively chemically etched using the emitter metal as a mask, and this etch is followed by a selective InGaP emitter etch. This etching sequence provides an approximately 0.2 μm undercut beneath the emitter metal, which is necessary for the deposition of self-aligned, non-alloy base contact (Fig. 9.49c). After the base metallization, the base and collector are chemically etched, leaving the subcollector exposed (Fig. 9.49d). Before the collector metallization, an etch of approximately 1000 Å into the substrate is performed to electrically isolate the HBT (Fig. 9.49e). Next, the collector contacts are deposited on the collector mesa and overlap the edge of the isolation mesa (Fig. 9.49f). By allowing this overlap, the active collector contact area near the edges of the mesa is maximized and the collector contact resistance is reduced. The HBT device structure has now been defined. Figure 9.50a shows the scanning electron microscope (SEM) top-view micrograph of a fully fabricated single-finger GaAsSb/InP DHBT with an emitter size of 0.25 × 5 μm^2.

At this point, the HBT has a non-planar topology. In order to make the HBT structure more planar to facilitate both via etching and the formation of planar probe pads for high-frequency testing, a planarization process (Fig. 9.49g) is required. The spin-on polyimide followed by a deposited Si_3N_4 layer completes the planarization process. Contact vias are defined using standard photolithography process (Fig. 9.49h). After the vias have been etched, a 1 μm layer of overlay metallization is deposited to form the probe pads and interconnect metallization. In the last step, an airbridge process is utilized to produce a second level of plated interconnect metallization allowing emitter fingers to connect to common-emitter probe pads. The cross-sectional SEM micrograph of a GaAsSb/InP DHBT with an emitter size of 0.35 × 5 μm^2 is shown in Fig. 9.50b.

9.8 III–V Metal-Oxide-Semiconductor Field-Effect Transistors

Silicon metal-oxide-semiconductor field-effect transistor (MOSFET)-based integrated circuit (IC) technology is the dominant semiconductor technology used by all electronics industries with a 2018 worldwide sales of over $470 billion. Miniaturization of the feature size of ICs to increase the density of components has been the foundation of the over 50-year-long steady advance of silicon technology. Following Moore's law, the exponential increase of transistor count per unit area as a function of time has been faithfully delivered by the industry since 1970. The minimum transistor gate length also decreases with time exponentially by incorporating strained SiGe channel material, high-κ gate dielectrics, and non-planar multigate structures (see Fig. 1.4). Although 5-nm-node CMOS circuits are on the horizon, further reduction of the feature size to below 2 nm will push the device structure into the quantum regime. At that moment, the silicon technology will reach the point at which significant materials and device innovations will be required to further technology developments. One possible solution for the future is the hybrid material system in which silicon, germanium, and III–V compound semiconductors are integrated together on silicon wafers. For example, taking advantage of the very high electron mobility in GaInAs and high hole mobility in germanium, advanced CMOS circuits consisting of n-GaInAs channel and p-Ge channel MOSFETs might be fabricated together on the silicon wafer.

Because of their high bulk electron mobility, III–V MOSFETs have long been pursued in hopes of achieving performance superior to that of their Si counterpart. However, due to the high surface state density, the realization of unpinned surface Fermi level in III–V compound semiconductors had been elusive until two decades ago. In searching for a low defect density, thermodynamically stable gate dielectric, native oxide and deposited oxide using a variety of deposition techniques have been investigated. Unlike the SiO_2/Si system, the native surface species in III–V materials are complicated. For example, the stable oxides of GaAs consist of As_2O_3, As_2O_5, Ga_2O_3, Ga_2O, and $GaAsO_4$ with different thermodynamic properties. Among all native oxides in this system, the As-oxides are the least stable while Ga_2O_3 is the most stable oxide. When the low-temperature-grown oxides are annealed at higher temperature, the composition of the oxide layer changes significantly. As temperature increases, the less stable As-oxides either evaporate away or are converted to the most stable oxide in the system (Ga_2O_3) along with elemental arsenic located at the interface. The pinning of the Fermi level at the GaAs-oxide interface has been unambiguously correlated with significant amounts of As_2O_3, As_2O_5, and elemental As present in native oxides. In order to fabricate functional III–V MOSFETs, it is necessary to completely remove the surface native oxide first before a stable and robust dielectric can be deposited. Thus, the successful development of III–V MOSFET technology relies strongly on the surface passivation and interface control technology.

9.8.1 III–V Alloy Surfaces and Semiconductor-Oxide-Metal Interfaces

The semiconductor surface state model was first proposed by John Bardeen in 1947 to explain the Fermi level pinning phenomenon at the metal–semiconductor (M–S) interface. It is assumed that the distribution of surface states becomes charge-neutral if the states are filled to a particular energy level, ϕ_0 (above $E_V = 0$), inside the forbidden gap. In the strong pinning limit, the metal Fermi level is pinned by the interface states at ϕ_0. The pinning energy level ϕ_0 is then the charge neutrality level of the interface states. Since then, two principal models, the intrinsic model of metal-induced gap states (MIGS) and the extrinsic model of defect states, have been developed. In the intrinsic case, a semiconductor in contact with a metal contains intrinsic states inside its band gap which are the evanescent tails of the metal wave function decaying into the semiconductor. These states are the dangling-bond states of the broken surface bonds of the semiconductor and called MIGS. On the other hand, in the extrinsic case, the interface states are introduced near the surface by processing-induced bond disorder (e.g., disorder caused by metal deposition, etching, or chemisorption of oxygen). In general, both types of interface states, MIGS and defects, could cause pinning. However, there is no universally accepted model for the M-S interface. Nevertheless, these studies of M-S contact have enriched our understanding of the chemistry and physics of III–V surfaces, such that III–V-oxide interface models can be developed.

(a) **Empirical CNL model of M-S interfaces**

In semiconductors, the dangling-bond energy is typically located inside the band gap and the interface trap density (D_{it}) increases exponentially in the energy ranges close to the band edges. The D_{it} distribution forms a U-shaped continuum of donor- and acceptor-type states with its minimum, or the pinning energy, located at the charge neutrality level (CNL). The CNL energy level represents a weighted average value over the density of states. CNL is pushed away by the large DOS of the conduction and valence bands from the intrinsic energy level E_i, as shown in Fig. 4.21. Therefore, the semiconductor surface pinning energy CNL is located inside the forbidden gap for most semiconductors. If the Fermi level E_F is above CNL, the states are of acceptor type and negatively charged if the states are occupied. If the Fermi level E_F is below CNL, the states are of donor type and positively charged if the states are occupied.

The CNL energies can be calculated from their pseudopotential band structures. The CNL values (above $E_v = 0$) of III–V binaries are determined by averaging values derived from various theoretical models as shown in Table 4.5. The experimental CNL values of the semiconductors can be extracted from Schottky barrier heights. In general, the calculated CNL is in close agreement with Schottky barrier heights (ϕ_{bv}) on p-type III–V binary compounds. Thus, one can determine the surface pinning energy of III–V binary and ternary compound semiconductors empirically based on the valence band alignment method developed by S. Tiwari and D. J. Frank (Sect. 6.1). Figure 9.51 shows the complete room-temperature conduction band-edge

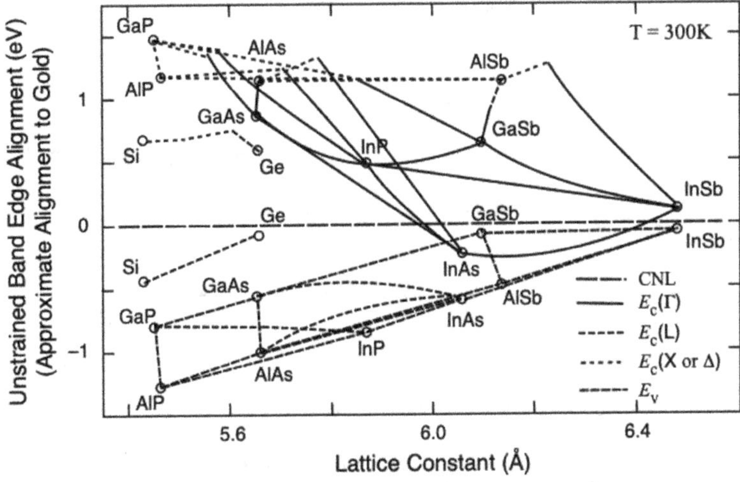

Fig. 9.51 Energy alignment of CNL with the band edges of elemental and III–V semiconductors at MOS interfaces plotted as a function of the lattice constant of semiconductors. For III–V compounds, the circles indicate the band edges of binaries and the lines show the band edges of ternaries

and valence band-edge energies as a function of the lattice constant for unstrained III–V compound semiconductors. This is the same figure that Tiwari and Frank derived (Fig. 6.5) for band discontinuity determination except the zero point of the energy is now representing the charge neutrality level (CNL) or trap neutral level (ϕ_0) for each lattice constant group. In the case of Si, the state-of-the-art SiO_2/Si interface has very low interface trap density of 10^9–10^{10}/cm^2–eV. Although the energy separation between CNL is ~0.6 eV and ~0.5 eV from the conduction band minimum (CBM) and valence band maximum (VBM), respectively, the Fermi level is unpinned. Thus, high-performance *n*-MOSFETs and *p*-MOSFETs are routinely fabricated. In the case of GaAs, CNL is far away from CBM (~0.8 eV) and VBM (~0.6 eV) so that both GaAs *n*-MOSFET and *p*-MOSFET are difficult to realize when significant interface traps presented. It is interesting to note that CNL is inside the conduction band of InAs. This indicates that InAs-based *n*-MOSFETs are easy to realize. For example, T. Brody and H. Kunig, in 1966, demonstrated both enhancement- and depletion-mode MOSFETs using evaporated SiO_2 on InAs film prepared by co-evaporation method. One interesting ternary system is InGaAs where the CBM of In-rich InGaAs intersects CNL at x_{InAs} ~ 0.75. Thus, In-rich InGaAs *n*-MOSFET is relatively easy to achieve.

For device applications, the next important issue after unpinning the Fermi level is achieving strong electron inversion in III–V MOSFETs. This can be explained using the distribution of the density of the surface states D_{it} as a function of the energy as shown in Fig. 9.52. It is assumed that the D_{it} distribution from VBM to CBM (with energy E_{CBM}) is nearly parabolic in a logarithmic scale due to the significant interface traps at III–V interfaces. The minimum value of D_{it} depends on the surface processing techniques. For simplicity, the D_{it} value at CBM is fixed at 10^{14} cm^2–eV

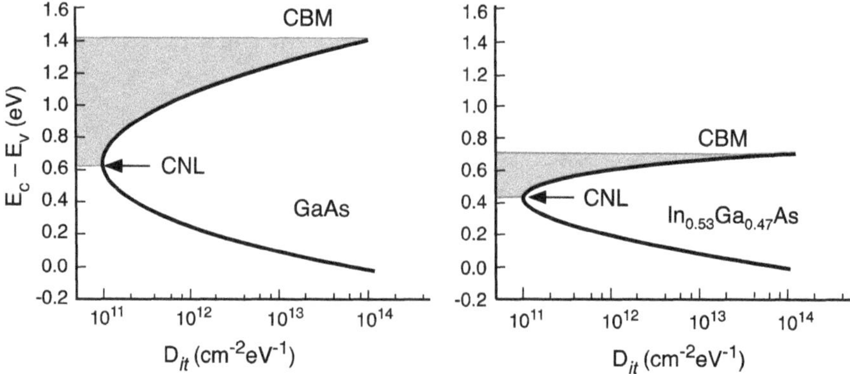

Fig. 9.52 Schematic for the parabolic D_{it} distribution within energy band of GaAs and $In_{0.53}Ga_{0.47}As$. The CNL is aligned 0.8 eV and 0.27 eV below CBM for GaAs and $In_{0.53}Ga_{0.47}As$, respectively. The shaded area shows the built-up negative charges in interface traps after Fermi level moves from CNL to CBM. Reprinted with permission from [19], copyright AIP Publishing

for all III–V compounds and has a minimum value located at CNL (with energy E_{CNL}). Assuming the strong electron inversion occurs when the Fermi level reaches CBM, the number of interface-trapped negative charges Q_{it} can be calculated by integrating D_{it} from E_{CNL} to E_{CBM}. It is obvious that the dominant factor is the energy difference between E_{CNL} and E_{CBM}. The built-in negative charges Q_{it} to prevent a strong inversion charge to participate in transport are larger for the material with a deep CNL below CBM. For example, in GaAs, with the CNL located ~0.8 eV below CBM, it builds up ~10^{-6} C/cm^2 negative charges to prevent a strong inversion charge to participate in transport. In contrast, the CNL and CBM potential difference for $In_{0.53}Ga_{0.47}As$ lattice-matched with InP is only 0.27 eV. The built-up negative charge is only ~3.4×10^{-7} C/cm^2, a factor of three smaller than that in GaAs. Therefore, it is much easier to realize an inversion-mode $In_{0.53}Ga_{0.47}As$/InP MOSFET than a GaAs MOSFET.

(b) **High-κ dielectrics and oxide-semiconductor (O-S) interfaces**

Scaling of the silicon complementary metal-oxide-semiconductor (CMOS) FET technology is at the center of Moore's law. The exponential increase of transistor count per unit area as a function of time has been faithfully delivered by the industry since 1970. The minimum transistor gate length also decreases with time exponentially. However, when the feature size is reduced to below 65 nm, the SiO_2 gate dielectric layer becomes so thin (<2 nm) that the gate leakage current due to direct tunneling of electrons through SiO_2 becomes very high. The excessively large leakage current leads to an unacceptably high circuit power dissipation level. The FET is a capacitance-coupled device, where the source–drain current of the device depends on the gate capacitance,

$$C = \epsilon_0 \kappa A / t \tag{9.98}$$

where ϵ_0 is the permittivity of free space, κ is the relative permittivity (dielectric constant), A is the area, and t is the thickness of the dielectric SiO_2. Since the tunneling current decreases exponentially with increasing the dielectric thickness, a new dielectric with a larger dielectric constant than that of SiO_2 ($\kappa = 3.9$) has to be introduced while maintaining the same capacitance. These non-native new gate oxides are called *high-κ dielectrics*. In selecting suitable high-κ dielectrics for microelectronics applications, there are several important requirements about their physical properties. First, the κ value must be high enough that it will be used for a reasonable number of years of scaling. A value of $\kappa \geq 20$ is preferred. Second, the oxide must have large bandgap energy and act as an insulator, by having oxide-Si band offsets over 1 eV to minimize leakage current. Third, the oxide must be thermodynamically as well as kinetically stable such that it is compatible with extreme processing conditions. Above all, the selected oxide must form a good electrical interface with the semiconductor, which is dependent on the deposition method. The oxides that satisfy most of these criteria are Al_2O_3, HfO_2, ZrO_2, Y_2O_3, La_2O_3, and various lanthanides (e.g., Gd_2O_3).

For III–V MOSFETs, due to the lack of high-quality native oxides, high-κ dielectrics used in current Si technology are implemented as gate oxides. In order to use these dielectrics as the gate oxide of III–V MOSFETs, the conduction band (CB) offset, ΔE_C, between the dielectric and III–V alloy must be over 1 eV to achieve a low leakage current. The CB offset at the M-S interface has previously been determined by the MIGS method using CNL as a reference energy level. The MIGS model is extended to determine band offsets between oxides and semiconductors. The CB offset is given by

$$\Delta E_c = \left(\chi_o - \Phi_{CNL,o}\right) - \left(\chi_s - \Phi_{CNL,s}\right) + S\left(\Phi_{CNL,o} - \Phi_{CNL,s}\right) \qquad (9.99)$$

Here Φ_{CNL} is the charge neutrality level measured from the vacuum level, χ is the electron affinity (EA), and the subscripts o and s represent the oxide and semiconductor, respectively. S is a dimensionless pinning parameter where $S = 1$ describes the Schottky limit of no pinning and $S = 0$ describes the Bardeen limit of strong pinning. Empirically, S depends only on the optical dielectric constant of the oxide, ϵ_∞, and follows the trend

$$S = \frac{1}{1 + 0.1(\epsilon_\infty - 1)^2} \qquad (9.100)$$

The CNL energies for high-κ oxides have been derived from the bands calculated by either the tight-binding or pseudopotential method. The calculated CNL energies (above the valence band edge) and measured bandgap energies, electron affinity, and optical dielectric constants of these high-κ oxides are listed in Table 9.3. The calculated band offsets of various high-κ oxides on III–V semiconductors are listed in Table 9.4. Most of these high-κ oxides deposited on III–V semiconductors have a large enough (>1 eV) conduction band offset for MOSFET applications.

Table 9.3 Experimental bandgap energy (E_g), dielectric constant (κ), electron affinity (EA), and optical dielectric constants (ϵ_∞) of various high-κ oxides along with the calculated charge neutrality level energy above the valence band edge (E_{CNL}) [20–22]

	E_g (eV)	κ	E_{CNL} (eV)	EA (eV)	ϵ_∞
a-Al_2O_3	6.3	7.5 [21]	3.2	3	3.2
HfO_2	6	25	3.7	2.2	4
ZrO_2	5.8	25	3.6	2.4	4.8
Y_2O_3	5.7	15	2.4	1.84	4.4
La_2O_3	6.0	30	2.4	1.9	4
Gd_2O_3	6	10 [22]	2.4	2.5	3.8

Table 9.4 Calculated conduction band offsets (ΔE_c, in eV) of various high-κ oxides on III–V semiconductors [20]

	Al_2O_3	HfO_2	ZrO_2	Y_2O_3	La_2O_3	Gd_2O_3
GaP	0.67	0.77	0.73	1.65	1.65	1.2
InP	1.7	1.72	1.64	2.6	2.6	2.1
AlAs	0.97	1.05	1.0	1.9	1.9	1.5
GaAs	1.46	1.52	1.42	2.4	2.4	1.9
InAs	2.38	2.54	2.44	3.4	3.4	2.9
GaSb	1.50	1.58	1.53	2.46	2.46	2.0
InSb	2.08	2.18	2.14	3.06	3.06	2.62
GaN	0.86	1.15	1.08	2.02	2.02	1.57

9.8.2 *Atomic Layer Deposition (ALD)*

Atomic layer deposition (ALD) is a vapor-phase film growth method used to deposit ultra-thin, uniform, conformal, and pinhole-free films onto a substrate. It was invented in the 1970s for the production of large-area flat-panel displays. Strong interest in ALD by the semiconductor industry began in the mid-1990s stemming from the need for non-native oxides for continuous scaling of Si MOSFET devices. In 2007, Intel successfully integrated an ALD Hf-based high-κ dielectric process into its 45-nm-node mass production technology. In the meantime, efforts to find non-native oxides on GaAs with low interfacial states density have found a solution in UHV deposition of amorphous $Ga_2O_3(Gd_2O_3)$ and single-crystal Gd_2O_3 on GaAs surface. After this breakthrough in material science, both depletion-mode and enhancement-mode GaAs MOSFETs were demonstrated using $Ga_2O_3(Gd_2O_3)$ as a gate dielectric along with an ion implantation process. The investigation of ALD high-κ Al_2O_3 and HfO_2 on GaAs and other III–V materials for the fabrication of III–V-based MOSFETs started in 2001 at Bell Labs. Soon after in 2003, GaAs MOSFETs with ALD Al_2O_3 gate dielectric were successfully demonstrated. Since then, the research

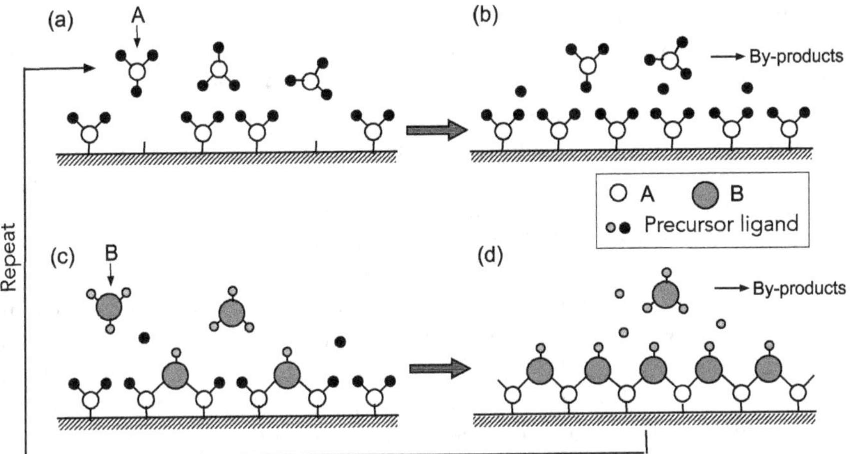

Fig. 9.53 Schematic representation of an ALD growth cycle leading to the formation of a binary dielectric film using self-limiting surface chemistry. **a** Precursor A is pulsed and reacts with surface, **b** excess precursor and reaction by-products are purged with inert carrier gas, **c** precursor B is pulsed and reacts with surface, and **d** excess precursor and reaction by-products are purged with inert carrier gas. Above procedures are repeated until the desired material thickness is achieved

of ALD high-κ dielectrics on III–V substrates has been extended to other material systems including $In_xGa_{1-x}As$, InP, InSb, GaSb, and GaN.

(a) **Principles of ALD**

An ideal ALD process consists of exposing the substrate surface alternatively to different precursors in a cyclic manner. It differs from CVD technique by keeping the precursors strictly separated from each other in the gas phase. As shown in Fig. 9.53, the process of one ALD growth cycle involves exposure of the substrate surface to alternating precursors A and B introduced sequentially without overlapping. In each alternate precursor pulse, the precursor molecules fully react with the surface in a self-limiting fashion that leaves no more than one monolayer (ML) at the surface. This ensures that the chemisorption reaction stops once all of the reactive sites on the substrate have been occupied. This automatic control of the amount of material deposited is a key feature of ALD. After terminating the pulse of precursor A, an inert gas (typically N_2 or Ar) purge follows to remove the excess of unreacted precursor and gaseous by-products. The second precursor B is then introduced following an inert gas purge to complete one reaction cycle. It should be noticed that only in an ideal case is a monolayer of desired material formed per each reaction cycle. In practice, the material thickness is determined by the nature of the precursor-surface interaction and only a fraction of a monolayer is formed. The ALD cycle can be performed multiple times to increase the film thickness.

Typically, the process of ALD is performed at relatively low temperatures (<350 °C). The temperature range, where thin film growth proceeds by the irreversible and saturating surface control in an ALD mode, is referred to as the 'ALD-window'. Within this ALD-window, the deposition rate per cycle is a constant independent of the growth temperature. The process outside the ALD-window, however, generally results in non-ALD-type deposition due to precursor condensation or insufficient reactivity at low temperatures, and precursor thermal decomposition or rapid precursor desorption at high temperatures. Thus, it is necessary to carry out each deposition process within the designated ALD-window.

The main advantages of ALD are derived from the self-limiting and sequential deposition process. First, ALD offers an accurate and simple thickness control by utilizing saturating, irreversible reactions where the film thickness can be tailored by the number of growth cycles. Secondly, ALD provides excellent conformality over large and/or complex-shaped substrates of high aspect ratio and three-dimensionally structured materials. This is made possible by self-limiting, irreversible reactions that provide the same amount of material adsorbed on different parts of substrates. Furthermore, due to the self-limiting reaction nature of ALD, there is no need to control the reactant flux homogeneity and the deposition temperature dependence is weak. A wide range of materials can be deposited using ALD, including oxides, metals, and semiconductors, and there is a wide range of properties that these films can exhibit, depending on the application. However, due to its sequential monolayer deposition nature, ALD suffers from slow deposition rates.

(b) **ALD of high-κ dielectrics on III–V surfaces**

Since 2007, the Si semiconductor industry has adopted ALD to integrate non-native oxides, e.g., HfO_2-based high-κ dielectric, into its IC mass production lines to advance the technology node beyond 45 nm. Next, at the 22 nm node, a three-dimensional fin field-effect transistor (FinFET) structure is developed. In a FinFET, the fin-shaped gate with a high aspect ratio protruding above the bulk surface needs to be covered with a gate oxide of high compositional and thickness uniformity. This is a task tailored for the conformal ALD process. To further improve the device performance with higher speed and reduced power, it is necessary to look toward alternative semiconductors with higher carrier mobility, such as III–V semiconductors.

For non-native oxide deposited on III–V substrate surface using ALD process, the achieved interface state density is primarily decided by the precursor characteristics and surface preparation of the substrate surface. There are some general requirements for a suitable ALD precursor which include: (a) sufficient volatility but no self-decomposition at the deposition temperature (b) adsorption to the surface sites without etching the substrate, and (c) sufficient reactivity toward the second precursor. The metal precursors used in ALD on III–V surfaces can be divided into two groups, inorganic and organometallic. Halides are the most common inorganic precursors used for both CVD and ALD processes. Chlorides, a kind of halides, are reactive, stable at a broad temperature range, and available for many metals including $AlCl_3$ and $HfCl_4$. However, the deposited film may suffer from high residual chlorine and hydrogen content. On the other hand, alkyls, such as trimethyl aluminum (TMA), are ideal metalorganic precursors containing a direct metal carbon bond.

They are highly volatile and very reactive with water through hydrolysis. Other metal precursors including the chelation of C (β-diketronates), O (cyclopendienyls), and N (amidinates) with alkyls to a metal have also been used. The variety in non-metal reactants is rather less than in metal reactants. Water, due to its high stability and reactivity in a broad temperature range, is the most commonly used oxygen source providing a hydroxyl-terminated surface in metal oxide ALD. Ozone is often used for deposition of oxides from metal precursors having bulky ligands that are not reactive with water. However, due to its strong oxidation power, there are potential concerns regarding undesirable surface oxide formation on the substrate by prolonged ozone exposures.

The other important factor that influences the quality of non-native oxide/III–V interfaces is the pre-cleaning or surface passivation of the substrate before ALD. Unlike the SiO_2/Si material system, the III–V substrates lack a high-quality, natural insulator. Instead, a number of different native surface species including Ga–O bonds, In-O bonds, As–O bonds, and other III–V–O bonds are readily formed on III–V surfaces when exposed to ambient environment. To achieve a low interface state density on III–V surfaces, these oxides must be removed as much as possible before depositing high-κ dielectrics. Currently, a number of ex situ surface treatments of III–V (100) surfaces have been developed. The use of wet chemical treatment by soaking samples in NH_4OH (29%) solution has been found to be effective in removing higher-oxidation states of As and Ga oxides. To prevent re-oxidation of the III–V surface after the etch, the sample can be further protected by monolayers of sulfur using a $(NH_4)_2S$ treatment before loading into the ALD system. The sulfur is then blown off from the III–V surface during the temperature ramp-up in the ALD system. The key to cleaning up As-oxide and providing high-quality III–V/high-κ interfaces is the *self-cleaning* effect of the ALD process as described in the following section. However, the removal of Ga oxides by the ALD process is incomplete. Further detailed studies examining Al_2O_3 and HfO_2 deposition on GaAs by ALD indicated that the efficiency of surface oxides removal is affected by the ALD precursors used. An ~1 nm interfacial layer containing significant Ga_2O content was observed at the Al_2O_3/III–V interface deposited using TMA/water precursors, while a thicker interfacial layer (~2.5 nm) was observed from the HfO_2 deposition using $HfCl_4$/water precursors. The enhanced reactivity of $Al(CH_3)_3$ compared to $HfCl_4$ is responsible for this difference.

Currently, Al_2O_3, HfO_2, and their alloys constitute the major ALD dielectrics studied on III–V substrates. Other metal oxides including Gd_2O_3, La_2O_3, and Y_2O_3 have also been studied. The best achieved interface state densities (D_{it}) are between 10^{11} and low-10^{12}/cm^{2}eV, which are orders of magnitude higher than that of $(Gd_2O_3)Ga_2O_3$/GaAs deposited by in situ UHV MBE deposition (~5×10^{10}/cm^{2}eV) and far worse than the ideal SiO_2/Si system (~10^9–10^{10}/cm^{2}eV). However, these interface state density values have already been improved significantly to allow the demonstration of inversion-mode MOSFETs on different III–V substrates. To further improve the technology, additional ALD metal oxides need to be developed for gate dielectrics of III–V substrates. Surface passivation methods before ALD require further refinement to achieve the right surface conditions including the formation of particular species of interfacial oxides. Furthermore, to make high-κ dielectric/III–V a manufacturable III–V CMOS technology, it is also necessary to avoid surface and interfacial defect formation in every step of device fabrication.

9.8.3 Oxides Deposition on III–V Substrates

(a) **Substrate surface preparation**

In the process of fabricating III–V-based MOSFETs, the preparation of a suitable surface condition of the semiconductor substrate for oxide deposition is the most critical step. In the ideal case, the surface of III–V semiconductors is a perfect, flat, and homogeneous surface. In reality, the atomic bonds in the terminated plane are sheared off to become unsatisfied dangling bonds. These dangling bonds are fairly susceptible to chemical reactions with atmospheric gases leading to the formation of moderately thick native oxides and carbonaceous adsorbents. These native surface species including, e.g., Ga–O bonds, As–O bonds, elemental As, and As and Ga anti-sites on GaAs, are the culprits of the Fermi level pinning. Thus, it is very important to remove these native oxides and carbonaceous adsorbents on air-exposed III–V substrates such that further surface treatment or 'passivation' of the surface can be carried out before the deposition of high-κ dielectrics.

On the other hand, under the UHV conditions, in a clean surface without atmospheric gas contamination, those unpaired electrons of dangling bonds tend to form covalent bonds between adjacent atoms in a process termed 'dimerization.' These surface dimmers have a tendency to form highly ordered rows in order to minimize the free energy of the surface. Using reflection high-energy electron diffraction (RHEED) technique, the observed reconstructed surface structures of GaAs vary from Ga-rich (4×2) reconstruction through As-rich (2×4) reconstruction depending on the substrate temperature and the As-to-Ga beam flux ratio. The As-rich (2×4) surface is terminated with As dimers aligned along the $[\bar{1}10]$ direction, while the Ga-rich (4×2) surface has Ga dimers aligned along the $[\bar{1}\bar{1}0]$ direction. Due to the high diffusivity of surface Ga atoms, prolonged exposure of the GaAs surface under high temperature and low As flux promotes the island formation leading to uneven morphology on Ga-rich GaAs surface. In contrast, the As-rich (2×4) surface, obtained under medium (~550 to 600 °C) temperature and As-beam flux conditions, provides a smooth surface. Since the FETs are surface devices, for in situ deposition of oxides under UHV conditions, the As-rich (2×4) surface is the one most relevant for MOSFET applications.

(b) **Surface passivation and oxide deposition**

Over the past five decades, persistent efforts have been carried out in searching for surface treatment methods to passivate III–V surfaces such that MOSFETs can be demonstrated. However, the conventional practice of thermal, anodic, and plasma oxidation methods failed to achieve this goal. Using an in situ deposition approach, the first successful passivation of GaAs surface was demonstrated in 1995 by Minghwei Hong et al. at Bell Labs [23]. In this method, electron-beam evaporated $Ga_2O_3(Gd_2O_3)$ [GGO] dielectric film was deposited on an MBE-grown wafer in a UHV connected multiple chamber system. During the experiment, the freshly grown GaAs with an As-stabilized (2 × 4) surface was transferred under UHV condition from the III–V MBE chamber to another chamber for oxide deposition. Prior to deposition, the GaAs surface stoichiometry and atomic order were preserved, as monitored using RHEED, with extremely low oxygen exposure. Then the GGO film was deposited on the clean, atomically ordered (100) GaAs surface at substrate temperatures below 600 °C to avoid chemical reactions with GaAs substrates. The oxide molecules were supplied by electron-beam evaporation of a single-crystal $Gd_3Ga_5O_{12}$ source. MOS structures on GaAs with a low D_{it} on the order of mid 10^{10} cm^{-2} eV^{-1} were demonstrated for the first time. However, low D_{it} values were not obtained in other oxide/GaAs systems using Al_2O_3, SiO_2, and MgO prepared by a similar approach. Furthermore, pure Ga_2O_3 films cannot passivate GaAs surfaces effectively and show very high leakage current with poorly defined breakdown characteristics.

Further studies indicated that the unpinning of the GaAs Fermi level using electron-beam evaporated GGO dielectric film results from the Gd_2O_3 epilayer restoring the surface As and Ga atoms to near-bulk charge. An understanding of the bonding between Gd_2O_3 and GaAs is important for the development of new dielectrics used in III–V-based MOSFETs. Based on the consideration of Gibbs free energies of formation of all possible pairs in Ga, As, Gd, and O, the growth initiates from bonding a layer of oxygen atoms to Ga (Ga_2O_3: −998 kJ/mole) by taking up the As site (GaAs: −70 kJ/mole) in GaAs structure. Then the oxygen atoms would proceed to bind to Gd atoms of the next row due to the very large Gibbs energy for Gd_2O_3 formation (−1739 kJ/mole). Thus, the initial growth of GGO on GaAs contains only Gd_2O_3, despite the fact that incoming fluxes consist of various other species including Ga_2O_3. In addition, this Gd_2O_3 film is grown in a single-crystal form on GaAs. The Gd_2O_3 film has a cubic structure isomorphic to Mn_2O_3 and is (110)-oriented in single domain on the (100) GaAs surface. Once a Gd_2O_3/GaAs interface is formed, however, the competition for oxygen is significantly reduced and inclusion of Ga_2O_3 becomes possible, resulting in an oxide mixture of Gd_2O_3 and Ga_2O_3. Thus, the high interface quality GGO/GaAs heterostructure consists of a transition from a single-crystal GaAs substrate, to an epitaxial coherent Gd_2O_3 oxide layer of two to three monolayers in thickness, and then to the amorphous mixed oxides. This result matches well with the model of high-quality SiO_2/Si interface, where the crystalline-Si to amorphous-SiO_2 transformation takes place via an ordered crystalline oxide layer ~5 Å thick. The in situ UHV deposition of Gd_2O_3 approach can be extended to other rare earth oxides due to their chemical similarity. For example, using Y_2O_3 as the high-κ dielectric, low trap densities of $(3 - 5) \times 10^{11}$ $eV^{-1}cm^{-2}$ have been achieved at the Y_2O_3/GaAs (001) interface. Similar to Gd_2O_3, the Y_2O_3 film is also grown in a single-crystal form on GaAs with Y_2O_3 (110) planes parallel to GaAs (001) planes.

In contrast to the in situ deposition approach, MOSFETs were also successfully demonstrated in 2003 using ex situ ALD high-κ Al_2O_3 films as the gate dielectrics on GaAs and other III–V materials. Metalorganic trimethyl aluminum and water were used as the metal precursor and oxygen source for the ALD process, respectively. Since the ALD process was conducted on the GaAs surface exposed to the ambient air after epitaxial growth, a thin native oxide layer consisting of Ga and As oxides as well as spurious organic contamination is formed on the GaAs surface. To achieve a low interface state density on III–V surfaces, these oxides must be removed as much as possible before depositing high-κ dielectrics. Currently, a number of ex situ surface treatments of III–V (100) surfaces using a combination of wet chemical etch in NH_4OH (29%) solution and $(NH_4)_2S$ treatment before loading into the ALD system have been developed as outlined in the previous section. The self-cleaning effect of the ALD process further cleans up the As-oxide and provides high-quality III–V/high-κ interfaces. For selected ALD precursors, the detailed process parameters are mainly decided by the precursor characteristics such as decomposition behavior, sticking coefficient, vapor pressure, etc. So far, a number of suitable metal precursors have been reported for ALD deposition of high-κ dielectrics on various III–V surfaces with Al_2O_3, HfO_2, and their alloys constitute the most studied systems.

To understand interface passivation on III–V surfaces by ALD requires the knowledge of the surface oxides likely to be present on the surface in the course of device fabrication. In situ film deposition and analysis of the chemical state of the oxide interface can give a clear picture of the detailed reactions. An example of interface reactions of Al_2O_3 and HfO_2 ALD processes on the NH_4OH-treated GaAs surface using in situ X-ray photoelectron spectroscopy (XPS) is shown in Fig. 9.54. The air-exposed starting material is covered with native Ga and As oxides and excess As. The As 2*p* spectra show two As oxidation states, As^{3+} and As^{5+}, corresponding to As_2O_3 and As_2O_5, respectively. The Ga–O signal of the Ga 2*p* spectra has multiple oxidation states including Ga^{3+} and Ga^{5+} as well as other oxide species. It is noted that the weaker bonded, higher-oxidation As states, e.g., As_2O_5 or As^{5+} states, are removed completely by wet NH_4OH-treatment, leaving behind As^{3+} oxidation As states, reduced GaO_x species as well as As–As bonding. The reduction of the As^{3+} oxidation state occurs during the TMA precursor pulse, whereby the Al^{3+} in the TMA precursor more efficiently replaces the As in the As^{3+} oxidation state, resulting in AlO_x formation and presumably volatile $As(CH_3)_3$ reaction products. However, the reaction with As^{3+} oxidation state is less efficient for Hf originating from the tetrakis (ethylmethylamino) hafnium (TEMA-Hf) precursor. It is also noted that As–As bonding and GaO_x oxidation states persist throughout the growth process in both Al_2O_3 and HfO_2 ALD samples. These detectable residual surface oxides at the interface result in a considerable interface state density $D_{it} \geq 2 \times 10^{12}$ cm^{-2} eV^{-1}. In contrast, the interface state density achieved for in situ deposited GGO grown by MBE on GaAs is as low as $D_{it} \sim 5 \times 10^{10}$ cm^{-2} eV^{-1}.

As shown in Fig. 9.54, the presence of surface defects such as As-As dimers and AsO_x persists in ex situ ALD fabricated high-κ dielectric/III–V semiconductor structures, and has been speculated to be the cause for the high D_{it}'s, or D_{it} peaks in the mid-bandgap of III–V materials. These defects mainly originate from the native oxide

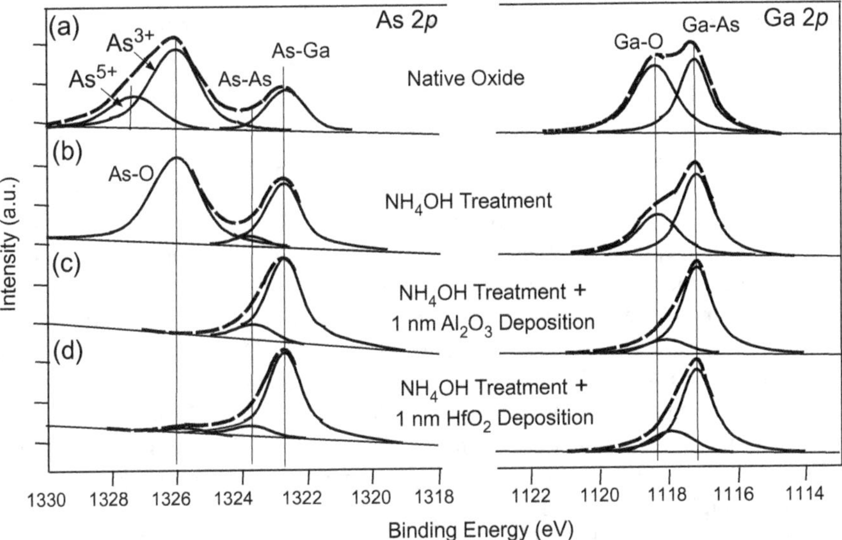

Fig. 9.54 As 2*p* and Ga 2*p* X-ray photoelectron spectroscopy (XPS) spectra of **a** native oxides on GaAs, and interface reactions after **b** NH_4OH-treatment of GaAs, **c** additional Al_2O_3 ALD process, and **d** additional HfO_2 ALD process. A substantial reduction of the As- and Ga-oxides is noted. Reprinted with permission from [24], copyright AIP Publishing

layer formed during the exposure of semiconductors to the ambient environment in ALD. To improve the high-κ dielectric/III–V semiconductor interface quality, one approach is to combine ALD with in situ UHV deposition, or in situ ALD process, such that the surface exposure to ambient is completely avoided during ALD processes. Direct comparisons between in situ and ex situ deposited ALD-Al_2O_3 on $In_{0.53}Ga_{0.47}As$ have revealed that the in situ method shows no detectable AsO_x component at the dielectric–semiconductor interface, whereas some AsO_x residue is detected using the ex situ approach. A reduction of D_{it} by more than half is observed for the in situ method. Furthermore, as detailed in the following section, the fabricated inversion-channel $In_{0.53}Ga_{0.47}As$ MOSFETs show much improved device performances for the in situ method than the ex situ method. The intrinsic drain current (I_d), transconductance (g_m), and effective mobility (μ_{eff}) are all improved by more than threefold over that of the ex situ method. $In_{0.53}Ga_{0.47}As$ MOSFETs with similar performances have also been achieved using in situ ALD HfO_2 and UHV-Y_2O_3 as gate dielectrics. Similar improvements of D_{it} using in situ ALD of Al_2O_3, HfO_2, and Y_2O_3 on GaAs have also been demonstrated. A low interface trap density D_{it} of ~2×10^{11} cm^{-2} eV^{-1} has been achieved in in situ ALD Y_2O_3 and HfO_2 on *n*-GaAs(001) and *p*-GaAs(001), respectively. The low interface trap density is key to the high performances of III–V-based MOSFETs.

9.8.4 High-κ Dielectric/III–V MOSFET Development

Using in situ deposition, and in situ and ex situ ALD approaches, various high-κ/III–V inversion-mode MOSFETs using GaAs, InGaAs, and GaN as channel materials have been demonstrated. Among them, the $In_{0.53}Ga_{0.47}As$ MOSFET is the most extensive studied due to its high electron mobility, high saturation velocity, and smaller charge neutrality level below the conduction band minimum of the channel material. Figure 9.55 benchmarks performances of state-of-the-art $In_{0.53}Ga_{0.47}As$ enhancement-mode planar and non-planar MOSFETs incorporated with various high-κ dielectrics. The non-planar structures include tri-gate, multigate, and gate-all-around (GAA) devices. The UHV deposited GGO and Y_2O_3, in situ ALD HfO_2 and Al_2O_3, and ex situ ALD Al_2O_3 are used as the gate dielectrics. The drain currents for all devices are compared at the same bias conditions of $V_{DS} = 1$ V and $V_{GS} - V_T =$ 1 V. The general trend is that the 1-μm gate length MOSFETs using in situ high-κ dielectrics show better I_{DS} and maximum g_m compared to smaller gate length devices with ex situ high-κ dielectrics, even though the bias conditions are not favorable for 1-μm gate length MOSFETs. The 1-μm gate length $In_{0.53}Ga_{0.47}As$ MOSFETs with in situ ALD-Al_2O_3 dielectrics show superb I_{DS} of 0.72 mA/μm, which compares favorably with other planar and non-planar InGaAs MOSFETs with ex situ ALD-Al_2O_3 dielectrics and significantly smaller gate lengths. The in situ ALD-Al_2O_3 MOSFETs also have the best maximum g_m of 0.98 mS/μm at $V_{DS} = 2$ V. Other MOSFETs with UHV deposited GGO and Y_2O_3, and in situ ALD HfO_2 high-κ dielectrics, also show excellent I_{DS} and maximum g_m. The low interface trap density associated with in situ high-κ dielectrics is key to the high performances of these devices.

During the past two decades, tremendous progress has been made in realizing both enhancement-mode and depletion-mode high-κ dielectrics/III–V MOSFETs.

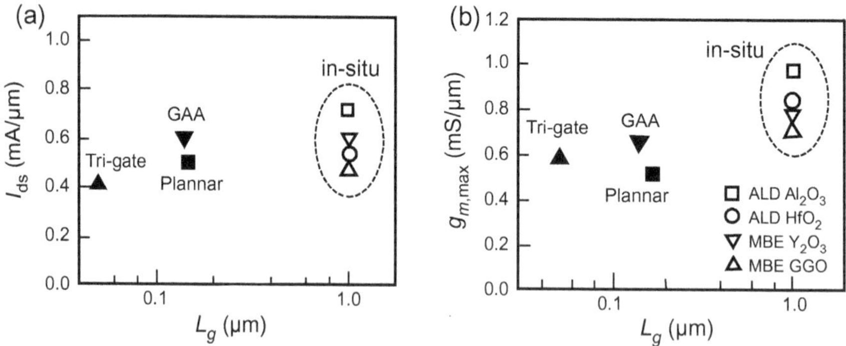

Fig. 9.55 Benchmark of **a** drain current at $V_{DS} = V_{GS} - V_T = 1$ V and **b** peak transconductance of in situ high-κ dielectrics/$In_{0.53}Ga_{0.47}As$ MOSFETs and other enhancement-mode planar and non-planar devices using ex situ (solid symbols) high-κ dielectrics. Reprinted with permission from [25], copyright AIP Publishing

However, there exist several difficult challenges to overcome before III–V materials become appropriate for future high-speed, low-power logic applications. First, for III–V MOSFETs to be competitive against scaled Si MOSFETs, their physical gate length needs to be scaled to 30 nm and below with acceptable I_{ON}/I_{OFF} ratio (>10^4) at $V_{CC} = 0.5$ V. Recent demonstration of high-performance devices described above has already laid a solid foundation in achieving this goal. Second, for CMOS logic applications, there is a need for *p*-channel MOSFETs with very high hole mobility. Although III–V materials show 10–20 times better electron mobility compared to Si, their hole mobility is comparable to that of Si. To improve the hole mobility in III–V materials, one approach is to use compressively strained III–V channels in MOSFETs. Alternatively, other novel materials exhibiting high hole mobility, e.g., strained Ge quantum wells, can be used as the *p*-channel materials. Finally, to take advantage of the advanced development of Si IC technologies and the high-speed, low-power capabilities of III–V devices, it is desirable to monolithically integrate III–V devices selectively onto the Si platform. To achieve this goal, one has to solve two significant challenges, namely the lattice-mismatch and thermal mismatch between Si and III–V's.

Problems

1. Refer to the experimental results of an $Al_{0.3}Ga_{0.7}As/GaAs$ HEMT reported in [6].

   (a) Estimate the threshold voltage V_T for three samples (R-96A, R-73A, and R-72A).

   (b) The 2DEG in the triangular quantum well has a sine-wave-like distribution perpendicular to the conduction channel. What is the average concentration peak of the 2DEG located ΔW away from the heterojunction interface? The results can be calculated using the equation

   $$N_S = \frac{\epsilon_2}{q(W_d + \Delta W)}(V_{GS} - V_T - V_x)$$

   Assume $V_{DS} = 0$ and comment on your results in terms of whether they are realistic.

   (c) Estimate ΔW again using Howard–Fang approximation for the triangular QW at the interface. Also discuss your results.

2. In an $Al_{0.2}Ga_{0.8}As/Ga_{0.85}In_{0.15}As$ pseudomorphic high-electron-mobility transistor (pHEMT), the $Al_{0.2}Ga_{0.8}As$ barrier is uniformly doped at a level of 2×10^{18} cm^{-3}. At the $Al_{0.2}Ga_{0.8}As/Ga_{0.85}In_{0.15}As$ interface, there is an undoped spacer layer on the $Al_{0.2}Ga_{0.8}As$ side.

   (a) Calculate the sheet carrier concentration in the triangular $Ga_{0.85}In_{0.15}As$ quantum well for different spacer layer thickness (W_{sp}) of 25 Å and 100

Å. Assume the donor level in $Al_{0.2}Ga_{0.8}As$ is sufficiently deep ($E_d = 50$ meV) so that the Fermi level is pinned there. The dielectric constant of $Ga_{0.85}In_{0.15}As$ is $13.41\varepsilon_0$.

(b) Plot the equilibrium energy band diagram of the whole pHEMT structure, including metal contact, for the $W_{sp} = 25$ Å case using *SimWindows*. Assume the total $Al_{0.2}Ga_{0.8}As$ barrier has a thickness of 150 Å and the undoped $Ga_{0.85}In_{0.15}As$ 2DEG channel is 100 Å thick. The whole structure is grown on top of an undoped GaAs substrate. The metal-$Al_{0.2}Ga_{0.8}As$ work function $q\phi_m$ equals ~0.9 eV.

(c) Repeat part (a) for a modulation-doped $Al_{0.3}Ga_{0.7}As$/GaAs heterostructure with a 25 Å spacer layer and $\Delta E_c = 0.2$ eV. This HEMT structure has the same doping level in the $Al_{0.3}Ga_{0.7}As$ barrier. The dielectric constant of GaAs is $12.85\epsilon_0$.

(d) Now the 2DEG channel material of the pHEMT is replaced with a $GaAs_{0.8}Sb_{0.2}$ layer, but keep all other materials unchanged. For $W_{sp} = 25$ Å, calculate the sheet carrier concentration in the triangular GaAsSb QW. Will this material system outperform the $Al_{0.2}Ga_{0.8}As/Ga_{0.85}In_{0.15}As$ pHEMT structure? Why?

The relevant material parameters are listed below:
For $Ga_{0.85}In_{0.15}As$: $m_e^*/m_0 = 0.025(1-x) + 0.71x - 0.0163x(1-x)$
For $GaAs_{0.8}Sb_{0.2}$: $m_e^*/m_0 = 0.00634 - 0.0483x - 0.0252x^2$.

3. The 2DEG with a sheet carrier density of 10^{13} cm^{-2} is obtained at the interface of an $Al_xGa_{1-x}N$/GaN HEMT structure grown by MOCVD. The $Al_xGa_{1-x}N$ barrier is undoped and has a thickness of 30 nm. Using parameters provided in the reference article [9], verify that the Al-composition (x) of the $Al_xGa_{1-x}N$ barrier is about 0.2 as shown in Fig. 11 of the referenced article.
4. Repeat Problem 3 with up-to-date material parameters of $Al_xGa_{1-x}N$ shown below:

$\Delta E_c(x) \approx 2.12x$,
$E_g(x) = xE_g(\text{AlN}) + (1-x)E_g(\text{GaN}) - 0.7x(1-x)$.
Dielectric constant $\epsilon(x) = 9.5 - x$,
$m^*(x) = 0.22m_e$,
$q\phi_m(x) = 1.3x + 0.84$ (eV).

$$P_{SP}(Al_xGa_{1-x}N/GaN) = -0.090x - 0.034(1-x) + 0.021x(1-x)\ \text{c/m}^2$$

$$P_{PE}(Al_xGa_{1-x}N/GaN) = -0.0525x + 0.0282x(1-x)\ \text{c/m}^2$$

Assuming the Al-composition of the $Al_xGa_{1-x}N$ barrier in the $Al_xGa_{1-x}N$/GaN HEMT is $x = 0.2$, calculate the 2DEG sheet carrier density at the $Al_xGa_{1-x}N$/GaN interface.

5. The active region of a p-channel pseudomorphic high-hole-mobility transistor (pHHMT) has a compressively strained InSb channel material and an $Al_{0.35}In_{0.65}Sb$ barrier. For a 5 nm InSb QW sandwiched between $Al_{0.35}In_{0.65}Sb$ barrier, the type-I band discontinuities, ΔE_c and ΔE_v, are determined as 0.342 and 0.254 eV, respectively. Assume the 200 Å thick $Al_{0.35}In_{0.65}Sb$ barrier consists of a uniformly Be-doped layer of $p = 10^{18}$ cm^{-3} and an undoped spacer layer, $W_{sp} = 70$ Å.

   (a) Calculate the sheet carrier concentration in the triangular InSb quantum well. Assume the acceptor level in $Al_{0.35}In_{0.65}Sb$ is sufficiently deep ($E_a =$ 40 meV) so that the Fermi level is pinned there.
   (b) Plot the equilibrium energy band diagram of the whole pHHMT structure, including metal contact, using *SimWindows*. The plot should extend into the AlInSb buffer layer for 200 Å. Assume the metal-$Al_{0.35}In_{0.65}Sb$ work function $q\phi_m$ equals ~0.16 eV.

6. In pursuing high-speed HBTs, InP-based material systems are the most promising. One of the systems being actively studied is based on using $GaSb_{0.5}As_{0.5}$ lattice-matched to InP as the base material.

   (a) Find the band discontinuity values between lattice-matched $GaSb_{0.5}As_{0.5}$ and InP. Is this a type-I or type-II heterojunction?
   (b) In $Ga_{0.47}In_{0.53}As$/InP *N-p-N* DHBTs, the valence band discontinuity between InP emitter and $Ga_{0.47}In_{0.53}As$ base is 0.36 eV. The ΔE_c in this type-I heterojunction is 0.252 eV. By replacing the $Ga_{0.47}In_{0.53}As$ with lattice-matched $GaSb_{0.5}As_{0.5}$, the maximum current gain can be improved. Calculate the maximum current gain improvement.
   (c) Compare the properties of $GaSb_{0.5}As_{0.5}$/InP and $Ga_{0.47}In_{0.53}As$/InP HBTs.

7. $Ga_xIn_{1-x}P$ is a useful material for visible light-emitting and HBT applications.

   (a) Calculate the composition x and band gap energy of $Ga_xIn_{1-x}P$ lattice-matched to GaAs.
   (b) Using model-solid theory, calculate the band discontinuities between lattice-matched $Ga_xIn_{1-x}P$ and GaAs.
   (c) In $Al_{0.3}Ga_{0.7}As$/GaAs HBTs, the valence band discontinuity between $Al_{0.3}Ga_{0.7}As$ emitter and GaAs base is 0.132 eV. By replacing the $Al_{0.3}Ga_{0.7}As$ with lattice-matched $Ga_xIn_{1-x}P$ the maximum current gain can be improved. Calculate the magnitude of current gain improvement.
   (d) Does $Ga_xIn_{1-x}P$ offer any advantages over $Al_{0.3}Ga_{0.7}As$ as the barrier layer in an $Ga_xIn_{1-x}P$/GaAs high-electron mobility transistor? Why? Note, the conduction band discontinuity in the $Al_{0.3}Ga_{0.7}As$/GaAs system is 0.263 eV.

8. **Mini project**: During the last 60 years, the density and feature size of Si ICs have followed Moore's law faithfully through device scaling. However, the current Si

technology will hit a feature size limit in the near future. New channel materials for increased electron and hole mobility, well above those achievable in strained Si, are needed. As gate dielectric scaling becomes increasingly difficult, carrier transport enhancement will become essential for current enhancement in the advanced FETs of future CMOS generations. One of the major problems encountered in developing FETs based on materials other than Si is the large disparity between electron and hole mobility. In order to design complementary FETs (i.e., one *n*-channel and one *p*-channel, and not limited to MOSFET), a material system with high hole mobility is required.

In this problem, the goal is to design a *p*-channel quantum well FET (QWFET) similar to the high-hole-mobility QWFET reported by Intel in 2008. (M. Radosavljevic et al., 2008 IEEE IEDM Tech. Dig., https://doi.org/10.1109/iedm.2008.4796798)

Your work should meet the following requirements:

- Select a material system among all available III–V materials for high hole mobility. Explain why you select it. However, you cannot copy the structure reported by Intel. Recall that in Chap. 6, we learned that under compressive strain, the heavy-hole band is lifted above the light-hole band in the growth direction (perpendicular to surface). But in the direction parallel to the surface (perpendicular to the growth direction), the effective mass becomes lighter. (See S. L. Chuang, Phys. Rev., **B43**, 9649, 1991.)
- The selected material system should be able to form QWFET structures with high crystal quality. Of course, high-quality heterostructures prepared on metamorphic substrate structures are allowed. (That means you can use materials with any lattice constant as the substrate).
- Design the layer structure of a QWFET based on the material system selected. We restrict the selection of FET structure to QWFETs (including pseudomorphic structures). Provide detailed material and device parameters such as ΔE_{v}, N_{S} and layer structures along with full energy band diagram. For heterostructures, the energy band discontinuities should be calculated. The portion of gate contact should also be included.
- To simplify the calculation, set the spacer layer thickness as $W_{sp} = 25$ Å and the acceptor level in the barrier layer as $E_a = 40$ meV. The Schottky barrier height of the barrier layer can be determined following the method reported by Tiwari and Frank (Appl. Phys. Lett., **60**, 630, 1992).
- Your final device layer structure should be a depletion mode FET at zero gate bias.

References

1. R. Dingle, H.L. Stormer, A.C. Gossard, W. Wiegmann, Appl. Phys. Lett. **33**, 665 (1978)
2. T. Mimura, S. Hiyamizu, T. Fujii, K. Nanbu, Jpn. J. Appl. Phys. **19**, L225 (1980)

3. S. Hiyamizu, J. Saito, K. Nanbu, T. Ishikawa, Jpn. J. Appl. Phys. **22**, L609 (1983)
4. H. Morkoc, T.J. Drummond, R. Fischer, J. Appl. Phys. **53**, 1030 (1982)
5. L. Pfeiffer, K.W. West, H.L. Stormer, K.W. Baldwin, Appl. Phys. Lett. **55**, 1888 (1989)
6. K. Hirakawa, H. Sakaki, J. Yoshino, Appl. Phys. Lett. **45**, 253 (1984)
7. E.F. Schubert, K. Ploog, Phys. Rev. **B30**, 7021 (1984)
8. A. Kastalsky, R.A. Kiehl, IEEE Trans. Electron. Dev. **33**, 414 (1986)
9. O. Ambacher, J. Smart, J.R. Shealy, N.G. Weimann, K. Chu, M. Murphy, W.J. Schaff, L.F. Eastman, R. Dimitrov, L. Wittmer, M. Stutzmann, W. Rieger, J. Hilsenbeck, J. Appl. Phys. **85**, 3222 (1999)
10. I.P. Smorchkova, C.R. Elsass, J.P. Ibbetson, R. Vetury, B. Heying, P. Fini, E. Haus, S.P. DenBaars, J.S. Speck, U. Mishra, J. Appl. Phys. **86**, 4520 (1999)
11. H. Kroemer, Proc. IRE **45**, 1535 (1957)
12. W. Snodgrass, W. Hafez, N. Harff, M. Feng, IEDM Tech. Dig. (11–13 Dec 2006)
13. N. Holonyak, Jr., M. Feng, IEEE Spectrum, **43**(2), 50 (2006)
14. H.K. Gummel, Proc. IEEE **57**, 2159 (1969)
15. A.W. Hanson, S.A. Stockman, G.E. Stillman, I.E.E.E. Electron, Dev. Lett. **14**, 45 (1993)
16. M.T. Fresina, Q.J. Hartman, D.A. Ahmari, N.F. Gardner, G.E. Stillman, J. Appl. Phys. **77**, 5437 (1995)
17. W. Hafez, J.W. Lai, M. Feng, I.E.E.E. Electron, Dev. Lett. **24**, 436 (2003)
18. B.W. Wu, B.F. Chu-Kung, M. Feng, K.Y. Cheng, J. Vac. Sci. Technol., B **24**, 1564 (2006)
19. P.D. Ye, J. Vac. Sci. Technol., A **26**, 697 (2008)
20. J. Robertson, J. Vac. Sci. Technol., A **31**, 050821 (2013)
21. M.D. Groner, F.H. Fabreguette, J.W. Elam, S.M. George, Chem. Mater. **16**, 639 (2004)
22. M. Hong, J. Kwo, A.R. Kortan, J.P. Mannaerts, A.M. Sergent, Science **283**, 1897 (1999)
23. M. Passlack, M. Hong, J.P. Mannaerts, Appl. Phys. Lett. **68**, 1099 (1996)
24. C.L. Hinkle, A.M. Sonnet, E.M. Vogel, S. McDonnell, G.J. Hughes, M. Milojevic, B. Lee, F.S. Aguirre-Tostado, K.J. Choi, H.C. Kim, J. Kim, R.M. Wallace, Appl. Phys. Lett. **92**, 071901 (2008)
25. M. Hong, H.W. Wan, K.Y. Lin, Y.C. Chang, M.H. Chen, Y.H. Lin, T.D. Lin, T.W. Pi, J. Kwo, Appl. Phys. Lett. **111**, 123502 (2017)

Further Reading

1. C. Weisbuch, B. Vinter, *Quantum Semiconductor Structures* (Academic Press, 1991)
2. J.H. Davies, *The Physics of Low-Dimensional Semiconductors* (Cambridge, 1998)
3. B.G. Streetman, S. Banerjee, *Solid State Electronic Devices*, 6th edn. (Prentice-Hall, 2006)
4. S.M. Sze, *High-Speed Semiconductor Devices* (Wiley, 1990)
5. C.W. Wilmsen (ed.), *Physics and Chemistry of III–V Compound Semiconductor Interfaces* (Plenum Press, New York, 1985)
6. S. Oktyabrsky, P.D. Ye (eds.), *Fundamentals of III-V Semiconductor MOSFETs* (Springer, New York, 2010)
7. S.M. George, Chem. Rev. **110**, 111 (2010)

Chapter 10
Heterostructure Photonic Devices

Abstract The light emitters made of III–V alloys are the most important and widely used photonic devices because Si and Ge do not emit light due to their indirect band structures. The semiconductor laser is similar to the solid-state ruby laser and Ar^+ gas laser in that the emitted radiation is highly monochromatic and produces a highly directional beam of light. However, the semiconductor laser is much smaller (on the order of 0.25 mm long) than other lasers and is easily modulated at high frequencies simply by modulating the injected current. Because of these unique properties, the semiconductor laser is one of the most important light sources for optical-fiber communication. It also has many applications in consumer electronics such as optical reading, laser printing, and face ID, to mention just a few. In addition, semiconductor lasers have significant applications in many areas of basic research and technology, such as high-resolution gas spectroscopy and atmospheric pollution monitoring. A related important photonic device is the light-emitting diode (LED), which has a device structure very similar to the semiconductor injection laser but without a resonant optical cavity. Today, visible LEDs play a leading role in displays and solid-state lighting applications. In this chapter, the device physics, structures, and characteristics of many types of heterostructure lasers are discussed. The current status and challenges of visible LEDs are briefly reviewed. In addition, the quantum cascade laser based on unipolar inter-subband transitions for long-wavelength applications is introduced. Also discussed is the mid-infrared quantum-well infrared photodetector, operating based on the same unipolar inter-subband optical transition. Finally, the concept of optoelectronic integration is demonstrated in the transistor laser where both transistor and laser operation are realized simultaneously in the same device.

10.1 Device Physics of Heterostructure Lasers

10.1.1 Basic Diode Laser Structure

The basic structure of a semiconductor injection laser is similar to that of a *p-n* junction diode as shown in Fig. 10.1. Like all other types of laser, it has a gain medium, an optical cavity, and an excitation source. The *p-n* junction, which usually

K. Y. Cheng, *III–V Compound Semiconductors and Devices*, Graduate Texts in Physics,
https://doi.org/10.1007/978-3-030-51903-2_10

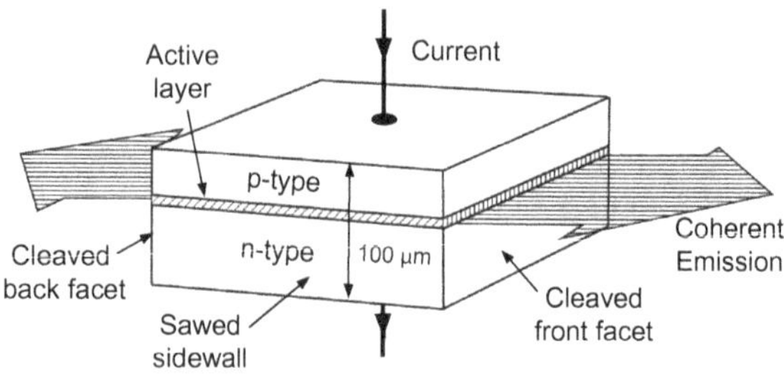

Fig. 10.1 Basic structure of a laser diode in the form of a Fabry–Perot cavity

lies in the (100) plane, forms the active region (gain medium) of the laser diode, which is a part of the optical cavity. A current flowing through the *p-n* junction supplies electrons and holes for light generation through radiative recombination. A pair of parallel (110) planes is cleaved perpendicularly to the (100) plane of the junction to form cavity mirrors. The two remaining sides of the diode are either roughened or electrically isolated to eliminate lasing in directions other than the main one. The structure is called a Fabry–Perot cavity.

When a forward bias is applied to the laser diode, a current flows through the *p-n* junction as the excitation source. Initially at low current, there is spontaneous emission in all directions. As the bias is increased, eventually a threshold current density J_{th} is reached at which the stimulated emission occurs and a monochromatic and highly directional beam of light is emitted from the junction (Fig. 10.1). As discussed later, in order to achieve efficient and reliable operation of laser diodes, it is necessary to design devices with a low threshold current density J_{th}.

In 1962, when laser diodes were first demonstrated in diffused GaAs and GaP_xAs_{1-x} *p-n* homojunction structures, the threshold current density J_{th} was about 5.0×10^4 A/cm^2 at room temperature. Most studies were done at liquid nitrogen temperature (77 K) or lower. Such a large current density imposes serious difficulties in operating the laser continuously at 300 K.

To reduce the threshold current density, heterostructure lasers have been proposed and built using epitaxial techniques. Figure 10.2 shows schematic representations of the bandgap under forward-biased conditions, the refractive index change, and the optical field distribution of light generated at the *p-n* junction for a homostructure and a double heterostructure (DH). The lack of carrier confinement under forward bias condition in the homojunction structure laser is evident from the low potential barriers. Consequently, only under very high injection conditions are there enough carriers for the laser to reach population inversion. The very small index change (<1%) induced from injected carriers near the junction is not enough to provide a strong optical confinement of the optical emission within the waveguide. Therefore, J_{th} is extremely high in homostructure lasers. For the double-heterostructure laser, the carriers are confined in the active region *d* by the heterojunction potential barriers on both sides, and the optical field is also confined within the active region by the abrupt reduction of the refractive index (~5%) outside the active region. These confinements can enhance the stimulated emission and substantially reduce the threshold current density. Since J_{th} for DH lasers can be as low as 10^3 A/cm^2 or less at 300 K,

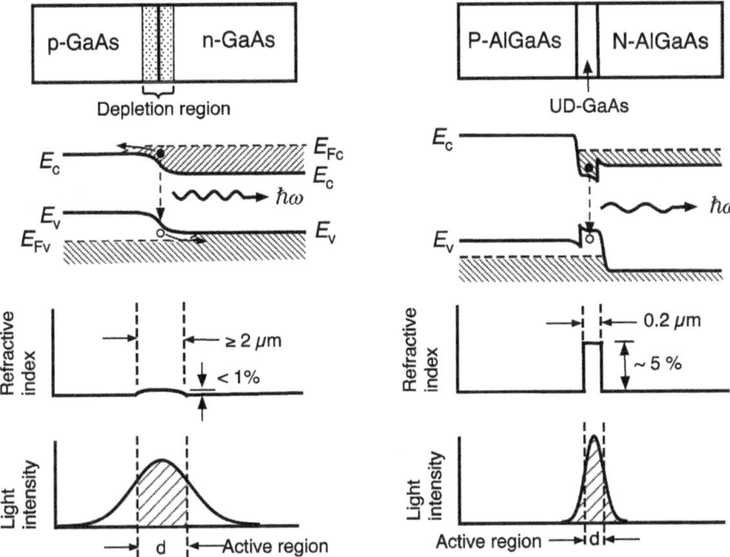

Fig. 10.2 Comparison of homojunction (left) and double-heterostructure (right) lasers. The second row shows energy band diagrams under forward bias. The refractive index change across the junction is shown in the next row. The confinement of light is shown in the bottom row

continuous room-temperature operation has been achieved. This achievement has led to increased applications of semiconductor lasers, especially for optical-fiber communication systems, laser printers, and optical memory storage applications. In the discussions followed, we will be concerned primarily with DH lasers.

10.1.2 Wave-Guiding and Confinement Factor

In a DH laser, the light is confined and guided by the dielectric waveguide. Figure 10.3 shows a three-layer symmetric dielectric waveguide with refractive indices, $\bar{n}_1$, $\bar{n}_2$, and $\bar{n}_3$, where an active layer is sandwiched between two cladding layers. Under the condition

$$\bar{n}_2 > \bar{n}_1 = \bar{n}_3 \tag{10.1}$$

the electromagnetic radiation with a ray angle θ at the active layer/cladding layer interface exceeding the critical angle $\theta_c = \sin^{-1}(\bar{n}_1/\bar{n}_2)$ is totally reflected into the active layer. Therefore, when the refractive index in the active region is larger than the index of its surrounding cladding layers, (10.1), the propagation of electromagnetic radiation is guided in a direction parallel to the layer interfaces along the z-direction. For idea DH lasers, the refractive index steps at each heterojunction can be made as

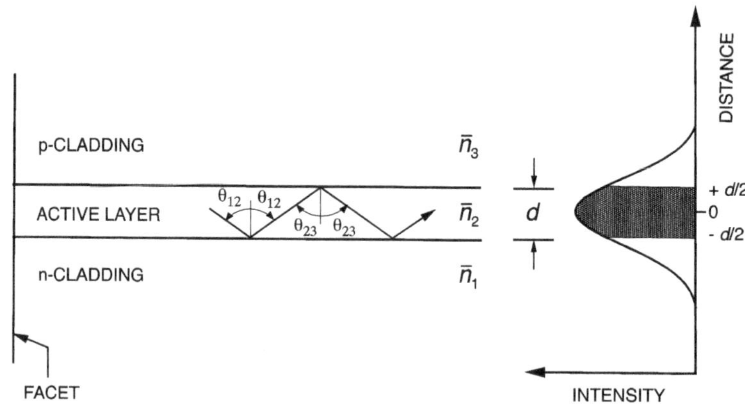

Fig. 10.3 Ray trajectories of the guided wave in a three-layer dielectric waveguide of a semiconductor laser. The intensity distribution of the optical waveguide mode is also shown. The shaded region represents the fraction of the mode within the active region

large as 10% or more and provide a well-defined waveguide. However, due to the finite difference in refractive indices in most structures, a portion of the electromagnetic radiation leaks out of the active region into cladding layers. The evanescent wave in the cladding layer will not participate in lasing action in the active layer and becomes a portion of the cavity loss. The ratio of the optical energy within the active layer to the total energy both within and outside the active layer defines a *confinement factor* Γ.

Consider the transverse electric (TE) wave $E_y(x, z, t)$ which is polarized transversely to the direction of propagation (z-direction) along the y-direction in a three-layer symmetric dielectric waveguide. The even mode of $E_y(x, z, t)$ is given by

$$E_y = A\cos(\kappa x)\exp[i(\omega t - \beta z)] \tag{10.2}$$

where $\beta = 2\pi/\lambda$ is the propagation constant, $\kappa^2 = \bar{n}_2^2 k_0^2 - \beta^2$, $k_0 = 2\pi/\lambda_0$, and $\lambda = \lambda_0/\bar{n}_2$. The even mode magnetic field $H_x(y, z, t)$ can be found from

$$H_x = -(i/\omega\mu_0)\left(\partial E_y/\partial z\right) \tag{10.3}$$

The average power propagating in the waveguide is determined by the Poynting vector, $P \propto E_y H_x^* \propto |E_y|^2$, where H_x^* is the complex conjugate. The confinement factor for the symmetrical three-layer dielectric waveguide with a thickness of d can be obtained for the even TE waves as

$$\Gamma = \frac{\int_0^{d/2}\left|E_y(x)\right|^2 \mathrm{d}x}{\int_0^{\infty}\left|E_y(x)\right|^2 \mathrm{d}x} = \frac{\int_0^{d/2}\cos^2(\kappa x)\mathrm{d}x}{\int_0^{d/2}\cos^2(\kappa x)\mathrm{d}x + \int_{d/2}^{\infty}\cos^2(\kappa x)\exp\left[-2\gamma(x - d/2)\right]\mathrm{d}x}$$

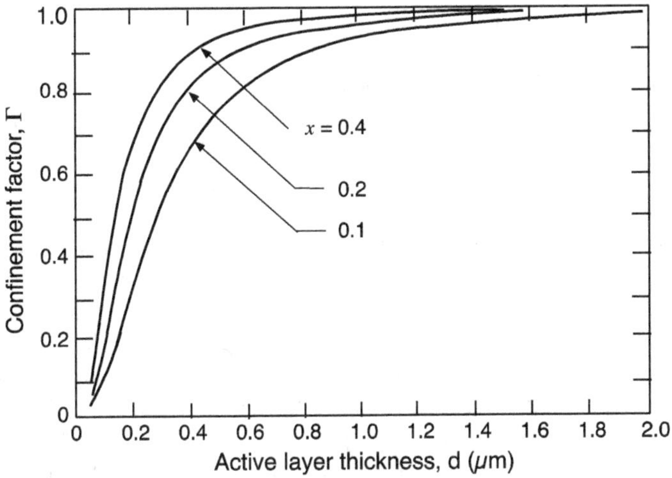

Fig. 10.4 Confinement factor for the fundamental mode as a function of active layer thickness for the $Al_xGa_{1-x}As/GaAs$ symmetric three-layer slab dielectric waveguide

$$= \left\{1 + \frac{\cos^2(\kappa d/2)}{\gamma[(d/2) + (1/\kappa)\sin(\kappa d/2)\cos(\kappa d/2)]}\right\}^{-1} \tag{10.4}$$

where $\gamma^2 = \beta^2 - \bar{n}_1^2 k_0^2$ in cladding layers. Similar expressions may be obtained for odd TE waves as well as for the transverse magnetic (TM) waves.

Figure 10.4 shows the variation of the calculated confinement factor Γ in an $Al_xGa_{1-x}As/GaAs$ DH structure for the fundamental mode with active layer thickness d. It can be seen that Γ decreases rapidly for $d < \lambda/\bar{n}_2$ (≈0.25 μm), where the active layer thickness becomes less than the wavelength of the radiation. Representing the fraction of the propagating mode within the active layer by Γ is an essential concept for understanding the influence of the active layer thickness on the gain coefficient necessary to reach the lasing threshold.

10.1.3 Threshold Condition in Fabry–Perot Cavity

In Chap. 8, it was shown that when the excitation satisfies the necessary condition of population inversion for stimulated emission, the absorption coefficient $\alpha(E)$ reduces to zero. Also, the gain coefficient $g(E)$ is related to $\alpha(E)$ as

$$g(E) = -\alpha(E) \tag{10.5}$$

where $g(E)$ is defined as positive when radiation is emitted and $\alpha(E)$ is negative when radiation is absorbed. At this point, $g(E) = \alpha(E) = 0$, the system reaches the transparency condition, which is determined solely by the material property of the active layer. However, since there are other losses in a practical laser diode, additional gain is needed for the laser diode to reach the lasing threshold.

Consider a Fabry–Perot laser diode cavity, which consists of a gain medium truncated by a pair of parallel reflection surfaces. The lasing condition can be obtained by considering the plane wave reflection between parallel, partially reflecting mirror surfaces as shown in Fig. 10.5. The cavity length is L. The transmission and reflection coefficients of the left mirror facet are indicated by t_1 and r_1, respectively. These parameters of the right mirror are indicated by t_2 and r_2, respectively. A plane wave electric field E_i is incident on the left cavity mirror so that E_x is t_1E_i inside the left boundary and $t_1E_i\exp(-KL)$ just inside the right boundary, where K is the complex propagation constant ($= \gamma + i\beta$). The first path of the field transmitted through the right mirror is $t_1t_2E_i\exp(-KL)$, and the reflected field is $t_1r_2E_i\exp(-KL)$. The next path of the wave transmitted through the right mirror becomes $t_1t_2r_1r_2E_i\exp(-3KL)$ and so on. The sum of these transmitted fields through the right mirror gives

$$\begin{aligned} E_t &= t_1t_2E_i\exp(-KL)\left[1 + r_1r_2\exp(-2KL) + r_1^2r_2^2\exp(-4KL) + \cdots\right] \\ &= E_i\left[\frac{t_1t_2\exp(-KL)}{1 - r_1r_2\exp(-2KL)}\right] \end{aligned} \tag{10.6}$$

The oscillation condition is reached when the denominator of (10.6) goes to zero, where a finite transmitted wave E_t is obtained with zero E_i. Thus

$$r_1r_2\exp(-2KL) = 1 \tag{10.7}$$

Using $K = i(\bar{n} - i\bar{\kappa})k_0$ in (10.7) with $k_0 = 2\pi/\lambda_0$ and $\bar{\kappa} = \alpha\,\lambda_0/4\pi$, it becomes

$$r_1r_2\exp(-\alpha L)\exp[-2i(2\pi\bar{n}/\lambda_0)L] = 1 \tag{10.8}$$

The above equation represents a wave making a round trip of $2L$ inside the cavity to the starting point with the same amplitude and phase. It of course contains two parts representing the amplitude and the phase of the oscillating wave. The amplitude requirement for oscillation is

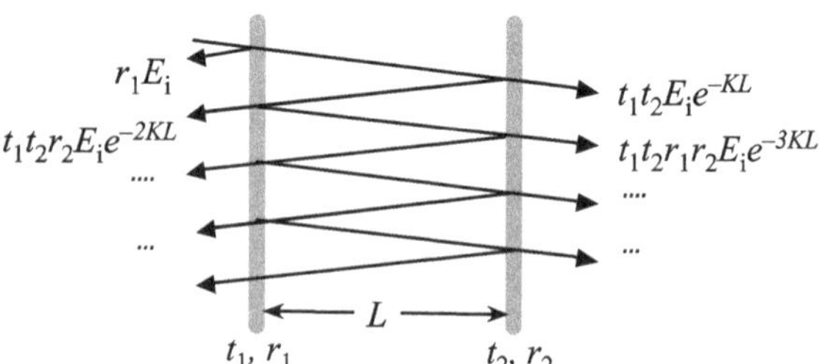

Fig. 10.5 Oscillation condition for a gain medium with parallel reflecting surfaces in a Fabry–Perot cavity

$$r_1 r_2 \exp[(\Gamma g - \alpha_i)L] = 1 \quad (10.9)$$

The negative α in (10.8) has been replaced by the net gain, which equals the difference between the model gain Γg and the total internal losses α_i. In the gain term, the confinement factor Γ represents the fraction of the mode energy contained within the active region that contributes to gain. The gain can be rearranged as the sum of the total internal loss (α_i) and mirror loss (α_m):

$$\Gamma g = \alpha_i + \alpha_m = \alpha_i + (1/2L)\ln(1/R_1 R_2) \quad (10.10)$$

and

$$\alpha_m = (1/2L)\ln(1/R_1 R_2) \quad (10.11)$$

The reflection coefficients are replaced with the power reflectances $R_1 = r_1 r_1^* \approx r_1^2$ and $R_2 = r_2 r_2^* \approx r_2^2$. Intrinsic loss in semiconductors with good crystal quality lies in the range $\alpha_i = 5$–$10\ \text{cm}^{-1}$.

According to discussions in Chap. 8, the peak of the gain curve increases and shifts to higher energy with the injection level. Empirically, the gain peak position g_p or $g_{\max}$ follows the injection carrier density in a nearly linear fashion as illustrated in Fig. 10.6. One can write

$$g_p = a(n - n_{\text{tr}}) \quad (10.12)$$

where g_p is the peak gain, $a = \partial g_p/\partial n$ is the differential gain coefficient, and n_{tr} is the transparency carrier density. For $n > n_{\text{tr}}$, there is a positive gain. However, to reach the threshold, more gain achieved by a higher carrier injection is needed to

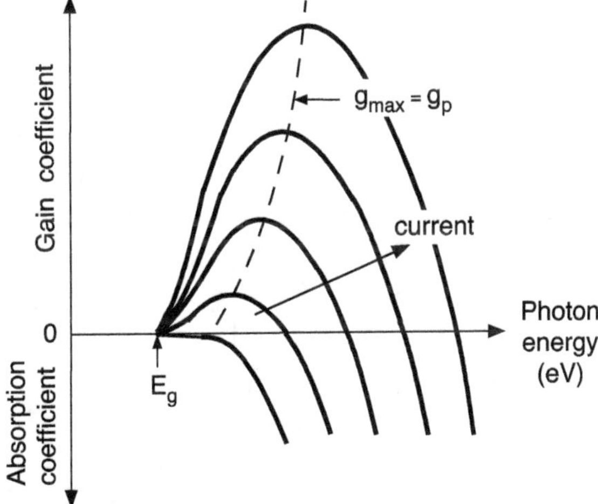

Fig. 10.6 Schematic of the change in laser gain spectra as a function of carrier injection of an InGaAsP/InP laser. Reprinted with permission from [1], copyright AIP Publishing

overcome internal and mirror losses. At the threshold, $g_p = g_{th} = a(n_{th} - n_{tr})$ and n_{th} is the threshold carrier density.

The phase condition of oscillation is

$$4\pi \bar{n} L/\lambda_0 = 2m\pi \quad m(\text{mode number}) = 1, 2, 3, \ldots \tag{10.13}$$

or

$$m\lambda_0 = 2\bar{n}L \tag{10.14}$$

For a typical laser diode, the mode number m can be very large since $\lambda_0 \ll L$. Figure 10.7 shows the lasing spectrum and the gain curve of a typical laser diode operating just above the threshold. The profile of the gain curve is usually broad, and multiple modes, whose gain is larger than the total loss, are allowed to lase simultaneously. This lasing mode reflecting a light wave situation in the oscillation direction of the laser cavity is called the *longitudinal mode*. Equation (10.14) can be used to derive the longitudinal mode spacing. Differentiation of this equation yields

$$\lambda_0 dm + m d\lambda_0 = 2L d\bar{n} \tag{10.15}$$

For adjacent mode, $dm = -1$, and (10.14) can be substituted for m to give

$$d\lambda_0 = \frac{\lambda_0^2}{2\bar{n}L[1 - (\lambda_0/\bar{n})(d\bar{n}/d\lambda_0)]} \tag{10.16}$$

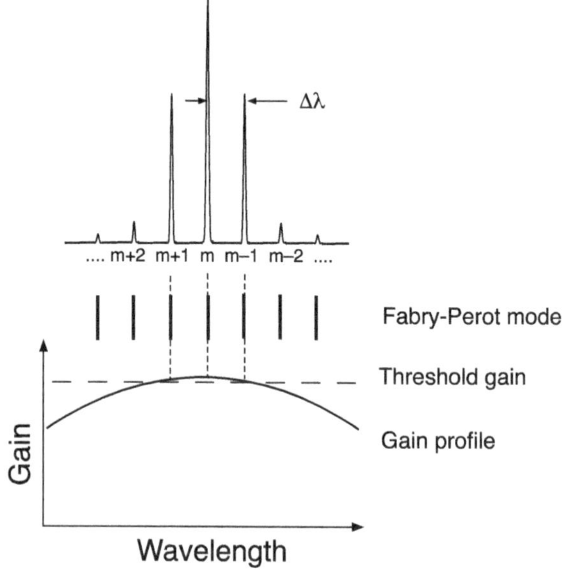

Fig. 10.7 Longitudinal mode emission spectrum and the gain profile of a FP laser diode

for the spacing between adjacent longitudinal modes. Since the refractive index change is near zero for adjacent modes, the mode spacing $d\lambda_0$ can be simplified as

$$d\lambda_0 = \Delta\lambda_0 \cong \lambda_0^2/2\bar{n}L = \lambda_0/m \tag{10.17}$$

For a GaAs laser diode with $L = 250\ \mu\text{m}$, $\lambda_0 = 0.85\ \mu\text{m}$, and $\bar{n} \sim 3.5$, the calculated mode number is ~2060 with a mode spacing of ~0.4 nm.

10.1.4 Threshold Current Density and Photon Density

To analyze the interaction between injected carriers and photons in a semiconductor laser so that the threshold current density and other dynamic properties can be derived, it is necessary to use the coupled rate equations which determine the time variation of the density of the two species. For the derivation of threshold current density, only the gain–loss relation, without the phase condition, is considered during the lasing action. For the active layer structure shown in Fig. 10.8 (d = thickness, L = length, and w = width), the rate of change of the injected electron density n is

$$\frac{\partial n}{\partial t} = D_n \nabla^2 n + \frac{J}{qd} - R(n) \tag{10.18}$$

where D_n is the carrier diffusion coefficient, $J = I/wL$ is the injected current density, and $R(n)$ is the rate of recombination loss.

The first two terms on the right-hand side represent carrier injection rate increase through diffusion and injection, respectively. The total recombination rate is expressed as

$$R(n) = A_{\text{nr}}n + B_r n^2 + Cn^3 + R_{\text{stim}} \tag{10.19}$$

where A_{nr}, B_r, C, and R_{stim} are the non-radiative, radiative, Auger and stimulated recombination rate coefficients, respectively. Under steady state and $n = p$,

$$\frac{\partial n}{\partial t} = D_n \nabla^2 n = 0 \tag{10.20}$$

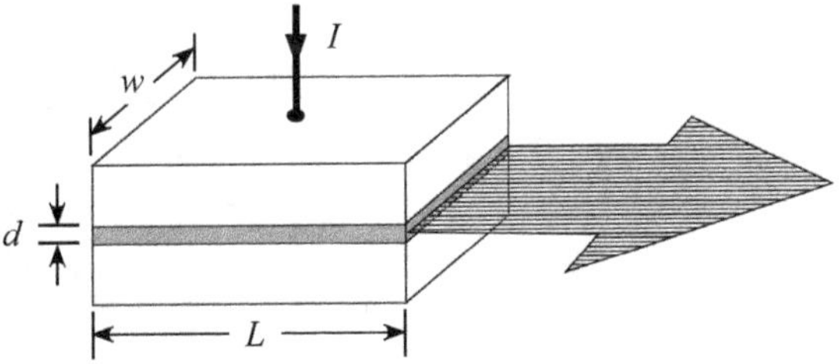

Fig. 10.8 Geometry and dimensions of a FP cavity laser diode

This leads to

$$J = qdR(n) \tag{10.21}$$

Near the threshold condition, $R_{\text{stim}} \approx 0$, and $A_{\text{nr}} = C = 0$ for the laser with high material quality. Therefore, $R(n) = B_r n^2 \approx n/\tau_s$, and, below threshold, the current density is related to the injected carrier density n by

$$J = \frac{qdn}{\tau_s} \tag{10.22}$$

where τ_s is the carrier lifetime.

When the injection current density at or just above threshold, $J = J_{\text{th}}, n = n_{\text{th}}$, and $g = g_{\text{th}} = a(n_{\text{th}} - n_{\text{tr}})$. Replacing g_{th} with $(\alpha_i + \alpha_m)/\Gamma$, the threshold carrier density becomes

$$n_{\text{th}} = n_{\text{tr}} + (\alpha_i + \alpha_m)/a\Gamma \tag{10.23}$$

This leads to the threshold current density of

$$J_{\text{th}} = \frac{qd}{\tau_s}\left(n_{\text{tr}} + \frac{\alpha_i + \alpha_m}{a\Gamma}\right) \tag{10.24}$$

This relationship between J_{th} and d is illustrated in Fig. 10.9 for 300 μm long InGaAs/InP lasers emitting at 1.3 μm. The curve shows a minimum J_{th} at an optimal

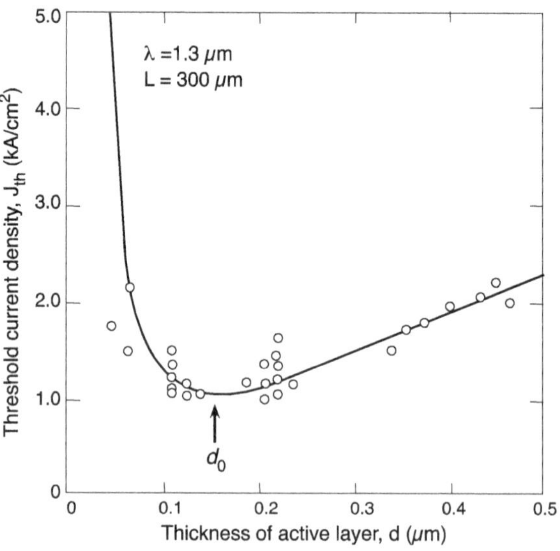

Fig. 10.9 Measured threshold current density at room temperature as a function of the active layer thickness for 1.3 μm InGaAsP DH lasers. Reprinted with permission from [2], copyright Japanese Society of Physics

active layer thickness d_0. When the active layer thickness is large, $d > d_0$, the confinement factor becomes saturated close to unity. The value in the parenthesis turns into a constant and J_{th} proportions to d in a linear fashion. For $d < d_0$, Γ is a small number and decreases rapidly with reducing active layer thickness. Therefore, the second term containing Γ, which represents the cavity loss, dominates the threshold current density. For thin active layer, the threshold current density increases rapidly with deceasing d.

At the same time, the rate of change in photon density N of the oscillating mode in the laser cavity can be described by the following photon rate equation:

$$\frac{\partial N}{\partial t} = \Gamma \beta R_{sp} + \Gamma R_{st} - \frac{N}{\tau_{ph}} \tag{10.25}$$

where R_{sp} and R_{st} are the spontaneous emission rate and stimulated emission rate, respectively, and τ_{ph} is the photon lifetime. β is the spontaneous emission factor that approximately equals the inverse of the optical mode number of the cavity. Therefore, the first term on the right-hand side represents the photon increase due to spontaneous recombination of the oscillation mode. The last term takes account of the decrease of the photon density in the mode by escape of photons from the resonator, as well as by scattering and other losses not directly connected with radiating transitions. The spontaneous emission rate can be expressed in the form

$$R_{sp} = n/\tau_s \tag{10.26}$$

The stimulated emission rate has the same form as the net absorption rate derived in Chap. 8 (8.77) and is expressed as

$$R_{st} = v_g g_p N = \frac{c}{\bar{n}} a(n - n_{tr}) N \tag{10.27}$$

When the laser is operated just above threshold, under steady-state condition, $\partial N/\partial t = 0$ and $\beta \ll 1$, the rate equation reduces to

$$N\left[\Gamma a(n - n_{tr})\frac{c}{\bar{n}} - \frac{1}{\tau_{ph}}\right] = 0 \quad \text{and} \quad N \neq 0 \tag{10.28}$$

Through the above equation, the photon lifetime can be determined as a function of the differential gain, carrier density, and the confinement factor.

$$\frac{1}{\tau_{ph}} = \Gamma a(n - n_{tr})\frac{c}{\bar{n}} \tag{10.29}$$

The optical emission processes in the laser cavity also change the number of injected excess carriers, which is controlled by the current injection following

(10.22). Under steady-state lasing conditions, the electron rate equation for high carrier injection, neglecting spontaneous emission, is

$$R(n) = \frac{J}{qd} = \frac{n}{\tau_s} + R_{\mathrm{st}} = \frac{n}{\tau_s} + \upsilon_g g_p N \tag{10.30}$$

and

$$N = \left(\frac{J}{qd} - \frac{n}{\tau_s}\right)\frac{1}{\upsilon_g g_p} \tag{10.31}$$

Using $\upsilon_g g_p = (c/\bar{n})a(n - n_{\mathrm{tr}})$ and $\Gamma \upsilon_g g_p = 1/\tau_{\mathrm{ph}}$, N can be expressed as

$$N = \frac{\Gamma \tau_{\mathrm{ph}} J}{qd} - \frac{1}{\tau_s}\left(n_{\mathrm{tr}}\Gamma\tau_{\mathrm{ph}} + \frac{\bar{n}}{ac}\right) = \frac{\Gamma\tau_{\mathrm{ph}}}{qd}(J - J_{\mathrm{th}}) \tag{10.32}$$

where $J_{\mathrm{th}} = (qd/\tau_s)(n_{tr} + \bar{n}/ac\Gamma\tau_{\mathrm{ph}})$ is the threshold current density. It is clear that

$$N \propto J - J_{\mathrm{th}} \tag{10.33}$$

Figure 10.10 shows the variations of the injected carrier density n and the photon density N as functions of the injected current density J. Following the assumption made above, the spontaneous emissions are neglected. Below the threshold current density J_{th} the photon density vanishes, and the electron concentration increases linearly with the pumping current (10.22). Above the threshold, the photon density increases linearly but the injected carrier density remains constant or pinned at n_{th}. This is because of the rather long carrier lifetime (~10^{-9} s) compared with the short photon lifetime (~10^{-12} s) such that, at $J \geq J_{\mathrm{th}}$, all carriers recombine immediately once injected into the cavity. If spontaneous emissions are not omitted, there is a minor deviation from the idealized behavior below threshold, as indicated by the dashed line in Fig. 10.10.

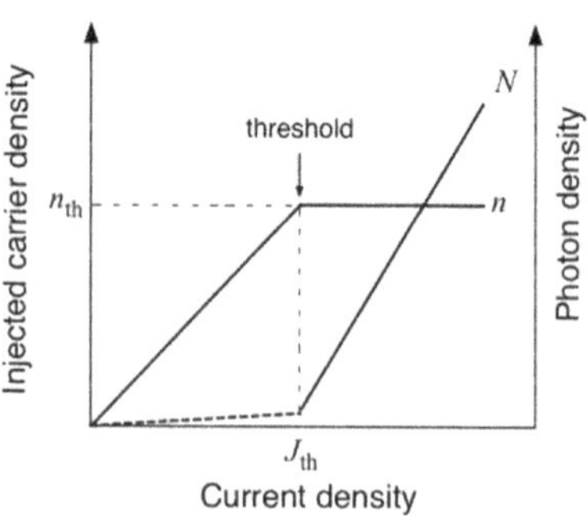

Fig. 10.10 Variations of injected carrier density (n) and generated photon density (N) as functions of injection current density

10.1.5 Laser Output Power and Efficiency

Before lasing, the light generated from the laser takes the form of spontaneous emission. At the threshold, a change from spontaneous emission to stimulated emission is observed, and light output is emitted with a high efficiency after lasing. The output power is formulated as follows:

$$P = \begin{pmatrix}\text{photon}\\ \text{energy}\end{pmatrix}\begin{pmatrix}\text{photon}\\ \text{density}\end{pmatrix}\begin{pmatrix}\text{effective volume}\\ \text{of the cavity}\end{pmatrix}\begin{pmatrix}\text{photon}\\ \text{escape rate}\end{pmatrix}$$
$$= (\hbar\omega)(N)\left(\frac{wdL}{\Gamma}\right)(\upsilon_g\alpha_m)$$

$$P = \frac{\hbar\omega}{q}\upsilon_g\alpha_m\tau_{\text{ph}}(J - J_{\text{th}})wL \tag{10.34}$$

Since $(J - J_{\text{th}})wL = (I - I_{\text{th}})$ and $1/\tau_{\text{ph}} = \upsilon_g(\alpha_i + \alpha_m)$, the output power is rewritten as

$$P = \frac{\hbar\omega}{q}(I - I_{\text{th}})\frac{\alpha_m}{\alpha_i + \alpha_m} \tag{10.35}$$

Clearly, the output power is a linear function of injection current above the threshold. Figure 10.11 shows the typical light output power from one mirror facet of the laser diode as a function of the injection current or the *L–I* curve. However, the linear relation deteriorates by Joule heating at high-current densities. The slope of the laser output characteristic *L–I* curve defines the external differential quantum efficiency η_d, which is expressed by a ratio of the photon number emitted out of a laser cavity to injected carrier number.

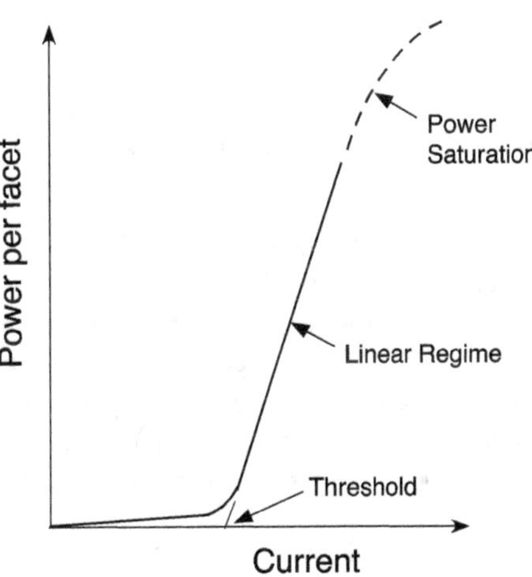

Fig. 10.11 Light–current (*L–I*) curve for a laser diode. The dashed line represents the onset of power saturation

$$\eta_d = \frac{\Delta P/\hbar\omega}{\Delta I/q} = \frac{\alpha_m}{\alpha_i + \alpha_m} \tag{10.36}$$

A typical η_d of a laser diode is about 50–80%, where the emission from both mirror facets is taken into account.

In the above discussion, injected carriers are assumed completely converting into photons inside the laser diode without any loss. If the spontaneous emission is no longer neglected, only a fraction of the total photons created goes into the laser mode. In addition, there are non-radiative transitions and leakage currents. An internal quantum efficiency η_i, which indicates a ratio of generated photon number to injected carrier number in the active layer, is introduced to show that only part of the injected carriers contribute to photon generation. The optical output power of the laser diode is now shown as

$$P = \frac{\hbar\omega}{q}(I - I_{\mathrm{th}})\frac{\alpha_m}{\alpha_i + \alpha_m}\eta_i \tag{10.37}$$

It follows that the single facet differential external quantum efficiency is

$$\eta_d = \left(\frac{\alpha_m}{\alpha_i + \alpha_m}\right)\eta_i = \frac{\eta_i}{1 - \alpha_i L/\ln R} \tag{10.38}$$

for the case of $R = R_1 = R_2$. This equation can be expressed in the following form:

$$\frac{1}{\eta_d} = \frac{1}{\eta_i}\left(1 - \alpha_i \frac{L}{\ln R}\right) \tag{10.39}$$

One can plot $1/\eta_d$ as a linear function of cavity length L of lasers which are otherwise the same. The slope of the straight line determines the loss parameter $\alpha_i/\ln R$. The internal quantum efficiency η_i can be deduced from the extrapolated η_d value at $L = 0$.

10.1.6 Characteristic Temperature

Light output power of a laser diode is ideally expressed by (10.37) which is a linear function of the injection current above threshold. In reality, lasers often show power saturation under cw operation, as recognized by the high-temperature range of Fig. 10.12. The power saturation tends to be large in proportion to the ambient temperature as a result of the temperature dependence of stimulated emission. This means that the threshold current also increases according to the rise in temperature. The temperature dependence of the threshold current is empirically given by

$$I_{\mathrm{th}} = I_{\mathrm{th}}(0)\exp(T_j/T_0) \tag{10.40}$$

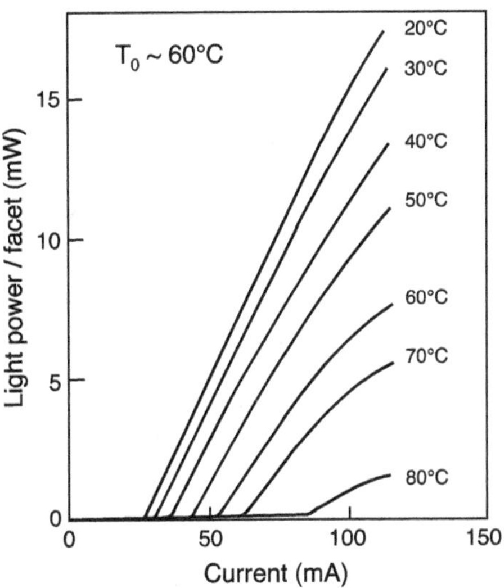

Fig. 10.12 Temperature dependence of *L–I* curves in a 1.3 μm InGaAsP/InP buried-heterostructure laser

where $I_{\text{th}}(0)$ is the threshold current at 0 °C and T_j is the junction (active layer) temperature. The T_0 term is called the *characteristic temperature* of the laser and indicates the degree of lasing characteristic change with temperature. A high characteristic temperature indicates a small change in lasing characteristics with temperature. The characteristic temperature of an AlGaAs/GaAs laser is usually between 100 and 200 K and is determined primarily from overflow of injected carriers from the active layer into cladding layers. For InGaAsP/InP lasers, the characteristic temperature has a maximum for an active layer emitting 1.3-μm-wavelength light, and it gradually decreases accordingly as the lasing wavelength shifts toward shorter or longer ranges. In the lasing mode at shorter wavelengths than 1.3 μm, where the composition of the active layer approaches InP, the characteristic temperature is reduced primarily by overflow of injected carriers due to reduced band discontinuities at the heterojunction. In contrast, when lasing wavelengths longer than 1.3 μm, Auger recombination and intra-valence band absorption between the split-off band and the heavy-hole band strongly reduce the characteristic temperature. The characteristic temperature is less than 100 K in such lasers.

10.2 Structures and Properties of Injection Lasers

In this section, the electronic and photonic characteristics of different semiconductor lasers are discussed. The active region of the laser forms a slab dielectric waveguide and is terminated by cleaving as shown in Fig. 10.13. The cavity length, characteristics of the gain medium, and lateral dielectric discontinuities determine the longitudinal mode behaviors inside the cavity. The width of the slab waveguide controls the lateral mode of the near-field pattern which can be observed on the mirror facet

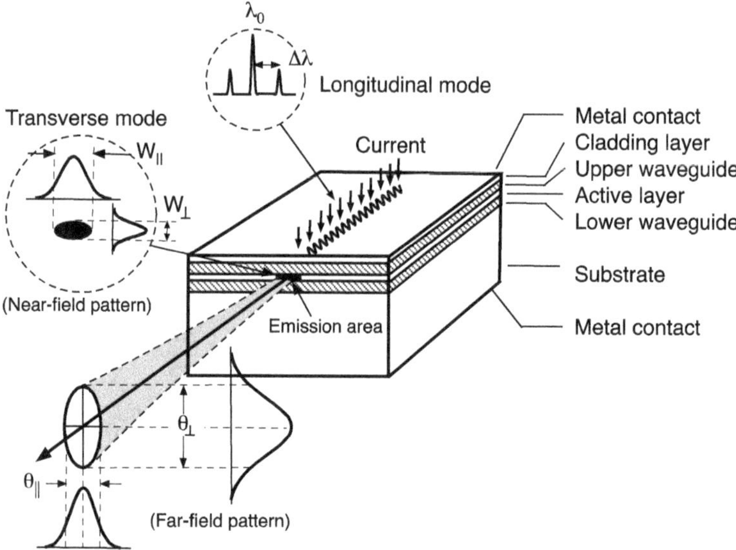

Fig. 10.13 Schematic representation of the near-field emission pattern, far-field emission pattern, and longitudinal mode pattern of a typical DH laser

under an infrared image converter microscope. When the optical waveguide is narrow ($\leq$10 μm), a fundamental transverse mode is usually achieved. On the other hand, the laser oscillates in a high-order transverse mode. The conditions to achieve fundamental transverse mode perpendicular to the junction plane depend on the index difference between the active and cladding layers and the active layer thickness. For laser structures with a thin active layer of $\leq$0.2 μm, a fundamental transverse mode is usually guaranteed. This is the situation of all lasers discussed in the following. As the modal field strikes the cleaved facet, it produces a radiation field outside the waveguide, called the far-field pattern. Due to the rectangular cross-sectional shape of the active region, the laser generates an elliptical far-field pattern with the long-axis perpendicular to the plane of the active layer.

10.2.1 Stripe-Geometry Lasers

The laser structure shown in Fig. 10.14 (left) is a broad-area DH laser because the entire area along the junction plane can emit radiation. There are serious problems associated with this design for practical applications. First, due to the large device cross section, the operation current is too high for room-temperature cw operation. Second, the large junction capacitance associated with the large device limits the operation speed. Third, it is difficult to achieve fundamental mode emission from the broad-area lasers. Figure 10.15 shows the near-field and far-field patterns of stripe-

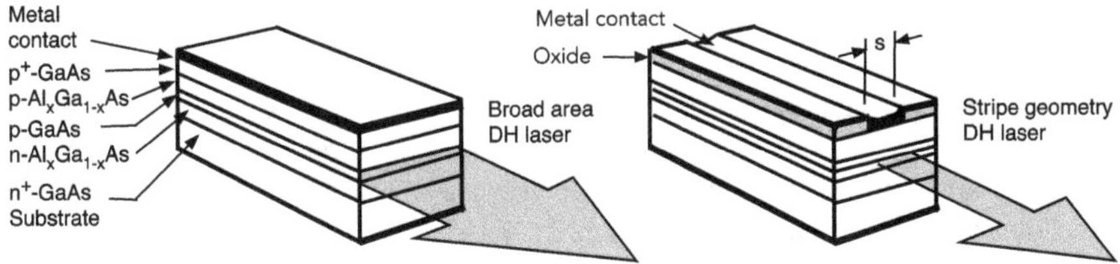

Fig. 10.14 Broad-area (left) and stripe geometry (right) double-heterostructure lasers in the Fabry–Perot cavity configuration

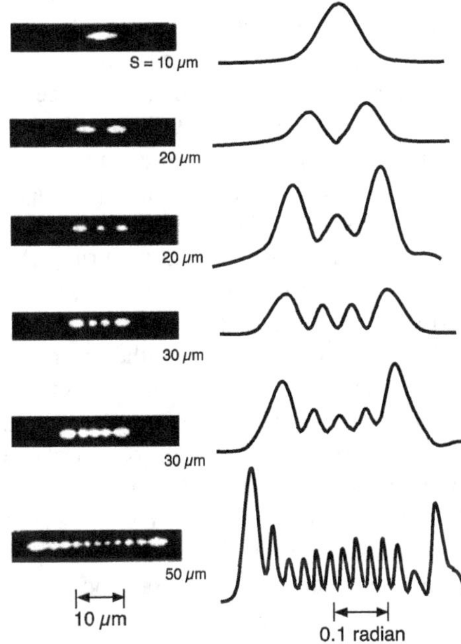

Fig. 10.15 Transverse modes on the laser facet as functions of stripe width (10–50 μm) for AlGaAs planar-stripe double-heterostructure lasers. Left: near-field patterns. Right: far-field patterns. Reprinted with permission from [3], copyright Japanese Society of Physics

geometry $Al_xGa_{1-x}As$ DH lasers with different waveguide widths between 10 and 50 μm. The fundamental mode operation can only be achieved in the device with a stripe width of $\leq$10 μm. As the stripe width is increased above 12 μm, higher-order modes along the junction plane are observed. The distinct and localized regions of stimulated emission shown in the near-field pattern are referred to as *filaments*. This spatial peak intensity of the emission can cause catastrophic laser mirror damage and limits the lifetime of the laser diode. Thus, most heterostructure lasers used today are made in narrow stripe-geometries (Fig. 10.14, right). Over the years, many different stripe-geometry laser structures have been developed. They can be divided into two categories: gain-guided lasers and index-guided lasers.

(a) **Gain-guided lasers**

The simplest gain-guided laser structure is an oxide-stripe geometry device (Fig. 10.16) where the oxide layer confines the injected current flow to a small region through a narrow stripe opening in the dielectric. Lasing from this device is restricted to the narrow region under the stripe contact. A variation of the above structure is the proton-bombardment-stripe laser, where the implanted protons or deuterons create a region of high resistivity that restricts the current flow to the center region, which is not bombarded. In these lasers, gain guiding is generated by carrier injection into a narrow region of the active layer, shown in Fig. 10.17, which is planar and continuous. Lasing occurs only in the gain region corresponding to the carrier injection region along the junction plane, and light propagates along the narrow gain region. Since the optical mode distribution along the junction is determined by the gain, rather than the refractive index difference, these lasers are called *gain-guided lasers*.

In contact stripe lasers, the current flow through the contact and upper cladding layers is by a majority carrier drift current which spreads laterally before reaching the active layer. Because of the spreading, the current flows through an area in the active region that is larger than the stripe contact area. Within the active layer, there is a lateral diffusion current of minority carriers due to the recombination, which also broadens the current distribution. These two effects greatly increase the threshold current of the stripe-geometry lasers. Generally, the threshold current of a 5–10 μm wide contact stripe laser is on the order of 100 mA. Figure 10.18 shows the typical *L–I* characteristics of a gain-guided InGaAsP laser. For small currents below the threshold, spontaneous emission is dominant. Stimulated emission is responsible for the steep increase in light output above the threshold. In all laser diodes, due to the higher mirror losses for the TM modes, only the TE-mode oscillates above threshold. In the neighborhood of the laser threshold, many longitudinal modes occur in the emission spectrum. With increasing injection current, side modes are strongly suppressed relative to the central mode and the emission spectrum narrows. However, the gain-guided laser does not emit light in a single longitudinal mode even at high injection levels.

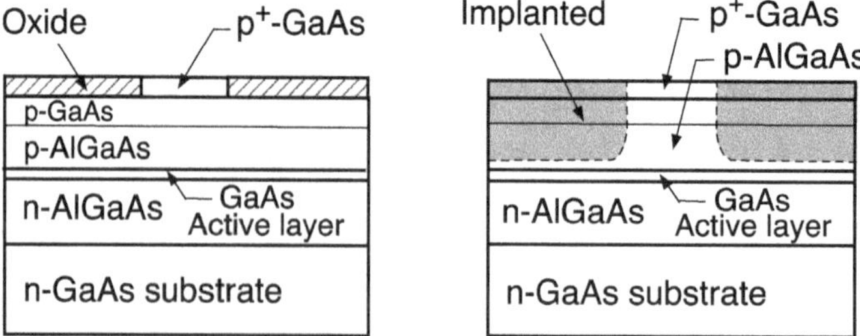

Fig. 10.16 Schematic cross-sectional view of different types of gain-guided laser structures: (left) oxide-stripe, and (right) proton- or deuteron-stripe

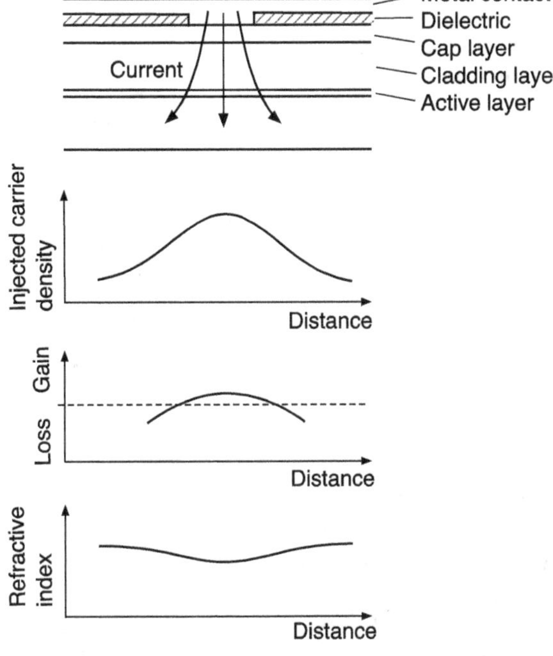

Fig. 10.17 Lateral current distribution, carrier distribution, gain, and refractive index in the active layer of a gain-guided stripe laser

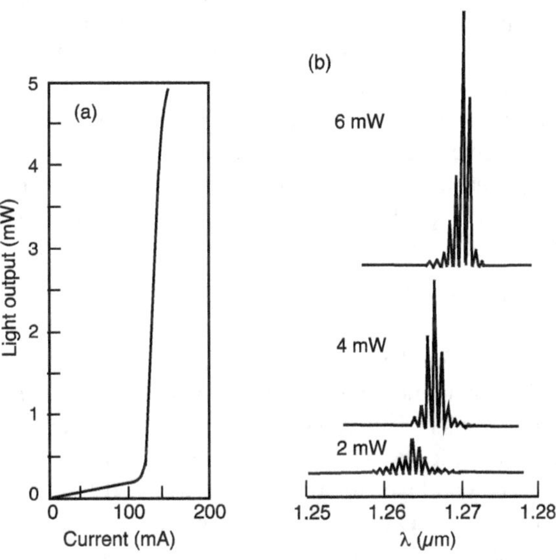

Fig. 10.18 **a** Measured cw *L–I* curve and **b** emission spectra at several power levels of a 1.3 μm deuteron-stripe gain-guided InGaAsP laser. Reprinted with permission from [4], copyright IEEE

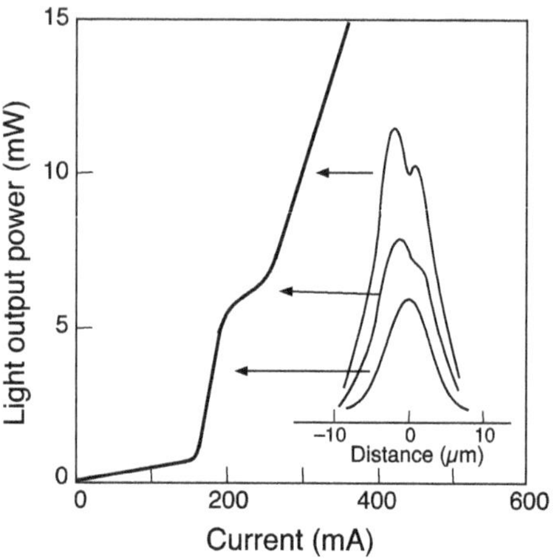

Fig. 10.19 'Kink' developed in the *L–I* curve in a gain-guided laser and the corresponding near-field patterns on the facet

At relatively high light output power, the lateral transverse mode becomes unstable due to the lateral index anti-guiding (Fig. 10.17). The effective index depression caused by carriers at the waveguide center is ~5×10^{-3}. Under high injection conditions, the carrier density of the center part of the light-emitting region decreases because of the high rate of stimulated emission occurring there. This carrier density reduction leads to a small increase in refractive index, and the emitted light is crowded within the center part of the waveguide. This confinement becomes quite unstable if there is any asymmetry in the refractive index distribution. The light-emitting area easily shifts toward the high-refractive index (low carrier density) region. In this situation, light output barely increases with current injection because the light confinement region does not geometrically correspond to the high-gain region. These phenomena result in the 'kink' or nonlinearity in the current–light output characteristics (Fig. 10.19). This nonlinearity can also be associated with a transition to higher-order modes, or a transition from TE to TM mode.

(b) **Index-guided lasers**

To overcome the index anti-guiding problems caused by the carrier-induced index reduction in gain-guided lasers, an effective index step of ~10^{-2} to ~0.2 along the junction plane is introduced to control the lateral mode in index-guided lasers. The added index step is achieved by introducing a lateral thickness non-uniformity in at least one layer for weakly index-guided lasers ($\Delta\bar{n} \cong 10^{-2}$). One example is the ridge-waveguide laser shown in Fig. 10.20. A narrow ridge ($\leq$5 μm) is etched through the upper cladding layer, and its edges are passivated with dielectric to restrict the injection current flow. When lasing, a fraction of the mode overlaps with the dielectric, which has a considerably lower refractive index than the cladding layer ($\bar{n} \sim 3.5$). This overlap introduces an effective lateral index step for lateral confinement. However, the magnitude of the lateral index step is affected by the thickness of the waveguide layer. If the waveguide layer is very thick, the evanescent wave in

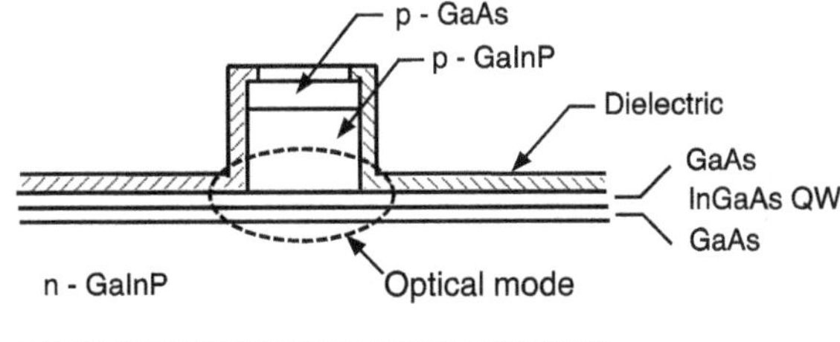

Fig. 10.20 Schematic cross section of a weakly index-guided ridge-waveguide laser and its optical mode

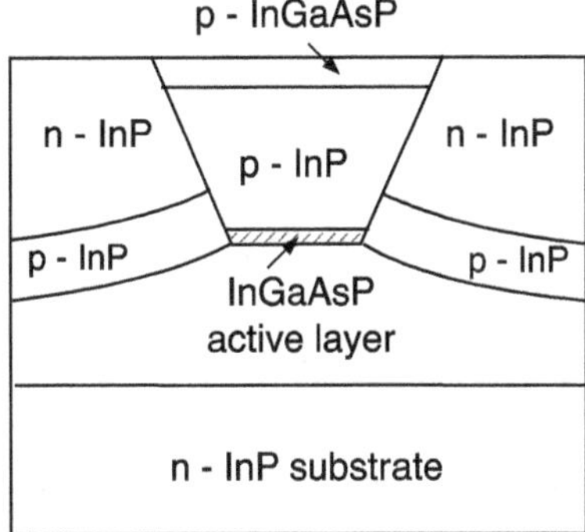

Fig. 10.21 Schematic cross section of a strongly index-guided buried-heterostructure laser

the dielectric becomes too weak to provide significant lateral mode confinement and the device may become very similar to a gain-guided laser.

In strongly index-guided lasers ($\Delta\bar{n} \cong 0.2$), the active region is buried in higher bandgap layers vertically and laterally formed by first and second epitaxial growths, respectively, to control the optical modes. Thus, this type of laser is called a buried-heterostructure (BH) laser (Fig. 10.21). Since the lateral index step along the junction plane in these laser structures is about two orders of magnitude larger than the carrier-induced effects, the lasing characteristics of BH lasers are primarily determined by the rectangular waveguide that confines the mode inside the buried active region. The current flow in a BH laser is limited to the active layer by current blocking layers deposited during the second epitaxy. The *p-n* junction in the current blocking layer is reverse-biased when a forward bias voltage is applied to the laser diode, thus confining the current flow within the active region. A highly resistive single layer such as Fe-doped InP has also been used as a current blocking layer instead of the reverse-biased *p-n* junction. As a result of the tight current confinement, the threshold current can be reduced. When compared to gain-guided lasers, BH lasers show some excellent characteristics. In addition to a stable transverse mode, BH lasers have a low threshold current, high external efficiency, and the capability of high output power operation. Figure 10.22 shows the *L–I* characteristics and lasing spectra of a typical 1.3 μm BH laser. It is clear that multiple longitudinal modes exist even at high injection current levels.

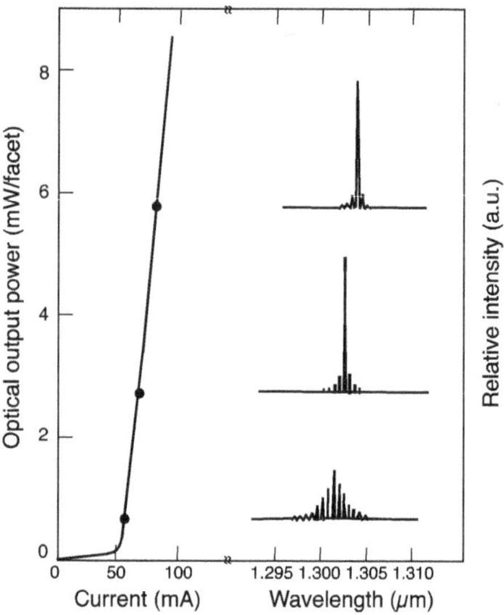

Fig. 10.22 Optical power spectra at three different power output levels as indicated in the L–I curve of a 1.3 μm InGaAsP buried-heterostructure laser. Reprinted with permission from [5], copyright IEEE

10.2.2 Distributed-Feedback Laser

All the laser structures described above use cavity facets that are formed by cleaving to obtain the feedback necessary for lasing. In these Fabry–Perot (FP)-type semiconductor lasers, the magnitude of facet reflections is the same for all longitudinal modes. The only longitudinal-mode discrimination in such a laser is provided by the gain spectrum. Since the gain spectrum is usually much wider than the longitudinal mode spacing, the resulting mode discrimination is poor. Therefore, FP-type semiconductor lasers do not emit light in a single longitudinal mode.

One way of improving the mode selectivity is to make the feedback frequency-dependent so that the cavity loss $(\alpha_i + \alpha_m)$ is different for different longitudinal modes. This can be achieved by incorporating a grating, i.e., a periodic variation of the refractive index, within the optical waveguide, which is generally produced by corrugating the interface between two dielectric layers (Fig. 10.23a). The feedback necessary for the lasing action in such a laser structure is not localized at the cavity facets but is distributed throughout the cavity length. Therefore, lasers that utilize this type of grating structure are called *distributed-feedback* (DFB) *lasers*. A variation of the DFB laser, where the grating is etched near the cavity ends and distributed feedback does not take place in the central active region, as shown in Fig. 10.23b, is called the *distributed Bragg reflector* (DBR) *laser*. The unpumped corrugated end regions act as effective mirrors whose reflectivity is of DFB origin and is therefore wavelength-dependent.

Assume the refractive index of the grating in a DFB laser has a sinusoidal variation of

$$\bar{n}(z) = \bar{n} + \bar{n}_a \cos(2\pi z/\Lambda) \tag{10.41}$$

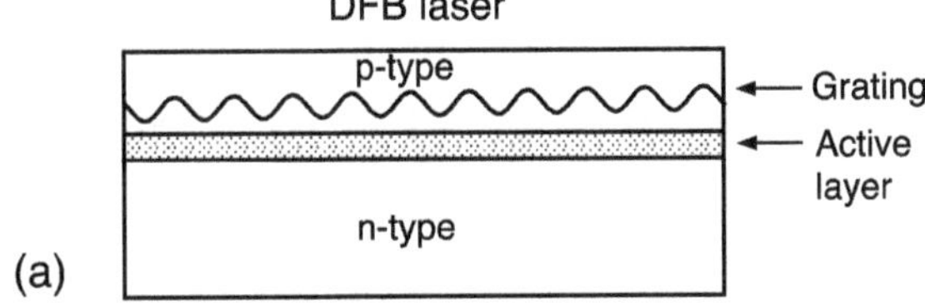

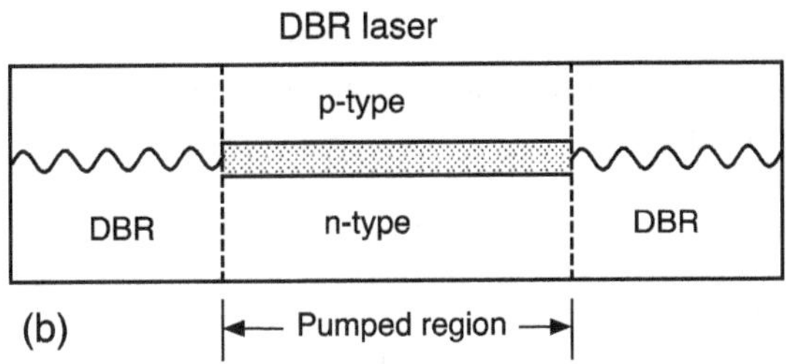

Fig. 10.23 Schematics of distributed-feedback (DFB) and distributed Bragg reflector (DBR) semiconductor lasers. Shaded area shows the active region of the device

where Λ is the period of the spatial variation of the refractive index and $\bar{n}_a$ is the amplitude of variation. It is close to the shape of gratings fabricated in real DFB lasers. This periodic perturbation of $\bar{n}$ can generate an infinite set of diffraction orders. However, only two counter-running phase-synchronizing waves near the Bragg wavelength can have significant amplitude. The Bragg condition that determines the Bragg wavelength is

$$\lambda_b = 2\bar{n}\Lambda/l \quad \text{for } l = 1, 2, 3, \ldots \tag{10.42}$$

where λ_b is the wavelength in free space that satisfies the Bragg condition and l is the grating order. Here, the grating pitches corresponding to $l = 1, 2,$ and 3 are called the first-order, second-order, and third-order gratings, respectively. The reflected wave intensity due to scattering with the periodic gating is determined by the grating height through the magnitude of refractive index change ($\bar{n}_a$) and by the distance between the grating and the active layer. Because of the presence of gain, these waves grow as they travel along the dielectric layer and transfer energy between each other due to Bragg scattering. The magnitude of the optical feedback depends on the structure length L and coupling constant $\kappa_c(= \pi\bar{n}_a/\lambda_0)$, which is determined by the grating structure. Without going into detail, the total electric field within the periodic (sinusoidal) dielectric layer can be expressed as the sum of the two counter-running waves:

$$E_x(z) = R(z)\exp(-i\beta_b z) + S(z)\exp(+i\beta_b z) \tag{10.43}$$

where $R(z)$ and $S(z)$ are complex amplitudes of the two counter-running waves and $\beta_b = 4\pi\bar{n}/\lambda_b$. The solutions of the coupled-wave equations determine a set of discrete eigenvalues γ_c shown as

$$\kappa_c = \pm i\gamma_c / \sinh(\gamma_c L) \tag{10.44}$$

For each value of γ_c there is a corresponding threshold gain and resonant wavelength. The resonance wavelength of the DFB laser is given as

$$\lambda_0 = \lambda_b \pm \left[\left(q + \frac{1}{2} \right) \frac{\lambda_b^2}{2\bar{n}L} \right] \tag{10.45}$$

with $q = 0, \pm 1, \pm 2, \ldots$ This result shows that the resonances are spaced $\lambda_b^2/2\bar{n}L$ apart and λ_0 never equals λ_b for any integer value of q. Once λ_0 is determined, the threshold gain $g_F = -\alpha_F$ can be obtained for each q from the following equation:

$$(\pi\bar{n}_a/\lambda_0)^2 \exp(-2\alpha_F L) \approx 4\alpha_F^2 \tag{10.46}$$

The emission spectrum is symmetric with respect to the Bragg wavelength so that two optical modes occur for the same gain coefficient. Also, the threshold increases with deviation from λ_b and therefore provides spectral selectivity. However, the grating pitch Λ and lasing wavelength λ_0 cannot be set randomly. The wavelength λ_b in (10.42) must be adjusted to the neighborhood of the peak material gain by controlling the grating pitch or composition of the active layer. The DFB and FP modes (gain profile) are displayed in the lasing spectra as shown in Fig. 10.24. The DFB mode and the gain profile differ because they are determined by grating pitch and bandgap energy, respectively. The temperature coefficient of the peak gain shift is about 0.25 nm/°C for 0.85 μm AlGaAs lasers and about 0.4 to 0.6 nm/°C for InGaAsP/InP lasers and is determined by the temperature dependence of the bandgap

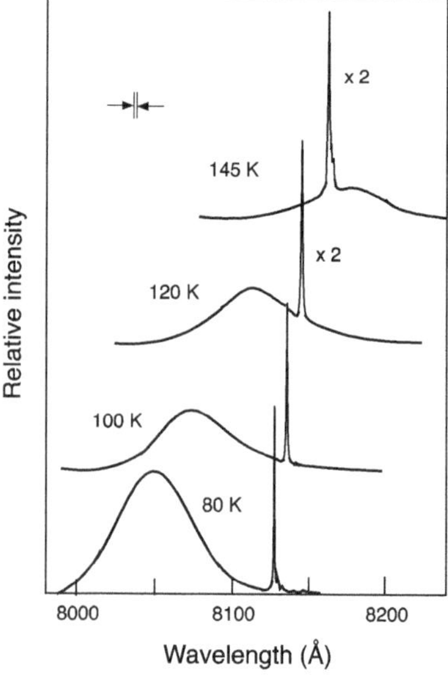

Fig. 10.24 Lasing and peak gain (FP mode) spectra of an AlGaAs/GaAs DFB laser at different temperatures. Reprinted with permission from [6], copyright IEEE

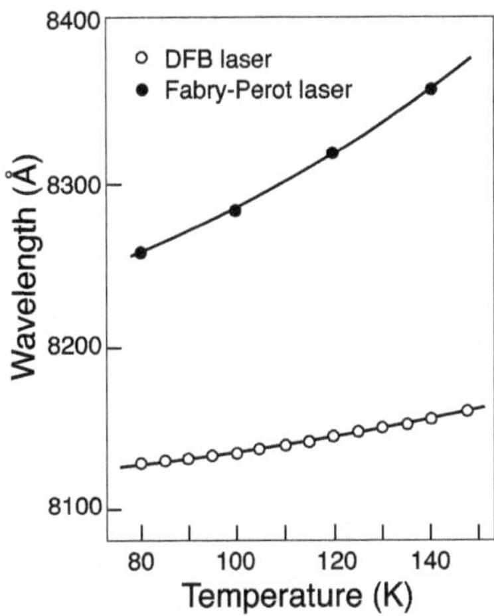

Fig. 10.25 Temperature dependence of the emission wavelength of an AlGaAs DFB laser and a FP laser. Reprinted with permission from [6], copyright IEEE

energy. In contrast, the temperature coefficient of the DFB mode is determined by the refractive index variation with temperature and is ~0.1 nm/°C. The wavelength change of a DFB laser as a function of ambient temperature is compared with that of a FP laser in Fig. 10.25.

As discussed above, the degeneracy in gain of the modes placed symmetrically around λ_b makes this idealized DFB laser unsuitable as a single-mode laser. One way to resolve the threshold gain degeneracy in DFB lasers is to use a Bragg grating with a quarter-wave shift ($\Lambda/2$) near the cavity center (Fig. 10.26). The phase of the grating is shifted by $\pi/2$ at the position of the shifter, which causes the phases of two waves propagating in the same direction in the cavity to coincide. Thus, the threshold gain can be further decreased and single-mode operation is established.

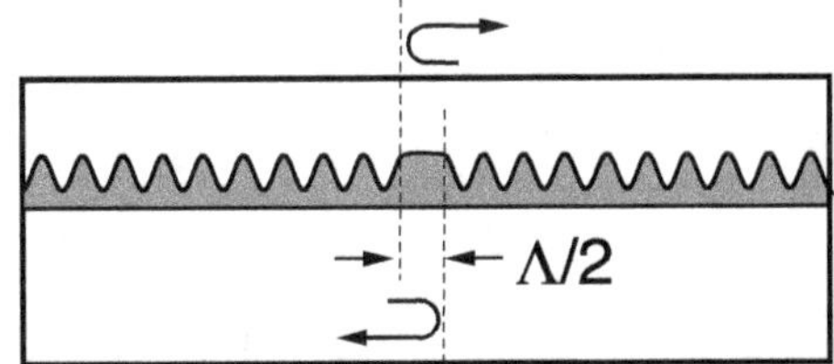

Fig. 10.26 Schematic of DFB laser diode with $\lambda/4$ phase shifting region between the grating portions. The added $\Lambda/2$ spacing corresponds to a $\lambda/4$ shift between two sections of grating

10.3 Laser with Quantum-Confined Active Region

10.3.1 Quantum-Well Laser

The history of semiconductor laser development is the history of the campaign to lower the threshold current, as is graphically illustrated in Fig. 10.27, using GaAs-based lasers as an example. The most significant changes in this endeavor did not take place until the concept of DH lasers had been introduced. The DH structure establishes an efficient way of collecting both electrons and holes as well as confining light in the same region for low threshold lasing. As the active layer is gradually thinned beyond the optimal active layer thickness d_0 in a DH laser, the optical confinement decreases and the threshold current increases according to (10.24). If the thickness of the active layer is further reduced to a few tens of nanometers, quantum effects arise. However, high-quality DH with ultrathin layers could not be attained until new epitaxy methods of MBE and MOCVD were developed in the mid-1970s for the growth of heterostructures. Current injection lasing in quantum wells with parameters to match those of standard DH lasers was first accomplished in 1978 by a group led by Holonyak at the University of Illinois in collaboration with Dupuis and Dapkus at Rockwell. The term *quantum-well* (QW) *laser* was coined in this work. Depending on whether they have a single active layer or multiple active layers, these devices are called *single quantum-well* (SQW) or *multiple quantum-well* (MQW) lasers, respectively.

Because of the thin QW active layer thickness, the optical confinement factor is very small in a SQW laser. This leads to a very large threshold current density. The remedy, as shown in Fig. 10.28, is to increase the confinement factor by adding extra wave-guiding layers, with higher refractive index, between QWs and cladding layers.

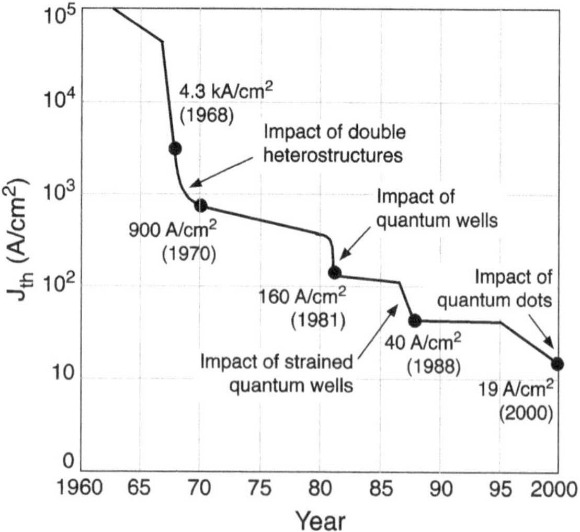

Fig. 10.27 Evolution of the threshold current density of semiconductor lasers. Reprinted with permission from [7], copyright IEEE

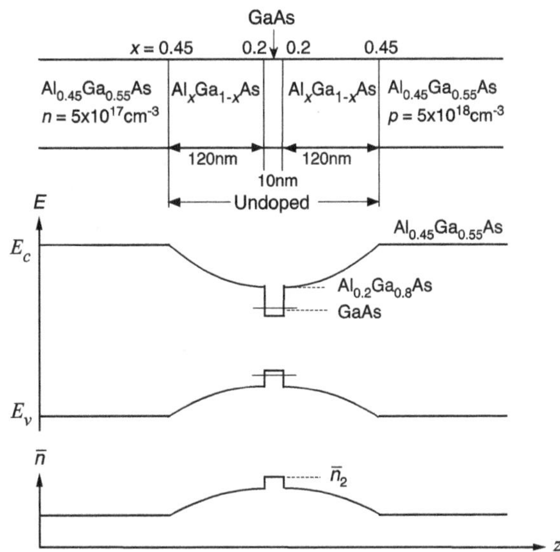

Fig. 10.28 AlGaAs graded-index separate-confinement heterostructure (GRIN-SCH) with GaAs quantum well and undoped $Al_xGa_{1-x}As$ wave-guiding layer, in which the Al content increases continuously from $x = 0.2$ to 0.45 with increasing distance from the quantum well

In this separate confinement heterostructure (SCH) design, the optical confinement is provided by the cladding layers while carriers in the QW are confined by wave-guiding layers. Furthermore, the refractive index of the wave-guiding layer can vary in linear or parabolic fashion, increasing from the cladding layer toward the QW, in the graded-index (GRIN) SCH structure. Injected electrons and holes would be more easily captured in the active QW material through a 'funnel' effect due to the funnel-shaped band-edge profiles around the QW.

The device characteristics of the QW laser can be understood on the basis of FP lasers with some modifications. In a QW, the electron and hole levels have quantum energies of confinement, so that the allowed interband transitions are not tied to the bandgap of the active layer material but can be tuned over a modest range by varying the thickness of the wells. The emission wavelength is always shorter than the bandgap energy of the QW material. The optical transitions have to obey the quantum mechanical selection rule; i.e., allowed transitions are between states in conduction and valence bands having the same quantum number. Because of the small thickness associated with the QW, there is no need to carefully match lattice parameters between the QW and surrounding materials. In fact, as long as the strained QW thickness is less than the critical thickness, the lattice-mismatch induced strain could enhance the device performance through energy band modification. This also allows a radical diversification of materials available for heterostructure components.

In a QW structure, the density of states (DOS) becomes two-dimensional (2D) as the injected electrons and holes are quantized only in the direction perpendicular to the active layer, as shown in Fig. 10.29. The step-like DOS introduces some excellent characteristics to lasers because the energy distribution of the injected carrier is narrower than that of a bulk DOS. In the bulk material (3D) shown in Fig. 10.30, the gain peak moves away from the bandgap as the density of injected carriers increases,

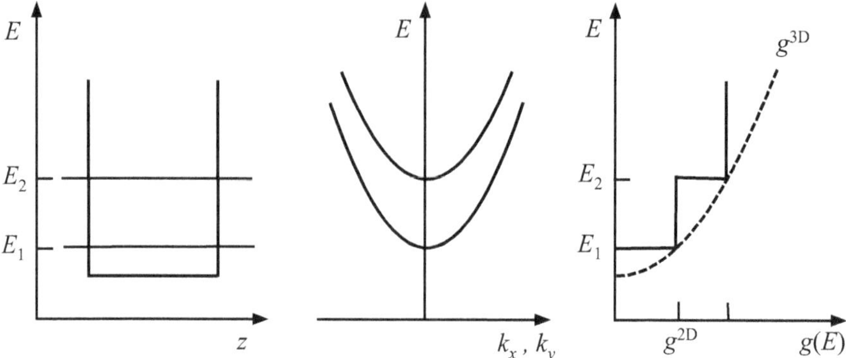

Fig. 10.29 Illustration of confined two states in a QW structure (left), the in-plane $E(k)$ dispersions (middle), and the resulting density of states (right)

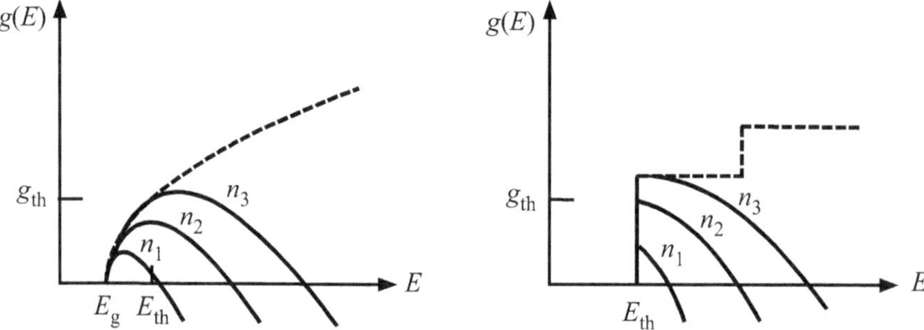

Fig. 10.30 Illustration of gain formation in bulk (left) and QW (right) structures for three different carrier densities, n_i. The dashed line is the peak gain profile for each structure

and all the carriers at lower energies are ineffective at contributing to the gain. The gain peak in the 2D structure stays at the effective bandgap, and the differential gain above threshold is much higher for carriers in the QW (2D) than in the bulk (3D). Furthermore, in the QW laser there are ~10^{12} cm^{-2} DOS within a temperature range of kT while there are ~10^{13} cm^{-2} DOS in a 1000 nm DH laser. Thus, the threshold is achieved sooner for the QW laser with fewer states to be inverted. Furthermore, the transparency condition is reached for lower injection currents as the QW width decreases, until very thin wells are reached. By then the quasi-Fermi level energies are so high in the QWs that some carriers are confined not in the QW, but in the wave-guiding layers. The threshold current density rises rapidly in very thin QW laser structures. It is clear that the use of the QW active layer greatly reduces the threshold current density below that of DH lasers, as shown in Fig. 10.27. In addition, other characteristics, such as spectral linewidth and high-speed modulation, are also improved by combination with the DFB structure.

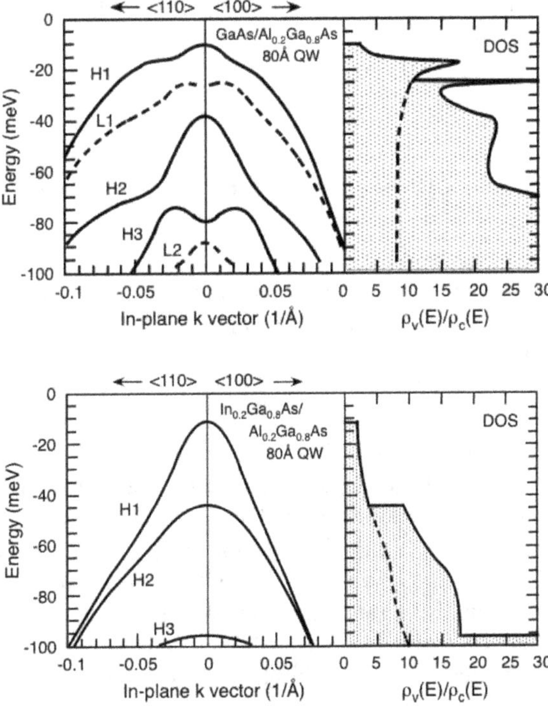

Fig. 10.31 Valence subband structure and DOS of an unstrained 80 Å GaAs/$Al_{0.2}Ga_{0.8}As$ QW structure (top), and a compressively strained 80 Å $In_{0.2}Ga_{0.8}As/Al_{0.2}Ga_{0.8}As$ QW structure (bottom). Reprinted with permission from [8], copyright AIP Publishing

In Fig. 10.27, the other steep drop in threshold current density comes from the incorporation of strain into the QW. As described in Chap. 6, the valence band of semiconductors can be modified using biaxial strain encountered in highly strained layers. The degenerated heavy- and light-hole bands are decoupled under the biaxial strain. In particular, under biaxial compression, the light-hole QW deepens and the highest confined state has light transverse hole character with a low DOS, as depicted in Fig. 10.31. Compared in this figure are the topmost valence bands and the corresponding DOS function of an unstrained GaAs/$Al_{0.2}Ga_{0.8}As$ QW and a compressively strained InGaAs/$Al_{0.2}Ga_{0.8}As$ QW. In the unstrained QW structure, due to the 1D quantum confinement, heavy-hole and light-hole bands are decoupled near the zone center. However, the valence band DOS is still dominated by the heavily hole band near the top of the valence band. In the compressively strained structure, the topmost in-plane valence band becomes light-hole-like and the DOS of hole states adjacent to the valence band edge is now narrowed and comparable to that of electrons near the conduction band edge. In unstrained lasers, the shift of quasi-Fermi level by carrier injection is much greater for electrons than holes due to the much higher valence band DOS. In strained QW lasers, the much reduced valence band DOS (lighter hole mass) decreases the transparency carrier density and threshold current density. In addition to device characteristic improvement, lasing is possible at a new wavelength that cannot be achieved in the lattice-matched QW structure.

With its superior characteristics, the strained QW structure has been incorporated into all types of semiconductor lasers and widely used in numerous photonic products including DVD players, laser printers, and optical-fiber communication systems.

10.3.2 Type-II Quantum-Well W-Type Lasers

Antimony-based semiconductor lasers emitting in the wavelength of mid-wave infrared (MWIR: 3–5 μm) are important for biomedical, environmental, and chemical sensing applications. Among all III–V compounds, InAsSb has the narrowest bandgap and is attractive for the active material of MWIR emitters. Strained InAsSb QW lasers with type-I heterostructure have been demonstrated with excellent performance but only at temperatures below ~200 K. Further, the performance of these QW lasers does not exceed that of their double-heterostructure counterparts because of the large leakage current due to inadequate carrier confinements by small conduction/valence band offsets and the increasing predominance of Auger recombination at high temperatures. Because the energy gap E_g and the split-off gap Δ_{so} of InAs-rich alloys are nearly equal, the CHHS Auger process dominates the non-radiative recombination through conduction-to-heavy-hole (CH) recombination coupled with a heavy-to-split-off-hole (HS) transition.

To circumvent these problems, new interband QW lasers with type-II band alignment have been developed. This new laser design has a W-type active region, named after the shape of the conduction band profile of the QW active region. Figure 10.32 shows the energy band profiles in the active region of a typical W-type laser which is lattice-matched to the InP substrate. The active region is made of multiple QW periods separated by barrier layers (e.g., InAlGaAs) and each period including a thin 'hole' QW (e.g., GaAsSb) sandwiched between two 'electron' QWs (e.g., InGaAs). Note that the type-II band alignment at each 'electron'-QW/'hole'-QW heterojunction has a broken-gap structure such that electrons and holes are confined in different

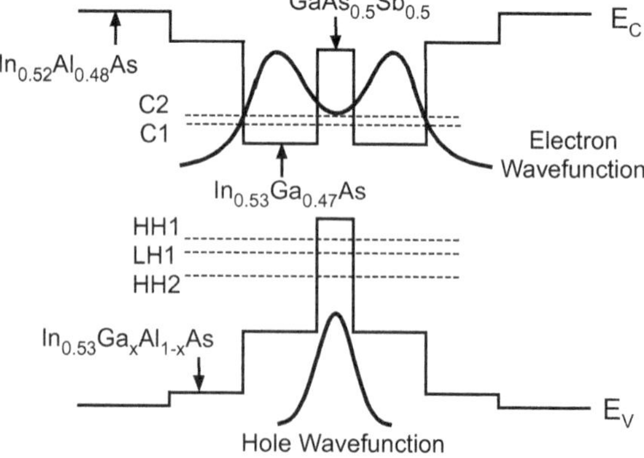

Fig. 10.32 Band alignment for a quantum-well laser structure lattice-matched to InP substrate with a W-type active region along with the ground-state electron and hole wavefunctions

but neighboring QWs. The symmetric (C1) and antisymmetric (C2) electron states are confined in the conduction band of two coupled InGaAs electron QWs, and heavy hole (HH) and light-hole (LH) subbands are formed in the valence band of GaAsSb hole QW. Although the wavefunctions of electrons and holes are centered in different QW materials, the symmetric wavefunction of two coupled electron QWs is sufficiently overlapped with the center hole wavefunction to yield high interband optical transitions. The fundamental C1-HH1 optical transition has an effective bandgap energy in the MWIR range, which is not limited by the bandgap energy of the constituent materials. Thus, the type-II band alignment effectively removes CHHS Auger processes by eliminating the resonance between E_g and Δ_{so}, even though it is presented in bulk InAs-rich alloys. At the same time, it also introduces large conduction and valence band offsets for enhanced electrical carrier confinements.

This W-type active region design can be incorporated in either diode lasers or cascade lasers for MWIR applications. In the cascade laser design, multiple stages of the type-II W-type QW active regions are connected in series in a staircase design such that the injected carriers are recycled for enhanced photon generation. The concept of cascade lasers will be discussed in more detail in Sect. 10.6. In type-II W-type diode lasers operating in the 3–4.5 μm spectral range, the derived room-temperature Auger coefficients are an order of magnitude lower than those of type-I lasers with a value less than mid-10^{-27} cm^6/s. On the other hand, the Auger coefficient of the cascade laser maintains a nearly constant value of ~5×10^{-28} cm^6/s in the 3–4.2 μm window. With reduced Auger non-radiative decay rates, better carrier confinements, and enhanced quantum efficiency (for the cascade laser), room-temperature pulse and cw operation of W-type diode lasers and cascade lasers have been demonstrated, respectively.

10.3.3 Quantum Dot and Quantum Wire Lasers

Multiple-dimension quantum confinement in the growth (junction) plane modifies the electronic band structure and DOS, as shown in Fig. 3.7 and spatially localizes the electrons and holes, resulting in improved laser characteristics. Even in Dingle and Henry's original patent of the QW laser in 1976, it was noted that additional modification of the DOS conducive to improved laser performance would occur with quantum wire (QWR) confinement [9]. At absolute zero temperature, the DOS in a QWR system is inversely proportional to $\sqrt{E - E_{mn}}$, where E_{mn} is a QWR energy eigenvalue. Thus to a first-order approximation, the DOS approaches infinity at each interband transition energy. In 1982, Arakawa and Sakaki further proposed that the performance of semiconductor lasers could be improved by reducing the dimensionality of the active regions of these devices from QW to quantum dots (QDs) [10]. Because of the 3D confinement of charge carriers in QDs, the conduction and valence bands of an ideal semiconductor QD nanostructure are split into completely discrete energy levels, similar to a single atom. The discrete energy levels in the QD nanostructure lead to a delta function-like DOS, and thus to a much narrower transition linewidth compared with bulk or QW semiconductors. The atom-like DOS in quantum dots should drastically improve the performance of optical devices, especially semiconductor lasers.

(a) Quantum dot formation

To date, the methods employed in QD fabrication experiments have included the post-growth lateral patterning of QW structures and regrowth on profiled substrates. Methods such as reactive ion etching have been employed in the above-mentioned techniques to etch mesas in QW heterostructures. Whether or not this process has been followed by epitaxial regrowth, it has inevitably been plagued by the pitfalls of non-radiative recombination via surface and interface states, and the volume-normalized photoluminescence (PL) intensities have invariably been lower than those obtained from the as-grown QW structures. These processes have yet to produce either useful devices or reliable fabrication processes. Further, these fabricated QD structures generally exhibit geometries far from ideal, such as large size, low density, and size non-uniformity. The non-ideal geometries of QD structures fabricated by these methods destroy the singularities in the DOS and eliminate their advantages completely.

A more promising way to fabricate QD nanostructures is through a strain-induced self-assembling technique. This process exploits the 3D island growth of highly lattice-mismatched heterostructures. During the growth of lattice-matched heterostructures, the growth mode is determined solely by the surface and the interface energies between the substrate and the epilayer. Figure 10.33 illustrates different epitaxial growth modes. If the sum of the epilayer surface energy γ_2 and the interface energy γ_{12} is smaller than the substrate surface energy γ_1, i.e., if the epitaxial material wets the substrate, the epitaxial growth occurs in a layer-by-layer mode resulting in a smooth epitaxial film. This is the Frank–van der Merve growth mode, as illustrated in Fig. 10.33a. On the other hand, if the sum of the surface energy of epitaxial layer γ_2 and interface energy γ_{12} is larger than the substrate surface energy γ_1, the epitaxial growth is then governed by Volmer–Weber growth mode, where 3D islands are formed on a bare substrate (Fig. 10.33b). However, in lattice-mismatched heterostructures, elastic strain relaxation becomes the major driving force for self-organized growth of dislocation-free high-density coherent islands on the substrate, accompanied by a thin wetting layer. The growth may initially proceed in a layer-by-layer mode until the critical thickness is reached. For InAs on GaAs, this transition occurs after about 1.8 monolayers (MLs) of InAs have been deposited, while for InAs

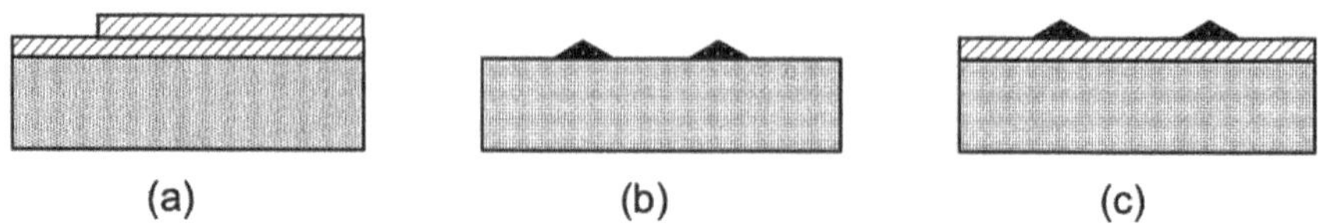

Fig. 10.33 Schematics of different growth modes: **a** Frank–van der Merve, **b** Volmer–Weber, and **c** Stranski–Krastanov growth modes

on InP, the critical thickness is about 3.2 MLs. Beyond that point, the accumulated elastic energy caused by the strain due to lattice-mismatch tends to be reduced via the formation of individual islands. In these islands, the elastic strains relax and the elastic energy decreases. This results in a Stranski–Krastanov (SK) growth mode, as illustrated in Fig. 10.33c.

The most widely studied III–V QD system via SK growth mode is (In,Ga)As on GaAs substrates grown either by MBE or MOCVD. Typical InAs self-assembled islands have a pyramid shape with a base width of about 20 nm and a height of a few nm. Thus the shape of these QDs is a flat pyramid, and their luminescence peaks depend mainly on the variation of the dot height. The lateral size of these islands is comparable to the exciton Bohr radius of the InAs-GaAs system and small enough to exhibit the 3D quantum confinement effect. Thus these self-assembled dots are suitable for QD optoelectronic device applications.

(b) **Self-assembled QD lasers**

In ideal QD lasers, most of the expected superior performance originates from their atom-like DOS, which is expressed by the delta function. Figure 10.34 displays theoretical room-temperature gain spectra for QD, QWR, QW, and bulk structures at a constant carrier density of 3×10^{18} cm^{-3} and an intraband relaxation time of $\tau_{in} = 0.1$ ps. The linewidth and peak intensity of these gain spectra are narrowed down and enhanced, respectively, with the increase of quantum confinement from QW to QWR and QD. The narrow linewidth of the gain spectrum is limited only by the homogeneous broadening caused by intraband relaxation in an ideal QD structure. Due to the large gain peak and narrow linewidth, a significant improvement of the differential gain, dg/dN, is expected for the QD laser. The large optical gain and differential gain result in a reduction of the threshold current density and improved

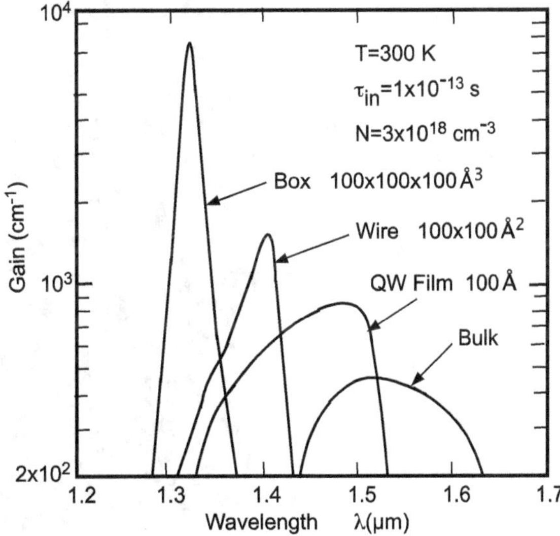

Fig. 10.34 Calculated gain spectra for $Ga_{0.47}In_{0.53}As$/InP (100 Å)^3 cubic quantum box, (100 Å)^2 quantum wire, 100 Å thick quantum well, and bulk crystal at $T = 300$ K. Reprint with permission from [11], copyright IEEE

differential quantum efficiency. The large differential gain also results in improved dynamic characteristics of QD lasers. Since the relaxation oscillation frequency of an ideal QD laser is proportional to the square root of the differential gain, a large differential gain enhances the modulation frequency of the laser. The other parameter important for high-speed modulation is the linewidth enhancement factor, which is inversely proportional to the differential gain. A large differential gain results in a small linewidth enhancement factor near zero, leading to a narrow spectral linewidth without spreading under high-frequency operation. Another important feature of ideal QD lasers is that, with infinite barriers for carrier confinement, the threshold current becomes independent of temperature, $T_0 = \infty$.

In practice, the edge-emitting QD laser shares the same double-heterostructure design as the QW laser. The active layer of a QD laser consists of one or multiple 2D arrays of QDs surrounded by a pair of larger but finite bandgap energy barriers to confine injected carriers. The active region is sandwiched by upper and lower cladding layers with larger bandgap energy and smaller refractive index for effective optical confinement of photons. In order to enhance the modal gain of the QD laser, high QD density with uniform dot size is necessary. Typically, the self-assembled QDs via SK growth mode can reach a dot density around 10^{11} cm^{-2} and show high luminescence efficiency. However, the self-assembled QD sizes are non-uniform with a typical size variation of more than ~10% (e.g., Fig. 10.35). A single-layer InAs-GaAs QDs sample shows a typical room-temperature PL emission peak at around 0.95–1.2 μm and a full-width at half-maximum (FWHM) of ~90 meV. The broad FWHM is caused by the fluctuation of the quantized energy among islands with non-uniform dot heights. When using these QDs as the active layer of the laser, the large inhomogeneous broadening of the spectrum indicates that only a fraction

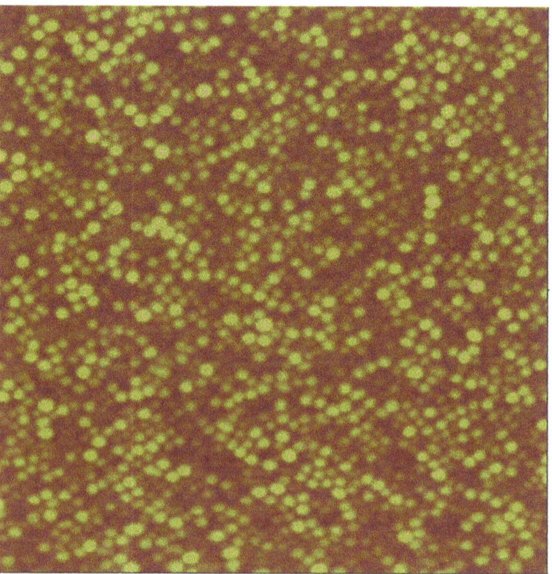

Fig. 10.35 Surface AFM (atomic force microscopy) image of the InAs QDs grown on InAlGaAs/InP deposited with 5.5 MLs of InAs by MBE. The imaged area is 2 μm × 2 μm and the dot density is ~5×10^{10} cm^{-2}

of QDs are participating in the laser action. The differential gain of the QD laser is reduced considerably, leading to a high threshold current density. To reduce the threshold current density of the QD laser, it is necessary to use long cavity structures and multiple stacked QD layers to increase the number of QDs in the active region.

In a multiple stacked QD system, the structural and optical properties of the InAs QDs are determined by the thickness of the intermediate GaAs barrier layers. The schematics of different stacked SK QD structures are shown in Fig. 10.36. If the GaAs barrier layer (thickness $t \geq 20$ nm) is much thicker than the strain field induced by the QDs, the average dot size and density of QDs in each layer are independent of each other. As the barrier layer thickness is reduced close to the average QD height (h), the QD-induced strain field in the lower layer provides the driving force for aligning self-assembled dots in the upper layers perpendicularly. The strain field also causes the upper layer islands to expand slightly as the number of stacked layers increases. Each closely stacked QD layer has a wetting layer indicating it is grown under the SK growth mode. Since the barrier layer between dot layers is thin, the electron wave functions of neighboring dots are overlapped such that QDs are coupled electrically through tunneling in the vertical (growth) direction. Therefore the vertically aligned dots act as a single quantum dot where the effective QD height is controlled by the total stacked dot layers used. Due to its large effective height, the emission wavelength becomes less sensitive to dot size fluctuations than that in a single QD layer. Thus a narrower emission spectrum FWHM of 25 meV is observed from the closely stacked QD structure. However, the accumulated strain in the structure could generate defects and degrade the luminescence intensity. To further improve the optical property of the closely stacked QD structure, the thickness of barrier layers is further reduced to a few (2–3) monolayers ($t \leq h$) and the QD material supply during the growth is progressively reduced. Under optimal growth conditions, the dots in the stacked structure are in contact with each other in the growth direction and form a columnar shape as a whole. In the columnar-shaped QD structure, both high luminescence intensity and narrow spectrum FWHM of $\leq$40 meV are obtained.

Through improved device technology and growth techniques, in 1993 realistic self-assembled QD lasers demonstrated certain predicted advantages of ideal QD lasers. However, the early models for ideal QD lasers were based on assumptions of lattice-matched heterostructures with infinite barrier height around QDs. In contrast, the realistic self-assembled QD laser structures demonstrated are based on strained heterostructures with finite energy barriers around QDs. In spite of their less ideal

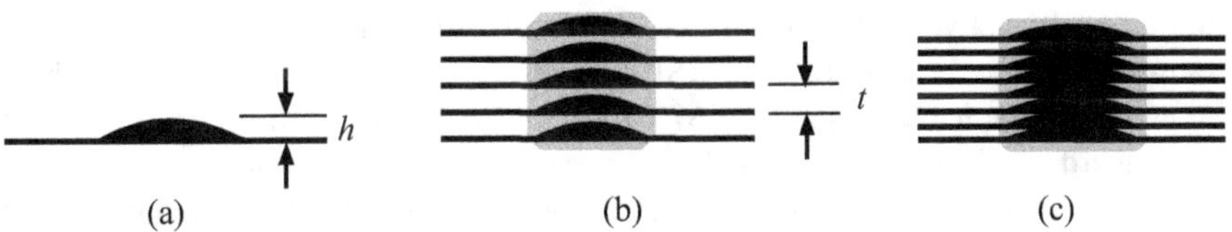

Fig. 10.36 Schematics of stacked multiple InAs QD layer structures of **a** isolated QD layers with GaAs barrier layer $t \geq 20$ nm, **b** closely stacked QD layers with a QD height $h \leq t < 20$ nm, and **c** columnar-shaped QD layers with $t \leq h$

structures, low threshold current density ($\leq$20 A/cm^2), large T_0 ($\geq$400 K) at temperatures below ~ 200 K, high internal quantum efficiency ($\geq$98%), and small linewidth enhancement factor (<1) have been achieved in In(Ga)As/GaAs self-assembled QD lasers. Nevertheless, it is necessary to improve the in-plane QD size homogeneity through innovative growth methods for further advancement of QD laser technology.

Furthermore, an individual dot in a QD laser can act as the recombination center for efficient capture and 3D confinement of injected carriers. This feature is useful in reducing the non-radiative recombination at defects or dislocations of III–V lasers grown on silicon substrates. Due to the high dislocation density generated in III–V on Si heterostructures, QW lasers grown on Si substrates usually show high threshold current and poor reliability. Using QDs as efficient recombination centers, the detrimental effect of dislocations is greatly diluted in QD lasers such that efficient and reliable laser operation can be achieved. Compared with QW lasers, the InAs/GaAs QD lasers grown on Si have shown high-temperature, high-power, and low threshold-current operation with over 2700 h of cw operation [12]. This is the longest lifetime of any GaAs-based laser grown on silicon.

(c) **Quantum wire (QWR) lasers**

The creation of high-density semiconductor QWR heterostructures requires the simultaneous variation of alloy composition in two orthogonal directions. The composition is readily controlled along the [100] (growth) axis using standard epitaxial crystal growth techniques, though controlling the alloy profile in an orthogonal direction ([110] or [$\bar{1}$10]) during growth is not straightforward. QWRs have been fabricated by the post-growth patterning of QW structures or regrowth on profiled substrates using reactive ion etching and ion milling techniques. However, these processes have yet to produce useful devices due to surface damage caused by ion bombardment. An improved method was to grow AlGaAs/GaAs heterostructures on patterned [001]-oriented GaAs substrates in which V-shaped grooves have been etched using wet chemicals. Because the epilayer growth rate depends on the crystallographic orientation, this scheme results in crescent-shaped GaAs QWR regions embedded in AlGaAs. These QWRs are thickest at the bottom of the V-groove, and taper on the slopes of the V. The lowest threshold current reported for this type of structure using single-wire active region has been 188 pA, which is comparable to that obtained from QW GRIN-SCH lasers, but hundreds of times greater than that predicted theoretically. In addition, due to the finite V-groove width ($\geq$0.5 μm) required to form a crescent-shaped QWR, it is difficult to produce a high-density coupled QWR active region using this and similar approaches. So far, the in situ strain-induced lateral-layer ordering (SILO) growth process has proven itself to be a simple and versatile technique for QWR formation. The in situ SILO technique can use standard (100) on-axis or vicinal substrates, creates lateral composition modulation perpendicular to the growth direction, and applies to a wide range of materials and growth conditions. It also avoids damage at interfaces because no pre- or post-growth processing is required.

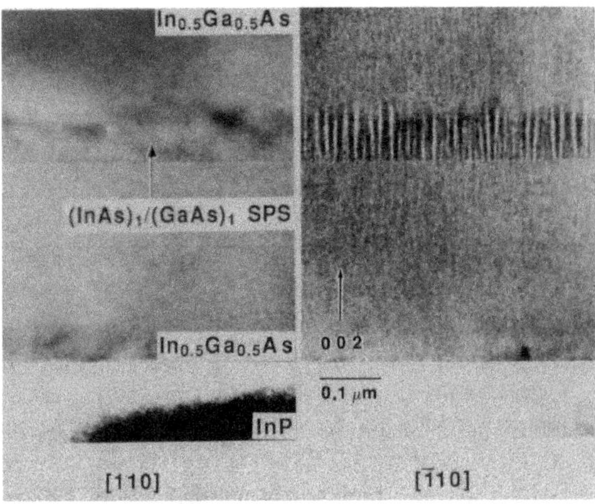

Fig. 10.37 Dark field cross-sectional TEM micrographs of the $(GaAs)_1/(InAs)_1$ short-period superlattices sandwiched between $Ga_{0.5}In_{0.5}As$ layers grown on InP. Within the SPS layer, a uniform image appears in the [110] cross section while strong dark and light fringes are observed in the [$\bar{1}$10] cross section. Reprint with permission from [13], copyright AIP Publishing

The principle of the SILO process is based on the deposition of alternating thin layers of two different materials to form a strained-layer supperlattice (SPS), where each of the layers is strained with respect to the growth substrate. This nearly strain balanced structure, however, is tensile for one of the layers and compressive for the other. Thus, the SPS as a whole has minimal strain, and defects due to lattice relaxation or critical thickness boundaries are not induced. During the growth of this SPS, the localized finite net strain that exists can trigger spinodal decomposition or induce a morphological instability on the surface. This will cause atoms in the alloy to migrate in a plane perpendicular to the growth direction. Once the composition modulation is initiated, it is stable and propagates as growth continues. In experiments, this modulation has always been observed to occur in the [110] direction, creating lateral superlattices (LSL). When combined with larger bandgap material in the growth direction [100] via standard QW growth methods, somewhat uniform arrays of QWRs are formed. Figure 10.37 shows the cross-sectional TEM (transmission electron microscopy) images from a $(GaAs)_m/(InAs)_n$ SPS structure, with $m \approx n \approx 1$, sandwiched between $Ga_{0.5}In_{0.5}As$ layers on an InP substrate. A uniform image appears in the [110] cross section, and a modulated image composed of dark and light fringes, each approximately 10 nm wide, in parallel to the growth direction appears in the [$\bar{1}$10] cross section. The salient feature is that the dark and light vertical fringes in the [$\bar{1}$10] TEM image represent In-rich and Ga-rich $Ga_xIn_{1-x}As$ alloys, respectively. As the composition varies in the [110] direction about the averaged SPS value of $x = 0.5$, so does the bandgap of the material, which is compressively strained in the In-rich well regions, and under tensile strain in the Ga-rich barrier regions. Energy-dispersive X-ray spectroscopy (EDS) confirmed that the contrast in the TEM image is due to lateral composition variations. In the orthogonal direction, the [110] image exhibits no such composition contrast, as this perspective constitutes a 'side view' of the LSL structure. The uniformity of this image indicates that the LSL interfaces are reasonably parallel to the (110) plane, and that the composition within each well and barrier region is approximately homogeneous in this plane. The LSL well and barrier regions form columns that propagate in the $[\bar{1}10]$ direction for considerable distances of ~400 nm. With this technology available to produce quantum confinement in the [110] direction, it becomes a relatively straightforward matter to sandwich a thin (~10 nm) laterally ordered layer between layers of bulk AlGaInAs in order to provide an additional degree of quantum confinement in the [001] direction, thereby creating QWRs that lie approximately 10 nm apart in the (001) plane. Based on the SILO process, a host of QWR heterostructures has been fabricated including $(GaP)_2/(InP)_2$ SPS on GaAs, $(GaP)_2/(InAs)_1$ SPS on GaAs, $(GaAs)_2/(InAs)_2$ SPS on InP, and $(GaP)_2/(InP)_1$ SPS on $GaAs_{0.66}P_{0.34}$, which cover the PL emission wavelength regimes from visible to near infrared [14].

Using the SILO process, both $Ga_xIn_{1-x}P/Al_{0.15}Ga_{0.35}In_{0.5}P$ visible (0.7 μm) laser structures and $Ga_xIn_{1-x}As/Al_{0.48}In_{0.52}As$ near IR (1.7 μm) laser structures with multiple quantum wire (MQWR) active regions have been formed in situ during MBE growth. Stripe-geometry Fabry–Perot lasers have been fabricated from these MQWR heterostructures in the same way as QW lasers, except that the contact stripe can be oriented in parallel or perpendicularly with respect to the QWRs. Stripe-geometry $Ga_xIn_{1-x}As/Al_{0.48}In_{0.52}As/InP$ SCH laser diodes with stripes oriented either in parallel ($[\bar{1}10]$ emission) or perpendicularly ([110] emission, Fig. 10.38) with respect to the QWR arrays exhibit anisotropic threshold current densities (J_{th}) varying in ratio by a factor of ~10. Liquid nitrogen temperature and room-temperature J_{th}'s as low as 200 A/cm^2 and 1 kA/cm^2 were obtained for lasers with stripes in the [110] direction, respectively. This behavior has been observed in both $Ga_xIn_{1-x}P/GaAs$ and $Ga_xIn_{1-x}As/InP$ MQWR laser structures. The dramatic anisotropy in J_{th} is due to the variation of the magnitude of the dipole moment with the direction of propagation

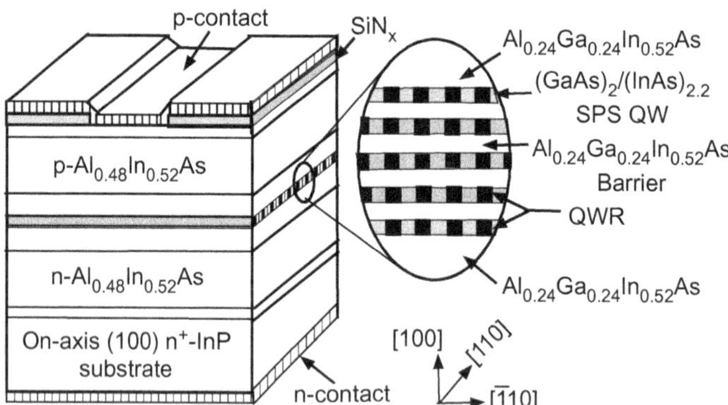

Fig. 10.38 Schematic diagram of a stripe contact $Ga_xIn_{1-x}As$ QWR Fabry–Perot laser diode grown using the SILO process. Notice that the contact stripe is perpendicular to the long axis of the QWRs. Reprint with permission from [15], copyright AIP Publishing

of the light wave in the QWR system. The magnitude of the dipole moment is maximized when the direction of the *E*-field coincident with that of the dipole moment vector, i.e., the wave propagation direction, is orthogonal to the QWR axis [16].

The improvement of J_{th} (>30%) has been seen in QWR lasers in comparison with their QW control sample counterparts. However, the improvement of J_{th} is several orders of magnitude smaller than theoretically predicted. Similar to self-assembled QDs, the self-assembled nature of SILO process generated QWRs does introduce non-uniformities to the wire dimensions. These non-uniformities serve to diminish theoretically expected improvements in lasing characteristics for laser diodes made from SILO-grown QWRs. Nevertheless, some unique properties have been observed in SILO-grown QWR lasers. Most importantly, the bandgap of GaInAs QWRs grown by the SILO process responds to temperature changes in ways that deviates from the norm for III–V semiconductors and leads to a temperature-invariant stable lasing wavelength in $Ga_xIn_{1-x}As$/InP MQWR laser system. As shown in Fig. 10.39, the 300 and 77 K lasing spectra of a $Ga_xIn_{1-x}As$ QWR laser that was grown using the SILO process are compared to those of a conventional SCH QW laser with lattice-matched $Ga_{0.47}In_{0.53}As$ QWs. The lasing wavelength shift of the $Ga_xIn_{1-x}As$ QWR laser between 300 and 77 K is ~0.9 Å/°C, which is comparable to that of a DFB laser and much less than the value of the conventional QW laser (~4.6 Å/°C). This effect is due to the temperature dependence of the multiaxial strain that exists in SILO-grown QWRs. This bandgap behavior with respect to temperature change of SILO-grown GaInAs QWRs is both physically interesting and potentially applicable to device applications.

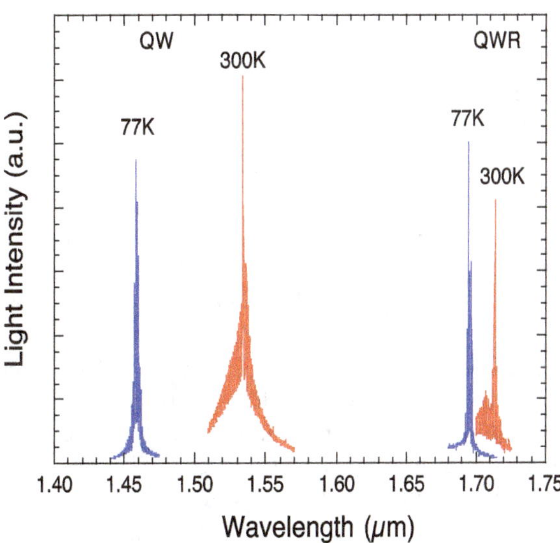

Fig. 10.39 Lasing spectra measured at 77 and 300 K from a GaInAs QW laser and from a GaInAs QWR laser that was grown using the SILO process. Notice that the lasing wavelength dependence on temperature is drastically reduced in the QWR laser diode. Reprint with permission from [15], copyright AIP Publishing

10.4 Vertical-Cavity Surface-Emitting Lasers

As discussed above, the optical cavity of a conventional edge-emitting laser lies in the plane parallel to the active layer. In contrast, the vertical-cavity surface-emitting laser (VCSEL) has an optical cavity perpendicular to the active layer plane. The consequences of this simple rearrangement of the orientation of the optical cavity are the improved output mode spectrum and manufacturability of semiconductor lasers. The VCSEL structure may provide a number of advantages over the conventional edge-emitting lasers. These include the following:

- VCSELs emit optical beams with a circular symmetric profile that improves the coupling efficiency with optical fiber and other optical components.
- Due to the very short optical cavity in VCSELs, dynamic single longitudinal mode operation is easily achievable.
- Ultralow threshold current operation of VCSELs is expected due to its small cavity volume. This also allows the formation of densely packed 2D laser arrays.
- Because of their vertical-cavity orientation, wafer-scale fabrication and probing can be performed before separating devices into discrete chips. This will reduce the production cost enormously.

Because of the difference in optical cavity orientation between VCSELs and edge-emitting lasers, some of the design parameters used for edge-emitting lasers, which have been discussed in preceding sections, may not be applied to the analysis of VCSELs. Hence, we need to develop a proper threshold current density model to identify important design parameters for VCSELs.

10.4.1 Threshold Current Density

Before we investigate the VCSEL structure with distributed Bragg reflector mirrors, the familiar Fabry–Perot structures are analyzed. The gain at the onset of the laser action in a Fabry–Perot cavity is generally written as

$$\Gamma g_{\text{th}} = \alpha_i + (1/2L)\ln(1/R_1R_2) \tag{10.47}$$

where α_i is the total internal loss, L is the effective length of the cavity, and R_1 and R_2 are the reflectivity of the mirrors. However, as shown in Fig. 10.40a, the gain in a surface-emitting laser is not distributed over the entire effective cavity length L; it is contained in a region of thickness d. The total gain in the cavity becomes $\Gamma g_{\text{th}}d$. The total loss in the cavity is divided into two components: one in the active region, α_a, and the other in the rest of the cavity, α_c. Thus the gain at the onset of the laser action must be modified as

$$\Gamma g_{\text{th}}d = \Gamma\alpha_a d + \alpha_c(L-d) + (1/2)\ln(1/R_1R_2) \tag{10.48}$$

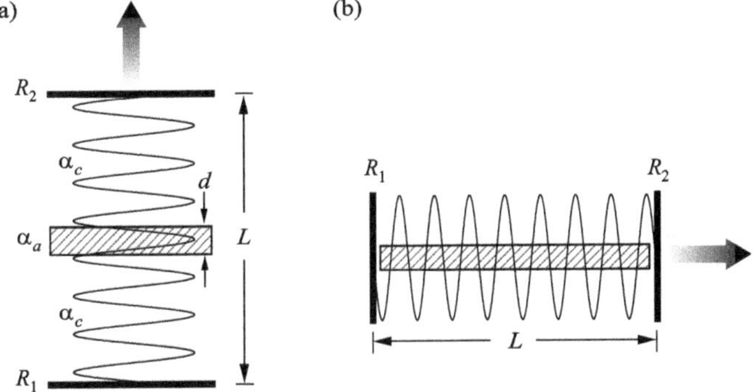

Fig. 10.40 Schematics of **a** surface-emitting laser and **b** edge-emitting laser showing the active region (shaded) and the standing waves inside the Fabry–Perot optical cavity

The confinement factor is modified as $\Gamma = \Gamma_L \Gamma_T$, where the transverse component is $\Gamma_T \approx 1$ and the longitudinal component represents the fill factor $\Gamma_L \approx d/L$. Equation (10.48) can be rearranged as

$$g_{\text{th}} = \alpha_a + \left(1/\Gamma^2\right)[\alpha_c(1-\Gamma) + (1/2L)\ln(1/R_1R_2)] \tag{10.49}$$

Empirically, the peak gain g_p is a linear function of the injected carrier density n and expressed as

$$g_p = a(n - n_{\text{tr}}). \tag{10.50}$$

As derived in Sect. 10.1, the threshold current density has the form of

$$J_{\text{th}} = \frac{qdn_{\text{th}}}{\tau_s} \tag{10.51}$$

where n_{th} is the injected carrier density at threshold and τ_s is the carrier lifetime. For lasers of high material quality, the dominant recombination process is radiative in nature.

The carrier lifetime can be written as

$$1/\tau_s = B_r n_{\text{th}} \tag{10.52}$$

where B_r is the radiative recombination coefficient. Combining (10.49) through (10.52), the threshold current density of the surface-emitting laser becomes

$$J_{\text{th}} = \frac{qdB_r}{a^2}\left\{\alpha_a + an + \frac{1}{\Gamma^2}\left[\alpha_c(1-\Gamma) + \frac{1}{L}\ln\left(\frac{1}{R}\right)\right]\right\}^2 \tag{10.53}$$

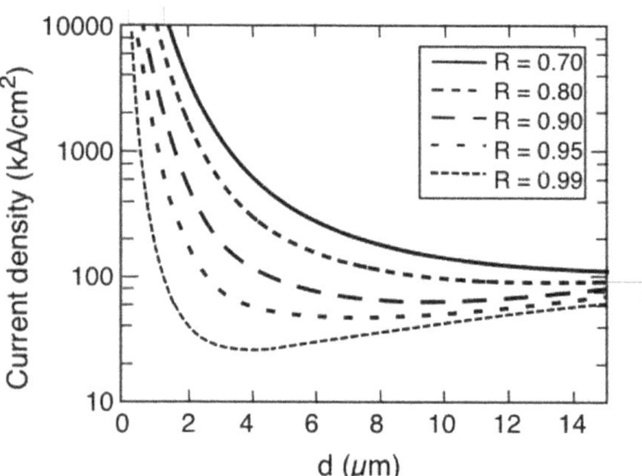

Fig. 10.41 Dependence of the threshold current density of a GaAs-AlGaAs VCSEL on the active layer thickness and mirror reflectivity. The parameters used for the calculation are: $a = 4 \times 10^{-16}$ cm^2, $B_r = 10^{-10}$ cm^3 s^{-1}, $\alpha_a = 25$ cm^{-1}, and $\alpha_c = 10$ cm^{-1}. Reprinted with permission from [17], copyright IEEE

Here we assumed that $R = R_1 = R_2$. Since $\Gamma = d/L$, for a fixed L, the threshold current density depends on d and R. Using a GaAs-AlGaAs laser with a short cavity of $L \leq 15$ μm as an example, (10.53) can be plotted as a function of d and R, as shown in Fig. 10.41. This figure clearly shows that to achieve low threshold current density in a surface-emitting laser, the active region must be thin and the mirror reflectivities must be extremely high, ideally $R \geq 99\%$!

10.4.2 Distributed Bragg Reflector Mirrors

The extremely high mirror reflectivity required for low threshold surface-emitting lasers can be achieved by using a distributed Bragg reflector (DBR) mirror design (Fig. 10.42). A DBR consists of N pairs of two dielectric, lossless materials with

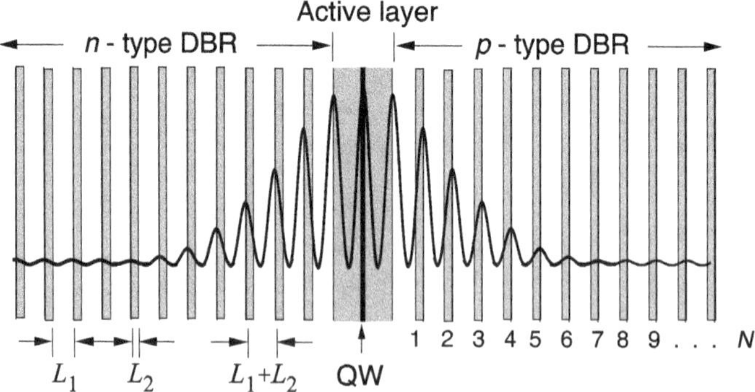

Fig. 10.42 Distribution of the optical power intensity inside a typical VCSEL. A DBR consists of N pairs of two materials of thickness L_1 and L_2 with refractive index $\bar{n}_1$ and $\bar{n}_2$, respectively

refractive index $\bar{n}_1$ and $\bar{n}_2$, where $\bar{n}_1 < \bar{n}_2$. The thickness of the two layers is a quarter of the Bragg wavelength, λ_b, that is,

$$L_1 = \lambda_b/4\bar{n}_1 \quad \text{and} \quad L_2 = \lambda_b/4\bar{n}_2 \tag{10.54}$$

The period of the DBR is $L = L_1 + L_2$. The reflectivity of this N-pair DBR at the Bragg wavelength is given by

$$R_b = \left[\frac{1 - (\bar{n}_1/\bar{n}_2)^{2N}}{1 + (\bar{n}_1/\bar{n}_2)^{2N}}\right]^2 \tag{10.55}$$

Since $\bar{n}_1/\bar{n}_2 < 1$, the reflection coefficient tends to unity as N increases. For a fixed number N, the reflection coefficient increases when the ratio $\bar{n}_1/\bar{n}_2$ is increased. The other important parameter of DBR design for VCSEL applications is the high-reflectivity band or stop band. The spectral width of the stop band is given by

$$\Delta\lambda_b \cong \frac{2\lambda_b \Delta\bar{n}}{\bar{n}_{\text{eff}}\pi} \tag{10.56}$$

where $\Delta\bar{n} = \bar{n}_2 - \bar{n}_1$ is the refractive index difference of the two constituent materials and $\bar{n}_{\text{eff}}$ is the effective index of the DBR mirror. The effective index can be calculated by requiring the same optical path length normal to the layers for the DBR and the effective medium, which leads to

$$\bar{n}_{\text{eff}} = 2\left(\frac{1}{\bar{n}_1} + \frac{1}{\bar{n}_2}\right)^{-1} \tag{10.57}$$

From (10.55) and (10.56), it is concluded that a large refractive index difference in DBR materials is advantageous in achieving high reflectivity and broader stop band with fewer pairs. The effect of refractive index difference on DBR performance is illustrated in Fig. 10.43, where the reflectances of the α (amorphous)-Si/Al_2O_3 and α-GaAs/α-AlAs DBRs are compared. Due to the large refractive index difference ($\Delta\bar{n} > 2$), the α-Si/Al_2O_3 dielectric DBR requires many fewer DBR pairs than an α-GaAs/α-AlAs DBR ($\Delta\bar{n} \sim 0.5$) to reach a reflectivity of >99% while maintaining a wide stop band. One additional advantage of the dielectric DBR reflectors is the absence of absorption loss over a wide optical spectrum. However, the dielectric DBRs are non-conductive and may pose some VCSEL design limitations.

10.4.3 Resonant Periodic Gain in VCSELs

From (10.53), a thin active region is required in VCSELs to reach a low threshold current density. However, a thin active region may not provide sufficient gain for the generation of ample optical power. As we have learned, enormous enhancement of

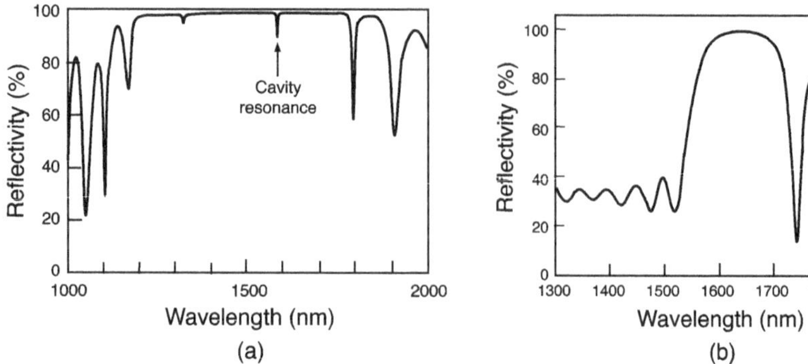

Fig. 10.43 Reflectance of **a** α-Si/Al_2O_3 and **b** α-GaAs/α-AlAs DBR. The α-Si/Al_2O_3 DBR consists of 6.5-pair lower DBRs, a 2λ active layer, and a 6-pair top DBR. The α-GaAs/α-AlAs structure has a single 20-pair DBR [18]

differential gain can be achieved by replacing the bulk active layer with QW material. The drawback of using a QW as the active layer is the reduction of longitudinal confinement factor Γ_L (~d/L) by which the threshold current density is increased. This problem can be resolved by using multiple quantum wells (MQWs) as the active layer of the VCSEL. Furthermore, to maximize energy transfer, a good overlap between the QW gain medium and the peak of the standing-wave pattern of the lasing mode is necessary.

Figure 10.44 illustrates the active region of a VCSEL with a periodic gain structure. The two DBRs are simplified and represented by two metal mirrors such that we can treat it as a Fabry–Perot resonator. For a mirror with high reflectivity, the standing wave of the optical field $E(z)$ can be approximated by

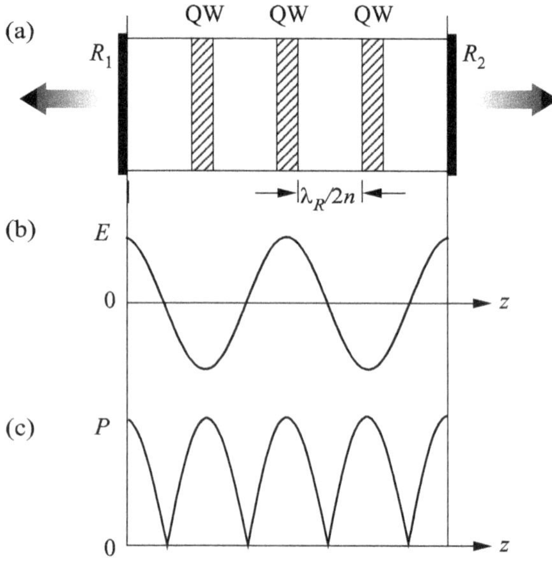

Fig. 10.44 **a** Schematic of the VCSEL cavity structure with periodic gain structure containing three QWs. **b** Magnitude of the electric field distribution. **c** The power density distribution

$$E(z) = E_0 \cos(kz) \tag{10.58}$$

The time-averaged optical power density in the standing wave is given by

$$P \sim EE^* \sim E_0^2 \cos^2(kz) \tag{10.59}$$

The positions of the very thin QW active layers, the gain medium, need to be aligned with the peak (antinode) of the standing-wave (power) pattern at a selected lasing wavelength λ_R to gain a maximum power transfer. This can be achieved by separating QWs with spacer layers of thickness $\lambda_R/2\bar{n}$, where $\bar{n}$ is the refractive index of the spacer layer at λ_R. Since the QW gain medium is anisotropic and wavelength selective, single longitudinal mode operation is favorable.

10.4.4 Current and Optical Confinements in VCSELs

In addition to providing highly reflective DBR mirrors and maximizing the optical field and gain overlap, a VCSEL design should contain some structures to effectively confine carriers in the active region, as in edge-emitting lasers. In Fig. 10.45, three commonly used current confinement schemes in VCSELs are shown. The air-post structure can be fabricated by etching to form a circular or rectangular mesa to confine current inside the small post. However, surface recombination may cause a large leakage current, especially for small devices. The proton-bombardment confinement can effectively limit the current into the center active region by producing insulating material in the surrounding region. Since the refractive index of the implanted region does not change after ion implantation, the lateral optical field is not well confined. To improve the optical confinement, a selectively oxidized VCSEL structure was developed (Sect. 5.6). This approach relies on the discovery of a native oxide for Al-containing III–V semiconductors by Nick Holonyak Jr. of the University of Illinois, which has a major impact on the way VCSELs are fabricated today.

The native oxide, formed by oxidizing Al(Ga)As by high-temperature ($\geq$400 °C) water vapor, is used to define the aperture out of which the light is emitted. A unique

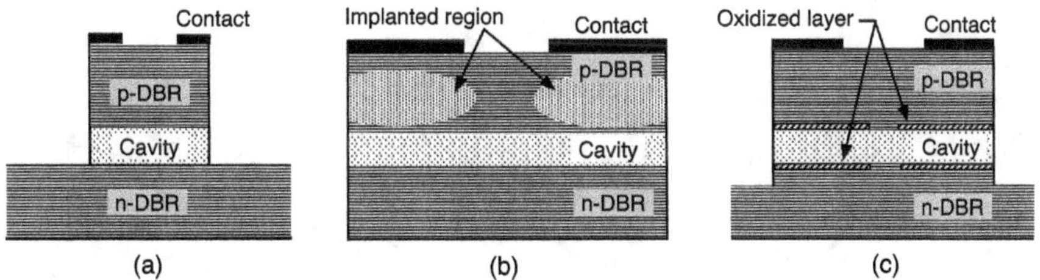

Fig. 10.45 Some of the common device geometries for VCSEL devices. The current confinement is achieved using **a** etched air-post, **b** ion implantation, and **c** selective oxidation

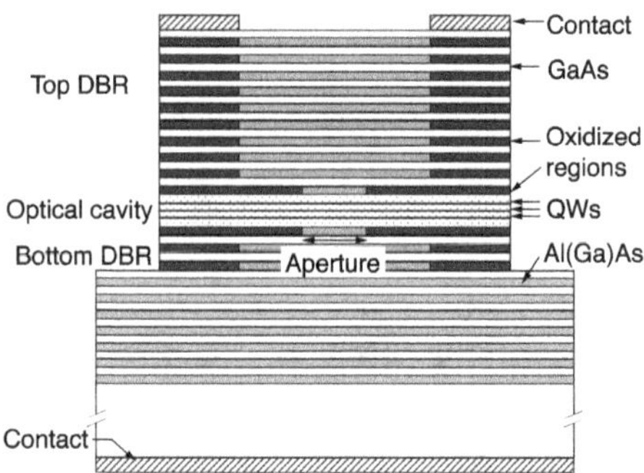

Fig. 10.46 Schematic of detailed structure of the oxide-confined GaAs/AlAs VCSEL. To achieve a tight optical aperture, the Al-composition of the two layers next to the optical cavity is higher than that of the DBR layers

feature of the oxide is the dependence of the oxidation rate on the Al-composition in the $Al_xGa_{1-x}As$ alloy (Fig. 5.20): the higher the Al-composition, the higher the oxidation rate. This gives one a high degree of control over the size and location of the oxide-defined aperture within the device structure. In addition to carrier confinement, the low refractive index of the oxidized AlAs (1.74) provides excellent optical confinement in a VCSEL. The detailed layer structure of an oxide-defined GaAs/AlAs VCSEL is shown in Fig. 10.46. In this case, the edge of the DBR mirrors is partially oxidized and forms a continuous longitudinal optical guide like a circular symmetric optical fiber.

10.4.5 Characteristics of VCSELs

As is the case with their edge-emitting laser counterparts, some of the important properties of VCSELs can be understood from their light–current–voltage (*L–I–V*) characteristics. Figure 10.47 shows the typical *L–I–V* curves of a GaAs/AlGaAs QW VCSEL with 20 and 30 pairs of epitaxially grown $Al_{0.15}Ga_{0.85}As$/AlAs top and bottom DBRs, respectively. There are two distinct characteristics in this VCSEL when compared with an edge-emitting laser. First, the maximum power output of the VCSEL, under continuous wave (cw) operation, has been severely limited by a sharp roll-over in the *L–I* characteristics. However, the device suffers no apparent damage

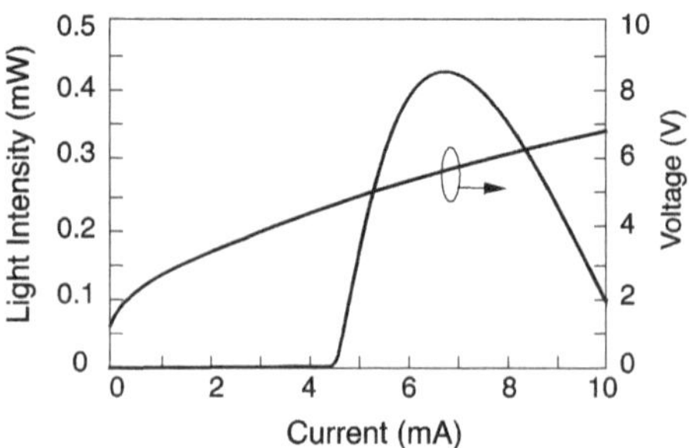

Fig. 10.47 Typical *L–I–V* curves of a 15 μm diameter GaAs VCSEL operating under cw condition at a heat sink temperature of 25 °C

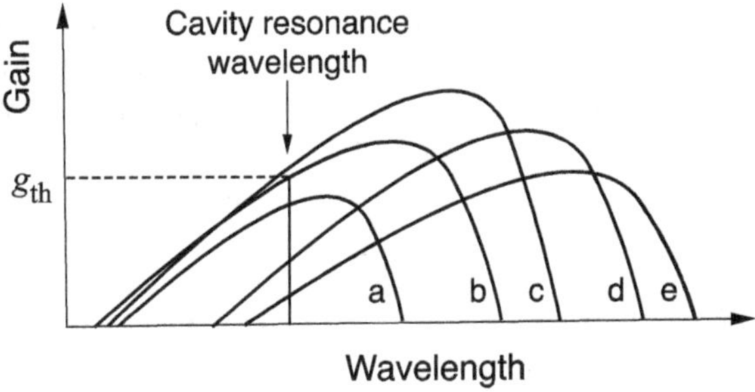

Fig. 10.48 Illustration of the effect of shifts in the gain spectrum of a VCSEL at different current injection levels. The shift in the Fabry–Perot cavity resonance is small in comparison and ignored for clarity. Each gain curve corresponds to an injection current **a** below I_{th}, **b** just above I_{th}, and **c** near the peak light output, before roll-over. In **d** and **e** the VCSEL ceases to lase ($I_d < I_e$)

even when the lasing is quenched at a current about twice the threshold value. Second, the voltage drop across the VCSEL is not clamped beyond the threshold value. The voltage across the VCSEL continues to increase with the injection current.

The sharp roll-over in the *L–I* characteristics can be qualitatively explained as the mismatch between the gain spectrum and the cavity resonance wavelength (Fig. 10.48), and the high series resistance of DBRs contributes to the continuous increasing of voltage above threshold. Since the VCSEL has an effective short cavity length of ~1 μm, the Fabry–Perot cavity resonance modes are spaced so far apart that there is only one such mode within the spectral gain bandwidth of the active material. The VCSEL lases in a single longitudinal mode as long as the gain at that wavelength exceeds the total cavity losses. Thus, the gain spectrum behaves similarly to that of an edge-emitting laser at low injection currents, where the gain peak increases with current and redshifts, as shown in Fig. 10.48. These gain spectra (a), (b), and (c) qualitatively correspond to the injection currents of $I < 4.5$ mA (I_{th}), $I \approx 5$ mA, and $I \approx 6$ mA, respectively, of the *L–I* curve shown in Fig. 10.47. However, we notice that the Fabry–Perot cavity resonance mode position is, in general, not coincident with the gain peak in a VCSEL. Furthermore, the position of the Fabry–Perot mode moves relative to the gain peak at a different rate as the temperature changes. For example, near room temperature, the Fabry–Perot mode wavelength of the AlGaAs DBR increases linearly with temperature by ~0.50 Å/°C while the gain peak wavelength of the GaAs active material increases with temperature at a faster rate of ~4.52 Å/°C. Therefore, the VCSEL power output (and threshold) is very sensitive to heating effects. Since current is injected into the active region through the thick multilayer heterostructure DBRs, a large series resistance is added to the laser. An evidence of the large series resistance in the VCSEL is the increasing voltage drop above threshold shown in Fig. 10.47. The voltage across an ideal laser diode beyond threshold should remain clamped at $h\nu/q$, where $h\nu$ is the lasing energy and q is the charge. The excess voltage across the VCSEL at high operation current reflects the high DBR resistance *R*. Thus, the active region temperature of the VCSEL increases quadratically with injection current since it is proportional to the electrical power dissipation (I^2R). Therefore, the resistive heating raises the active region temperature significantly, and the maximum gain will decrease and redshift at high currents as shown in curves (d) and (e) of Fig. 10.48. Although the peak gain may still be higher than the threshold, because of the mismatch between the gain peak and the cavity resonance wavelength, the VCSEL lasing is quenched resulting a roll-over in the *L–I* curve. These high-current injection levels qualitatively correspond to $I > 10$ mA in the *L–I* curve of Fig. 10.47.

If the Fabry–Perot cavity resonance mode is on the long-wavelength side of the gain spectrum, a rise in temperature brings the gain peak toward the cavity resonance mode. Hence, a decrease in threshold with increasing temperature can be observed in VCSELs provided the reduction in gain constant is not overwhelming. Because of this, the cavity resonance is often intentionally designed to be at a slightly longer wavelength relative to the peak laser gain at room temperature for optimum VCSEL performance.

10.5 Light-Emitting Diodes

Light-emitting diodes (LEDs) are *p-n* junctions designed to operate under forward-biased conditions to generate spontaneous emissions of radiation in the ultraviolet (UV), visible, and infrared (IR) regions of the electromagnetic spectrum. The basic structure of the LED is similar to that of a semiconductor injection laser, as shown in Fig. 10.1, except without a pair of cleaved facets. Under a forward bias condition, the current flow through the LED provides injected excess minority carriers —electrons and holes—necessary for the generation of electroluminescence by the radiative recombination of minority carriers in and near the semiconductor *p-n* junction. As discussed in Chap. 8, highly efficient radiative recombination of excess minority carriers can only be achieved in direct bandgap semiconductors. Although the luminescence efficiency of indirect bandgap materials can be improved by doping isoelectronic impurities, the external quantum efficiency is still relatively low. Therefore, today's high-brightness LEDs are all made in direct bandgap semiconductor materials.

The use of LEDs can be divided into different categories depending on their emission wavelengths; e.g., an IR light source for short-distance optical-fiber communications, and visible light sources for displays and solid-state lighting. Today the rapid technological advancements have expanded the usage of visible LEDs into many economically and technologically important applications including light sources in traffic signals, full-color large outdoor displays, automotive applications (signaling, tail light, and headlight), backlighting in liquid-crystal flat-panel displays, and as sources of general illumination. Thus, the discussion in this section will focus on LEDs that emit visible light (λ between 400 and 750 nm).

10.5.1 *Efficiency of LEDs*

When a *p-n* junction LED is under a forward bias condition, the recombination of injected excess minority carriers in the vicinity of the *p-n* junction can release the carriers' excess energy as photons. The photon flux generated inside the LED then must emerge from the device and provide desired optical stimulus to the eye. However, not all injected carriers will recombine radiatively near the *p-n* junction and a part of the generated photon flux may suffer internal reflections inside the LED die. Light may also be absorbed by the substrate or internally reflected back at the surface boundary and reabsorbed by the semiconductor. Therefore the overall device efficiency is always less than unity and is expressed as

$$\eta_e = \eta_i \eta_{\text{xtr}} \tag{10.60}$$

where η_i, η_{xtr}, and η_e are the internal, extraction, and external quantum efficiency, respectively.

The internal quantum efficiency is defined as

$$\eta_i = \frac{\text{number of photons generated per second}}{\text{number of injected electrons per second}} = \frac{P_a/h\nu}{I/q} \tag{10.61}$$

where P_a is the optical power emitted from the active region, I is the injection current, and q is the electron charge. The material quality and the *p-n* junction design of the LED play significant roles in determining the internal quantum efficiency. Non-radiative recombination can take place through mid-gap states (Shockley–Read–Hall recombination), surface states, and Auger processes. In high-quality direct bandgap semiconductors, only non-radiative surface recombination becomes important, but it can be minimized using heterojunctions. In homojunction devices, some injected electrons may penetrate deep and pass the *p-n* junction before recombination. By utilizing the properties of carrier and optical confinement of the heterojunction, these unwanted losses are greatly reduced in double-heterojunction devices.

The extraction quantum efficiency is defined as

$$\eta_{\text{xtr}} = \frac{\text{number of photons emitted into free space per second}}{\text{number of photons emitted from active region per second}} = \frac{P/h\nu}{P_a/h\nu} \tag{10.62}$$

where P is the optical power emitted into free space. In practice, not all the optical power emitted by the active region is emanated into free space. For example, light emitted toward the substrate of the LED can be reabsorbed if the bandgap of the substrate is narrower than the active layer. Light may also be absorbed by the metal contact surface. A major loss mechanism that limits extraction quantum efficiency is related to the total internal reflection at the surface–air interface of the LED. Approaches that involve either encapsulating or shaping the device die have been shown to be effective in alleviating the total internal reflection problem as discussed next.

The external quantum efficiency is defined as

$$\eta_e = \frac{\text{number of photons emitted into free space per second}}{\text{number of electrons injected into LED per second}} = \frac{P/h\nu}{I/q} \tag{10.63}$$

It is closely related to the power efficiency or wall-plug efficiency of the LED, which is defined as

$$\eta_{\text{pw}} = \frac{\text{optical power output}}{\text{input electric power}} = \frac{P}{IV} \tag{10.64}$$

Assume θ_1 is the angle of incidence with respect to the surface normal in the semiconductor at the semiconductor–air interface of the light emission. According to Snell's law, the incident angle of the refracted ray into air θ_2 can be expressed as

$$\bar{n}_1 \sin\theta_1 = \bar{n}_2 \sin\theta_2 \tag{10.65}$$

where $\bar{n}_1$ and $\bar{n}_2$ are the refractive indices of the semiconductor and air, respectively. When $\theta_2 = 90°$, the refracted light travels along the surface and no light can escape the semiconductor. Then $\theta_1 \equiv \theta_\text{c}$ is defined as the critical angle for total internal reflection.

$$\theta_c = \sin^{-1}(\bar{n}_2/\bar{n}_1) \tag{10.66}$$

This angle defines the light escape cone where light emitted into the cone can escape from the semiconductor. Otherwise, light outside the cone suffers total internal reflection. Since the refractive index of semiconductors is quite large (~3.5) compared to air ($\bar{n}_2 = 1$), the critical angle for total internal reflection is small. Thus most of the emitted radiation is trapped inside the semiconductor by total internal reflection.

One common practice to improve the extraction quantum efficiency of the LED is to encapsulate the device within a dome of transparent dielectric material of high-refractive index. Using a dielectric with a refractive index 1.5 to encapsulate the LED of $\bar{n}_1 = 3.5$, the light output is enhanced by more than 2.5 times. This method is useful for low-cost high-volume applications.

The other method to increase the extraction quantum efficiency is by shaping of LED dies. Ideally, a spherically shaped LED can eliminate the internal reflection problem since light is incident at a normal angle at the semiconductor–air interface. However, the high cost of shaping each die into a sphere makes this approach impractical. Instead, high luminous efficiency (>100 lm/W) and high external quantum efficiency (60%) AlGaInP/GaP red LEDs have been achieved in LED dies by shaping them into truncated inverted pyramids (TIPs) (Fig. 10.49). The TIP geometry improves light extraction by redirecting totally internally reflected photons from the sidewall interfaces toward the top surface of the chip near normal incidence,

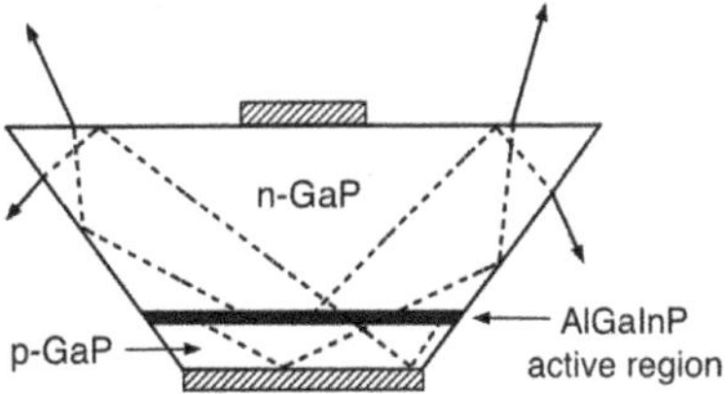

Fig. 10.49 Schematic diagram of a truncated inverted pyramid (TIP) AlGaInP/GaP LED. Reprinted with permission from [19], copyright AIP Publishing

allowing them to escape. In addition, totally internally reflected photons at the top surface are redirected for escape through the beveled sidewalls. These two processes provide the TIP device with a significant reduction in photon path length within the die compared to a conventional chip, thus reducing internal absorption losses. The light extraction efficiency can also be improved by using textured semiconductor surfaces formed by wet-chemical etching. This method has been found to be effective in GaInN LEDs. A substantial amount of surface roughness of pyramid-like structures with smooth side surfaces can be obtained by wet-chemical and photochemical etchings. This strong surface scattering structure can easily out-couple the light with a few scattering events.

10.5.2 Materials and Technology Evolution

For visible LED applications, the energy bandgap of semiconductors used must be larger than 1.8 eV (~700 nm) to match the luminosity function of the human eye, which is a function of wavelength. In the spectral range from red to yellow/orange, there are three ternary alloys, $GaAs_{1-x}P_x$, $Al_xGa_{1-x}As$, and $Ga_xIn_{1-x}P$, and one quaternary alloy, $Al_xGa_yIn_{1-x-y}P$, with suitable direct energy bandgaps. For LEDs with even shorter wavelength in the blue and green spectral range, the current choices are III-nitride material systems, specifically $Ga_xIn_{1-x}N$. The relevance of these alloys to the visible LED development is summarized in Fig. 10.50 where the evolution of LED performance in terms of luminous efficiency is plotted versus year. It reveals that the performance of visible III–V LEDs has improved dramatically over the past half century since the demonstration of the first GaAsP red LED in 1962 by Nick Holonyak Jr. at General Electric. Over this period of time, the LED performance increased roughly ten times every decade, mainly driven by the advancement of material technologies including material selections, epitaxy methods, and heterostructure designs to improve the internal quantum efficiency. The development of chip technologies with improved light extraction advances the external quantum efficiency performance of LEDs.

The GaAsP crystal used for the fabrication of the first red LED by Holonyak was grown using a closed tube vapor transport technique. Then the vapor-phase epitaxy (VPE) technique was developed for commercial production of GaAsP red LEDs. The green and red GaP LEDs were grown by LPE doped with isoelectronic impurities

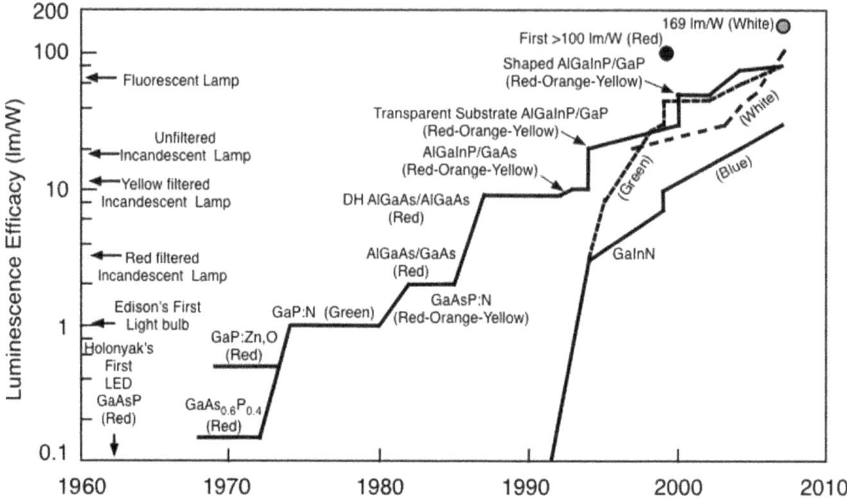

Fig. 10.50 Light output of visible LEDs during the last 50 years. The luminescence efficacy increased at a rate of ~10 ×/decade. At the moment, the luminous efficacy of laboratory white LEDs exceeds that of all other conventional light sources. Reprinted with permission from [20], copyright IEEE

N and Zn-O pairs, respectively. The VPE growth technique was then developed to incorporate a significant amount of N isoelectronic impurities into indirect bandgap GaAsP alloys creating relatively high-performance amber, yellow, and yellow-green LEDs. Nevertheless, the quantum efficiency of these indirect bandgap-based visible LEDs is limited by the fundamental physics. Furthermore, the ternary GaAsP alloy does not lattice-match to any III–V binary substrates except near the binary ends, which leads to high defect density and low luminescence efficiency. Consequently, it was necessary to search for alternative material systems. In 1970, cw operation of AlGaAs/GaAs DH injection laser diodes at room temperature demonstrated that LPE growth and DH design were useful for the fabrication of AlGaAs red LEDs. In addition, the light-absorbing GaAs substrate could be easily removed by the use of wet-chemical etching to enhance the luminescence efficiency of LEDs to over 10 lm/W, which was bright enough for outdoor applications including traffic signals, automobile stoplights, and alphanumeric signs. However, the high aluminum content (>60%) used in AlGaAs layers of red LEDs posed a serious reliability problem in hot humid outdoor applications where the reaction of Al with oxygen could lead to device failure. To solve this problem, the quaternary alloy AlGaInP was then developed. The potential of using AlGaInP as a light-emitting material from red to yellow/amber was recognized much earlier since it can lattice-match to GaAs substrates over almost the whole direct bandgap region. In spite of this, the epitaxy technologies of the time were LPE and VPE, both of which had insurmountable technology difficulties for AlGaInP growth. During the late 1970s to mid-1980s, the MOCVD epitaxy technology had progressed to a point that high-optical-quality III–V alloys could be produced even for aluminum-containing materials. Since 1990, MOCVD-grown AlGaInP LEDs became the dominant high-efficiency red, orange, and yellow LEDs. To enhance the light extraction efficiency of AlGaInP LEDs, additional chip technologies using transparent GaP window layer, removing the light-absorbing GaAs substrates, and utilizing shaped chip have been implemented. These innovations have led to a red LED with a performance of 100 lm/W!

Although the performance of red LEDs had improved steadily, high-performance blue and green LED technologies were not available until the early 1990s due to the lack of suitable material technologies. Since the early 1970s, there had been a number of reports of GaN-based blue/violet light-emitting devices using VPE method. The real breakthrough did not come until the late 1980s and early 1990s when Akasaki's team at Nagoya University, Japan, solved some key issues of GaN growth: The MOCVD-grown GaN layer quality was substantially improved by using an AlN low-temperature buffer layer on (0001) sapphire substrates. The *p*-type GaN was finally realized using Mg as a dopant coupled with a post-growth electron beam irradiation to activate acceptor atoms. The first GaN *p-n* junction LED was finally realized in 1992. In 1995, high-performance green and blue LEDs were demonstrated by a Nichia team using GaInN/GaN quantum-well structures.

The current state-of-the-art external quantum efficiency of high-performance visible LEDs as a function of wavelength is summarized in Fig. 10.51. Relative to the peak eye response wavelength (555 nm), AlGaInP technology and III-nitride technology dominate longer wavelength LEDs (red, orange, and amber) and shorter wavelength LEDs (green and blue), respectively. Although the external quantum efficiency of both red and blue LEDs exceeds 50%, there exists a 'green gap' where the efficiency drops quickly between 500 and 600 nm. For the AlGaInP system, it is the direct–indirect bandgap transition that limits the external quantum efficiency beyond 600 nm. For the GaInN system, the efficiency drop in green LEDs is due to effects associated with lattice-mismatch when a high InN mole fraction layer is used. The highly dense dislocations induced in the GaInN layer act as non-radiative recombination centers. Furthermore, the compressive strain-induced built-in electric

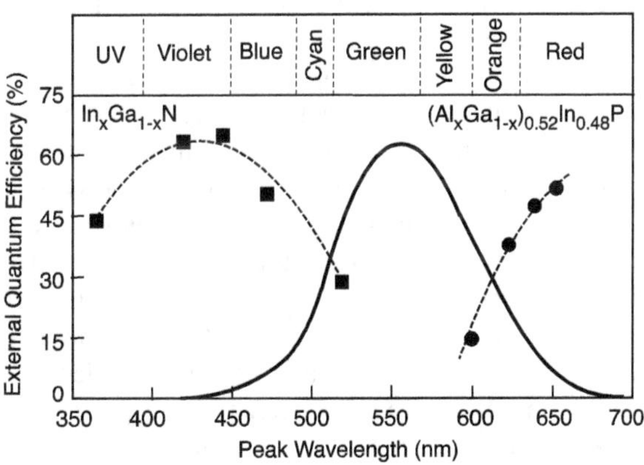

Fig. 10.51 External quantum efficiency of LEDs of different colors as a function of wavelength. InGaN devices produce green and blue, and AlInGaP devices produce red, red–orange, and yellow. CIE eye sensitivity function is also shown as the solid curve. Reprinted with permission from [20], copyright IEEE

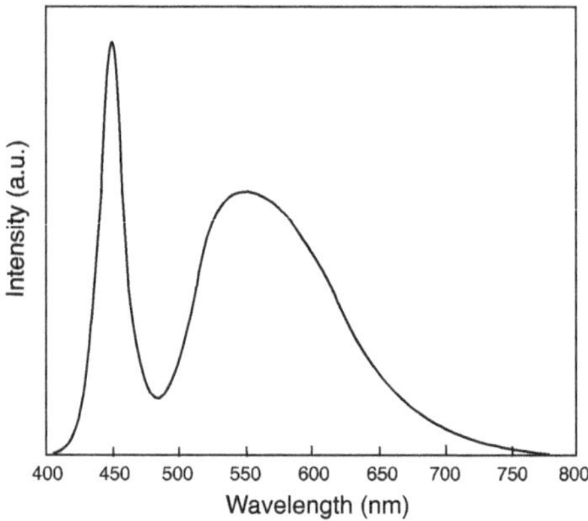

Fig. 10.52 Luminescence spectrum of blue (450 nm)-pumped phosphors for generating white light

fields along the *c*-axis of the wurtzite crystal cause electrons and holes to separate spatially within the same quantum well (e.g., Fig. 7.29), leading to an efficiency drop. At the moment, green LEDs are not efficient enough to mix with red and blue LEDs to generate white light. Instead, phosphor-converted LEDs are used. For instance, white light can be generated by a blue LED pump source coated with a layer of yellow-emission phosphor such as $Y_3Al_5O_{12}$:Ce^{3+} or YAG. The excitation spectrum of YAG at ~440–460 nm is well-matched to blue GaInN LEDs. The resulting light is a combination of blue (450 nm) and broadband yellow/orange (peaked at 550–600 nm), which appears white to the eye (Fig. 10.52). The current white LED performance has already reached 100 lm/W and exceeds the performance of conventional lighting technologies such as fluorescent and halogen.

10.6 Quantum Cascade (QC) Lasers

Lasing in semiconductor QW lasers takes place through radiative recombination of conduction band electrons and valence band holes across the direct energy gap of the QW, under the momentum conservation condition ($\Delta k = 0$), known as *interband transitions* (Fig. 10.53a). The transition energy is determined by the bandgap energy of the QW plus the confinement energies of the electron and hole. Consequently, the smallest transition energy is determined by the gap of the QW material. Alternatively, the transitions in a QW can also occur between confined states of the same band (conduction or valence). Since it relies on only one type of carrier making electronic transitions between subbands, the process is referred as *unipolar inter-subband transitions*. In this case, the smallest transition energy reduces toward zero as the well width is increased. This unique feature is central to the inter-subband transition because it enables the fabrication of long-wavelength devices over a wide frequency range (from mid-infrared to terahertz) using the same heterostructure materials. As shown in Fig. 10.53b, the initial and final states of subband within a QW have the

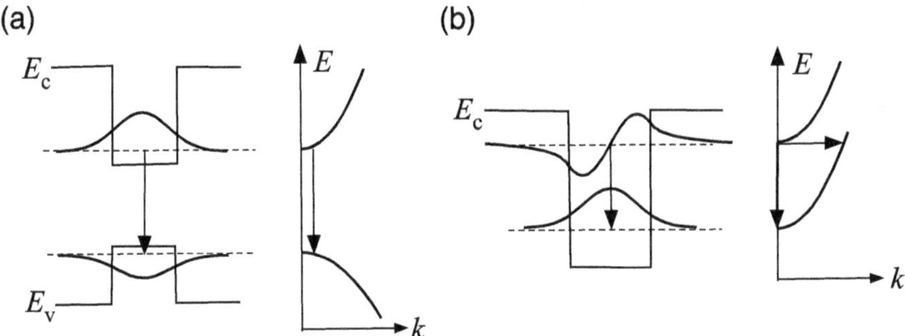

Fig. 10.53 Energy band diagram and density of state dispersion curves of a quantum well of **a** interband transition, and **b** inter-subband transition

same in-plane dispersion. This allows the scattering of an electron from the upper to lower state by any transition, elastic or inelastic, as long as it provides the necessary momentum exchange. Therefore, the electron lifetime is very short, of the order of picoseconds, and is dominated by non-radiative recombination process.

In 1971, R. Kazarinov and R. Suris proposed a unipolar inter-subband laser structure where electrons in the QW of a superlattice made transitions between subbands for light amplification [21]. In their designs, a strong electric field is applied to the superlattice such that electrons tunnel from the ground state of one QW to the excited state of the neighboring well, emitting a photon in the process via the photon-assisted tunneling process. The electrons are then non-radiatively relaxed to the ground state and tunneled into the excited state of the next stage ready for photon generation again. The process repeats sequentially for many stages to generate optical gain of the laser. This work prompted a great deal of interests, although mostly theoretical due to the lack of a proper thin film growth technique. Finally, in 1994, the first unipolar inter-subband laser, called the *quantum cascade* (QC) *laser*, was demonstrated at Bell Laboratories, thanks to the convergence of two key techniques: molecular beam epitaxy (MBE) and band-structure engineering. Using quantum wells, superlattices, and digitally composition graded alloys as building blocks, the band-structure engineering method allows the design of devices with arbitrary shape of energy band diagrams. Thus, QC laser structures can be designed and their properties tailored for specific needs. To construct such a complex device structure consisting of hundreds of thin-layer heterojunctions, MBE with its ability to grow atomically abrupt heterojunctions incorporating precisely designed composition and doping profiles, is the epitaxial method of choice. Thus far, QC lasers with emission wavelengths from mid-infrared (~3.5 μm) to terahertz (≥200 μm) have been demonstrated.

10.6.1 QC Laser Active Region Designs

The basic structure of a QC laser consists of a periodic arrangement of cascade periods, each one consisting a number of QWs and barriers with a complex heterostructure potential profile. Figure 10.54a shows a portion of the conduction band energy diagram of the first QC laser reported in 1994. The laser structure was grown by MBE with the $Al_{0.48}In_{0.52}As$-$Ga_{0.47}In_{0.53}As$ heterojunction material system lattice-matched to InP substrate. The injected electrons make diagonal radiative transitions between subbands of two neighboring QWs and are recycled from stage to stage, contributing in each stage to the gain and photon emission. Therefore, each electron injected above lasing threshold generates M photons leading to an optical power proportional to M, where M is the number of cascade stages. However, in order to achieve lasing, population inversion in the structure is a necessary condition. Unlike interband transition lasers, where carrier recombination involves band-to-band transitions, optical transitions in inter-subband lasers are taking place between discrete subbands as in a three-level laser system. Therefore, a careful design of scattering times of each subband and between subbands is necessary.

As noted in Fig. 10.54a, each stage of the QC laser is divided into a gain or active region and an injection/relaxation region. The gain region is the section where a population inversion between the two levels of the laser transition takes place. Following the gain region is an injection/relaxation region to raise the energy of the electron state, thus facilitating injection by resonant tunneling in the next period. This function is achieved by digitally alternating QWs and barriers with changing duty cycle to create a section of material with effective increasing bandgap. The

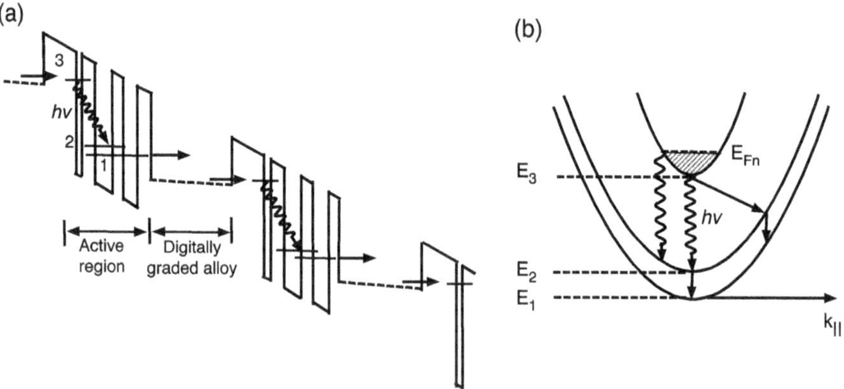

Fig. 10.54 **a** Conduction band energy diagram of a portion of the QC laser based on a diagonal transition between states 3 and 2. The dashed lines are the effective conduction band edges of the digitally graded electron-injecting region. Electrons are injected into the energy level state 3 of the active region. **b** E–k curves of the energy subbands from three QWs. The bottoms of these subbands correspond to the energy levels 1, 2, and 3 indicated in (**a**). Reprinted with permission from [22], copyright AAAS

middle part of the injection region is also doped to prevent the formation of strong space-charge domains and provide electron charge injection to the next stage.

Usually the active region contains two-to-three QWs to form a ladder of three states such that electrons are injected in the third state and the population inversion is maintained between the third and second states. Assuming perfect injection efficiency in the third state, the population inversion condition is simply $\tau_{32} > \tau_2$; i.e., the total lower state lifetime τ_2 is shorter than the electron scattering time from the third state to the second state τ_{32}. One way to shorten the transition lifetime between two states is by increasing the overlap of wave functions. In the first working QC laser reported in 1994, a diagonal transition between states 3 and 2 of two adjacent QWs was used where the reduced wave function overlap enhances the scattering time from state 3 to state 2. On the other hand, strong reduction of the relaxation time τ_{21} is obtained in the heavily overlapped and closely spaced states 2 and 1 of two adjacent QWs. Furthermore the separation between these two subbands is designed to be equal to the optical phonon energy. Thus the strong inelastic relaxation by means of resonant optical phonon emission between these two states will reduce the lifetime τ_{21} to sub-picosecond. Finally, the tunneling escape time out of state 1 is extremely short, further assisting the population inversion.

In addition to the diagonal transition active region design, laser structures based on a vertical transition (i.e., with the initial and final states placed in the same QW) for lower threshold current have also been developed. The main challenge in designing such QC lasers is to suppress electrons tunneling out of the state 3 into the continuum, such that the electron population in state 3 exceeds that of state 2. The population inversion is achieved if the relaxation time τ_{32} for the transition from state 3 to state 2 exceeds the electron lifetime τ_2 in state 2. To efficiently reduce the relaxation time τ_{21} by means of resonant optical phonon emission, the energy separating states 2 and 1 are designed to be equal to the optical phonon energy. To prevent accumulation of electrons in the lower state 1, a downstream superlattice injector is used where state 1 is aligned with the miniband of the superlattice (Fig. 10.55). The injector superlattice is also designed so that a minigap faces the upper-state 3, thus suppressing electron

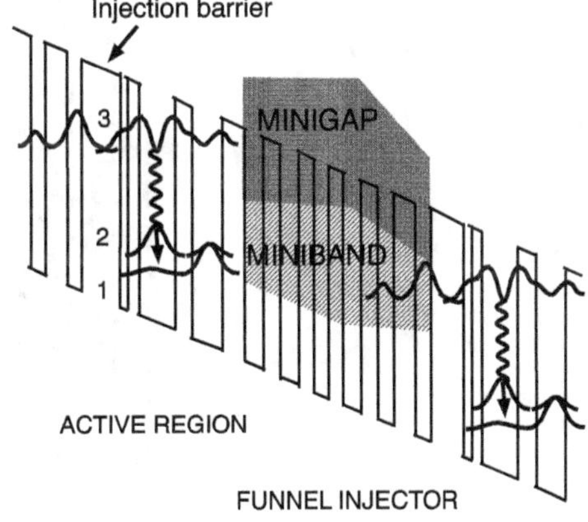

Fig. 10.55 Conduction band energy diagram of a portion of the QC laser based on a vertical transition between states 3 and 2 of the same QW. Electrons are injected into the energy level state 3 of the active region. Reprinted with permission from [23], copyright AIP Publishing

escape by tunneling. As a result, electron population builds up sufficiently to allow laser action at a reasonably low threshold current density.

High-performance unipolar inter-subband QC lasers with QW active regions have been demonstrated with emission wavelengths up to 11 μm. But further extension of the laser emission to even longer wavelength using the same active regions is difficult. Since inter-subband transitions are characterized by decreasing radiative efficiency and optical waveguide losses ($\propto\lambda^2$) with decreasing transition energy, conventional inter-subband QC laser structures show prohibitively high threshold current densities at longer wavelength. This problem can be solved by replacing the QW in the active region with semiconductor superlattices (SLs) that consist of a periodic stack of nanometer-thick layers of two materials (QWs and barriers). The advantages of SL QC lasers based on optical transitions between conduction minibands are the intrinsic population inversion associated with the large interminiband-to-intraminiband relaxation time ratio and the high oscillation strength of the laser transition at the SL Brillouin zone boundary. In addition, due to the wide energy minibands, SL QC lasers can be driven with very large current densities without compromising laser performance.

A schematic band diagram of the SL QC laser with graded (chirped) SL is shown in Fig. 10.56. In the SL active region, the well thicknesses are gradually decreased in the direction of electron motion. At zero bias, the states of the QWs are localized because the variation in the thickness of adjacent layers is such that the corresponding energy levels of the isolated wells are out of resonance. Minibands are formed when an electric field of the appropriate magnitude and polarity is applied to bring these states into resonance. The energy levels in each active SL are grouped together in two well-defined minibands, with the intermediate injector region bridging them together across the cascade stages. Laser action takes place at the edge of the minigap between minibands 2 and 1.

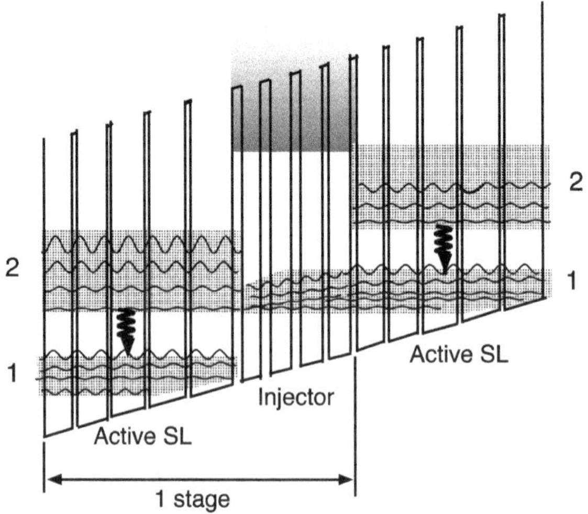

Fig. 10.56 Conduction band energy profiles of a portion of the chirped superlattice QC laser active regions and injector regions. The shaded areas indicate the extent of the electron minibands. Laser transitions are indicated by wavy arrows between level 2 and level 1. Reprinted with permission from [24], copyright AIP Publishing

As mentioned above, the SL QC laser exhibits attractive features of very large oscillator strength and a favorable ratio of lifetimes. However, when a strong electric field is applied on a periodic SL, it will break the miniband of the SL into a set of localized states, losing all the favorable features of the extended states. Alternatively the bound-to-continuum active region, which can be seen as an evolution of the SL design, is developed for maintaining a high population inversion and low threshold current densities [25]. Because of the localization of the upper state, the oscillator strength is spread between the upper-state and a number of lower-states in the miniband. This active region design offers many interesting features, including high characteristic temperature T_0 (as high as 190 K) and long-wavelength (16 μm) operation, including terahertz generation.

10.6.2 Threshold Current and Waveguide Designs

(a) Threshold current density

The lasing condition of a QC laser may be analyzed by considering a three-level system of the active region in a rate equation approach. As shown in Fig. 10.57, electrons are injected in state 3 from the ground state of the injector of the previous stage at a rate of J/q, where J is the current density (A/cm^2). Electrons may scatter from state 3 to states 2 and 1 with rates $1/\tau_{32}$ and $1/\tau_{31}$, respectively. The lifetime of state 3 may include an escape process into the continuum with a lifetime τ_{esc} and $1/\tau_3 = 1/\tau_{32} + 1/\tau_{31} + 1/\tau_{esc}$. Also included are electrons that are thermally activated to state 2 with an equilibrium population n_2^{th}. Its value is determined by n_g, the sheet doping density of the injector, Δ, the energy difference between the state 2 and the Fermi level of the injector and is expressed as

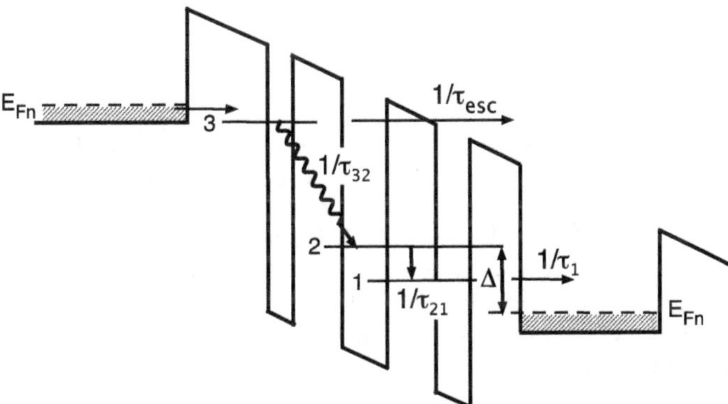

Fig. 10.57 Schematic diagram of a QC laser with various tunneling and relaxation processes used in the rate equations

$$n_2^{\text{th}} = n_g \exp(-\Delta/kT) \tag{10.67}$$

The rate of change of the population of each state may be formulated as

$$\frac{\mathrm{d}n_3}{\mathrm{d}t} = \frac{J}{q} - \frac{n_3}{\tau_3} - Ng(n_3 - n_2) \tag{10.68a}$$

$$\frac{\mathrm{d}n_2}{\mathrm{d}t} = \frac{n_3}{\tau_{32}} + Ng(n_3 - n_2) - \frac{n_2 - n_2^{\text{th}}}{\tau_2} \tag{10.68b}$$

$$\frac{\mathrm{d}N}{\mathrm{d}t} = \frac{c}{\bar{n}}\left\{[Ng(n_3 - n_2) - \alpha_T] + \beta\frac{n_3}{\tau_{\text{sp}}}\right\} \tag{10.69}$$

In above equations, n_i is the electron sheet density per period in state i, N is the photon flux per period and unit active region width, $\bar{n}$ is the refractive index, g is the gain coefficient, α_T is the total modal loss, β is the fraction of the spontaneous emission coupled into the laser mode, and τ_{sp} is the spontaneous emission lifetime. The gain for inter-subband transition between states 3 and 2 per *one stage* is expressed as

$$g = \Gamma\frac{4\pi q^2}{\epsilon_0\bar{n}\lambda}\frac{z_{32}^2}{(2\gamma_{32})L_p} \tag{10.70}$$

where z_{32} is the dipole matrix element, λ is the wavelength, $2\gamma_{32}$ is the linewidth or the full-width at half-maximum of the transition, Γ is the total confinement factor, and L_p is the length of one period.

Below threshold, under steady-state conditions, we can set $\mathrm{d}n/\mathrm{d}t = 0$ and $N = 0$. Then (10.68a) becomes

$$n_3 = J\frac{\tau_3}{q} \tag{10.71}$$

Inserting this result in the second rate equation, one obtains

$$\Delta n = n_3 - n_2 = J\frac{\tau_3}{q}\left(1 - \frac{\tau_2}{\tau_{32}}\right) - n_2^{\text{th}} = J\frac{\tau_{\text{eff}}}{q} - n_2^{\text{th}} \tag{10.72}$$

where $\tau_{\text{eff}} = \tau_3(1 - \tau_2/\tau_{32})$ is the effective lifetime. This equation links the population inversion Δn to the electric pumping, J. The population inversion can only exist if $\tau_{\text{eff}} > 0$, i.e., if $\tau_{32} > \tau_2$. At the threshold current density, the modal gain must equal to the total loss, i.e., $g\Delta n = \alpha_T$. Thus, the threshold current density per each stage of the QC laser is

$$J_{\text{th}} = \frac{q}{\tau_{\text{eff}}}\left(\frac{\alpha_T}{g} + n_2^{\text{th}}\right) = \frac{q}{\tau_{\text{eff}}}\left[\frac{\epsilon_0\bar{n}\lambda(2\gamma_{32})L_p\alpha_T}{4\pi q^2\Gamma z_{32}^2} + n_2^{\text{th}}\right] \tag{10.73}$$

Above threshold, the gain is fixed and N increases linearly similar to diode lasers described in Fig. 10.10. From (10.73), it is clear that a low threshold and high-efficiency QC laser will be obtained in the laser structure with a large ratio of τ_{32}/τ_2, a narrow transition linewidth γ_{32}, and a long upper-state lifetime τ_3.

(b) Mid-infrared waveguide designs

In (10.73), an important contribution to the laser threshold is the total loss α_T experienced by the light traveling in the waveguide. The major losses are the mirror loss α_m, and the waveguide loss α_w. The mirror loss can be reduced using high-reflection coating at the mirror facets. The major source of waveguide losses is free-carrier absorption in the doped semiconductor regions and the metallic contact layers. While the influence of the metallic layers can be suppressed, the free-carrier absorption loss in doped semiconductors is unavoidable. This semiconductor loss follows a simple Drude model and increases approximately quadratically with the wavelength in mid-infrared range. In shorter wavelength QC lasers, such as the first demonstrated QC laser ($\lambda = 4.3$ μm), the dielectric waveguide is a simple choice if the material system in use provides sufficiently large refractive index variations to build the waveguide. In such cases, the high-index core consisting of the active regions is sandwiched between low-index cladding layers leading to a relatively symmetrical optical intensity profile in the waveguide. Nevertheless, in order to decouple the mid-infrared wavelength light from the surface plasmon at the semiconductor/metal interface, the minimum thickness of the wave-guiding layer needed is of the order of a wavelength. This requirement is easily fulfilled for the bottom-cladding layer by using a suitable substrate material as a part of the cladding layer. For the top cladding layer, however, this requires a prohibitively long material growth time for QC lasers as the emission wavelengths increase beyond 15 μm. To limit the thickness of the top cladding layer, a plasmon-enhanced waveguide design has been developed. Figure 10.58 shows the optical mode and refractive index profiles of such an AlInAs/GaInAs/InP QC laser structure. At the top of the laser structure, a heavily doped (7×10^{18} cm^{-3}) GaInAs:Si layer is added to suppress the coupling between the fundamental waveguide mode and

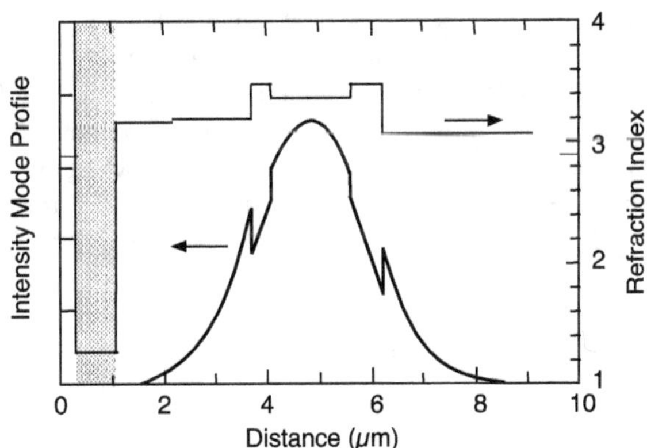

Fig. 10.58 Mode intensity profile and the refractive index profile as a function of layer thickness of a QC laser incorporating a plasmon-enhanced heavily doped semiconductor confinement layer. The surface is located at zero distance. Reprinted with permission from [26], copyright AIP Publishing

the surface plasmon mode propagating along the metal contact–semiconductor interface. This heavily doped GaInAs layer has a plasma frequency (9.4 μm) approaches but does not exceed that of the waveguide mode (8.4 μm) of the laser. Near the plasma frequency, the anomalous dielectric dispersion strongly depresses the refractive index of the layer at the laser wavelength from ~3.5 (value of an undoped layer) to ~1.2. The insertion of this low refractive index top layer beneath the metal contact completely decouples the waveguide mode from the interface plasmon and increases the confinement factor of the mode by pushing the mode away from the metal contact.

For a long-wavelength (≥15 μm) QC laser, rather than decoupling the guided wave and the interface plasmons, radically different waveguide designs utilizing the surface plasmons have been developed. Since metals and doped semiconductors behave like plasma in mid-infrared frequencies, the electromagnetic surface waves (interface plasmons) at a metal–semiconductor interface can be used as the main guiding mechanism. The metal layer is now deposited directly above the active material of the laser and used as the wave-guiding interface. The obvious advantage of this approach is that no additional cladding layer is needed since the amplitude of the optical wave decreases exponentially in the direction normal to the metal–semiconductor interface. In this type of waveguide, the modal losses are strongly dependent on the dielectric constants of the metal and semiconductor used. The loss of the plasmon waveguide at the metal–dielectric interface can be expressed as

$$\alpha = \frac{4\pi \bar{n}_m \bar{n}_d^3}{\kappa_m^3 \lambda} \tag{10.74}$$

where $\bar{n}$ and κ are the real and imaginary parts of the dielectric constant of the metal ($\sqrt{\epsilon_m} = \bar{n}_m + i\kappa_m$) and dielectrics ($\sqrt{\epsilon_d} = \bar{n}_d$). Therefore, it is apparent that the losses at the interface are inversely proportional to the wavelength λ and can be minimized by choosing metals having a refractive index with a large imaginary component. Palladium and gold are the most suitable metals for wavelengths around 10 μm and ≥15 μm QC laser applications, respectively. The normalized mode intensity depth profile of the surface plasmon waveguide for a QC laser with an emission wavelength of 11 μm is shown in Fig. 10.59. It is noted that the confinement factor is more strongly enhanced (≥70%) than that of a regular dielectric waveguide (~40%).

To further improve the waveguide design for the realization of QC lasers at even longer wavelength in the terahertz range, a double-surface plasmon waveguide was developed. In this waveguide structure, the lightly doped GaInAs lower cladding layer is replaced with a heavily doped layer acting as a bottom metal. The waveguide structure is essentially the same as a microwave metal waveguide, and the optical confinement is close to unity as shown in Fig. 10.59b.

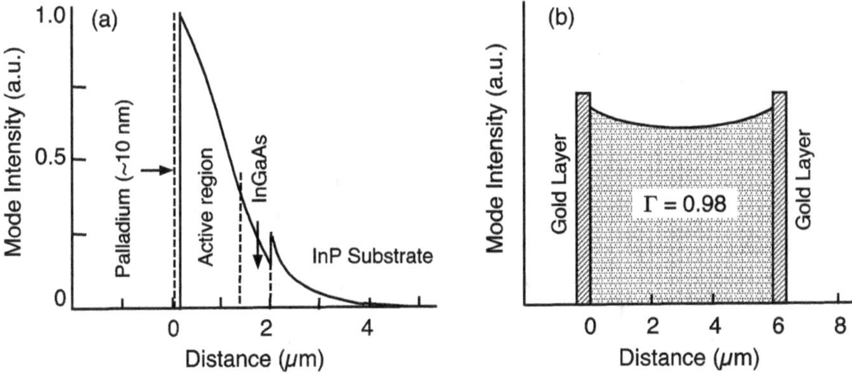

Fig. 10.59 Mode intensity profile as a function of layer thickness of a QC laser incorporating **a** a surface plasmon waveguide with $\lambda = 11\ \mu m$. Reprinted with permission from [27], copyright Optical Society of America; and **b** a double-surface plasmon waveguide with $\lambda = 21\ \mu m$. The surface is located at zero distance. Reprinted with permissions from [28], copyright AIP Publishing

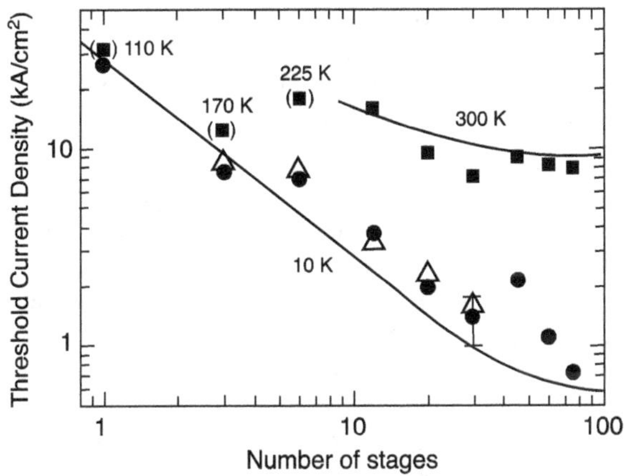

Fig. 10.60 Threshold current density as a function of the number of active region stages measured at 10 K in pulse mode (solid circles) and in cw mode (open triangles), and in pulse mode at room temperature (solid squares). Lasers with $M \leq 6$ stages did not lase at room temperature. The threshold current density at the highest operating temperature is displayed instead. Reprinted with permission from [29], copyright AIP Publishing

10.6.3 Quantum Cascade Laser Operation Characteristics

QC lasers are unique among semiconductor lasers in their operation principles which are based on inter-subband transitions and on the cascading scheme, where a large number (typically $16 \leq M \leq 35$) active regions are stacked upon each other separated by electron injection regions. Under an appropriate applied bias voltage, the localized states of M active regions merge into one continuous cascade of optical transitions, allowing each electron to create M photons. Consequently, QC lasers display an external quantum efficiency significantly larger than unity, which is unattainable for conventional interband lasers.

One important consequence of increasing the number of stages for waveguide consideration, which in turn increases the total active layer thickness, is the increase in confinement factor Γ with increasing M. For example, the confinement factors of identical QC lasers incorporated with different numbers of active regions of $M = 1$, 3, 6, 12, 20, 30, 45, 60, and 75 are calculated. These lasers have three-well vertical transition design of the active region emitting at ~8 μm wavelength. The confinement factor grows approximately linearly with the number of stages up to $M \approx 30$ ($\Gamma_{30} = 0.49$) and saturates for higher M ($\Gamma_{75} = 0.81$). The increased confinement factor for lasers with larger M leads to a larger modal gain coefficient and, thus, decreased laser threshold current.

Figure 10.60 shows the measured and calculated J_{th} as a function of M measured at room temperature and low temperatures (10 K). A low-temperature threshold current density $J_{\mathrm{th}} \leq 2$ kA/cm^2 is consistently achieved for $M \geq 15$. Devices with $M \geq 12$ stages displayed laser action in pulse mode at and above room temperature. For smaller M, J_{th} rises rapidly, at a rate proportional to $1/M$, as a direct consequence of the reduced confinement factor. In addition, lasers with $M \leq 6$ stages did not show laser action at room temperature and their highest operation temperatures are 225, 170, and 110 K for samples with six, three, and one stages, respectively.

In conventional interband lasers, the turn-on voltage V_{on} of the I–V characteristics is fixed and depends on the energy gap of the semiconductor material used. For QC lasers, the turn-on voltage of the device is not fixed but depends on the number of stages M. At zero bias, due to the highly localized subband states in each QW, the QC laser structure is highly resistive. Only when a suitable electric field is reached, merging all localized states into one continuous cascade of transitions allowing efficient injection and carrier transport between stages can large currents then flow through the structure. The energy drop per stage is determined by the sum of transition energies between subband states ($E_{32} + E_{21}$) in the active region, and the voltage drop across the entire M-stage cascade is given by

$$V_S(M) = \frac{E_{32} + E_{21}}{q} \cdot M \tag{10.75}$$

The turn-on voltage of the QC laser is simply expressed as $V_{\mathrm{on}} = V_S + V_{\mathrm{os}}$, where V_{os} is the offset voltage associated with the contact resistance and lightly doped waveguide layers. Thus the operation voltage of the device is approximately proportional to M which is a consequence of cascading. Figure 10.61 shows the measured I-V characteristics for all laser samples shown in Fig. 10.60.

Using the voltage drop across the entire M-stage cascade $V_S(M)$, the differential series resistance $\rho = \partial V/\partial J$, and the threshold current density J_{th}, the voltage at laser threshold V_{th} can be calculated as

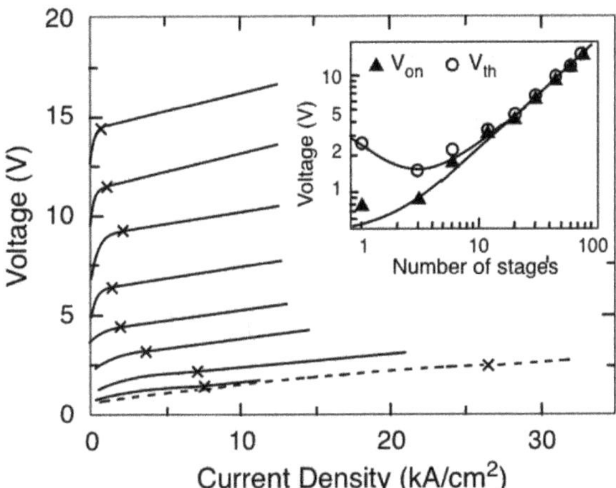

Fig. 10.61 Current–voltage characteristics of QC laser devices having different stages M between 1 and 75 with the uppermost trace corresponding to $M = 75$. The trace of $M = 1$ is drawn as a dashed line. The cross symbol on each curve denotes the threshold voltage, V_{th}. Inset: V_{on} (solid triangles) and V_{th} (open circles) as a function of M. Reprinted with permission from [29], copyright AIP Publishing

$$V_{\mathrm{th}}(M) = \frac{E_{32} + E_{21}}{q} \cdot M + V_{\mathrm{os}} + \frac{\partial V}{\partial J} \cdot J_{\mathrm{th}}(M) \tag{10.76}$$

The last term of the above equation accounts for the bias acquired above V_{on} by the current that is needed for the gain to overcome the losses. As shown in the inset of Fig. 10.61, V_{on} monotonically deceases with decreasing M. V_{th} follows this general trend except for devices of very low M (≤ 3) due to the associated high threshold current.

From the rate equations, by again setting the time derivatives to zero and differentiating over J, one obtains the slop efficiency per facet for the whole laser stack:

$$\frac{\mathrm{d}P}{\mathrm{d}I}(M) = \frac{Mh\nu}{2q}\left(\frac{\alpha_m}{\alpha_i + \alpha_m}\right)\left(1 - \frac{\tau_2}{\tau_{32}}\right) \tag{10.77}$$

where $h\nu$ is the photon energy. It is apparent that a large number of stages will result in a large slope efficiency and high-optical output. Figure 10.62 shows the

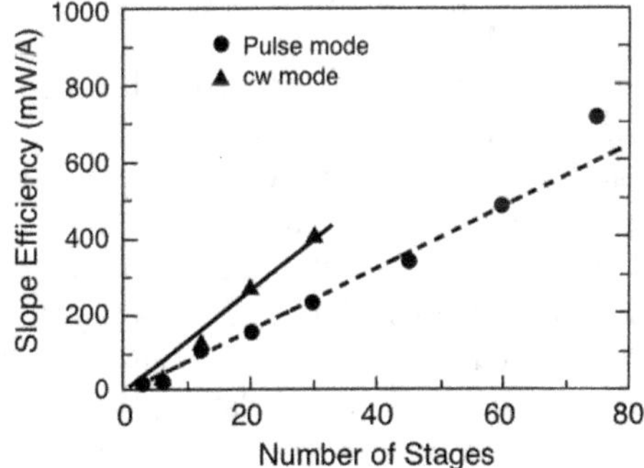

Fig. 10.62 Measured slope efficiency per facet of QC lasers as a function of M obtained under cw mode (triangles) and pulse mode (circles). The average slope efficiency per facet per stage of these samples is about 8 mW/A. Reprinted with permission from [29], copyright AIP Publishing

measured slope efficiency at cryogenic temperature for those samples with different M mentioned above.

From the slope efficiency the total external differential quantum efficiency per facet is calculated as

$$\eta_d(M) = \frac{q}{h\nu}\frac{\partial P}{\partial I}(M) \tag{10.78}$$

Using the data from Fig. 10.62, $\eta_d \sim 8.5\%$ per stage is obtained. Consequently, a value of $\eta_d(M)$ larger than unity is obtainable for QC laser structures with $M \geq 12$ stages.

10.6.4 Interband Cascade Lasers—Type-II QC Laser

The quantum cascade laser, which employs unipolar inter-subband optical transitions in the active region, is the first laser technology to achieve room-temperature cw operation beyond the MWIR (3–5 μm) milestone. However, for QCLs using the InGaAs/InAlAs material system, the shortest wavelength for which room-temperature cw lasing has been attained is 3.8 μm. This limitation is due to the excessive strain required to maintain sufficient conduction band offset in the active region. Other QCLs and type-I QW lasers based on different III–V material systems have yet to achieve room-temperature cw operation in the MWIR spectral range. On the other hand, room-temperature pulse operation has been achieved in W-type QW lasers with the type-II band alignment for the MWIR emission. Thus, it is possible to combine the interband W-type laser active region with the multistage QC structure for lower threshold and higher quantum efficiency such that room-temperature cw lasing can be realized. The resulting device design is a type-II bipolar QCL with interband optical transitions in the active region, or an *interband cascade laser* (ICL).

The single-stage band structure of an InAs-based ICL is illustrated in Fig. 10.63. Similar to QCLs, multiple stages of the InAs-based type-II W-type QW active regions are connected in series in a staircase ICL design. The injected electrons from the previous stage on the left tunnel through the AlSb barrier into the conduction band of the InAs QW and recombine with holes in the valence band of the neighboring GaInSb QW. The location of the electron subband energy in the W-type active region is positioned inside the bandgap of the GaSb layer to prevent electrons from tunneling into the following GaSb layer and force the electrons to recombine with holes in the GaInSb layer. When an electron transitions into the valence band, it gives up energy in the form of photon emission followed by tunneling through the AlSb barrier into the GaSb hole subband of the hole injector region. Separating the hole injector and electron injector regions is an AlSb/InAs heterostructure with a type-II interface. Under zero bias, the conduction subbands of InAs QW on the electron injector side and valence subbands of GaSb QW on the hole injector side are separated by a small energy. When properly biased, these states are energetically aligned to form a

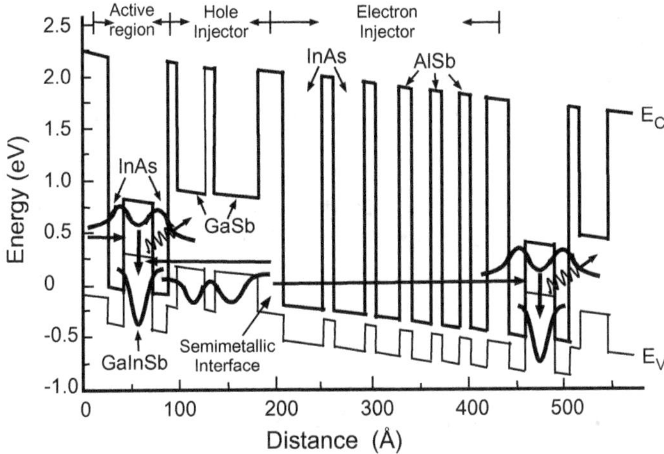

Fig. 10.63 Conduction band and valence band profiles for one stage of an ICL under bias, along with the active region of the succeeding stage. Some of the important electron and hole wavefunctions are also plotted. Reprinted with permission from [30], copyright IEEE

semimetal interface such that electrons in the valence band of GaSb can be scattered elastically back to the conduction band of InAs in the electron injector region which consists of InAs/Al(In)Sb chirped superlattices. Thus, similar to type-I QCL, an electron recycling through additional stages in ICL increases the injection quantum efficiency to produce multiple photons.

While the cascading concept of ICL was motivated in part by the earlier development of the inter-subband QCL, the intricate physical principles governing QCL and ICL operation are not completely equivalent. The major difference between ICL and QCL results from their vastly different carrier lifetimes. The short effective lifetime of carriers in the upper level of the QW in an inter-subband QCL is limited by a combination of optical phonon and interface roughness scattering and has an empirical value of $\leq$1 ps. In contrast, the interband transition in the QW of the ICL has a rather long carrier lifetime of ~1 ns. The relatively long carrier lifetime in the type-II QW of ICL leads to two orders of magnitude larger modal gain per stage than a QCL. The consequence is that it typically requires 30 or more stages in a QCL, and less than 10 stages in an ICL, to achieve room-temperature laser operation with a low threshold current density. For $\lambda \approx 4.6$ μm, the lowest reported threshold current densities at room temperature for QCLs with 30-40 stages remain at ~800 A/cm^2 and increase at shorter wavelengths. By contrast, the lowest room-temperature threshold current density reached for a 10 stage ICL emitting at ~3.6 μm is <100 A/cm^2. Due to the fewer stages used, the savings in bias voltage combined with the low threshold current density make the ICL an ideal MWIR source (3–4 μm) for applications that require low power dissipation.

The above discussions concerned only the drawback of the necessity of using a large number of stages in QCL to achieve low threshold current densities. However, the increased photon generation rate through increasing the number of stages in QCL leads to increased quantum efficiency and higher *L–I* curve slope efficiency above threshold. In addition, the threshold current density of QCLs has weak temperature dependence. For example, the typical characteristic temperatures T_0 just above room temperature of ICL and QCL are ~46 K and ~200 K, respectively. This feature is unique to the inter-subband transition of the QCL structure where the two subbands of the laser transition are nearly parallel and are insensitive to the thermal broadening of the electron distribution in the excited state (Fig. 10.54b). These features further enhance the ability of QCLs to produce high cw output power at wavelengths beyond 4 μm.

10.7 Quantum-Well Infrared Photodetectors (QWIPs)

The interband transition energy of intrinsic infrared photodetectors is determined by the bandgap energy of the semiconductor. Thus the spectral response of the detectors can be tuned by selecting the proper bandgap of the photosensitive material. This means detection of very long-wavelength radiation ($\geq$12 μm) requires small bandgaps down to 62 meV. However, it is well known that narrow bandgap materials (e.g., $Hg_xCd_{1-x}Te$) are more difficult to prepare and process than larger bandgap semiconductors such as GaAs. Alternatively, optical absorption transitions in semiconductor QWs can take place in the form of unipolar inter-subband transitions similar to quantum cascade lasers described in Sect. 10.5. In a QW, the inter-subband transitions can occur between confined states of the same band (conduction or valence) and the smallest transition energy reduces toward zero as the well width is increased. This unique feature enables the fabrication of long-wavelength photodetectors over a wide frequency range using the same heterostructure materials. In 1985, L. C. West and S. J. Eglash first observed the strong inter-subband absorption near 8–10 μm in GaAs/AlGaAs MQWs [31]. Subsequently B. F. Levine et al. demonstrated the first GaAs/AlGaAs *quantum-well infrared photodetector* (QWIP) at Bell Laboratories in 1987 [32]. Currently, highly uniform, large area QWIP arrays at low-cost are reproducibly produced using advanced epitaxial growth techniques such as MBE. Furthermore, the use of III–V materials offers the potential for monolithic integration of QWIP arrays with high-speed GaAs multiplexers and other electronics. As a result, these features allow them to compete successfully with the conventional IR detectors for several important applications, including infrared imaging.

10.7.1 Operation Principles of QWIPs

QWIPs utilize the photoexcitation of electrons (holes) between the ground and first excited state in the conduction (valence) band of the QW to generate a photocurrent. Figure 10.64 shows the conduction band diagram of a *bound-to-bound* (BTB) *n*-

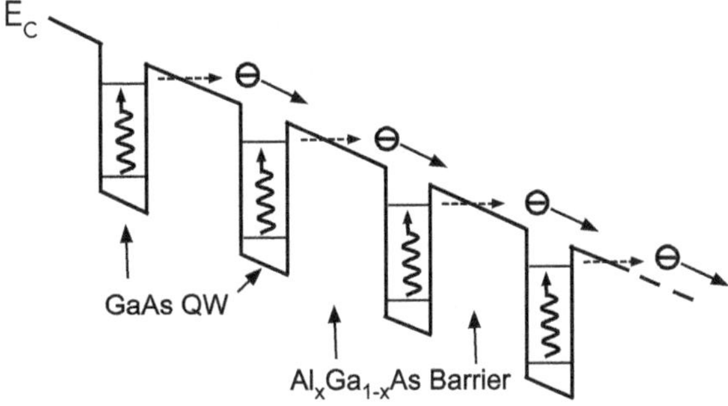

Fig. 10.64 Schematic conduction band diagram of a bound-to-bound *n*-type QWIP under a finite bias

type QWIP under a finite bias. Only photons having energies corresponding to the energy separation between the two states are absorbed, resulting in a detector with a sharp absorption spectrum. By varying the QW width, QW dopant, and barrier composition, and by using different heterostructure materials, the wavelength of the peak response can be continuously tailored beyond 15 μm. Because the photon energy at far infrared wavelength is insufficient to create carriers across the bandgap, carriers are introduced into QWs by doping inside the wells. In addition, stacked identical QW layers repeating many times (10–100) are used to enhance the photon absorption efficiency.

For BTB-type QWIP structures containing two bound states in quantum wells, the generated photoexcited carriers have to tunnel through the barrier for photocurrent generation. By reducing the quantum-well width in a QWIP, the excited state can be pushed upward to energy equal to the barrier height (quasi-bound state) and even into the continuum states, forming *bound-to-quasi-bound* (BTQB) and *bound-to-continuum* (BTC) transitions, respectively. The major advantage of these kinds of QWIPs is that the photoexcited electrons can escape from the quantum well without tunneling through the energy barrier. Thus, the bias voltage required to efficiently extract photoelectrons from the well can be drastically reduced, significantly reducing the dark current. In addition, the barrier thickness can be substantially increased, thereby further reducing the ground-state sequential tunneling by many orders of magnitude. As we will see later in this section, the optimum quantum-well shape design is that with the first excited state in resonance with the top of the barrier, i.e., the quasi-bound transition QWIP.

However, it should be noted that the quantum mechanical selection rules for the inter-subband transition require a component of the optical electric field along the growth (z) direction (i.e., TM polarization). For TE polarized light (in the x–y plane of quantum wells), the absorption is negligible such that the normal incident radiation will not be absorbed. To overcome this problem, the 45° edge facet light coupling geometry is commonly used as shown in Fig. 10.65a. Although this geometry wastes one half of the incident light, it is simple and convenient to use for characterizing the device performance. The majority of imaging applications of QWIPs require large area 2D arrays where the edge facet geometry is not suitable. To make a backside-illuminated *n*-type QWIP detector array, a grating-coupled design must be employed

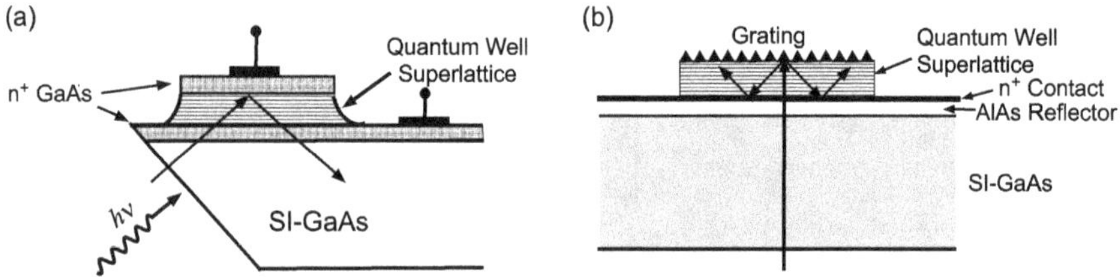

Fig. 10.65 Schematic cross-sectional views of **a** a 45° edge-coupled QWIP and **b** a two-dimensional grating-coupled QWIP

as shown in Fig. 10.65b. The fabricated gratings, either etched into the top contact or in the substrate, can refract normally incident IR radiation so that a component of the radiation electric field is parallel to the growth direction. On the other hand, for *p*-type QWIPs, there is no such selection rule due to the valence band mixing effects. Therefore, all polarizations of normal incident radiation are absorbed in *p*-type QWIPs.

10.7.2 Dark Current

Since the dark current contributes to the noise of the photodetector, it is vital to understand the dark current for the optimization of QWIPs. The dark current consists of two major components: thermionic emission and tunneling. Carriers that are thermally excited or scattered into states above the top of the well form the thermionic emission dark current. Carriers can tunnel into a neighboring well from densely populated states near the bottom of the well. Also, carriers can be thermally excited into states lying high in the confined states and tunnel through the triangular barrier into states above the top of the well to form the tunneling current as in the bound-to-bound case. To derive the dark current (I_d), the processes controlling the dark current paths of a simple one-well structure are modeled in Fig. 10.66.

The emission of electrons from the well contributes to the dark current and is denoted as I_{em}. This current tends to lower the electron density in the well and must be balanced by trapping or capture of electrons in the barrier into the well. The capture

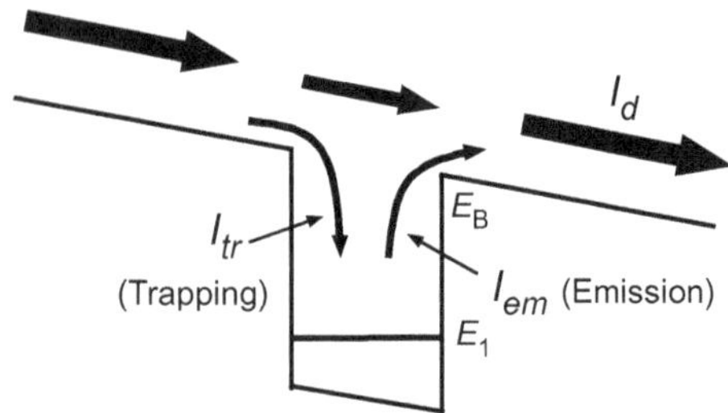

Fig. 10.66 Schematic representation of the processes controlling the dark current in a single well of a biased QWIP

process forms a trapping current $I_{\text{tr}} = p_c I_d$, where p_c is the capture probability. The capture probability p_c is given by

$$p_c = \frac{\tau_T}{\tau_L + \tau_T} \tag{10.79}$$

where τ_T is the transit time for an electron across one period of the MQW structure, and τ_L is the carrier lifetime associated with the trapping process. The electron lifetime of the relaxation process depends on the QW design. For a typical QWIP with bound-to-bound state transition, the relaxation time τ_L is ~1 ps. For the bound-to-continuum relaxation process, since the excited electron is extended in the barrier region, the relaxation should take more time (~5 ps). The transit time can be estimated by $\tau_T \approx L_p/\upsilon$, where L_P is the period of the MQW structure and υ is the electron drift velocity. For typical parameters of $\upsilon = 10^7$ cm/s and $L_P = 30$–50 nm, τ_T is estimated to be in the range of 0.3–0.5 ps. A capture probability of $p_c = 0.06$–0.1 is expected for the bound-to-continuum MQW structure.

For a dark current I_d flowing through the QWIP heterostructure, a fraction of the current $p_c I_d$ is trapped by the quantum well. The remaining current transmitted to the next stage is $(1 - p_c)I_d$. Under the steady-state condition, $I_{\text{tr}} = I_{\text{em}}$. Thus,

$$I_d = I_{\text{tr}} + (1 - p_c)I_d \tag{10.80}$$

and

$$I_d = I_{\text{em}}/p_c. \tag{10.81}$$

To calculate I_{em} associated with the escape of electrons from the ground state, we use the following expression:

$$I_{\text{em}} = qA\int_{E_1}^{\infty}[T(E, F)/\tau_S]\mathrm{d}n_{\text{2D}} \tag{10.82}$$

$$\mathrm{d}n_{\text{2D}} \equiv \left(m^*/\pi\hbar^2\right)f(E)\mathrm{d}E \tag{10.83}$$

$$f(E) = 1/\{1 + \exp[(E - E_F)/kT]\} \tag{10.84}$$

where E_1 is the ground-state energy, E is the total energy of an electron, F is the average field, $T(E, F)$ is the transmission coefficient, E_F is the Fermi energy, and τ_S is the scattering time to transfer electrons from the ground state to a state near or above the barrier. Since the two processes of carrier scattering out of the well and trapping (relaxation) into the well are inverse of each other, we take $\tau_S = \tau_L$. We then have

$$I_{\mathrm{em}} = \frac{qAN_{2\mathrm{D}}}{\tau_L} \tag{10.85}$$

$$N_{2\mathrm{D}} \equiv \int_{E_1}^{\infty} T(E, F)\mathrm{d}n_{2\mathrm{D}} \tag{10.86}$$

The total dark current can be calculated from $I_d = I_{\mathrm{em}}/p_c$ or

$$I_d = \frac{qAN_{2D}}{\tau_L}\left(\frac{\tau_L + \tau_T}{\tau_T}\right) \tag{10.87}$$

For QWIPs with the bound-to-continuum or bound-to-quasi-bound well design, $p_c \ll 1$ or $\tau_{\mathrm{T}} \ll \tau_{\mathrm{L}}$, the dark current becomes

$$I_d \approx \frac{qAN_{2D}}{\tau_T} = \frac{qA\upsilon N_{2D}}{L_p} = qA\upsilon \int_{E_1}^{\infty} \left(\frac{m^*}{\pi\hbar^2 L_p}\right) T(E, F) \frac{1}{1 + \exp[(E - E_F)/kT]} \mathrm{d}E \tag{10.88}$$

The field-dependent drift velocity is expressed as

$$\upsilon(F) = \frac{\mu F}{\sqrt{1 + (\mu F/\upsilon_{\mathrm{sat}})^2}} \tag{10.89}$$

where μ is the low-field mobility and υ_{sat} is the saturation drift velocity. We can further simplify the dark current expression by neglecting the tunneling contribution, i.e., setting $T(E, F) = 1$ for the bound-to-continuum transitions where $E > E_{\mathrm{B}}$ and E_{B} is the barrier height. The condition of $E > E_{\mathrm{B}}$ also implies $(E - E_{\mathrm{F}})/kT \gg 1$. The final expression of the dark current becomes

$$I_d = qA\upsilon\left(\frac{m^* kT}{\pi\hbar^2 L_p}\right) \exp[-(E_{\mathrm{B}} - E_1 - E_{\mathrm{F}})/kT] \tag{10.90}$$

10.7.3 Photoconductive Gain

The device operation of the QWIPs is similar to that of extrinsic semiconductor photoconductors except that the incident photons are only absorbed in discrete quantum wells. The photocurrent caused by inter-subband excitations in a QWIP can be estimated through the concept of photoconductivity gain, which is defined as the number of electrons flowing through the external circuit for each photon absorbed. For a simplified analysis of the photoconductivity gain, four approximations are made to

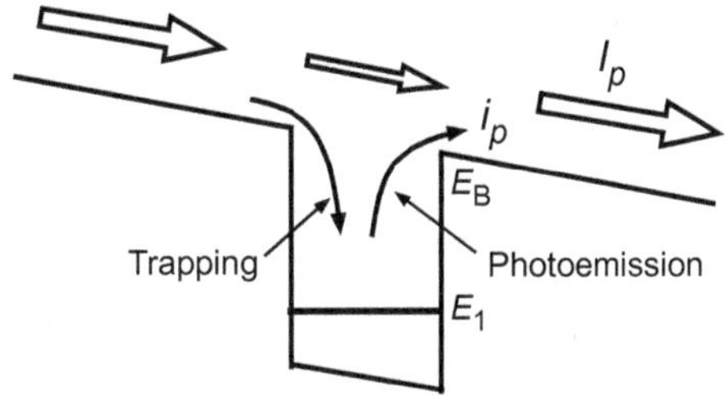

Fig. 10.67 Schematic representation of the photoconductive gain mechanism in a single well of a biased QWIP

define the physical regime: (a) The barriers are thick such that the inter-well tunneling is negligible. (b) The electron density in each well remains constant under applied bias. (c) The QW confines one bound state with the excited state either in resonance or very close to the top of the barrier. (d) The two-terminal metal contacts form ideal ohmic contacts. The physical process and the gain mechanism of a simple one-well structure are given in Fig. 10.67.

When infrared radiation is shone on the detector, there is a photemission of electrons from the well that contributes to the photocurrent I_p in the external circuit. The photoconductive gain is a result of the extra current injection from the contact necessary to balance the loss of electrons from the QW due to photoemission. Thus the total current I_p consists of contributions from the direct photoemission and the extra current injection. The photocurrent generated from one well is

$$i_p = qA\Phi\eta_w p_e = qA\Phi\left(\frac{\eta}{N_w}\right)p_e \tag{10.91}$$

where A is the detector illumination area, Φ is the incident photon number per unit time, η_w is the absorption quantum efficiency for one well, N_w is the total number of wells, and $\eta = N_w\eta_w$ is the total absorption quantum efficiency. The escape probability p_e for an excited electron from the well is given by

$$p_e \equiv \frac{\tau_R}{\tau_R + \tau_E} \tag{10.92}$$

where τ_E is the escape time and τ_R is the inter-subband relaxation time. For QWIPs with bound-to-continuum or bound-to-quasi-bound state transition case, once an electron is excited, it is already in the continuum and $\tau_E \approx 0$, $p_e \approx 1$.

For a photocurrent I_p injected from the contact, a fraction of the current $p_c I_p$ is captured by the quantum well, where p_c is the capture probability. The remaining current transmitted to the next stage is $(1 - p_c)I_p$. From current continuity, the injected current must equal to the sum of the transmitted current and photoemission current.

$$I_p = (1 - p_c)I_p + i_p \tag{10.93}$$

Therefore,

$$i_p = p_c I_p \tag{10.94}$$

Using (10.91), we get

$$I_p = qA\Phi\eta\left(\frac{p_e}{p_c N_w}\right) \equiv qA\Phi\eta g \tag{10.95}$$

and

$$g \equiv \frac{p_e}{p_c N_w} \tag{10.96}$$

is the photoconductive gain. Under the approximation $p_e \approx 1$ and $p_c \ll 1$ for the bound-to-continuum and bound-to-bound QWIP designs, the gain becomes $g \approx 1/(p_c N_w)$.

10.7.4 Detectivity

For single IR detectors, the most commonly used figure-of-merit is the detectivity, which is a normalized signal-to-noise ratio. The peak detectivity D^* is given by

$$D^* = R_p\sqrt{A\Delta f}/i_n \tag{10.97}$$

where R_p is the peak responsivity (A/W), A is the photosensitive area of the detector, i_n is the dark current noise ($A/\sqrt{\text{Hz}}$), and $\Delta f = 1$ Hz is the measurement bandwidth. For a photoconductor such as a QWIP with small QW capture probability $p_c \ll 1$, the noise current is given by

$$i_n = \sqrt{4qI_d g\Delta f} \tag{10.98}$$

The peak responsivity is defined as the ratio of the photocurrent I_p to the incident optical power given by

$$R_p = \frac{I_p}{Ah\nu\Phi} = \frac{q\Phi\eta g}{h\nu\Phi} = \left(\frac{q}{h\nu}\right)\eta g \tag{10.99}$$

where h is Planck's constant and ν is the frequency of the incident light. The absorption quantum efficiency of the QWIP can be expressed as

$$\eta = \frac{p_e}{2}\left[1 - \exp(-2\alpha l)\right] \tag{10.100}$$

where α is the absorption coefficient and $l = N_w L_p$ is the length of the multiple quantum-well structure. For most QWIP cases, $2\alpha l \ll 1$ and $\eta \approx p_e \alpha l$.

In QWIPs, the responsivity involves two processes: absorption of IR radiation by photoexcitation and transport of the photoexcited carriers to the output of the detector. The high absorption quantum efficiency, however, does not necessary result in high detector responsivity. To achieve a large photocurrent, the photoexcited electrons must escape the quantum wells efficiently with a higher value of g. For QWIPs, the photoconductive gain is given by

$$g = \frac{p_e}{p_c N_w} = \left(\frac{\tau_R}{\tau_R + \tau_E}\right)\left(\frac{\tau_L + \tau_T}{\tau_T}\right)\frac{1}{N_w} \tag{10.96}$$

where the first two fractions correspond to p_e and $1/p_c$, respectively. For the bound-to-continuum and bound-to-quasi-bound cases, the process of escape takes little time, $\tau_E \approx 0$ and $p_e \sim 1$. For the bound-to-bound case, the calculated and experimental value of τ_R is about 1 ps. To ensure $p_e \sim 1$, we must have $\tau_E \ll 1$ ps. This implies that the excited state in the bound-to-bound case must be close to the top of the barrier so that the tunneling escape time is much less than 1 ps.

10.7.5 Experimental Results

In the following, we compare the measured absorption coefficient spectra, responsivities, and detectivities of several $Al_xGa_{1-x}As/GaAs$ QWIPs with different quantum-well designs reported by Levine et al. The material parameters of these devices are listed in Table 10.1.

The IR absorption spectra for samples A–F were measured at room temperature and are shown in Fig. 10.68. The spectra of the bound-to-continuum QWIPs (samples A, B, and C) are much broader than the bound-to-bound QWIP (samples E) and bound-to-quasi-bound QWIP (samples F). The bound-to-continuum absorption can correspond to a transition from the bound state to numerous mini-subbands,

Table 10.1 Quantum-well structure parameters for five AlGaAs/GaAs QWIP samples A–F, including quantum-well width L_w, barrier width L_b, $Al_xGa_{1-x}As$ composition x, doping concentration N_d, and type of inter-subband transition: bound-to-continuum (BTC), bound-to-bound (BTB), and bound-to-quasi-continuum (BTQC) [33]

Sample	L_w (Å)	L_b (Å)	x_{AlAs}	N_d (10^{18} cm^{-3})	Inter-subband transition
A	40	500	0.26	1	BTC
B	40	500	0.25	1.5	BTC
C	60	500	0.15	0.5	BTC
E	50	500	0.26	0.42	BTB
F	50	50/500	0.3/0.26	0.42	BTQB

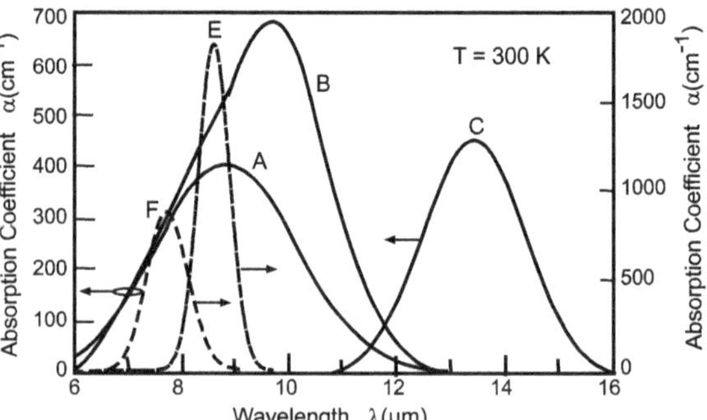

Fig. 10.68 Absorption coefficient spectra $\alpha(\lambda)$ versus wavelength measured at room temperature for QWIP samples A, B, C, E, and F. Reprint with permission from [33], copyright AIP Publishing

each having a small spectral width. The summation of these absorptions gives rise to a greater spectral width than a single bound-to-bound transition. Due to the conservation of integrated oscillator strength, the magnitude of the absorption coefficient for the bound-to-continuum QWIPs is significantly lower than that for the bound-to-bound or bound-to-quasi-bound QWIPs. Among all measured samples, the bound-to-bound QWIP (sample E) has the narrowest absorption spectral width and highest peak absorption coefficient. In the following, we will focus the discussion on samples A, E, and F due to the proximity of their peak absorption wavelengths.

The bias voltage-dependent peak responsivity R_p for the bound (E), quasi-bound (F), and continuum (A) QWIPs measured at $T = 20$ K is displayed in Fig. 10.69. The low bias responsivity of bound-to-continuum QWIP (A) is approximately linear and that it saturates at high bias. The quasi-bound QWIP (F) behaves quite similarly to the continuum QWIPs. However, the fully bound QWIP (E) has a significantly lower responsivity despite its high absorption coefficient. The responsivity does not start out linearly with applied bias but is zero for finite bias. There is a zero bias offset

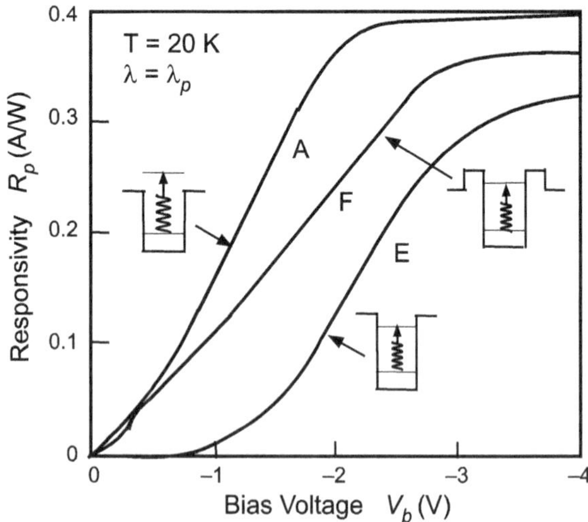

Fig. 10.69 Bias-dependent peak responsivity measured at $T = 20$ K for bound (E), quasi-bound (F), and continuum (A) QWIP samples. Reprint with permission from [33], copyright AIP Publishing

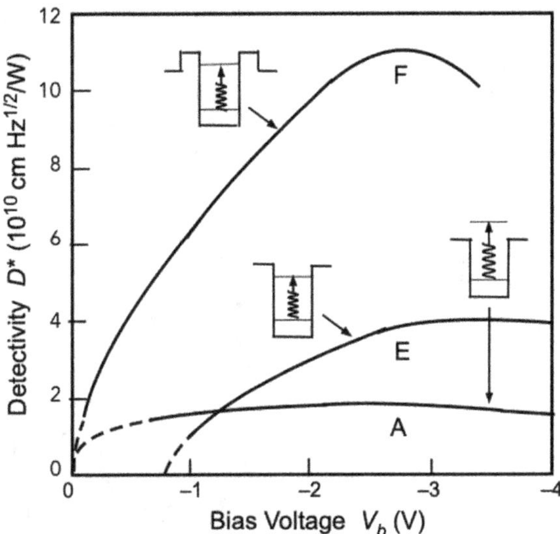

Fig. 10.70 Bias voltage-dependent detectivity D^* measured at 77 K for continuum (A), bound (E), and quasi-bound (F) QWIP samples. The dashed lines near the origin are extrapolations. Reprint with permission from [33], copyright AIP Publishing

of nearly 1 V, due to the necessity of field-assisted tunneling for the photoexcited carrier to escape from the well. Since the bias voltage-dependent photoconductive gains are quite similar for all three QWIP designs, the magnitude of responsivities is mainly determined by the absorption quantum efficiency ($\eta \propto p_e\alpha$). The measured escape probabilities p_e for continuum, quasi-bound, and bound QWIPs at zero bias are 34%, 28%, and 2.9%, respectively. These lead to the corresponding zero-bias absorption quantum efficiencies η of 5.6%, 5.8%, and 0.6% for continuum, quasi-bound, and bound QWIPs, respectively. It is clear that higher peak spectral responsivity is achieved in QWIPs with the upper state in the continuum or nearly in resonance with the top of the barrier.

Finally, the bias-dependent detectivities for a continuum (A), a quasi-bound (F), and a bound (E) QWIP are shown in Fig. 10.70. For all samples D^* has a maximum value at a bias between −2 and −3 V. Since these QWIPs all have different cutoff wavelengths, these maximum D^* values cannot be simply compared. However, the general behavior of the photodetector can be revealed from the detectivity expressed as a function of detector parameters. For quasi-bound and continuum QWIPs, $\tau_L \gg \tau_T$, and substituting (10.88), (10.98), and (10.99) into (10.97), we get

$$D^* = \frac{\lambda}{2hc}\frac{\eta}{\sqrt{N_w}}\sqrt{\frac{\tau_L}{N_{2\mathrm{D}}}} \tag{10.101}$$

where λ is the wavelength, η is the absorption quantum efficiency, N_w is the number of quantum wells, and $N_{2\mathrm{D}}$ is given by

$$N_{2D} = \left(\frac{m^*kT}{\pi\hbar^2}\right)\exp[-(hc/\lambda_c kT) + (E_F/kT)] \tag{10.102}$$

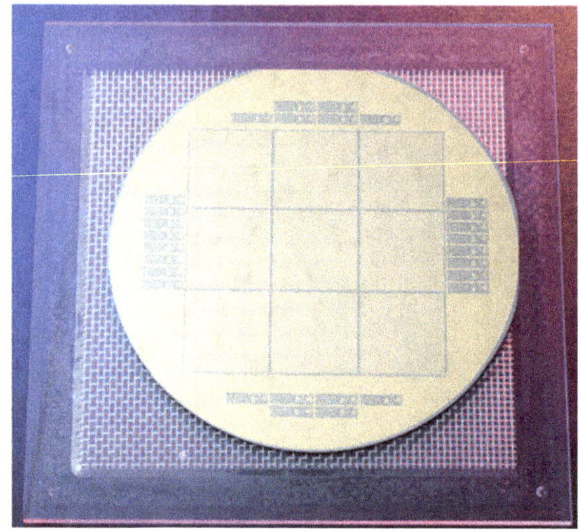

Fig. 10.71 Nine 1024 × 1024 QWIP detector arrays fabricated on a four-inch semi-insulating GaAs wafer. The process evaluation chips containing large and small area test detectors were placed at the sides of the detector arrays. Reprint with permission from [34], copyright IEEE

where m^* is the effective mass in the well, T is the temperature, c is the speed of light, λ_c is the cutoff wavelength, and $hc/\lambda_c = (E_B - E_1)$.

From the above two equations, it is clear that higher η, longer τ_{L}, shorter λ_c or lower T leads to a higher D^*. Among these detector parameters, λ_c and T are the most sensitive, being in the exponent. In addition, provided η can be made high, say close to 100%, one should use the least number of quantum wells in a QWIP for higher detectivity.

The continuous progress in QWIP device optimization, material growth, and device processing has culminated in the realization of large format focal plane arrays (FPAs) for long-wavelength (8–12 μm) infrared (LWIR) spectrum applications. The most common and commercially viable QWIP FPAs are based on lattice-matched GaAs/$\mathrm{Al}_x\mathrm{Ga}_{1-x}\mathrm{As}$ material systems grown on 4- and 6-inch GaAs semi-insulating substrates. This material system is robust and highly reproducible. Figure 10.71 shows nine megapixel QWIP detector arrays fabricated on a 4-inch semi-insulating GaAs wafer made at the NASA Jet Propulsion Laboratory (JPL). The LWIR bound-to-quasi-bound QWIP device structure consists of 50 pairs of 40 Å GaAs quantum wells and 600 Å $\mathrm{Al}_{0.27}\mathrm{Ga}_{0.73}\mathrm{As}$ barriers. The device structure was grown using MBE. 1024 × 1024 pixel detector arrays and process evaluation chips (PECs) were fabricated using standard photolithography, dry etching, and metallization processes. The responsivity of the large area test detectors in the PECs peaked at 8.4 μm, and the peak responsivity of 130 mA/W was achieved at bias $V_B = -1$ V. The QWIP detector array was then integrated with the silicon readout integrated circuits. These finished LWIR QWIP FPAs produced excellent images (Fig. 10.72) with 99.98% pixel operability.

Fig. 10.72 One image frame taken with a megapixel LWIR QWIP FPA. Pixel pitch is 19.5 μm and the FPA was cooled to 72 K. Reprint with permission from [34], copyright IEEE

10.8 Transistor Lasers

For high-speed operation of over 700 GHz, the downscaling HBT becomes a high-current-density device. As much as 10^4 A/cm^2 goes into the base, which is 10–100 times as dense as that needed for a state-of-the-art high-speed laser. For HBTs made of direct-gap III–V materials—in this case, InP/InGaAs—Milton Feng and Nick Holonyak, Jr. at the University of Illinois speculated that at these normally destructive current densities, light was probably being generated to remove the energy lost in recombination instead of heat generation. Their extensive research efforts led to the invention of the *light-emitting transistor* (LET), particularly, the *transistor laser* (TL). This HBT-based integrated optoelectronic device is, simultaneously, both an electrical and an optical signal source driven by the base current. It operates as a three-port device with an electrical input, electrical output, and a third optical output. The unique three-terminal configuration of the TL enables the realization of new electro-optical applications such as nonlinear signal mixing, frequency multiplication, and optoelectronic logic gates.

10.8.1 Basic Device Structure and Operation Principles

A transistor laser has a device layer structure essentially similar to a HBT with a few modifications. As shown in Fig. 10.73, a resonant cavity is created using reflective facets at opposite ends of the recombination (active) region to aid the stimulated

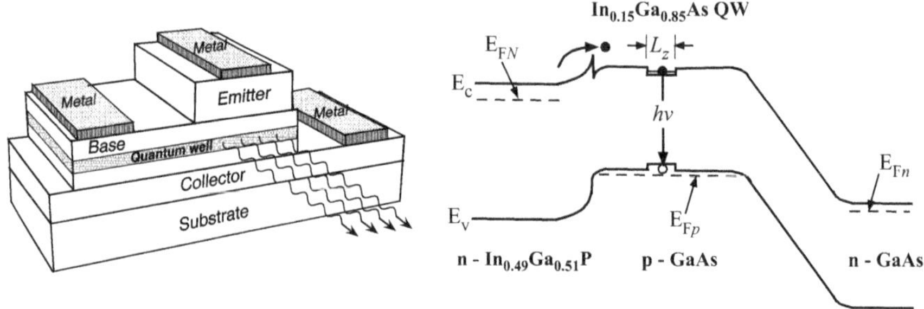

Fig. 10.73 Device structure and energy band diagram of an InGaP/GaAs transistor laser embedded with InGaAs QWs inside the base region

recombination and emission of photons. The other major departure from the conventional HBT structure is the insertion of a QW or MQWs inside the base region. Under forward normal operation of a *N-p-n* HBT, injected electrons from the emitter diffuse across the base. Because the base–collector junction is reverse-biased to block holes, the electrons coming from the emitter through the base sweep right into the collector without noteworthy recombination with holes inside the base. To enhance the recombination of electrons and holes in the base region of the transistor laser, the inserted QW acts like a special recombination center, or an 'optical collector,' to trap electrons that flow from emitter to collector. This enhanced recombination process produces a much stronger optical signal emerging from the base region and a complementary electrical signal emerging from the collector. The light intensity of the spontaneous emission from the base increases with increasing base current, which supplies the required amount of holes. Once the supplied base current reaches the lasing threshold, the intense directed stimulated emission of a coherent laser beam is emerging from the ends of the laser cavity.

In two-port devices such as the two-terminal diode laser, the light output power solely depends on the injection current. The three-terminal HBT is also a two-port device where output collector current (I_C) is a function of both injected base current (I_B) and the collector-to-base voltage ($V_{CB} = V_{CE} - V_{BE}$, V_{BE} is constant at given I_B). On the other hand, the three-port transistor laser, in common-emitter operation shown in Fig. 10.74, has an input port 1 (I_B), an output port 2 (I_C–V_{CE} characteristics), and uniquely in the base a QW 'optical collector' output port 3 (L–V_{CE} characteristics, L = coherent light output intensity). It has a unique set of current output (I_C–V_{CE}) family of curves and a distinct set of coherent light output (L–V_{CE}) family of curves. As an example, the (I_C–V_{CE}) and (L–V_{CE}) family curves of a *n*-InGaP/p^+-GaAs/n^+-GaAs transistor laser incorporated a single undoped 15 nm InGaAs base-region QW at wavelength $\lambda \approx 980$ nm are shown in Fig. 10.75. The electrical output of the transistor laser is examined first. For $I_B < 56$ mA the transistor laser exhibits the usual transistor (HBT) behavior with the differential current gain $\beta = \Delta I_C/\Delta I_B$ increasing with I_B. At base current $I_B = 56$ mA the gain decreases abruptly, corresponding to stimulated recombination and 'faster' base recombination, as will be discussed in the

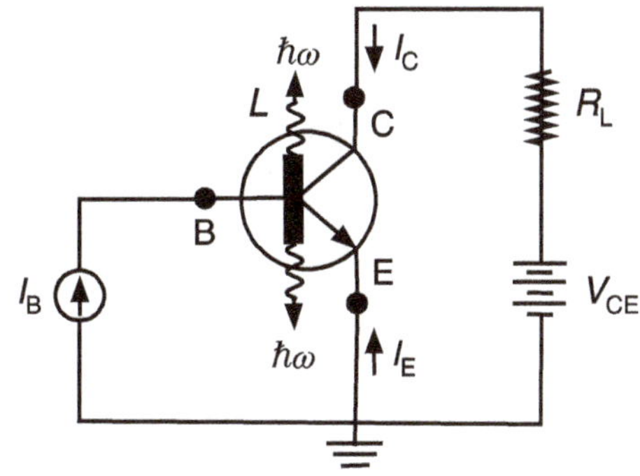

Fig. 10.74 Biasing diagram of a common-emitter operation of a transistor laser

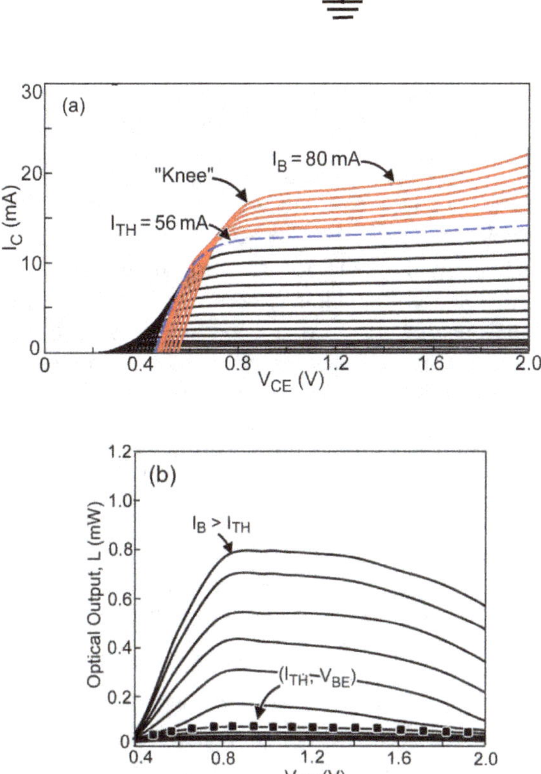

Fig. 10.75 **a** Collector I–V_{CE} characteristics of a transistor laser. Below the knee voltages, the transistor is biased in saturation. **b** The coherent light output L–V_{CE} characteristics of the transistor laser. Reprinted with permission from [35], copyright AIP Publishing

next section. Thus, it determines the threshold current (I_{th}) of the laser action. This abrupt compression of the I–V characteristics and reduction in β is unique to the transistor laser. This feature also eliminates the need to have an additional external feedback system, such as a photodetector, to verify that the device is operating as a laser. The coherent optical output characteristics (L–V_{CE}) of the transistor laser for $I_B \geq I_{th}$ are shown in Fig. 10.75b. For small V_{CB} where the BC junction has not reached strong reverse bias, the laser output increases with emitter electron injection into the base. The stimulated recombination then saturates shown as a constant output

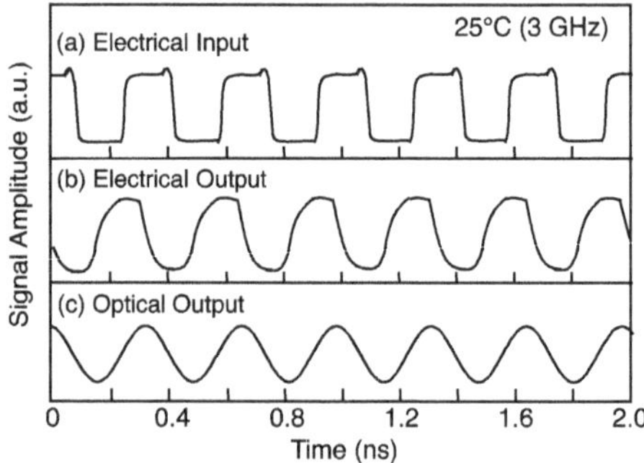

Fig. 10.76 Three-port laser operation of a transistor laser at room temperature in the common-emitter configuration. Curve (**a**) is a 3 GHz electric input voltage waveform, **b** a 3 GHz electrical output voltage signal, and **c** a 3 GHz photodetector output voltage waveform corresponding to the modulated laser output. Reprinted with permission from [36], copyright AIP Publishing

power limited by the hole supply at base current I_B. Once the BC junction reaches sufficient reverse bias, direct tunneling of electrons across the BC junction reduces the hole supply in the base region leading to a gradual drop in laser output power.

In a directly modulated three-port transistor laser, a coherent stimulated optical signal emerges from the base and, simultaneously, a complementary electrical signal emerges from the collector. As an example, a 3-GHz small signal sinusoidal voltage with peak-to-peak amplitude of 0.6 V is supplied at the base (input port) of the device. The input voltage V_{BE}, the electrical output voltage V_{CE}, and the laser output signal measured using a power meter are all displayed simultaneously on separate channels of a digital sampling oscilloscope and shown in Fig. 10.76. These data demonstrate that direct electrical and optical modulation of a transistor laser at high-speed is possible. However, the ultimate bandwidth of the transistor laser is limited by the modulation speed of the optical output signal, since the state-of-the-art switching speed of HBT has already reached 800 GHz.

10.8.2 Effective Minority Carrier Lifetime in Transistor Lasers

For semiconductor lasers, the modulation bandwidth is determined by how fast the injected carriers in the active region can be depleted when the device is turned off. In a double-heterostructure *p-i-n* diode laser, injected carriers are trapped in the active region, bounded by upper and lower cladding layers, and the depletion process is accomplished only through recombination processes (Fig. 10.77a). In laser diode structures, the minority carrier density and lifetime are about $\geq 10^{18}$ cm^{-3} and ~1 ns, respectively. The long minority carrier lifetime puts an upper limit on the maximum modulation speed of the diode laser at ~ 20 GHz.

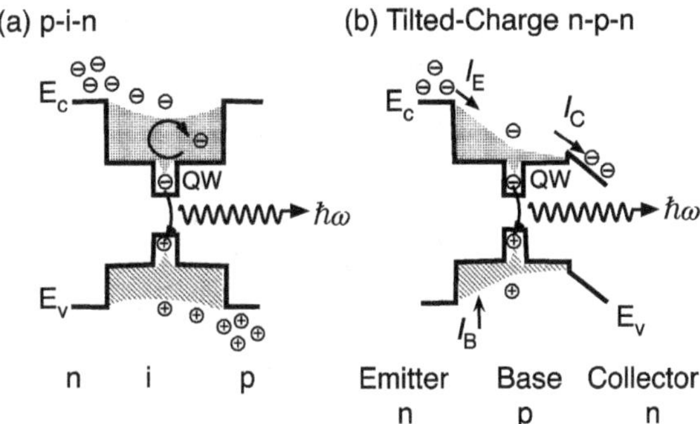

Fig. 10.77 Schematic energy band diagram and charge population distribution in **a** a *p-i-n* double-heterostructure laser, and **b** an *n-p-n* transistor laser. Reprinted with permission from [37], copyright AIP Publishing

In *N-p-n* transistor lasers, the active region is a heavily doped thin *p*-type base bounded by *n*-type emitter and collector. The base region can be described as a series of three layers: (a) *p*-type base doped N_A (cm^{-3}), (b) undoped QW, and (c) *p*-type base doped N_A (cm^{-3}). Under normal mode operation, the EB junction is forward-biased and the BC junction is reverse-biased creating a tilted injected minority charge distribution with zero charge density at the BC junction as shown in Fig. 10.77b. Furthermore, the base charge density profile of the transistor laser deviates from the simple HBT 'triangular' approximation. The straight-line profile is slightly tilted more steeply upward at the emitter end of the base than near the collector with a 'kink' at the QW position. This is due to the enhanced recombination of the QW compared to the bulk region and the reduced base distance from emitter to the QW. It is evident from the *I–V* characteristics (e.g., Fig. 10.75a) where, at the lasing threshold, the QW recombination lifetime decreases abruptly (current gain $\beta_{stim} < \beta_{spon}$) as stimulated recombination speeds up the overall rate of recombination in the QW.

As it becomes clear later that the recombination rate in bulk base regions is very small compared to that in the QW, the base charge distribution can be seen as a superposition of two triangular charge populations denoted by Q_1 and Q_2, as shown in Fig. 10.78. Q_1 is responsible for the diffusion of carriers to the QW for radiative recombination, while Q_2 is responsible for the diffusion of carriers from the emitter to the collector. The QW acts like an optical collector in addition to the regular electrical collector of the device. The populations Q_1 and Q_2 have the values of $Q_1 = q\Delta n_1 A W_{EQW}/2$ and $Q_2 = q\Delta n_2 A W_{EC}/2$, respectively, with total base charge $Q = Q_1 + Q_2$. In these expressions, A, Δn, W_{EC}, and W_{EQW} are the base cross-sectional area, excess minority carrier density at EB junction, base width, and emitter to QW distance, respectively. From the charge control model, it is obvious that $I_C = Q_2/\tau_{t,2}$ and $I_B = (Q_1 + Q_2)/\tau_{bulk} + Q_1/\tau_{t,1}$, where $\tau_{t,1} = W_{EQW}^2/2D$ is the transit time from emitter to QW, $\tau_{t,2} = W_{EC}^2/2D$ is the transit time across the whole base, and D is the diffusion constant. We note that for a GaAs base of 90 nm, $\tau_{t,2} \approx 1.56$ ps using a diffusion constant of 26 cm^2/s. Since $W_{EQW} < W_{EC}, \tau_{t,1}$ has a value also in the picosecond range. The bulk lifetime τ_{bulk} in GaAs layers can be obtained from the current gain of a comparison HBT without a base QW, which has

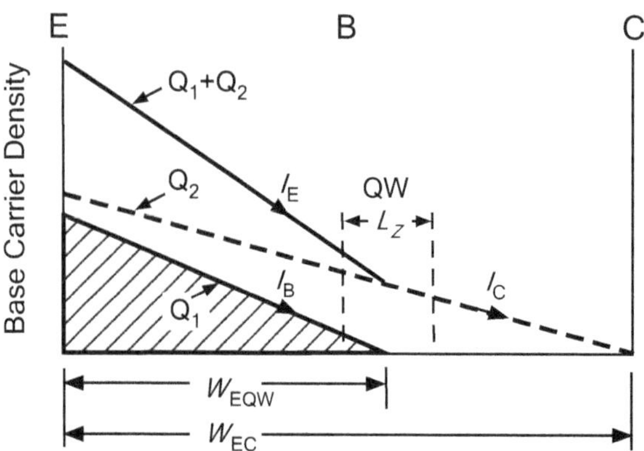

Fig. 10.78 Charge analysis of the base minority carrier density of a transistor laser showing the continuity condition applied to the minority carrier population at the base QW active region. Q_1 and Q_2 triangles correspond to carrier diffusion and collection at the QW and the electrical collector, respectively. Reprinted with permission from [38], copyright AIP Publishing

a value of ~135 ps. The overall effective base recombination lifetime τ_{B} of the entire base population, not including parasitic charging delays, can be expressed as

$$\frac{1}{\tau_{\mathrm{B}}} = \frac{1}{\tau_{\mathrm{bulk}}} + \left(\frac{Q_1}{Q_1 + Q_2}\right)\frac{1}{\tau_{t,1}} \tag{10.103}$$

Since $\tau_{\mathrm{bulk}} \gg \tau_{t,1}$, the effective base recombination lifetime reduces to $\tau_{\mathrm{B}} \approx (1 + Q_2/Q_1)\tau_{t,1}$. When the transistor laser is operated below threshold in spontaneous mode, $Q_2 \gg Q_1$ and (10.103) reduces to $\tau_{\mathrm{B}} = \tau_{\mathrm{B,spon}} \approx (Q_2/Q_1)\tau_{t,1}$. When the transistor laser is operated near and just beyond the threshold, Q_1 increases rapidly approaching the same order of magnitude of Q_2. Therefore, τ_{B} is larger but of the same order of magnitude as $\tau_{t,1}$ which is in the picosecond range. Beyond the lasing threshold τ_{B} decreases as stimulated emission speeds up the overall recombination rate in the base region. A larger proportion of the base carriers are transported to the now faster optical collector (QW) for stimulated emission leading to decreased gain β_{stim}.

In a semiconductor laser stimulated emission, the output intensity responds to either a current step or a small modulation signal that involves the interaction between the photon population in the cavity and the fraction of injected carriers that are in excess of the threshold density. Hence the lifetimes associated with both injected carriers and the photons are involved in the process. For diode lasers, the typical carrier lifetime and photon lifetime are on the order of 1 ns and 5 ps, respectively. The strong interaction between the populations of injected carriers and photons with very different lifetimes leads to pulsations during switch-on and a natural resonance during the sinusoidal modulation of the laser. These undesirable features result in bit error rate degradation and limit the useful bandwidth of the device for optical data communication links. For transistor lasers, the spontaneous recombination lifetime

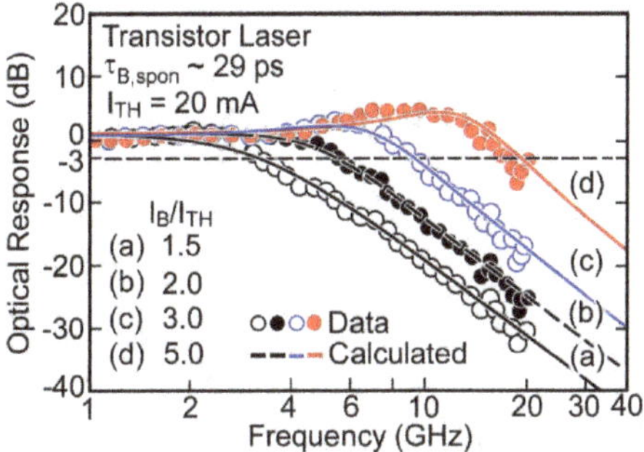

Fig. 10.79 Measured and fitted optical response of a transistor laser showing absence of carrier–photon resonance owing to fast spontaneous recombination lifetime, $\tau_{B,spon}$. Reprinted with permission from [37], copyright AIP Publishing

$\tau_{B,spon}$ is reduced by more than two orders of magnitude from the value of diode laser and on the same order of magnitude as the photon lifetime. The fast spontaneous recombination lifetime leads to a resonance-free frequency response at low biases and relatively small resonance appears only at large I_B/I_{th} as shown in Fig. 10.79. It should be noted that the fitted data of $\tau_{B,spon}$ ~ 29 ps quoted in the figure is an effective lifetime which subsumes all the parasitic charging delays. This number can be further reduced through the refinement of device designs to efficient utilization of the full device bandwidth up to –3 dB point with lower power consumption.

10.8.3 Voltage Modulation of a Transistor Laser

In transistor lasers, the collector characteristics can be used to directly extract the transport and recombination dynamics in the QW base due to its close proximity to the active base region. For example, the threshold current $I_{B,th}$ of coherent light generation can be identified from the current gain compression in current output (I_C–V_{CE}) family of curves. In Fig. 10.75b, it is also observed that the laser output drops at high reverse bias of V_{CE} due to increased carrier loss via tunneling. Thus, it is possible to use a collector junction design with efficient generation of tunneling current such that the laser operation can be more meritoriously controlled by collector bias voltage modulation in addition to current modulation. The major tunneling mechanisms at the reverse-biased BC junction of a transistor laser are the *intra-cavity photon-assisted* (ICP) *tunneling* process and the direct band-to-band tunneling process. While Franz–Keldysh effect is ordinarily referred as the electric field dependence of the fundamental absorption edge of a semiconductor (Sect. 7.5), the ICP tunneling process further includes the effect of electrical and optical cavity coupling. The ICP tunneling can be adjusted to lower voltage by narrowing the collector junction transition region with higher doping at the expense of shifting to more direct band-to-band tunneling. Thus, a collector incorporated with a tunnel junction provides an enhanced means for voltage-controlled modulation of the optical output of the transistor laser.

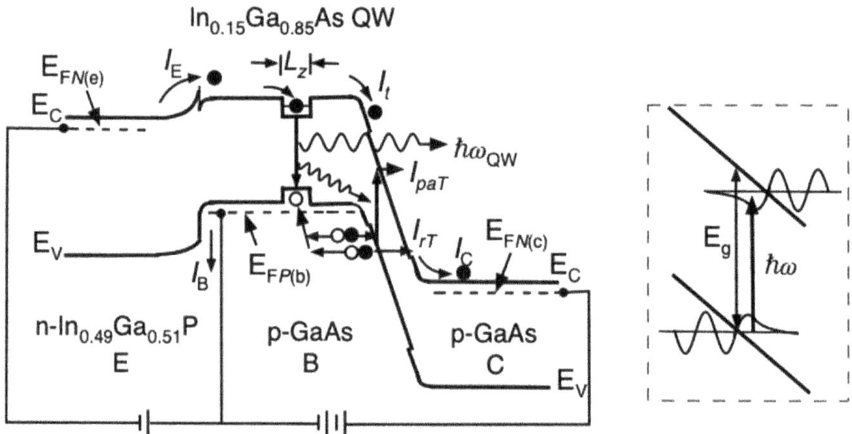

Fig. 10.80 Schematic band diagram of a TJ-TL shown with a generic resonator cavity. The inset shows the photon-assisted tunneling process at the tunnel junction. Reprinted with permission from [39], copyright AIP Publishing

Figure 10.80 shows the energy band diagram of a tunnel junction transistor laser (TJ-TL) where a highly doped p^+ (base-side) and n^+ (collector-side) TJ is incorporated around the BC junction. Under normal operation where the BC junction is reverse-biased, the collector current I_C consists of the usual transport component across the base I_t, the direct TJ current I_{rT}, and the ICP tunneling current I_{paT}, or

$$I_C = I_t + I_{rT} + I_{paT} \tag{10.104}$$

Comparing to a regular transistor laser with relative flat I_C-V_{CE} characteristics (Fig. 10.75a), the effects of collector tunneling of a TJ-TL are evident from the upward slope of the collector current I_C versus V_{CE} (Fig. 10.81a) due to various tunneling components. Under the condition of stimulated emission, the tunneling process occurs predominantly via ICP absorption, where $I_C \approx I_t + I_{paT}$. The coherent photons generated at the QW interact with the collector junction field (V_{CB}) by way of ICP absorption and assist electron tunneling from base to collector, resulting in hole resupply current (I_{paT}) into the base region and a reduction in the coherent optical field. Direct tunneling can be observed at higher V_{CE} biases of I_C–V_{CE} characteristics according to (10.104).

The optical output L-V_{CE} characteristics of a TJ-TL are shown in Fig. 10.81b. In laser operation of the TJ-TL under weak collector junction field (left region 1 of Fig. 10.81b), the laser output increases with emitter electron injection to the base. ICP tunneling ($I_{paT} > 0$) further enables the supply of holes to the QW and improves the laser optical output. The photon absorption resulting from the weak collector junction field is insufficient to overcome the optical gain established by emitter and base carrier injection. The stimulated recombination is then saturated limited by the hole supply to the base current I_B. However, under strong reverse-biased collector junction field

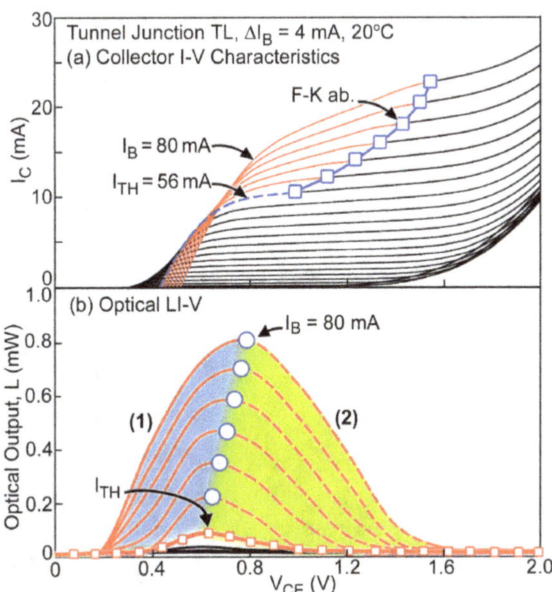

Fig. 10.81 **a** Collector I–V_{CE} characteristics and **b** optical output L–V_{CE} characteristics of a TJ-TL as shown schematically in Fig. 10.80. Reprinted with permission from [40], copyright AIP Publishing

(right region 2 of Fig. 10.81b), optical output is reduced and subsequently quenched by the field-dependent ICP absorption.

Figure 10.82 illustrates the ICP tunneling process at the BC junction that is singled out from the total collector current for clarity. The electron–hole pairs are generated at the collector junction under photon-assisted tunneling inside the optical cavity. Electrons drifting to the collector contribute to the excess collector current I_{paT}. Holes drifting to the base contribute as majority carrier injection and diffuse to the base QW for stimulated recombination (I_{r2}). Under steady-state condition $I_{paT} = I_{r2}$,

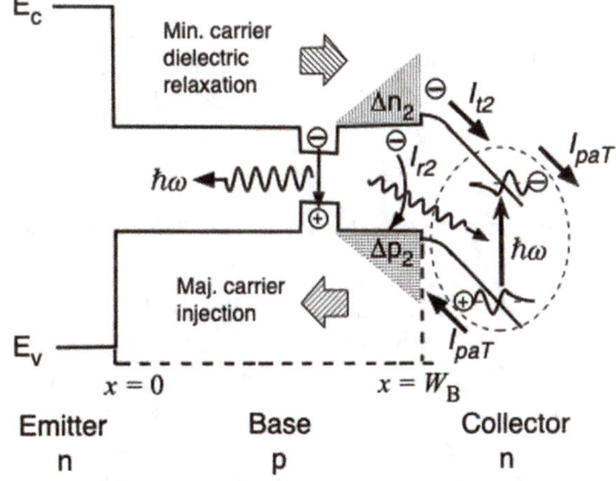

Fig. 10.82 Tunneling excess carrier distribution in the base of a transistor laser under the influence of intra-cavity photon-assisted tunneling (circled region). Minority electron injection from the emitter and excess majority hole supply from the base are omitted for clarity. Reprinted with permission from [41], copyright AIP Publishing

the excess hole distribution (Δp_2) is established. The excess hole charge ($+q\Delta p_2$) lowers the emitter junction energy barrier and induces minority carrier injection to establish the excess electron distribution (Δn_2) at the collector junction satisfying charge neutrality. The time constant that characterizes this transport process is quite small, on the order of femtoseconds, and is referred to as the *dielectric relaxation time*, $\tau = \epsilon_r \epsilon_0 \rho$ where ρ is the resistivity. The excess electrons that do not recombine will drift to the collector and contribute as additional collector current I_{t2}. These carrier transport and redistribution processes inside the transistor laser establish a tunneling current gain, β_2, and can be expressed as

$$\beta_2 = \frac{I_{t2}}{I_{r2}} = \frac{q\Delta n_2 v_{\text{drift}}}{W_{\text{CQW}} q \Delta p_2 / 2\tau_{\text{stim}}} = \frac{W_{\text{C}}/\tau_{\text{drift}}}{W_{\text{CQW}}/2\tau_{\text{stim}}} = \frac{2W_{\text{C}}}{W_{\text{CQW}}} \frac{\tau_{\text{stim}}}{\tau_{\text{drift}}} \approx \frac{\tau_{\text{stim}}}{\tau_{\text{drift}}} \tag{10.105}$$

W_{CQW} is the distance between the QW and the BC junction, and W_{C} is the collector junction thickness. τ_{stim} and τ_{drift} are the QW stimulated recombination lifetime and the carrier drift time, respectively. The tunneling current gain β_2 is approximately the ratio of τ_{stim} to τ_{drift}. As shown in Fig. 10.81a, the ICP tunneling current I_{paT} increases with the collector voltage V_{CE} due to the increased absorption of cavity photons. The cavity photon density reduction leads to the QW stimulated recombination lifetime increase and, thus, an increasing tunneling current gain β_2 at high bias V_{CE}.

As shown in Fig. 10.82, the tunneling collector current I_{C2} due to the stimulated recombination can be expressed as

$$I_{\text{C2}} = I_{t2} + I_{\text{paT}} = I_{t2} + I_{r2} = (\beta_2 + 1) I_{r2} = (\beta_2 + 1) I_{\text{paT}} \tag{10.106}$$

Once the currents I_{C2} and I_{paT} are determined from the measured collector I–V and optical L–V curves, the tunneling current gain $\beta_2 = \left(I_{\text{C2}}/I_{\text{paT}}\right) - 1$ can be derived experimentally. In a typical AlGaAs/InGaAs QW/GaAs transistor laser, the tunneling current gain as a function of V_{CB} is experimentally determined and shown in Fig. 10.83. As expected, for a given base current, β_2 increases with collector V_{CB}. Thus, the photon-field enhanced optical absorption augments ICP tunneling at high collector voltages in a transistor laser and leads to a unique property of voltage modulation and switching of a transistor laser.

10.8.4 Excited State Operation of a Transistor Laser

In the diode laser, the charge densities in the undoped QW are largely pinned at and beyond lasing threshold, and the lasing transition is in the ground states. This is not the case for transistor lasers. Since the base region of the *N-p-n* transistor laser is heavily doped with acceptors except the QW, the occupation of the QW states is primarily by minority carrier electrons. It is governed by the boundary conditions imposed by the charge transport from the emitter to the collector, i.e., a tilted emitter-to-collector base population distribution. Therefore the charge population in the QW base does not

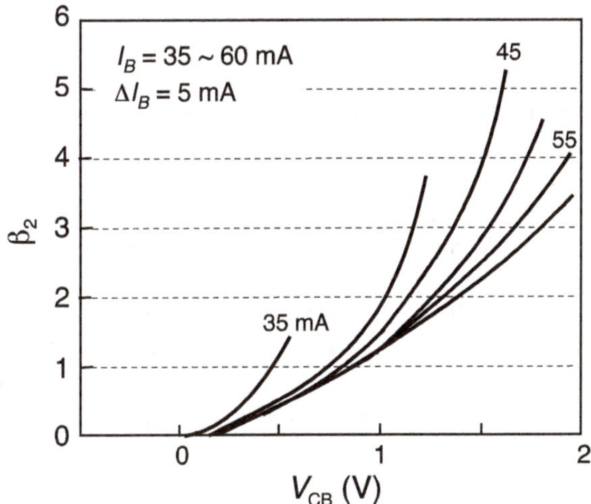

Fig. 10.83 Tunneling current gain β_2 of an AlGaAs/InGaAs QW/GaAs TJ-TL at different base currents and collector bias voltages. Reprinted with permission from [41], copyright AIP Publishing

saturate but continuously increases beyond lasing threshold, causing the lasing action to evolve from the ground-state transition at low I_B to the excited state transition at high base current. Figure 10.84 shows the I–V characteristics of a transistor laser where the I–V curves are distorted in the region of stimulated recombination ($I_B > I_{B,th}$) when compared to the region of spontaneous recombination. In the region of high bias voltage ($V_{CE} > 2.7$ V) the stimulated recombination is quenched into spontaneous recombination even at $I_B > I_{B,th}$. The corresponding lasing spectra at the base current bias points of $V_{CE} = 1.5$ V at $I_B = 30$, 50, 70, and 90 mA are also shown in Fig. 10.84. At 30 mA the spontaneous spectral peak developed into laser modes with the major lasing peak at $\lambda \approx 1002$ nm. The intensity of ground state stimulated radiation at ~1000 nm increases proportionately as I_B is increased further to 50 mA. At $I_B = 70$ mA lasing occurs on both the broad ground state and on the

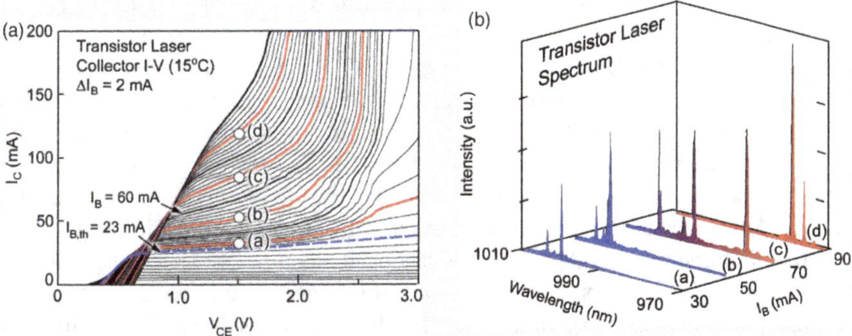

Fig. 10.84 Transistor laser collector I–V characteristics (left) and laser spectra (right) at V_{CE} = 1.5 corresponding to bias points: **a** $I_B = 30$ mA, **b** $I_B = 50$, **c** $I_B = 70$, and **d** $I_B = 90$. Reprinted with permission from [42], copyright AIP Publishing

next higher state of the QW ($\lambda \approx 978$ nm). At $I_B = 90$ mA the laser operation is observed exclusively on the higher state ($\lambda \approx 978$ nm).

The 3-dB bandwidth of the transistor laser operating on the ground state increases with increasing base current and begins to saturate (at 7 GHz) around 50 mA because of the ground-state band-filling and gain saturation. Continuous increase of the charge population it shifts the operation from the ground state to the first excited state at 70 mA. The 3-dB bandwidth of the transistor laser operates on the first excited state continue to rise with increasing I_B. Further increase in the base current causes the first excited state to saturate (at 13.5 GHz) due to gain saturation and heating. The limited bandwidth measured arises mainly from large input base capacitance and base resistance of the current device. With proper device geometry and improvement in contact resistance, it should be possible to achieve transistor laser operation near the intrinsic bandwidth of greater than 40 GHz.

10.8.5 Light-Emitting Diodes with Tilted Base-Charge Distribution

Similar to diode lasers, the modulation speed of double-heterostructure light-emitting diodes (LEDs) is limited by the carrier recombination lifetime of the device. As shown in Fig. 10.77a, the buildup of confined excess carriers in the active region can only be depleted through the slow radiative recombination process which limits the maximum modulation speed of the LED to about 1 GHz. On the other hand, a tilted base charge population in the transistor laser design offers a unique solution to select and keep fast recombination carriers in the base by collecting slow recombination carriers at the reverse-biased collector junction. Thus, the action of the collector in a transistor laser may be incorporated into the design of a carrier drain in LEDs to enable high-speed operation.

Figure 10.85a shows the cross section of a tilted-charge LED. The device layer structure is similar to an *N-p-n* transistor laser with some modifications to achieve two-terminal operation. Emitter current I_E is injected from the top into the emitter and transported to the QW base active region. The injected minority carriers that recombine in the base form the base current I_B. To empty carriers that are too slow to recombine within the transit time (τ_t) across the base layer, a reverse bias field at the lower base boundary is needed. In the tilted-charge LED design the collector of the transistor laser is replaced by an undoped drain layer which forms a *p-n* junction with the base. A reverse built-in field is maintained at the base–drain junction by the zero potential difference obtained via a common contact metallization extending from the drain to the base. The zero base–drain potential difference ensures that there is no base charge population at the base–drain boundary, hence establishing a tilted emitter-to-drain population distribution in the base. The excess minority carriers removed from the base by the built-in field forms the drain current I_D. The tilted-charge LED can be biased as a usual two-terminal device and displays an *I-V* characteristic resembling

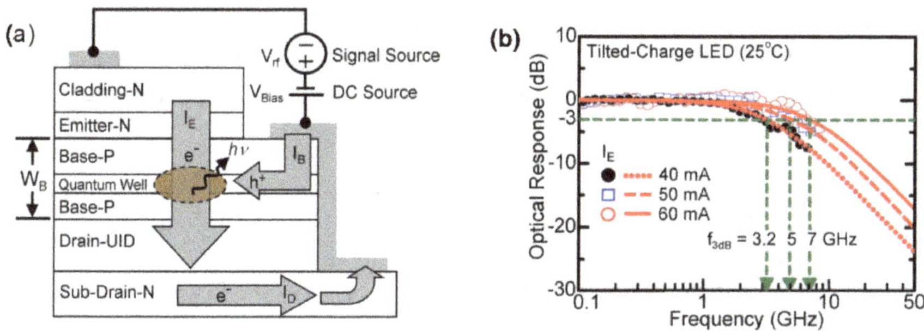

Fig. 10.85 **a** Cross section of an *N-p-n* tilted-charge LED with a *p*-type base QW active region. **b** The optical output frequency response of the tilted-charge LED at bias currents $I_E = 40$, 50, and 60 mA showing the -3 dB frequency of 3.2, 5, and 7 GHz, respectively. Reprinted with permission from [43], copyright AIP Publishing

that of a *p-n* junction diode. The turn-on voltage is determined by the emitter–base potential difference since the base and drain have a common potential. The optical output frequency responses of the tilted-charge LED biased at different I_E are shown in Fig. 10.85b. The tilted-charge LED operates similar to a *p-n* junction LED with reasonable efficiency but has a considerably higher modulation speed. The maximum 3-dB bandwidth of the device achieved is 7 GHz at $I_E = 60$ mA.

Problems

1. A GaAs/$Al_{0.3}Ga_{0.7}As$ symmetric double-heterostructure laser with a cavity length of 250 μm has the following device parameters: $\alpha_i = 10\ \text{cm}^{-1}$, $\bar{n}$(GaAs) = 3.6 at the lasing wavelength, and an active layer thickness of 0.15 μm. Find the gain coefficient. Note, the mirror power reflectivity can be calculated from

$$R = \left|\frac{\bar{n}_0 - \bar{n}_1}{\bar{n}_0 + \bar{n}_1}\right|^2$$

where $\bar{n}_0(\text{air}) = 1$ and $\bar{n}_1 = \bar{n}(\text{GaAs})$.

2. From an article [*App. Phys. Lett.* **61**, 255 (1992)] on strained 0.98 μm InGaAs single QW lasers using different cladding layers (GaAs and GaInAsP), extract the following parameters for two types of lasers. Assume the mirror reflectivity is 32%.

   (a) α_i
   (b) η_i
   c. Transparency current density.

3. The peak gain of a 1.55 μm InGaAsP laser diode can be approximated by

$$g_p = a(n - n_{\text{tr}}) = 3 \times 10^{-16}\left(n - 2 \times 10^{18}\right)$$

Assume the cavity length is 250 μm, the refractive index of the active layer is 3.4, the internal loss is 25 cm^{-1}, the carrier lifetime is 1.5 ns, the internal quantum efficiency is 80%, the active layer thickness is 0.1 μm, and the optical confinement factor is 0.4.

(a) Calculate the threshold current density of the laser.
(b) What is the threshold current of the laser if the active region is 2.5 μm wide?
(c) What is the differential quantum efficiency η_d of the laser?
(d) The external quantum efficiency η_e or slope efficiency relates to η_d by

$$\eta_e = \frac{\Delta P}{\Delta I} = \frac{\hbar\omega}{q}\eta_d \, (W/A).$$

Find η_e of the laser.

4. For a GaAs/AlGaAs laser with a lasing wavelength of 780 nm, calculate the mode number and mode spacing of the laser for a cavity length of 250 μm and $\bar{n}$ (active layer) = 3.54.
5. Calculate the photon lifetime of a laser with the following parameters: cavity length = 250 μm, internal loss = 10 cm^{-1}, and the refractive index of cleaved mirrors is 3.4.
6. Calculate the output power of a GaAs/AlGaAs laser operated at an injection current level of 30 mA with a lasing wavelength of 870 nm and the following device parameters: cavity length = 250 μm, internal loss = 20 cm^{-1}, the refractive index of cleaved mirrors is 3.65, the internal quantum efficiency is 0.8, and the threshold current is 18 mA.
7. The figure below is the measured optical spectrum of a GaAs laser just below the threshold. For both mirrors, $\bar{n}$(active region) = 3.58, calculate and plot the peak gain between 7960 and 8030 Å under this current injection level.

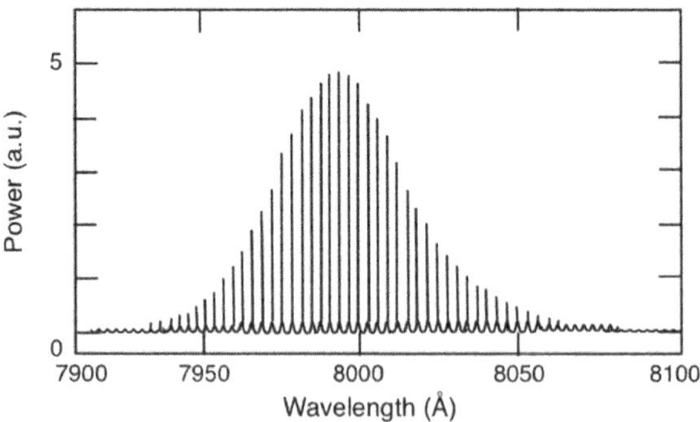

8. This problem concerns DBR mirror designs.

   (a) Plot the mirror reflectance R for InP/InGaAsP lattice-matched 1.55 μm DBR mirror as a function of the number of mirror pairs. Assuming $\bar{n}$(InP) = 3.17, $\bar{n}$(InGaAsP) = 3.40, and the adjacent layer to the DBR is air.
   (b) How many pairs of InP/InGaAsP DBR mirrors are required to reach $R =$ 99%?
   (c) How many mirror pairs are required for AlAs/GaAs mirror stacks to reach the same reflectance of 99%?
   (d) At $R = 99\%$, calculate the stop-band width for both material systems.

9. In an InGaAs/GaAs/$Al_{0.2}Ga_{0.8}As$ 980 nm VCSEL, the active region consists of 3 InGaAs quantum wells, each 8 nm thick, separated by two 8-nm-thick GaAs barriers, and clad on each side by another 8-nm-thick layer of GaAs. The rest of the cavity (active region) consists of an $Al_{0.2}Ga_{0.8}As$ spacer layer. The DBR mirrors are composed of quarter wavelength stacks of AlAs/GaAs. The DBR mirrors begin with an AlAs quarter-wave layer next to the cavity. The cavity between the DBR mirrors has to be one optical wavelength to maximize the gain.

   (a) For $\bar{n}$(GaAs) = 3.52, $\bar{n}$(AlAs) = 2.95, $\bar{n}$(InGaAs) = 3.60, and $\bar{n}$($Al_{0.2}Ga_{0.8}As$) = 3.39, determine the thickness of the two AlGaAs spacer layers.
   (b) Assuming the averaged refractive index of the cavity is 3.56, determine the number of the AlAs/GaAs period in the top and bottom DBR mirrors if the required power reflections are 99.5% and 98.5%, respectively. Consider contributions by AlAs/GaAs DBR pairs only.
   (c) Determine the threshold current density of a VCSEL with the following device parameters: $a = 4 \times 10^{16}$ cm^2, $B_r = 10^{-10}$ cm^3/s, $\alpha_a = 25$ cm^{-1}, α_c = 10 cm^{-1}, $n_{tr} = 1.5 \times 10^{18}$ cm^3, and $\Gamma \approx$ (cavity length, d)/(total device length, L).

10. **Mini project**: As the transistor has evolved from point contact, to *p-n* junction, to III–V semiconductor heterojunction bipolar (HBT), to particularly high-current density high-speed HBT, we arrive at the possibility that the HBT can be modified and operated as a three-port light-emitting device (an electrical input, and electrical and optical outputs). The three-port light-emitting transistor (LET) can be substantially modified and improved by incorporating a quantum well, or multiple quantum wells, into the *p*-type base region of a N-p^+-n HBT in order to tailor the base recombination and thus the transistor electrical and optical properties. The light-emitting HBT can be further designed to support stimulated recombination and to be operated as a laser—a transistor laser.
    Design a transistor laser with an emission wavelength near 1.3 μm, which is important for short-haul light wave communications. The HBT used in your design can be either an N-p^+-n or P-n^+-p structure. Following are two related publications on long-wavelength transistor lasers:

- F. Dixon, et al., 'Transistor laser with emission wavelength at 1544 nm,' Appl. Phys. Lett. **93**, 021111 (2008)
- Y. Huang, et al., 'InAlGaAs/InP light-emitting transistors operating near 1.55 μm,' J. Appl. Phys. **103**, 114505 (2008).

Your work should meet the following requirements:

- Select a material system among all available III–V materials for the 1.3 μm emission wavelength. Both strained and lattice-matched systems can be used. Explain why you select it.
- The selected material system should be able to form transistor laser structures with high crystal quality. However, heterostructures prepared on metamorphic substrate structures are not allowed.
- The material selected for the base region should accommodate doping to mid-10^{19} cm^{-3} for low base sheet resistance and high-speed operation. Specify the dopant used in the base region.
- Design the layer structure of a transistor laser based on the material system selected. Provide detailed layer structures along with a full energy band diagram. For heterostructures, the energy band discontinuities should be calculated.

References

1. K.N. Dutta, J. Appl. Phys. **51**, 6095 (1980)
2. Y. Itaya, Y. Suematsu, S. Katayama, K. Koshino, S. Arai, Jpn. J. Appl. Phys. **18**, 1795 (1979)
3. H. Yonezu, I. Sakuma, K. Kobayashi, T. Kamejima, M. Ueno, Y. Nannichi, Jpn. J. Appl. Phys. **12**, 1585 (1973)
4. B. Schwartz, W.W. Focht, N.K. Dutta, R.J. Nelson, P. Besomi, IEEE Trans, Electron. Dev. **31**, 841 (1984)
5. R.J. Nelson, R.B. Wilson, P.D. Wright, P.A. Barnes, N.K. Dutta, IEEE J. Quantum Electron. **17**, 202 (1981)
6. M. Nakamura, K. Aiki, J. Umeda, A. Katzir, A. Yariv, H.W. Yen, IEEE J. Quantum Electron. **11**, 436 (1975)
7. Z. Alferov, IEEE J. Sel. Topics Quantum Electron. **6**, 832 (2000)
8. S.W. Corzine, R.H. Yang, L.A. Coldren, Appl. Phys. Lett. **57**, 2835 (1990)
9. R. Dingle, C.H. Henry, Quantum effects in heterostructure lasers, U.S. Patent 3982207, Sept. 21, 1976
10. Y. Arakawa, H. Sakaki, Appl. Phys. Lett. **40**, 939 (1982)
11. M. Asada, Y. Miyamoto, Y. Suematsu, IEEE J. Quantum Electron. **22**, 1915 (1986)
12. A.Y. Liu, S. Srinivasan, J. Norman, A.C. Gossard, J.E. Bowers, Photonics Res. **3**, B1 (2015)
13. K.Y. Cheng, K.C. Hsieh, J.N. Baillargeon, Appl. Phys. Lett. **60**, 2892 (1992)
14. A.M. Moy, A.C. Chen, K.Y. Cheng, L.J. Chou, K.C. Hsieh, J. Cryst. Growth **175/176**, 819 (1997)
15. D. Wohlert, K.Y. Cheng, S.T. Chou, Appl. Phys. Lett. **78**, 1047 (2001)
16. M. Asada, Y. Miyamoto, Y. Suematsu, Jpn. J. Appl. Phys. **24**, L95 (1985)
17. E. Towe, R.F. Leheny, A. Yang, IEEE J. Sel. Topics Quantum Electron. **6**, 1458 (2000)

18. H.C. Lin, *Design and Fabrication of Long-Wavelength Vertical-Cavity Surface-Emitting Lasers using Wafer Bonding Technologies*, Ph.D. Thesis, ECE Dept., University of Illinois at Urbana-Champaign, 2002
19. M.R. Krames et al., Appl. Phys. Lett. **75**, 2365 (1999)
20. M.G. Craford, Proc. IEEE **101**, 2170 (2013)
21. R. Kazarinov, R. Suris, Sov. Phys. Semicond. **5**, 707 (1971)
22. J. Faist, F. Capasso, D.L. Sivco, C. Sirtori, A.L. Hutchinson, A.Y. Cho, Science **264**, 553 (1994)
23. J. Faist, F. Capasso, D.L. Sivco, J.N. Baillargeon, A.L. Hutchinson, S.-N.G. Chu, A.Y. Cho, Appl. Phys. Lett. **68**, 3680 (1996)
24. A. Tridicucci, C. Gmachl, F. Capasso, D.L. Sivco, A.L. Hutchinson, A.Y. Cho, Appl. Phys. Lett. **74**, 638 (1999)
25. J. Faist, M. Beck, T. Aellen, E. Gini, Appl. Phys. Lett. **78**, 147 (2001)
26. C. Sirtori, J. Faist, F. Capasso, D.L. Sivco, A.L. Hutchinson, A.Y. Cho, Appl. Phys. Lett. **66**, 3242 (1995)
27. C. Sirtori, C. Gmachl, F. Capasso, J. Faist, D.L. Sivco, A.L. Hutchinson, A.Y. Cho, Opt. Lett. **23**, 1366 (1998)
28. K. Unterrainer, R. Colombelli, C. Gmachl, F. Capasso, H.Y. Hwang, A.M. Sergent, D.L. Sivco, A.Y. Cho, Appl. Phys. Lett. **80**, 3060 (2002)
29. C. Gmachl, F. Capasso, A. Tredicucci, D.L. Sivco, R. Kohler, A.L. Hutchinson, A.Y. Cho, J. Sel. Topic Quantum Electron. **5**, 808 (1999)
30. I. Vurgaftman, W.W. Bewley, C.L. Canedy, C.S. Kim, M. Kim, J.R. Lindle, C.D. Merritt, J. Abell, J.R. Meyer, IEEE J. Sel. Topics Quantum Electron. **17**, 1435 (2011)
31. L.C. West, S.J. Eglash, Appl. Phys. Lett. **46**, 1156 (1985)
32. B.F. Levine, K.K. Choi, C.G. Bethea, J. Walker, R.J. Malik, Appl. Phys. Lett. **50**, 1092 (1987)
33. B.F. Levine, A. Zussman, S.D. Gunapala, M.T. Asom, J.M. Kuo, W.S. Hobson, J. Appl. Phys. **72**, 4429 (1992)
34. S.D. Gunapala, S.V. Bandara, J.K. Liu, J.M. Mumolo, S.B. Rafol, D.Z. Ting, A. Soibel, C. Hill, IEEE J. Sel. Topics Quantum Electron. **20**, 3802312 (2014)
35. M. Feng, N. Holonyak Jr., H.W. Then, C.H. Wu, G. Walter, Appl. Phys. Lett. **94**, 041118 (2009)
36. M. Feng, N. Holonyak Jr., G. Walter, R. Chan, J. Appl. Phys. **87**, 131103 (2005)
37. M. Feng, H.W. Then, N. Holonyak Jr., G. Walter, A. James, Appl. Phys. Lett. **95**, 033509 (2009)
38. H.W. Then, M. Feng, N. Holonyak Jr., Appl. Phys. Lett. **94**, 013509 (2009)
39. M. Feng, J. Qiu, C.Y. Wang, N. Holonyak Jr., Appl. Phys. Lett. **119**, 084502 (2016)
40. H.W. Then, C.H. Wu, G. Walter, M. Feng, N. Holonyak Jr., Appl. Phys. Lett. **94**, 101114 (2009)
41. M. Feng, J. Qiu, C.Y. Wang, N. Holonyak Jr., J. Appl. Phys. **120**, 204501 (2016)
42. H.W. Then, M. Feng, N. Holonyak Jr., Appl. Phys. Lett. **91**, 183505 (2007)
43. G. Walter, C.H. Wu, H.W. Then, M. Feng, N. Holonyak Jr., Appl. Phys. Lett. **94**, 231125 (2009)

Further Reading

1. H.C. Casey, M.B. Panish, *Heterostructure Lasers* (Academic, New York, 1978)
2. C. Weisbuch, B. Vinter, *Quantum Semiconductor Structures* (Academic, New York, 1991)
3. M. Fukuda, *Reliability and Degradation of Semiconductor Lasers and LEDs* (Artech House, Boston, 1991)
4. G.P. Agrawal, N.K. Dutta, *Semiconductor Lasers*, 2nd edn. (Van Nostrand Reinhold, New York, 1993)
5. T.E. Sale, *Vertical Cavity Surface Emitting Lasers* (Research Studies Press, Taunton, 1995)
6. E.F. Schubert, *Light-Emitting Diodes*, 2nd edn. (Cambridge University Press, Cambridge, 2006)
7. J. Faist, F. Capasso, C. Sirtori, D.L. Sivco, A.Y. Cho, Quantum cascade lasers, in *Semiconductors and Semimetals*, **66**, chap. 1 (2000), 1
8. F. Capasso, Opt. Eng. **49**, 111102 (2010)

9. J. Faist, *Quantum Cascade Lasers* (Oxford University Press, Oxford, United Kingdom, 2013)
10. I. Vurgaftman, R. Weih, M. Kamp, J.R. Meyer, C.L. Canedy, C.S. Kim, M. Kim, W.W. Bewley, C.D. Merritt, J. Abell, S. Höfling, J. Phys. D Appl. Phys. **48**, 123001 (2015)
11. B.F. Levine, J. Appl. Phys. **74**, R1 (1993)
12. H.C. Liu, Quantum well infrared photodetector physics and novel devices, in *Semiconductors and Semimetals*, **62**, chap. 3 (2000), 129
13. N. Holonyak, Jr., M. Feng, IEEE Spectrum, February (2006), pp. 50–55
14. H.W. Then, M. Feng, N. Holonyak Jr., Proc. IEEE **101**, 2271 (2013)

Appendix A
Values of Important Physical Constants

Quantity	Symbol	Value	Unit (MKS)
Bohr radius	a_B	5.2917×10^{-11}	m
Velocity of light	c	2.9979×10^{8}	m/s
Electron volt	eV	1.6022×10^{-19}	J
Permittivity of free space	ϵ_0	8.8542×10^{-12}	F/m
Planck's constant	h	6.6262×10^{-34}	J s
	$\hbar$	1.0546×10^{-34}	J s
Boltzmann's constant	k	1.3807×10^{-23}	J/K
Thermal voltage at $T = 300$ K	kT/q	2.585×10^{-2}	V
Electron rest mass	m_0	9.1095×10^{-31}	kg
Permeability of free space	μ_0	$4\pi \times 10^{-7}$	H/m
Avogadro's number	N	6.0220×10^{23}	1/mole
Electronic charge	q	1.6022×10^{-19}	C
Wavelength of 1 eV quantum	λ	1.2398×10^{-6}	m
Wavenumber of 1 eV quantum	k	8065.6	1/cm
Frequency of 1 eV quantum	f	241.8	THz

K. Y. Cheng, *III–V Compound Semiconductors and Devices*, Graduate Texts in Physics,
https://doi.org/10.1007/978-3-030-51903-2

Appendix B
Important Physical Properties of Some Indirect Semiconductors[a]

		Si	Ge	AlP	AlAs	AlSb	GaP
Density (g/cm^3)		2.329	5.323	2.40	3.76	4.26	4.138
Lattice constant (Å)		5.43102	5.6579	5.4672[b]	5.660[b]	6.1355[b]	5.4505[b]
Melting temperature (°C)		1412	937	2550	1740	1065	1457
Indirect bandgap energy @ 300 K (eV)		1.1242	0.664	2.488	2.164	1.616	2.273
Direct bandgap energy @ 300 K (eV)				3.553	3.003	2.30	2.991
Intrinsic carrier concentration (cm^{-3})		1.02×10^{10}	2.33×10^{13}				2
Dielectric constant		11.9	16.2	9.8	10.06	12.04	11.11
Refractive index @ E_g		3.6	4.0	2.95	3.18	3.445	3.42
Optical phonon energy (meV)		64	37	62	50	42	50.7
Thermal conductivity (W/cm °C)		1.32	0.6	0.9	0.91[c]	0.57[c]	1.0
Electron effective mass (m_0)	Longitudinal	0.9163	1.59	3.67	1.1	1.8	0.254
	Transverse	0.1905	0.0823	0.212	0.19	0.259	4.8
Hole effective mass (m_0)	Heavy hole	0.537	0.284	0.513[d], 1.372[e]	0.409[d], 1.022[e]	0.336[d], 1.372[e]	0.67[e]
	Light hole	0.153	0.043	0.211[d], 0.145[e]	0.153[d], 0.109[e]	0.123[d], 1.372[e]	0.17[e]
Mobility @ 300 K (cm^2/V s)	Electron	1450	3900	60	294	200	160
	Hole	370	1800			400	135

[a]O. Madelung, Ed., *Semiconductor: Group IV Elements and III–V Compounds*, in R. Poerschke, Ed., *Data in Science and Technology*, Springer, Berlin, 1991
[b]I. Vurgaftman, J. R. Meyer, and L. R. Ram-Mohan, *J. Appl. Phys.*, **89**, 5815 (2001)
[c]H. C. Casey, Jr, and M. B. Panish, *Heterostructure Lasers*, Part B, Academic Press, New York, 1978
[d]B ∥ [100]
[e]B ∥ [111]

K. Y. Cheng, *III–V Compound Semiconductors and Devices*, Graduate Texts in Physics,
https://doi.org/10.1007/978-3-030-51903-2

Appendix C
Important Physical Properties of Direct III–V Binary Semiconductors[a]

		GaAs	GaSb	InP	InAs	InSb
Density (g/cm^3)		5.3176	5.6137	4.81	5.667	5.7747
Lattice constant (Å)		5.65325	6.09593	5.8687	6.0583	6.47937
Melting temperature (°C)		1240	712	1062	942	527
Bandgap energy $E_g(\Gamma)$ @ 300 K (eV)[b]		1.424	0.726	1.353	0.354	0.174
Indirect bandgap $E_g(X)$ @300 K (eV)[b]		1.899	1.033, 0.753 (*L*)	2.273	1.37, 1.07 (*L*)	0.63 (0 K)
Intrinsic carrier concentration (cm^{-3})		2.1×10^6		1.2×10^8	1.3×10^{15}	2×10^{16}
Dielectric constant		12.85	15.69	12.56	15.15	16.8
Refractive index @ E_g [c]		3.655	3.82	3.45	3.52	4.0
Optical phonon energy (meV)		36	28.8	42.7	29.6	23.7
Thermal conductivity (W/cm °C)		0.44	0.33	0.68	0.27[b]	0.17
Electron effective mass (m_0)		0.067	0.0412	0.077	0.022	0.014
Hole effective mass (m_0)	Heavy hole	0.50	0.28	0.56	0.35	0.45
	Light hole	0.076	0.050	0.12	0.026	0.0158
Mobility @ 300 K (cm^2/V s)	Electron	9200	3750	4400	25,000	77,000
	Hole	400	680	150	450	800

[a]O. Madelung, Ed., *Semiconductor: Group IV Elements and III–V Compounds*, in R. Poerschke, Ed., *Data in Science and Technology*, Springer, Berlin, 1991
[b]I. Vurgaftman, J. R. Meyer, and L. R. Ram-Mohan, *J. Appl. Phys.* **89**, 5815 (2001)
[c]H. C. Casey, Jr, and M. B. Panish, *Heterostructure Lasers*, Part B, Academic Press, New York, 1978

K. Y. Cheng, *III–V Compound Semiconductors and Devices*, Graduate Texts in Physics,
https://doi.org/10.1007/978-3-030-51903-2

Appendix D
Important Physical Properties of Wurtzite III–Nitride Semiconductors[a]

		GaN	AlN	InN
Density (g/cm^3)		6.15	3.257	6.81–6.89
Lattice constant (Å)	a	3.189	3.112	3.54
	c	5.185	4.982	5.70
Melting temperature[b] (°C)		2500	3000	1100
Band gap energy @ 300 K (eV)		3.438[c]	6.138[c]	0.64[d]
Dielectric constant		8.9–9.5	8.5	15
Refractive index @ E_g		2.29	2.15	2.9
Optical phonon energy (meV)		91.8	113	73
Thermal conductivity (W/cm °C)		>2.1	3.3	0.45
Breakdown field strength (V/cm)		5×10^6	1.8×10^6	
Electron effective mass (m_0)		0.2–0.22	0.3	0.07–0.26
Hole (heavy) effective mass (m_0)		1.0	$k_z - 3.53$ $k_x - 10.42$	1.63
Mobility @ 300 K (cm^2/V s)	Electron	1000	135	2700
	Hole	≤200	14	

[a]S. L. Rumyantsev, M. S. Shur, and M. E. Levinshtein, in *GaN-Based Materials and Devices*, M. S. Shur and R. F. Davis, Ed., World Science, New Jersey, 2004
[b]O. Madelung, Ed., *Semiconductor: Group IV Elements and III–V Compounds*, in R. Poerschke, Ed., *Data in Science and Technology*, Springer, Berlin, 1991
[c]I. Vurgaftman, J. R. Meyer, and L. R. Ram-Mohan, *J. Appl. Phys.*, **89**, 5815 (2001)
[d]J. Wu, W. Walukiewicz, W. Shan, K. M. Yu, J. W. Ager III, S. X. Li, E. E. Haller, H. Lu, and W. J. Schaff, *J. Appl. Phys.*, **94**, 4457 (2003)

K. Y. Cheng, *III–V Compound Semiconductors and Devices*, Graduate Texts in Physics,
https://doi.org/10.1007/978-3-030-51903-2

Appendix E
Bandgap Energy of III–V Semiconductor Ternary Alloys[a]

Alloy	$E_g(x)$	Cross-over	Reference
$Al_xGa_{1-x}P$	$E_g(X) = 2.273 + 0.215x$		
$Al_xGa_{1-x}As$	$E_g(\Gamma) = 1.424 + 1.247x\ (x < 0.4)$ $E_g(\Gamma) = 1.424 + 1.102x + 0.477x^2$ $E_g(X) = 1.899 + 0.210x + 0.055x^2$ $E(L) = 1.707 + 0.645x$	$x_c(\Gamma - X) = 0.39$	b a
$Al_xGa_{1-x}Sb$	$E_g(\Gamma) = 0.726 + 1.129x + 0.368x^2$ $E_g(L) = 0.799 + 0.746x + 0.334x^2$ $E_g(X) = 1.020 + 0.492x + 0.077x^2$	$x_c(\Gamma - L) = 0.23$ $x_c(L - X) = 0.4$	c
$Al_xIn_{1-x}P$	$E_g(\Gamma) = 1.353 + 2.68x - 0.48x^2$ $E_g(X) = 2.273 - 0.165x + 0.38x^2$	$x_c(\Gamma - X) = 0.36$	
$Al_xIn_{1-x}As$	$E_g(\Gamma) = 0.354 + 1.949x + 0.7x^2$ $E_g(X) = 1.37 + 0.794x$	$x_c(\Gamma - X) = 0.64$	
$Al_xIn_{1-x}Sb$	$E_g(\Gamma) = 0.1737 + 1.696x + 0.43x^2$ $E_g(X) = 0.63 + 1.455x$	$x_c(\Gamma - X) = 0.65$	
$Ga_xIn_{1-x}P$	$E_g(\Gamma) = 1.353 + 0.774x + 0.65x^2$ $E_g(X) = 2.273 - 0.2x + 0.2x^2$	$x_c(\Gamma - X) = 0.71$	
$Ga_xIn_{1-x}As$	$E_g(\Gamma) = 0.358 + 0.589x + 0.477x^2$		
$Ga_xIn_{1-x}Sb$	$E_g(\Gamma) = 0.174 + 0.138x + 0.415x^2$		
AlP_xAs_{1-x}	$E_g(X) = 2.164 + 0.104x + 0.22x^2$		
GaP_xAs_{1-x}	$E_g(\Gamma) = 1.424 + 1.163x + 0.19x^2$ $E_g(X) = 1.899 + 0.134x + 0.24x^2$	$x_c(\Gamma - X) = 0.47$	
InP_xAs_{1-x}	$E_g(\Gamma) = 0.354 + 0.899x + 0.1x^2$		
$AlSb_xAs_{1-x}$	$E_g(X) = 2.164 - 0.828x + 0.28x^2$		

(continued)

K. Y. Cheng, *III–V Compound Semiconductors and Devices*, Graduate Texts in Physics,
https://doi.org/10.1007/978-3-030-51903-2

(continued)

Alloy	$E_g(x)$	Cross-over	Reference
$GaSb_xAs_{1-x}$	$E_g(\Gamma) = 1.424 - 2.127x + 1.43x^2$		
$InSb_xAs_{1-x}$	$E_g(\Gamma) = 0.354 - 0.85x + 0.67x^2$		

[a]I. Vurgaftman, J. R. Meyer, and L. R. Ram-Mohan, *J. Appl. Phys.*, **89**, 5815 (2001)

[b]H. C. Casey, Jr, and M. B. Panish, *Heterostructure Lasers*, Part B, Academic Press, New York, 1978

[c]K. Y. Cheng, G. L. Pearson, R. S. Bauer, and D. J. Chadi, *Bull. Am. Phys. Soc.*, **21**, 365 (1976)

Appendix F
Bandgap[a, b] and Polarization Parameters[c] of Wurtzite III-Nitride Semiconductor Ternary Alloys

Alloy	$E_g(x)$, $P^{SP}(x)$, $P^{PE}(x)$
$Al_xGa_{1-x}N$	$E_g(x) = 3.438 + 2.020x + 0.7x^2$ (eV)
	$P^{SP}(x) = -0.090x - 0.034(1-x) + 0.021x(1-x)$ (C/m^2)
	$P^{PE}_{AlGaAs/InN}(x) = -0.28x - 0.113(1-x) + 0.042x(1-x)$ (C/m^2)
	$P^{PE}_{AlGaAs/GaN}(x) = -0.0525x + 0.0282x(1-x)$ (C/m^2)
	$P^{pE}_{AlGaAs/AlN}(x) = 0.026(1-x) + 0.0248x(1-x)$ (C/m^2)
$Al_xIn_{1-x}N$	$E_g(x) = 0.64 + 3.028x + 2.5x^2$ (eV)
	$P^{SP}(x) = -0.090x - 0.042(1-x) + 0.070x(1-x)$ (C/m^2)
	$P^{PE}_{AlInAs/InN}(x) = -0.28x + 0.104x(1-x)$ (C/m^2)
	$P^{PE}_{AlInAs/GaN}(x) = -0.0525x + 0.148(1-x) + 0.0938x(1-x)$ (C/m^2)
	$P^{PE}_{AlInAs/AlN}(x) = 0.182(1-x) + 0.092x(1-x)$ (C/m^2)
$In_xGa_{1-x}N$	$E_g(x) = 3.438 - 4.198x + 1.4x^2$ (eV)
	$P^{SP}(x) = -0.042x - 0.034(1-x) + 0.037x(1-x)$ (C/m^2)
	$P^{PE}_{InGaAs/InN}(x) = -\ 0.113(1-x) - 0.0276x(1-x)$ (C/m^2)
	$P^{PE}_{InGaAs/GaN}(x) = 0.148x - 0.0424x(1-x)$ (C/m^2)
	$P^{PE}_{InGaAs/AlN}(x) = 0.182x + 0.026(1-x) - 0.0456x(1-x)$ (C/m^2)

[a] I. Vurgaftman and J. R. Meyer, *J. Appl. Phys.*, **94**, 3675 (2003)
[b] J. Wu, W. Walukiewicz, W. Shan, K. M. Yu, J. W. Ager III, S. X. Li, E. E. Haller, H. Lu, and W. J. Schaff, *J. Appl. Phys.*, **94**, 4457 (2003)
[c] O Ambacher, J Majewski, C Miskys, A Link, M Hermann, M Eickhoff, M Stutzmann, F Bernardini, V Fiorentini, V Tilak, B Schaff, and L F Eastman, *J. Phys.: Condens. Matter* **14**, 3399 (2002)

K. Y. Cheng, *III–V Compound Semiconductors and Devices*, Graduate Texts in Physics, https://doi.org/10.1007/978-3-030-51903-2

Index

K. Y. Cheng, *III–V Compound Semiconductors and Devices*, Graduate Texts in Physics,
https://doi.org/10.1007/978-3-030-51903-2

C

K

L

GPSR Compliance
The European Union's (EU) General Product Safety Regulation (GPSR) is a set of rules that requires consumer products to be safe and our obligations to ensure this.

If you have any concerns about our products, you can contact us on

ProductSafety@springernature.com

In case Publisher is established outside the EU, the EU authorized representative is:

Springer Nature Customer Service Center GmbH
Europaplatz 3
69115 Heidelberg, Germany

www.ingramcontent.com/pod-product-compliance
Lightning Source LLC
Chambersburg PA
CBHW071606290626
47387CB00006B/1036